Plant Gene Isola

PRINCIPLES AND PRACTICE

THE LIB

Uni

CONTENTS

Plant Gene Isolation

PRINCIPLES AND PRACTICE

Edited by
GARY D. FOSTER
and
DAVID TWELL
Department of Botany, University of Leicester, UK

JOHN WILEY & SONS

Chichester · New York · Brisbane · Toronto · Singapore

Other Wiley Editorial Offices

John Wiley & Sons, Inc., 605 Third Avenue,
New York, NY 10158-0012, USA

Jacaranda Wiley Ltd, 33 Park Road, Milton,
Queensland 4064, Australia

John Wiley & Sons (Canada) Ltd, 22 Worcester Road,
Rexdale, Ontario M9W 1L1, Canada

John Wiley & Sons (Asia) Pte Ltd, 2 Clementi Loop #02-01,
Jin Xing Distripark, Singapore 0512

Library of Congress Cataloging-in-Publication Data

Plant gene isolation : principles and practice / edited by Gary D.
    Foster and David Twell.
        p.   cm.
    Includes bibliographical references (p.   ) and index.
    ISBN 0-471-95538-8 (cased). — ISBN 0-471-95539-6 (pbk.)
    1. Plant gene isolation.   I. Foster, Gary D.   II. Twell, David.
QK981.P5   1996
581.87'322'072—dc20   95–49528
                                                                 CIP

British Library Cataloguing in Publication Data

A catalogue record for this book is available from the British Library

ISBN 0-471-95538-8 (case)
ISBN 0-471-95539-6 (paper)

Typeset in 10/12pt Palatino by Keytec Typesetting Ltd, Bridport, Dorset
Printed and bound in Great Britain by Bookcraft (Bath) Ltd
This book is printed on acid-free paper responsibly manufactured from sustainable forestation,
for which at least two trees are planted for each one used for paper production.

Contents

PART II: LIBRARY SCREENING

5   Heterologous and Homologous Gene Probes   103
N.G. Halford

6   Differential Screening   125
P.A. Sabelli

PART III: MAP-BASED CLONING

9 Classical Mutagenesis and Genetic Analysis   215
I. Vizir, G. Thorlby and B. Mulligan

10 Chromosome Walking   247
O. Leyser and C. Chang

PART IV: INSERTIONAL MUTAGENESIS

11   T-DNA-mediated Insertional Mutagenesis   275
K. Lindsey and J.F. Topping

12   Transposon-mediated Insertional Mutagenesis   301
M. Roberts

PART V: PCR-BASED CLONING

13 PCR Techniques 331

C. *Thomas*

14   cDNA and Genomic Subtraction   369
M.P. Bulman and S.J. Neill

PART VI: SEQUENCING PROJECTS

15   EST and Genomic Sequencing Projects   401
R. Cooke, R. Mache and H. Höfte

Contributors

Mike P. Bulman
Department of Biological Sciences, Washington-Singer-Laboratories, University of Exeter, Perry Road, Exeter EX4 4QG, UK
Tel: 01392 263263   Fax: 01392 263700

Caren Chang
Department of Plant Biology, University of Maryland, College Park, MD 20742, USA
Tel: 301 405 1643   Fax: 301 314 9082

Richard Cooke
Laboratoire de Physiologie et Biologie Moléculaire Végétales, CNRS URA 565, Université de Perpignan, Avenue de Villeneuve, 66860 Perpignan-Cedex, France

Richard B. Flavell
John Innes Centre, Norwich Research Park, Colney, Norwich NR4 7UH, UK

Nigel G. Halford
Institute of Arable Crops Research, Long Ashton Research Station, Department of Agricultural Sciences, University of Bristol, Long Aston, Bristol BS18 9AF, UK
Tel: 01275 392181   Fax: 01275 394281   E-mail: nigel.halford@bbsrc.ac.uk

Herman Höfte
Laboratoire de Biologie Cellulaire, INRA Versailles, Route de Saint-Cyr, 78026 Versailles Cedex, France

Patrick J. Hussey
School of Biological Sciences, Royal Holloway, University of London, Egham Hill, Egham, Surrey TW20 0EX, UK
Tel: 01784 443539   Fax: 01784 434326   E-mail: p.hussey@rhbnc.ac.uk

John Hunsperger
Haplogenetics, P.O. Box 2217, Gilroy, CA 95021-2217, USA

Ottoline Leyser
Department of Biology, University of York, Heslington, York YO1 5YW, UK
Tel: 01904 434333   Fax: 01904 434336

Keith Lindsey
*Department of Biological Sciences, University of Durham, South Road, Durham DH1
3LE, UK*
Tel: 0191 374 7059   Fax: 0191 374 2417

Régis Mache
*Laboratoire de Biologie, Moléculaire Vegetale, CNRS URA 1178, B.P. 53X, 38041
Grenoble Cedex, France*

Graham Moore
John Innes Centre, Norwich Research Park, Colney, Norwich NR4 7UH, UK

Bernard Mulligan
*Department of Life Sciences, University of Nottingham, University Park, Nottingham
NG7 2RD, UK*
Tel: 0115 9513255   Fax: 0115 9513251

James A.H. Murray
*Institute of Biotechnology, University of Cambridge, Tennis Court Road, Cambridge
CB2 1QT, UK*
Fax: 01223 334162   E-mail: j.murray@biotech.cam.ac.uk

Steve J. Neill
*Department of Biological Sciences, University of the West of England, Bristol,
Coldharbour Lane, Bristol BS16 1QY, UK*
Tel: 01272 656261   Fax: 01272 763871

Michael Roberts
*Plant Laboratory, Department of Biology, University of York, Heslington, York YO1
5YW, UK*

Paolo A. Sabelli
*Institute of Molecular and Cell Biology, National University of Singapore, 10 Kent
Ridge Crescent, Singapore 0511*

Alison G. Smith
*Department of Plant Sciences, University of Cambridge, Downing Street, Cambridge
CB2 3EA, UK*
Fax: 01223 333953   E-mail: as24@mole.bio.cam.ac.uk

Imre E. Somssich
*MPI für Züchtungsforschung, Abteilung Biochemie, Carl-von-Linné-Weg 10, 50829
Köln, Germany*
Tel: 49 221 5062 310   Fax: 49 221 5062 313
E-mail: somssich@vax.mpiz-koeln.mpg.d400.de

Chris Thomas
Advanced Technologies (Cambridge) Ltd, Cambridge Science Park, Cambridge CB4 4WA, UK

Glenn Thorlby
Department of Life Sciences, University of Nottingham, University Park, Nottingham NG7 2RD, UK
Tel: 0115 9513255   Fax: 0115 9513251

Jennifer F. Topping
Department of Biological Sciences, University of Durham, South Road, Durham DH1 3LE, UK

Igor Vizir
Department of Life Sciences, University of Nottingham, University Park, Nottingham NG7 2RD, UK
Tel: 0115 9513255   Fax: 0115 9513251

Simon A.J. Warner
Department of Botany, University of Leicester, University Road, Leicester LE1 7RH, UK
Fax: 0116 252 3381

Bernd Weißhaar
MPI für Züchtungsforschung, Abteilung Biochemie, Carl-von-Linné-Weg 10, 50829 Köln, Germany

Rod A. Wing
Soils and Crop Sciences Department, Crop Biotechnology Center, Texas A & M University, College Station, Texas 77843-2123, USA
Tel: 409 862 2244, Fax: 409 862 4790   E-mail: rodwing@tam2000.tamu.edu

Sung-Sick Woo
Soils & Crop Sciences Department, Crop Biotechnology Center, Texas A & M University, College Station, Texas 77843-2123, USA

Hong-Bin Zhang
Soils & Crop Sciences Department, Crop Biotechnology Center, Texas A & M University, College Station, Texas 77843-2123, USA

Preface

Developments in the technologies enabling the manipulation and improvement of a variety of plant processes by genetic manipulation have burgeoned in recent years. This is reflected in the plethora of reviews and texts in the general area of plant genetic engineering. In this subject area, the primary enabling technology continues to be the isolation of the 'gene of interest', which for many applications represents a formidable task—similar to looking for 'a needle in a haystack'. With the advent, however, of new and advanced molecular and genetic techniques the process of gene isolation may often be reduced to a routine procedure.

In this book our aim is to provide an accessible and comprehensive guide to both the background and theory of current experimental methods needed to plan and execute gene isolation from plants. Our approach in designing the style and content of the book was to achieve a balance somewhere between a textbook and a laboratory manual. In other words, we neither wanted a review of current techniques nor a list of detailed protocols. Our intentions are to: first, provide a set of chapters which give both the research student and the more experienced investigator insight into a variety of techniques and approaches; and second, to enable the reader to get a 'feel' for what is involved experimentally, thereby allowing the selection of the most appropriate methods to isolate their gene of interest.

We also hope that this book will be useful as a teaching aid for advanced students in that each chapter provides a historical and well-referenced introduction explaining the development and principles on which different techniques and approaches are based. This has required lengthy discussion on the background and intricacies of a particular approach in some chapters, whilst in others working experimental protocols are provided to allow the reader to assess exactly what is involved in order to carry out a selected technique. In all chapters a comprehensive list of references has been provided and the authors have highlighted five key references which are strongly recommended for further reading.

The book is structured into six parts preceded by an introductory chapter which provides an overview of plant genome constituents and their organisation. The introductory chapter provides the molecular and genetic context to the problem of identifying and isolating specific genes and as to why some approaches are more or less effective in particular species.

Parts I and II cover the most common and now traditional approaches for plant gene isolation, namely the generation of cDNA or genomic libraries and screening for the gene of interest using different strategies. Part I describes how to isolate plant RNA and DNA and generate cDNA and genomic libraries in different cloning vectors. Part II describes several approaches to screen for the gene of interest once a library has been generated. These range from direct hybridisation using defined labelled probes, through differential screening, to antibody screening for expressed proteins and/or genetic complementation where functional assays are possible.

Parts III and IV deal with the techniques involved in the identification and isolation of genes using a molecular genetic approach. These techniques rely on the phenotypic identification of the gene of interest *in planta* following mutagenesis and subsequently the use of specific cloning methods for the recovery of the gene from plant DNA. Part III includes a discussion of the relative merits of chemical and physical methods of mutagenesis and approaches towards classical methods of genetic mapping. This section also provides access to the more recent and increasingly used technologies involved in chromosome walking using yeast artificial chromosomes. Part IV describes two increasingly important direct approaches for identifying genes, insertional mutagenesis involving T-DNA tagging and the use of plant transposable elements.

Part V contains two chapters describing the application of PCR-based techniques to plant gene isolation, which often allows the direct isolation of genes from cDNA or genomic DNA, thus bypassing the majority of methods described in Parts I–IV. However, in many cases PCR technology has become an integral part of the methods in Parts I–IV and is therefore used in combinational fashion.

Part VI covers one of the most recent advances in plant gene isolation, which in the long term may have a profound effect on the way we go looking for genes. This part deals with the technique of automated sequencing of large numbers of randomly selected cDNA clones and ordered sequencing of selected plant genomes. Sequence information generated in this way is analysed by comparison with DNA databases to identify matches with previously identified genes from plants and other organisms. This chapter deals with the value and technology behind this approach and explains how the information from such projects is made available to scientists worldwide.

Overall, we hope that this book will be a useful source book for plant molecular biologists, geneticists and biotechnologists considering gene characterisation or manipulation, providing them with explanation and guidance when considering which route to follow on the way to the isolation of their gene of interest. Thus, we hope that this book will help make the process of 'looking for a needle' a less daunting task.

The editors would like to thank all the authors for a tremendous effort on this project, even those busy contributors who caused no end of worry for the editors; the chapters were definitely worth the wait. We would also like to thank BBSRC and The Royal Society for their support of our fellowship awards which have allowed us to devote ourselves full-time to research and scholarship. Thanks also go to our academic colleagues who have provided advice and encouragement over the years at Leicester. GF would also like to thank Diana,

James and Kirsty, who have tolerated the out-of-hours extra work required to proofread and edit the book, and all the Foster clan for support over the years. DT would like to thank Tracy for thinking of Remedios (never to be) and for valuable literary advice.

Happy cloning!

Gary Foster and David Twell
Department of Botany, University of Leicester
October 1995

1 Plant Genome Constituents and Their Organisation

RICHARD B. FLAVELL
GRAHAM MOORE
John Innes Centre, Norwich Research Park, Colney, Norwich, UK

The genomes of higher plants are remarkably diverse in size and nucleotide sequence content. Yet from studies over the past twenty-five years or so a framework of understanding has emerged that enables genome organisation to be interpreted and underlying principles to be described. It is the aim of this chapter to summarise this understanding and the underlying principles to provide a background for the subsequent chapters which deal with the many aspects of gene isolation.

1 INTRODUCTION

We are now in the era of 'genome projects' aimed at identifying all the genes of a few species and sequencing the complete genomes of, for example, *E. coli*, yeast, human, the monocot rice and the dicot *Arabidopsis*. The gathering and interpretation of all the new data on genomes are proceeding very rapidly and information on one genome is used to interpret that gained on another. This unification of genome research, driven by molecular genetics, is a powerful development in biology. Research on plant genomes is now becoming better recognised because the exploitation of plant genes and genomes via plant breeding is likely to contribute far more to the quality of life on the planet than is, for example, the exploitation of human genes, which benefits individual families and not whole societies. Therefore, it is especially important to advance the understanding of plant genomes as rapidly as possible.

Several units of sequence organisation can be recognised in plant genomes. The largest is the chromosome. At the sub-chromosomal level are segments with conserved gene complements. Next, with the dimensions of multimegabases, are satellite arrays or specific repetitive sequences concentrated within genomic segments. At the sub-megabase level are blocks of amplified sequences defined by undermethylated sites. Well-characterised repetitive elements and genes are units of a few kilobases. Finally, many other regulatory sequences are composed of relatively few base pairs. This review of plant genome organisation has been constructed to describe these sorts of units.

Plant Gene Isolation: Principles and Practice. Edited by G. D. Foster and D. Twell.
© 1996 John Wiley & Sons Ltd.

However, two general features relevant to understanding genome organisation across the plant kingdom are reviewed first: variation in genome size and cytosine methylation.

2   VARIATION IN GENOME SIZE

Plant genomes vary in DNA content over 1500-fold (Bennett & Smith, 1991). Estimates of genome size have been gained by microdensitometry of feulgen-stained nuclei, reassociation kinetics, nuclear volume and flow cytometry of 4,6-diamidino-2-phenylindole (DAPI)-stained nuclei. A defined genome has usually been used as an internal standard to obtain comparable estimates. Near absolute DNA contents of a few genomes, e.g. those of *E. coli*, *Saccharomyces cerevisae*, and *Arabidopsis thaliana*, will be revealed by DNA sequencing within the next decade or so. These DNA contents will then serve as more accurate internal standards. The different methods have produced sometimes substantially different values for genome sizes. For example, *Arabidopsis thaliana* has been estimated to contain from 70 to 190×10^6 base pairs (Bennett & Smith, 1991).

A substantial part of the variation in plant DNA content is due to polyploidy. Polyploidy can also occur during plant development. It has been estimated that 50% of angiosperms are polyploids resulting from the doubling of chromosomes within a species or the hybridisation of two species without chromosome reduction. However, recent molecular mapping data (Helentjaris et al., 1988; Whitkus et al., 1992; Moore et al., 1995a) have revealed that some species (e.g. maize) which behave genetically as diploids contain two nearly complete copies of a basic genome (Bennett, 1983). Thus the haploid chromosome complement does not contain the 'primary genome' but that of a cryptic polyploid. Evidence for duplicated segments in *Arabidopsis* and *Brassica* species genomes has also emerged from molecular mapping (Kowalski et al., 1994; Lydiate et al., 1993). These important discoveries have major implications for the genetic characterisation of plant genomes, the classification of genes and gene families and the understanding of the variation of DNA contents of plant genomes.

The second major source of variation in nuclear DNA content comprises families of repeated sequences. In the 1960s and 1970s, studies of the reassociation kinetics of short, denatured DNA fragments in solution led to quantification of the proportion of plant DNAs that consisted of short sequences (> 50 bp) present in one or a few copies per haploid genome and the proportion consisting of short sequences present in families of 50 up to several million copies per haploid genome (Flavell, 1980; Thompson & Murray, 1981). Families of repeated sequences were defined in these studies as related sequences differing from one another by up to about 25% and thus are presumed to have evolved from a common ancestral sequence. The amounts of repetitive DNA account for a large proportion of the variation in genome size, thereby indicating that mutations resulting in DNA amplification and the ability to tolerate the products of amplification are major sources of variation in genome evolution. The proportions of repetitive DNA exceed 50% in genomes larger than 1.5 pg and approach 95% in very large genomes (Flavell, 1980). High proportions of repeats have major implications for exploring genomes via molecular genetics.

Reassociation kinetics revealed that most plant genomes contain a huge number of families of repeated sequences (Flavell, 1980; Flavell et al., 1977; Rimpau et al., 1978; Thompson & Murray, 1981). The subsequent characterisation of cloned copies of individual sequences from the families has revealed a much more complex situation, because families of repeats often share some common segments while other segments are specific to a family or subset of a family (Bedbrook et al., 1980b; Grandbastien, 1992). It is therefore important to understand the evolutionary and functional units of DNA in the genome rather than fragments of common evolutionary origin to comprehend genome organisation. Many members of families of repeated sequences are apparently disposable, because such families expand and contract repeatedly in evolution (Flavell, 1985). This behaviour of repeats also leads to variation in DNA content, even within species (Flavell, 1982). Total nuclear content is related to minimum cell cycle times and maximal developmental rates in diploids (Bennett, 1973). If DNA contents increase, maximum developmental rates decrease. Therefore, it is concluded that natural selection sustains genome sizes at optimum levels.

3   CYTOSINE METHYLATION

In plants, both CpG and CpXpG motifs can have methyl groups on the cytosine residues as a result of nuclear methylases (Finnegan et al., 1993). Some adenine and other cytosine residues are also methylated. Methyl groups are added to DNA post-replication. The patterns of methylation form an important feature of genome organisation. Addition of methyl groups to a newly replicated DNA strand usually follows the pattern on the old complementary strand. Errors occur, but de novo methylases re-establish the pattern. What specifies which CpG residues are methylated and which are not is not fully understood. Small genomes of *Arabidopsis* and rice (Leutwiler et al., 1984; Deshpande & Ranjekar, 1980; Wu & Tanksley, 1993) appear to be relatively undermethylated compared to larger genomes such as those of wheat, barley (Swartz et al., 1962; Gruenbaum et al., 1981; Moore et al., 1993a), maize (Bennetzen et al., 1994; Springer et al., 1994) and pea. The increased methylation appears to correlate with the increased accumulation of repetitive sequences in the genome. It has been proposed that the highly methylated vertebrate genome evolved from the unmethylated invertebrate genome by the accumulation of repetitive DNA, which became differentially methylated, a process which is associated with transcriptional silencing and chromatin condensation (Bird, 1987). Many of the genes of vertebrates reside in undermethylated islands, protected from methylation, perhaps to preserve function and the environment in which they originally evolved. Genes in cereals such as wheat, barley and maize also appear to reside in such islands, with long stretches of highly hypermethylated repetitive DNA between them (Antequera & Bird, 1988; Moore et al., 1993b; Bennetzen et al., 1994; Springer et al., 1994; Finnegan et al., 1993). In mammalian genomes, it can be inferred that the methylated CpGs in the repetitive sequences have frequently undergone transitions to TpGs. Thus the repetitive sequences have become CpG-poor in contrast to the CpG-rich gene regions. In plant genomes such as wheat and barley, there is no evidence for large-scale transitions of CpGs (Finnegan et al., 1993; Moore et al., 1993b). Thus they appear to have escaped the mutational load associated with methylated CpG residues.

The reduced levels of cytosine methylation in islands close to active genes provide ways of isolating genomic fragments with active genes. Numerous laboratories have used DNA fragments isolated by the restriction endonuclease *Pst*I, which cleaves the motif only when the cytosines are unmethylated. *Hpa*II has been used similarly. Short fragments released by cleavage with such enzymes are often highly enriched for single- or few-copy sequences, because repetitive sequences are especially highly methylated (Moore et al., 1993a). Restriction endonucleases which cleave genomic DNA only rarely and only where cytosines are unmethylated, such as *Not*I, have also been used to isolate sequences that are in few copies and so detect the distribution of single/few-copy sequences in the genome (Moore et al., 1993a).

The role and organisation of methyl residues is worthy of special consideration, not only because it can be used as a tool to dissect the genome. It has been argued that high-density cytosine methylation suppresses gene expression due to the binding of 'silencing' proteins and stimulation of chromatin conformations that prevent transcription, either directly or indirectly (Finnegan et al., 1993).

4 ORGANISATION OF THE GENOME INTO CHROMOSOMES

The nuclear genome is organised into chromosomes, the number of which varies between species and occasionally within species, where supernumerary B chromosomes are present (Jones & Rees, 1982; Beukeboom, 1994). Chromosomes consist of essentially one long DNA helix wound around nucleosomes; the nucleoprotein chain is differentially coiled throughout its length at interphase, reflecting the consequences of the presence of specific proteins and the activity of specific regions. At metaphase, when the genome is relatively inactive, the chromosomes are most condensed and therefore most easily observed cytologically, counted or separated from one another using a micromanipulator. Single chromosomes purified by micromanipulation have been used for the construction of chromosome-specific DNA libraries. The chromosomal unit has special importance because it is the basis of genetic linkage descriptions of the genome. Chromosomes provide the means by which the genome constituents are replicated and segregated regularly in mitosis and meiosis. Regular segregation depends upon centromeres that facilitate attachment of the spindle fibres. The faithful replication of chromosome ends requires special telomeric DNA structures that have been highly conserved during evolution (Blackburn, 1994). In meiosis, chromosomes recombine following the intimate pairing of homologous chromosomes, the formation of double-stranded chromosome breaks and hybrid DNA and religation of the DNA strands in new combinations. All of these properties have implications for the organisation of DNA sequences within chromosomes. Centromeric and telomeric DNA sequences are essential, as are any sequences that promote replication, chromosome pairing and recombination and chromosome condensation. Much more remains to be learnt about such sequences for plant chromosomes. The amount of DNA in individual chromosomes and chromosomal arms is likely to be under selection pressure, since chromosome arm length may determine the position of the chromosomes in the nucleus relative to other chromosomes (Bennett, 1982; Heslop-Harrison & Bennett, 1990). However, many quantitatively small changes in DNA content occur frequently during evolution, as do translocations of segments between chromosomes.

A schematic representation of a chromosome is given in Fig. 1 as an aid to understanding the general organisation of sequences as described in this chapter.

5 LARGE GENOME SEGMENTS DEFINED BY THEIR CONSERVED LINEAR ORDER OF CONSTITUENT GENES

The momentum of mapping genes on plant chromosomes has increased substantially over the past few years, as the techniques of placing molecular markers on recombination maps have become facile and gene sequences have become available as cDNAs. For genomes such as maize, wheat, rice, *Arabidopsis* and soybean, over 1000 molecular markers (restriction fragment length polymorphisms (RFLPs), random amplified polymorphic DNA (RAPD) or microsatellites) exist on the recombination maps but, of course, not all markers register cDNAs or genes.

In a number of cases, markers positioned on the genetic map of one species have been located by reciprocal mapping on the genetic map of another (Bonierbale et al., 1988; Gebhardt et al., 1991; Whitkus et al., 1992; Kurata et al., 1994; Ahn & Tanksley, 1993; Kowalski et al., 1994). Some of these comparisons have important implications for plant genetics/biology, particularly those between *Arabidopsis* and *Brassica*, rice and wheat, and rice and maize. Although the genomes of rice and *Arabidopsis* are significantly smaller than those of the other species in these comparisons, the gene order along genome segments is similar in the small and large genome of the related species. In cereals, these comparisons now include barley, rye, sorghum, sugarcane, fox-tail millet, rice, wheat and maize (Moore et al., 1995b). They have revealed a series of genomic units containing genes whose order is

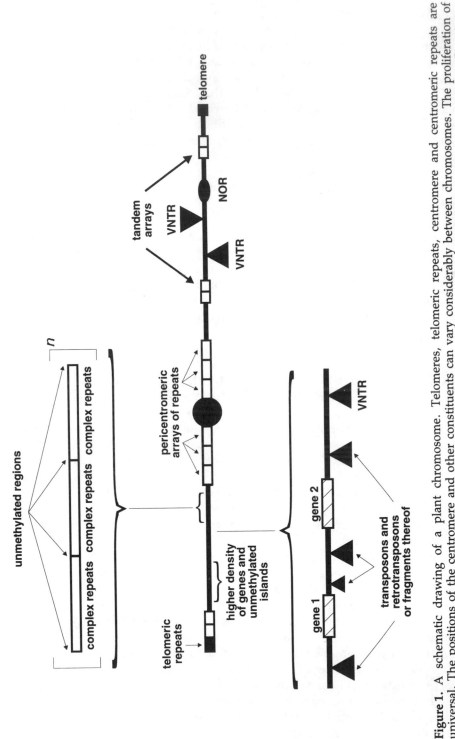

Figure 1. A schematic drawing of a plant chromosome. Telomeres, telomeric repeats, centromere and centromeric repeats are universal. The positions of the centromere and other constituents can vary considerably between chromosomes. The proliferation of 'megabase' units to complex repeats defined by undermethylated regions has been included because of observations on Gramineae (Moore et al., 1993a). A region containing a higher density of genes and undermethylated islands has been included for similar reasons (Moore et al., 1993b). NOR, nucleolus organising region: VNTR, variable number tandem repeats

conserved in all these species' genomes. Integrating these units results in a predicted ancestral genome from which these cereals speciated some 60 million years ago (Moore et al., 1993b; Moore et al., 1995a, b). This is illustrated in Fig. 2.

From comparative mapping of today's species it is possible to deduce how the ancestral genomic segments have become restructured to form the present-day chromosomes of each species. The chromosomes of rice, maize and wheat are illustrated in Fig. 2 as compilations of these units. The 12 chromosomes of rice have evolved from 20 ancestral segments whose gene complements have been conserved amongst many grass species. The order of these units in the 10 maize chromosomes is shown, revealing the 'tetraploid' nature of maize due to the duplications of the ancestral segments.

It is important to note that recognition of these component genomic units is based only on conserved 'single-copy' cDNA sequences. This conserved DNA framework may contribute only a small fraction of the DNA but its recognition is proving extremely valuable because it includes the genes. Consequently, all the genetic mapping information which has been generated by studying these cereal species separately in the last 50 years can now be collated onto composite comparative maps and exploited with greater efficiency (Moore et al., 1993b). Knowledge of where a gene or quantitative trait locus (QTL) maps in one cereal species predicts where the homologous genes map in the other species. This will be of great benefit in plant breeding. By characterising the homologous genes in crops grown in different parts of the world, it will be possible to trace the selection of allelic variation during speciation and agriculture and so comprehend better the genetic basis of developmental differences within and between species.

The gramineae segments with conserved gene synteny are large. The sizes of the segments depend upon the number of translocations, inversions or other sorts of rearrangements that have occurred during species divergence and presumably the evolutionary distance between the species. In other groups of species assayed to date, the conserved synteny is extensive but not over whole chromosome arms: Lee Menancio-Hautea et al., (1993) and Weeden et al., (1992) (legumes); and Kowalski et al. (1994) (Brassica/Arabidopsis). It can be expected that as more detailed fine-structure genetic maps are generated, conserved synteny of genes along small chromosomal units will be discovered across very wide evolutionary distances, so providing new tools for taxonomy and the construction of evolutionary trees.

Linear gene order is conserved during evolution by accident or by selection. Gene order may be conserved as a result of selection for a large number of genetic linkages or for other reasons we do not understand. Alternatively, the conserved linkage observed may simply reflect a common ancestral genome and relatively few segmental rearrangements. Conserved gene synteny is in contrast to the variable amount and sequence of the DNA lying between the genes. The rice genome is only one-tenth that of the diploid wheat genome, and the Arabidopsis genome is only one quarter of the diploid Brassica genome. The much larger amounts of DNA lying between homologous genes in wheat

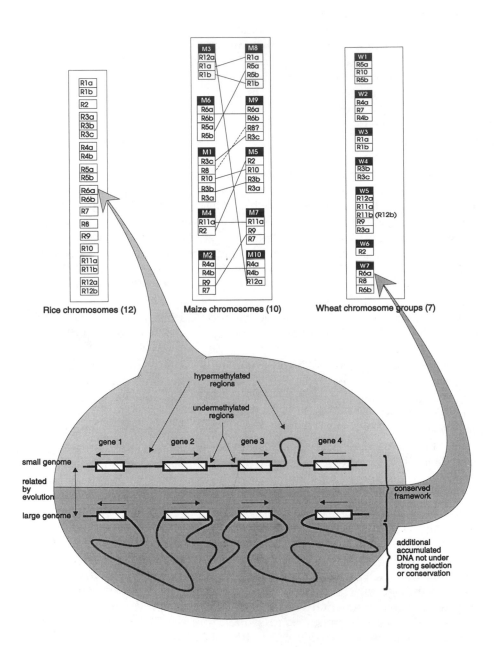

| R1a |
| R1b |
| R2 |
| R3a |
| R3b |
| R3c |
| R4a |
| R4b |
| R5a |
| R5b |
| R6a |
| R6b |
| R7 |
| R8 |
| R9 |
| R10 |
| R11a |
| R11b |
| R12a |
| R12b |

Rice chromosomes (12)

M3	M8
R12a	R1a
R1a	R5a
R1b	R5b
	R1b
M6	M9
R6a	R6a
R6b	R6b
R5a	R8?
R5b	R3c
M1	M5
R3c	R2
R8	R10
R10	R3b
R3b	R3a
R3a	
M4	M7
R11a	R11a
R2	R9
	R7
M2	M10
R4a	R4a
R4b	R4b
R9	R12a
R7	

Maize chromosomes (10)

| W1 |
| R5a |
| R10 |
| R5b |
| W2 |
| R4a |
| R7 |
| R4b |
| W3 |
| R1a |
| R1b |
| W4 |
| R3b |
| R3c |
| W5 |
| R12a |
| R11a |
| R11b (R12b) |
| R9 |
| R3a |
| W6 |
| R2 |
| W7 |
| R6a |
| R8 |
| R6b |

Wheat chromosome groups (7)

hypermethylated regions

undermethylated regions

gene 1   gene 2   gene 3   gene 4

small genome

related by evolution

large genome

conserved framework

additional accumulated DNA not under strong selection or conservation

compared with rice and in *Brassica* compared with *Arabidopsis* are principally due to the accumulation of much more repetitive DNA in the larger genomes. Thus the unit structures in plant genomes perceived by comparative mapping of conserved genes may be based on a framework of little more than the genes themselves when species which have diverged over long evolutionary time-scales are compared. This is illustrated schematically in the lower part of Fig. 2.

The gene synteny framework will be an extremely valuable aid for gene isolation. In species with large genomes, gene isolation by chromosome walking is extremely time-consuming, because markers mapped close to the desired gene by recombination mapping may be a long way in base pairs from the gene. Also, the high concentration of repetitive DNA inhibits genome walking in libraries. In such a situation, localising the exact position of the gene between known markers may be extremely difficult. Where the homologous gene can be located in the homologous segment of a small genome, however, e.g. in rice or *Arabidopsis*, and flanking markers mapped physically close by, in regions without repeats, the problems shrink. Following isolation of the gene from the small genome species, its homologue can then be isolated directly from the relevant species. Thus, as the contiging of the complete genomes of rice and *Arabidopsis* into ordered DNA fragments provides unlimited mapping molecular markers, then gene isolation will be facilitated for all closely related species and perhaps all monocots and dicots (Moore et al., 1993b; Dunford et al., 1995; Kilian et al., 1995). Many research groups are now following these strategies.

6   MULTIMEGABASE UNITS

6.1   Tandem arrays of repeats

Tandem arrays of essentially the same sequence are clustered at many locations in a genome. Examples have been characterised in many plant species (Bedbrook et al., 1980a; Grellet et al., 1986; Martinez-Zapater et al., 1986; Schweizer

---

Figure 2. Genome evolution: the restructuring of chromosomal units during Gramineae divergence and the fine-scale divergence of chromosome regions due to the differential accumulation of repetitive DNA. The chromosome units of the 12 rice chromosomes, defined by the genes they contain, are shown alongside the organisation of the equivalent homologous units in the maize and wheat chromosomes. The figure shows how the order and number of units has changed during species divergence. Because the homologous units are defined by the conserved gene synteny, they may differ in total DNA content due to major changes in non-genic, methylated repetitive DNA. This is illustrated at the bottom of the figure. Four genes are conserved between two species that have diverged from a common ancestor. The genome of the upper species has not accumulated much intergenic repetitive DNA relative to the lower species. The conserved framework based on gene synteny is proving to be extremely valuable in comparative gene mapping and the isolation of genes by map-based cloning, as described in the text

et al., 1988; Dennis et al., 1980; Deumling & Greilhuber, 1982). Such arrays were originally called 'satellites' because they were recognised as separate peaks in density gradient centrifugation—a consequence of their low sequence complexity. In some genomes these arrays consist of up to millions of repeat units. These arrays are commonly clustered around the centromeres, in telomeric and subtelomeric locations and also in some interstitial sectors (Fig. 1) (Bedbrook et al., 1980a; Deumling & Greilhuber, 1982; Gerlach & Peacock, 1980). Such arrays may have arisen via a rolling circle replication mechanism. However, once an array exists, however short, it can expand or contract via unequal crossing over or intrastrand recombination to produce individual chromosomes with arrays of varying lengths (Flavell, 1982, 1985). The amplified repeat units can have lengths of a few base pairs up to many hundreds or thousands. Such arrays undergo continual reamplification because of their reiterated organisation, and a later amplified unit length may differ from a previously amplified unit length. Thus within arrays of essentially the same repeat, much longer compound repeat units can be discerned, when fine-structure sequence analysis is performed (Bedbrook et al., 1980b). Specific variants of a repeat unit can become dispersed over a genome via physical movement or gene conversion. At any point in evolution, however, chromosome- or region-specific variants can be found (Bedbrook et al., 1980b).

The clustering of major arrays of repeats around centromeres and telomeres and at specific interstitial locations raises the question of whether this is where they are tolerated most easily or whether they are focused there as a consequence of the forces of natural selection. Probably both alternatives are true, but what advantages tandem arrays of repeats confer on a chromosome or a genome remain speculative. They often adopt a heterochromatic chromatin conformation and may be late-replicating (Flavell, 1980; Gerlach & Peacock, 1980; Bedbrook et al., 1980a). Whether any single-copy genes lie within such arrays or clusters of arrays is an important issue for understanding genome organisation. Their cloning might prove difficult except as cDNAs initially.

Ribosomal 18S, 5S, 25S and 5S RNA genes are highly reiterated in clusters, the former at sites called nucleolus organisers (NORs), because the extended rDNA chromatin, upon transcription, catalyses nucleolus formation (Fig. 1). Because the units are tandemly reiterated there is within-species variation for the number of copies in allelic arrays, and because there are tandemly arrayed subrepeats in the intergenic regions of many rDNA repeats there is much variation in the length of rDNA repeat units within and between species (Flavell, 1986b; Ellis et al., 1988; Lapitan et al., 1991). Variations in the lengths of rDNA repeat units have been used since the 1970s as sources of RFLPs, making rDNA loci convenient molecular chromosomal markers (Flavell, 1985). The high frequency of rDNA and 5S RNA gene reiteration led to their cloning in the 1970s from some plant species (Gerlach & Bedbrook, 1979; Flavell, 1986b). However, the tandem arrays are unstable in many E. coli hosts, thus enhancing the difficulty of cloning them. This is probably true for all long tandem arrays of repeats. This implies that most genomic libraries are incomplete. In mammalian studies, other multigene families have been found in large repeat arrays, although whether such arrays occur in plants has yet to be shown.

Because unequal crossing over and other mechanisms lead to variation in copy number within arrays, cleavage with restriction enzymes which do not cut within the array generates RFLPs. These can be recognised using pulse field gels to fractionate the long arrays but much more convenient RFLPs are generated by this method where the arrays are short. This is why 'microsatellites' or variable number tandem repeats (VNTRs) are now being actively sought as dominant markers in plant genomes to aid molecular mapping—the variation is high and the visualisation of the variation is facile (Hillel et al., 1990). Microsatellites are, for example, arrays of di-, tri- or tetranucleotides, and are common because of the aberrations in the replication processes and the common occurrence of unequal crossing over which probably develop them. These microsatellites are localised at many sites (Fig. 1) within the genome (Bell & Ecker, 1994; Morgante & Olivieri, 1993). This further supports the hypothesis that such arrays can be generated anywhere but very long arrays are tolerated preferentially near centromeres, telomeres or other particular sites.

Tandem arrays of sequences evolve in concert, due to homogenisation processes such as gene conversion and rounds of unequal crossing over (Flavell, 1982, 1985). Because arrays in different chromosomes evolve in concert, there must be exchange of sequences or gene conversion between arrays such as occurs during replication of telomeric arrays. However, units within one array are usually more identical than units from different arrays. Thus restriction endonucleases able to discern specific variant sequences within arrays are useful for identifying chromosome-specific variants of tandem arrays of sequences common to many chromosomes.

6.2 Telomeres

In plant studies, the term telomere has often been used loosely. It has been used to describe heterochromatin associated with the terminal regions of chromosomes and regions containing telomere-like repeat sequences. However, several studies in *Arabidopsis thaliana* (Richards & Ausubel, 1988), tomato, *Lycopersicon esculentum* (Ganal et al., 1991) and wheat (Cheung et al., 1994) have demonstrated that telomeres consist of repeat units with the sequence (TTTAGGG)$_n$ in *Arabidopsis* and wheat and (TT(T/A)GGG)$_n$ in tomato (Fig. 1). The length of the array of telomeric repeats in these species, as in other eukaryotes, is heterogeneous, reflecting variation created in the process of telomere replication (Burr et al., 1992).

In a few organisms, the sequences that lie immediately adjacent to telomeric arrays have been cloned and studied: yeast (Horowitz et al., 1984); human (Cross et al., 1990); and *Arabidopsis* (Richards et al., 1991). These subtelomeric sequences are frequently repetitive, being present at one or more telomeres and occasionally at interstitial sites (Fig. 1); they reveal a high level of RFLP. In plants, satellite arrays have also been located in the terminal regions (Bedbrook et al., 1980a, b). Most of these satellite sequences appear to be separated by intervening sequences of at least 100 kb from the telomeric array (Ganal et al., 1991).

Although in barley some sequences homologous to the *Arabidopsis* telomeric repeat are found adjacent to members of the satellite repeat, most homologous sequences do not lie on the same pulsed field gel fragments as the satellite repeats (Roder et al., 1993). A family of related sequences associated with $(TTTAGGG)_n$ motifs has been cloned from wheat. These sequences reveal a high level of polymorphism (Cheung et al., 1994). However, these sequences are located at interstitial sites on wheat chromosomes, suggesting that TTTAGGG motifs occur at interstitial sites as well as at the termini of plant chromosomes.

6.3   Centromeres

Centromeres have been isolated from *Saccharomyces* and *Schizosaccharomyces*. Among higher eukaryotes only a human centromere has been cloned in its entirety (Tyler-Smith et al., 1993). It is largely composed of tandem arrays of the alphoid repeat spanning approximately 2.5 megabases. There is no such detailed analysis of a plant centromere, although the ordered yeast artificial chromosomes (YACs) of *Arabidopsis* chromosomes now make this possible. Various sequences have been isolated which are specific to the pericentromeric regions (Fig. 1) (e.g. Hutchinson & Lonsdale, 1982; Maluszyska & Heslop-Harrison, 1991; Alfenito & Birchler, 1993; Gerlach & Peacock, 1980). However, the importance of these sequences to centromere function is unclear.

7   MEGABASE UNITS REVEALED BY CYTOSINE-METHYLATION-SENSITIVE RESTRICTION ENDONUCLEASES

Most repetitive sequences in wheat and barley genomes are highly methylated, but studies have indicated the presence of unmethylated recognition sites for the restriction enzymes *Mlu*I and *Nru*I at non-random intervals in or around repeats (Moore et al., 1993a). Cleavage of the genomes of rice, barley, rye and wheat with these enzymes yields DNA fragments of discrete size classes, ranging from hundreds of kilobases to several megabases in length. The generation of these size classes suggests that unmethylated recognition sites for *Mlu*I are non-randomly distributed within these genomes. Furthermore, some of these fragments appear to be multicopy. Thus the spacing of un-methylated *Mlu*I sites along the genome may be confined to a limited number of distances. This may reflect the fact that very long pieces of DNA in these regions have been amplified. If amplification has occurred, it could help explain the presence of 'isochores' observed in plant genomes (Montero et al., 1990). In either case the preservation of the pattern of undermethylated sequences is noteworthy and has been included in the schematic chromosome in Fig. 1. The bulk of the DNA, if not all, in these cereal units is formed from highly complex combinations of many repeats. Perhaps these unmethylated regions define chromatin domains, matrix attachment regions or origins of replication.

8   KILOBASE UNITS

8.1   Transposable elements

Most of the repetitive DNA not consisting of tandem arrays probably consists of one kind or another of transposable element or sequences derived or mobilised by such elements. Transposable elements are units of DNA which are predisposed to move to another location, sometimes involving replication of the unit, with the help of products of genes on the element or on a related element. These properties can lead them to being fixed in populations relatively rapidly, even though they reduce fitness through being mutagenic (Flavell, 1985).

8.1.1   Elements encoding a transposase

The first major class contains elements characterised by inverted terminal repeats and that move by a multistep chromosome breakage and religation process, sometimes involving replication of the element. Examples include *Ac*, *Spm* and *Mu* in maize (Döring & Starlinger, 1986), *Tam* elements in *Antirrhinum* (Coen & Carpenter, 1986), *Tph*I in petunia (Gerats et al., 1990) and *Tag*I in *Arabidopsis* (Tsay et al., 1993). Autonomous versions of these elements, i.e. those which contain a full gene complement for promoting transposition, contain an open reading frame encoding for a transposase (Coen et al., 1989; Fedoroff, 1989). This enzyme product interacts with the inverted repeats of the element to facilitate transposition to a new site. The transposition mechanism leads to a high frequency of transposition to close-by sites; the insertion frequency at a particular site reduces as the distance from the original location increases. This presumably leads to the clustering of specific transposon variants in the genome. Transposase molecules from an autonomous element can interact with the ends of other elements in the genome lacking coding sequences for transposase, provided the inverted repeats possess recognition sites for the transposase. Thus a large number of mutant elements, often carrying internal deletions, are propagated in genomes by the presence of one or a few active autonomous elements. These sorts of elements—as autonomous or deleted forms—are usually present in a few to a few hundred copies per genome.

Transposition results in duplication of the sequence motif at the insertion site, one copy residing at either end of the element (Schwarz-Sommer et al., 1985). The number of duplicated bases (between 2 and 8) is characteristic of the element. Upon excision of the element, the target site duplication remains, sometimes with a few bases in between. Such sequence motifs, characteristic of the one-time presence of a transposable element at the locus, are extremely common in many plant genomes (e.g. Lucas et al., 1992).

Recently, two other families of transposable elements in this class have been described which are in large numbers, at least in some genomes. The *Tourist* elements in the grass family are in high abundance (> 10 000 copies) in maize (Bureau & Wessler, 1992), while the *Stowaway* family has a wide distribution in monocotyledonous as well as dicotyledonous species. They are small, have a conserved terminal inverted repeat and insert into TA sites. Since a *Tourist* element is present on average every 30 kb of the maize genome, one is likely to be present on most YACs in a maize genome YAC library.

8.1.2   Retroelements

The second class of transposable elements comprises active and inactive derivatives of retrotransposable elements. These are the most abundant in plant genomes, many being present in thousands of copies, and the dispersed *Bis*1 element in barley occupies several per cent of the total DNA (Moore et al., 1991a). Their abundance is probably because they replicate via mRNA transcript intermediates, and consequently many progeny elements can emanate from one parent copy (Grandbastien, 1992).

Many plant retroelements have now been characterised in detail and range from about 1 to 13 kb in length (Grandbastien, 1992). Retroelements differ from retroviruses in lacking the sequence encoding the structural envelope protein, essential for cell-to-cell movement. The other sequences encoded within the element are typical constituents of retroviruses, namely: the *gag* gene (encodes a structural protein), *prot* gene (protease), reverse transcriptase gene, RNase H gene and integrase gene. Most of the elements are also defined by having long terminal repeats (LTRs) that possess promoter and transcription termination sequences. The duplications are generated in the replication cycle. The LTRs are characteristically bordered by short inverted repeats typically terminating in 5'-TG. . .CA-3'. Inside from the left LTR is a primer binding site complementary to the 3' end of a host tRNA used for the synthesis of the first (−) DNA strand. Inside from the right LTR is a polypurine tract used for synthesis of the second (+) DNA strand. Each element is flanked by very short direct repeats of the host target DNA, created upon insertion. These characteristics are usually easily recognised in DNA sequences, even though there is substantial sequence divergence within and between families of such elements, within and between species (Moore et al., 1991b). Within the category of element, two subgroups can be identified: those in which the organisation of the coding domains is the same as in retroviruses, and those in which the endonuclease domain is found upstream of the reverse transcriptase domain.

Related retroposons without LTRs have also been identified. Elements from this class are the Cin4 elements from maize, which terminate in poly(A) tails and vary in size from 1 to 6.5-kb, the smaller ones being deletion derivatives of the 6.5 kb element. No master, complete copy of the Cin4 element has yet been identified. The del2 elements from *Lilium speciosum* are also of this class (Grandbastien, 1992).

8.2 Cytoplasmic organelle DNAs

The cytoplasmic organelles of mitochondria and chloroplasts originated as symbionts within a primitive cell. During eukaryote evolution many of the genes have been transferred from the organellar to the nuclear genomes. However, longer sectors of organellar DNA have been transferred to nuclear chromosomes during more recent evolution. These sectors are readily detected by hybridisation to cloned chloroplast DNA (Steele Scott & Timmis, 1984).

8.3 Genes

There are perhaps 20 000–30 000 different gene families in the higher plant genomes. Many genes are present in only copy per haploid genome but a high proportion are present in multigene families, although the genes whose RNA products are required in large numbers, e.g. ribosomal RNAs, are highly reiterated. The failure to find other genes in high numbers implies that they are rarely amplified or, perhaps more likely, such products are selected against.

To define the functional unit of the gene is difficult because we know relatively little about long-range regulatory elements. However, there is much information accumulated in databases to enable promoter regions to be interpreted and open reading frames, introns, polyadenylation and transcription termination sites to be recognised. These topics have been reviewed elsewhere.

Surprising results came from the sequencing of regions from the human and mouse genomes containing the T cell receptor complex. Not only are the genes conserved, but also large blocks of the repetitive sequences flanking the genes (Koop & Hood, 1994). This long-range conservation occurs despite these species having diverged from one another more than 60 million years ago. It suggests that sequences (repetitive) flanking genes may be important for gene function. A number of studies have identified sequences which may organise chromatin and therefore indirectly affect function. Members of multigene families are often localised close together in the genome but not necessarily so. They may have diverged in function or in their regulatory pattern of expression during development.

The average G + C content of monocot genes is higher than that of dicot genes; the majority of monocot genes have a G + C content of 60–70%, while dicot genes average around 46%. This is a consequence of a higher G + C content at the third codon. For this reason, synonymous codons are used differently by monocots and dicots. In fact, monocots and mammals share the most commonly used codon for 13 out of 18 amino acids while dicots share only 7 out of 18 amino acids. Because of the sequence divergence between homologous genes in dicots and monocots, it is not a straightforward exercise to use dicot DNA sequences to locate and map homologous sequences in large complex Gramineae genomes or vice versa. This is currently slowing up gene isolation in plants. However, monocot and dicot gene sequences are accumulating from major gene-sequencing projects and these facilitate dicot—monocot comparisons. From the Japanese Rice Genome programme, by the end of 1994, 13 000 rice cDNA clones, and 8400 green shoot, etiolated shoot and root libraries, had been sequenced by single-run partial DNA sequencing and 7800 sequences deposited in the DDBJ (DNA Databank of Japan) (Sasaki, 1994). A similar programme is underway to sequence *Arabidopsis* genes (Newman et al., 1994; Höfte et al., 1993). Over 18 000 *Arabidopsis* partial gene sequences have been submitted to databases. A large number of the partial gene sequences have been searched for similarity to known genes of all organisms in the Protein Information Resources (PIR) database. Such searches reveal that 20% of the genes have significant sequence similarly to previously sequenced genes in this database. The genes can be placed into various categories: those that show homology at the nucleotide or amino acid level with genes from many eukaryotic species and are therefore probably involved with basic cellular function; those genes which reveal nucleotide or amino acid level homology to only plant genes; and those that show nucleotide or amino acid level homology to genes derived from only other monocots or other dicot plants. Until most of the rice and *Arabidopsis* genes have been sequenced and compared, it will remain unclear what proportion of genes will occupy each of these categories.

The mapping of gene sequences onto the genetic map has now become much simpler, given the large number of markers (RFLP, minisatellites, VNTR, RAPDs) that can be obtained to cover an entire genome. Genetic distances between genes depend upon the frequency of recombination between the genes. If recombination does not occur at random over the genome, then the genetic map will not represent the physical distance between genes. It is clearly important to know the physical distance between genes to interpret genetic maps and to produce a viable strategy for map-based cloning.

Major deviations between physical and genetic maps have been noted. In wheat and barley, the nucleolus organiser is located towards the distal end of the chromosome arm, whereas on the genetic map it is located close to the centromere (Snape et al., 1985; Gustafson et al., 1990). This is due to a relatively low level of recombination around the centromeres; recombination occurs preferentially in the terminal one-third of these cereal chromosomes.

Comparisons of physical versus genetic distance have been made for a number of loci in maize (Civardi et al., 1994). The interval between the $\alpha 1$ and $sh1$ genes shows 1750 kb per centiMorgan (cM), but within the $\alpha 1$ and $bz1$ loci it is 217 kb/cM and 14 kb/cM respectively. These studies suggest that the genes themselves are the sites of recombination. In tomato, greater than 10-fold differences between genetic and physical distances in different genome sections have been noted (Ganal et al., 1989; Segal et al., 1992). In *Arabidopsis*, sevenfold differences have been found (Dean & Schmidt, 1995). Thus recombination map distances should not be equated with physical map distances.

Clarification of the physical versus genetic maps will reveal whether genes are clustered in large genomes or spread more or less randomly through the genome. For wheat and barley, recent studies suggest that there is a higher density of genes in regions closer to the telomeres. The information comes from the mapping of random cDNAs (Devos & Gale 1993a, b; Devos et al., 1993; Snape et al., 1985), from the physical mapping of cDNA sequences in stocks carrying terminal deletions of different sizes (Werner et al., 1992), and from the distribution of unmethylated sites cleaved by the restriction endonuclease NotI which appear to be preferentially associated with single- or few-copy sequences (Moore et al., 1993a). Consequently, for illustrative purposes, a higher density of genes nearer the telomeres has been shown in Fig. 1.

Gene density would be expected to be very low in the pericentromeric heterochromatin. Any other regions consisting of tandem arrays of a single sequence or of very long, complex sequences would also be expected to have few genes. Where such arrays are common in large genomes, then major differences between genetic and physical maps and gene clustering might be expected. The co-location of regions of higher gene density and recombination frequencies is consistent with the genes being the sites of recombination (Civardi et al., 1994).

The clustering of genes into gene-rich sectors in large genomes has important implications for how genomic libraries might be made and screened in order to minimise effort and maximise efficiency. It should be noted that gene-rich sectors appear not to be amplified in genomes (but duplications are clearly stable) but they are rich in short repeats. This is because they are the sites of insertion of transposable elements. Intergenic regions can tolerate large amounts of additional DNA in many instances (Fig. 2). Such repeats have therefore been shown interspersed with genes in Figs 1 and 2.

8.4   Matrix attachment regions

There is now strong support for the organisation of chromatin in nuclei into a series of loops, 5–200 kb, attached at their bases to a proteinaceous network, the nuclear matrix. The points of attachment, about 1 kb, are termed matrix attachment regions, MARs (or scaffold attachment regions, SARs) (Laemmli et al., 1992).

There are many reports of plant sequences which interact with the nuclear matrix, including some which lie close to genes (Slatter & Gray, 1991; Laemmli et al., 1992; Hall et al., 1991; Allen et al., 1993; Breyne et al., 1992). They are AT-rich and are up to 1000 bp. The sites of attachment are defined using matrix preparations isolated by using detergent or salt. A number of recent reports have questioned whether such experiments accurately reflect the in vivo situation (Jackson et al., 1990; Eggert & Jack, 1991). Recently, however, MARs isolated from tobacco and soybean have been shown to stabilise expression of a transgene, implying that such sequences adjacent to a gene have an effect in vivo (Allen et al., 1993; Breyne et al., 1992). It is likely that MARs are vital components for facilitating access to the genome and allowing regulated transcription, replication and other processes to be activated. MARs would therefore be expected to be widely distributed—every 5–200 kb. It will be important to explore this prediction as long genome sequences emerge. It is likely that the overall structural features of AT-rich DNA rather than the sequence per se determine the affinity of MARs for the matrix (Hall et al., 1991; Thompson et al., 1995).

8.5 Origins of replication

Initiation of DNA replication along eukaryotic chromosomes occurs at a large number of sites. Studies in yeast provide evidence that there are specific sequences (origins of replication), some of which, when cloned into plasmids, confer autonomous replication on them. In contrast to the studies on yeast, little is known on the origins of replication found in higher eukaryotes. Two different approaches to studying them using the dihydrofolate reductase domain in CHO cells have given different results (review: DePamphis, 1993). The first approach indicates that there are short, specific DNA regions within which replication is initiated, and the second indicates initiation of replication across a broad region. Little similar work has been undertaken in plants, and that which has concentrates on studying the mode of replication of the ribosomal RNA genes (Van't Hof & Lamm, 1992; Hernandez et al., 1993). Nevertheless, it is likely that signals exist in chromatin that provoke chromosome replication at a very large number of sites.

9 EVOLUTION OF REPETITIVE SEQUENCES— IMPLICATIONS FOR GENOME ORGANISATION

Genomes are products of evolution. They reflect the kinds of molecular events that change DNA sequences and the consequences of selection upon the variation created. Families of repetitive sequences evolve rapidly in plant genomes relative to the coding sequences. This is because some sequences, such as tandem arrays and transposable elements, are predisposed to change in frequency in the genome and species as a consequence of their sequence and the fact that they are repeated (Flavell, 1982, 1985). They become caught up in unequal crossing over, transposition, gene conversion and other processes that lead to changes in sequence and copy number. Transposition also leads to the insertion of short pieces of DNA into new sites in the genome and hence sequence scrambling. Insertion of transposable elements with the same LTRs close by but in opposite orientation can, for example, together with transposable elements with inverted repeats, explain why 10% of cereal fragments a few kilobases long contain inverted repeats (Flavell, 1980). DNA segments of one tandem array may become parts of other tandem arrays, and similarly parts of one transposable element may become part of another (Moore et al., 1991b; Lucas et al., 1992). Such events greatly enhance genomic complexity but can be interpreted when primary repetitive units are known.

Repeated sequences are easily lost, by unequal crossing over or intrastrand recombination between homologous sequences. Thus the repetitive DNA complement is continuously being amplified and lost. Because of the continual reamplification of individual members or subsets of repeated sequence families, ancient families are continually being recycled as new subfamilies. These are unlikely to be distributed uniformly across the genome—as has been shown for several families in tandem arrays, where chromosome-specific variants have been discovered. This may also be the explanation for why subfamilies of the retrotransposon *Bis*1 and *Hi*-10 repeats in cereals are not distributed uniformly across the 'megabase' repeat units revealed after cleavage of DNA with *MluI*, which recognises unmethylated sites (Abbo et al., 1995).

Where related species have been shown to possess varying amounts of repetitive DNA, the quantitative differences are frequently found distributed over the genome and not all concentrated in one chromosome arm (Flavell, 1986a). Therefore, it can be envisaged that changes in genome organisation are occurring continuously all over the genome, especially in species with a high proportion of repetitive DNA. The amount of DNA accumulated within any genome segment is presumably the outcome of the random mutational processes and genetic drift, and their interaction with the forces of selection on the individuals carrying the variation. All of these events involving repeated sequences lead to chromosome divergence, and possibility speciation, as populations diverge (Flavell, 1982).

10   FUTURE PERSPECTIVES

Plant genome characterisation via mapping, genome sequencing and cDNA sequencing is progressing alongside the human genome project and the sequencing of the yeast, *E. coli* and other genomes. The functions of many plant genes will be gained from databases of their counterparts in bacteria, fungi and animals. Nevertheless, it will be essential to make in vivo mutants in the plant genes, by transposons or, in transgenic plants, using sense or antisense constructs, to understand the role of a gene in a particular host. The same goes for all the *cis*-acting regulatory sequences and the *trans*-acting factors regulating gene expression.

The richness of within-species plant diversity emanates from combinations of specific alleles. Crop plant improvement programmes also rely on such allelic variation. Therefore, whatever knowledge about gene function is gained from comparative databases, there will be the need to study particular alleles isolated from individual plants. This is why gene mapping and efficient gene isolation methodologies are so important. The discovery of conserved frameworks in plant genome segments, based on gene syntenies, is going to make gene mapping and isolation of specific alleles in more species much more efficient.

While the advances in DNA-sequencing techniques are providing much knowledge about the genomes, an understanding of the three-dimensional structure of chromatin and the consequences of genome coiling for the control of gene expression and the behaviour of chromosomes lags far behind. A reminder of this is the variable expression of transgenes when inserted into plant chromosomes. They can be silent, or change activity during somatic growth or between generations (Flavell, 1994; Matzke & Matzke, 1995). Understanding such events is likely to give us new insight into the links between genome structure and genome behaviour.

The stage is therefore set for a golden era in the characterisation of genomes and the isolation of plant genes, from many species. What will be the value of such research to societies worldwide? Much has been made of the value of recognising and isolating human genes that cause disease, but the benefit is restricted to the relatively few families carrying the inherited diseases. In contrast, if the fruits of plant genome research are put to use via crop improvement programmes, especially in making transgenic crops, the benefits could be enormous for all societies. The importance of plants to food production, as sources of raw materials and for sustaining the environment, surely means that plant genome studies will be of greater significance to the future of our planet than even those of the human genome.

ESSENTIAL READING

Dean, C. and Schmidt, R. (1995) Plant genomes: a current molecular description. *Annu. Rev. Plant Physiol. Plant Mol. Biol.*, **46**, 395–418.

Finnegan, E.J., Brettell, R.I.S. and Dennis, E.S. (1993) The role of DNA methylation in the regulation of plant gene expression. In *DNA Methylation: Molecular Biology and Biological Significance* (eds J.P. Jost and H.P. Saluz), pp. 218–261. Birkhauser Verlag, Basel.

Flavell, R.B. (1985) Repeated sequences and genome change. In *Advances in Plant Gene Research*, Vol. 2 (eds E. Dennis and B. Hohn), pp. 139–156. Springer Verlag, Vienna.

Moore, G., Gale, M.D., Kurata, N. and Flavell, R.B. (1993b) Molecular analyses of small grain cereal genomes. Current status and prospects. *Bio/Technology*, **11**, 584–589.

Moore, G., Devos, K., Wang, Z. and Gale, M. (1995b) Cereal genome evolution: grasses line up and form a circle. *Curr. Biol.*, **5**, 737–739.

REFERENCES

Abbo, S., Dunford, R.P., Foote, T.N. et al. (1995) Organisation of retroelement and stem-loop repeat families in the genomes and nuclei of cereals. *Chromosomal Res.*, **3**, 5–15.

Ahn, S. and Tanksley, S.D. (1993) Comparative linkage maps of the rice and maize genomes. *Proc. Natl. Acad. Sci. USA*, **90**, 7980–7984.

Alfenito, M.R. and Birchler, J.A. (1993) Molecular characterisation of a maize B chromosome centromeric sequence. *Genetics*, **135**, 589–597.

Allen, G.C., Hall, G.E., Childs, L.C. et al. (1993) Scaffold attachment regions increase reporter gene expression in stably transformed plant cells. *Plant Cell*, **5**, 603–613.

Antequera, F. and Bird, A.P. (1988) Unmethylated CpG islands associated with genes in higher plant DNA. *EMBO J.*, **7**, 2295–2299.

Bedbrook, J.R., Jones, J., O'Dell, M. et al. (1980a) A molecular description of telomeric heterochromatin in *Secale* species. *Cell*, **19**, 545–560.

Bedbrook, J.R., O'Dell, M. and Flavell, R.B. (1980b) Amplification of rearranged sequences in cereal plants. *Nature*, **288**, 133–137.

Bell, C.J. and Ecker, J.D. (1994) Assignment of 30 microsatellite loci to the linkage map of *Arabidopsis*. *Genomics*, **19**, 137–144.

Bennett, M.D. (1973) Nuclear characters in plants. *Brookhaven Symp. Biol.*, **25**, 344–366.

Bennett, M.D. (1982) Nucleotypic basis of the special ordering of chromosomes in eucaryotes and the implication of the order for genomic evolution and phenotypic variation. In *Genome Evolution* (eds G.A. Dover and R.B. Flavell), pp. 239–261. Academic Press, London.

Bennett, M.D. (1983) The spatial distribution of chromosomes. In *Kew Chromosomes Conference II* (ed. P.E. Brandham), pp. 71–79. George Allen and Unwin Ltd, London.

Bennett, M.D. and Smith, J.B. (1991) Nuclear DNA amounts in angiosperms. *Phil. Trans. R. Soc. Lond. (Biol.)*, **334**, 309–345.

Bennetzen, J.L. and Freeling, M. (1993) Grasses as a single genetic system: genomic composition, collinearity and compatibility. *Trends Genet.*, **9**, 259–261.

Bennetzen, J.L., Schrick, K., Springer, P.S. et al. (1994) Active maize genes are unmodified and flanked by diverse classes of modified highly repetitive DNA. *Genome*, **51**, 565–576.

Beukeboom, L.W. (1994) Bewildering Bs: an impression of the 1st chromosome conference. *Heredity*, **73**, 328–336.

Binelli, G., Gianfranceschi, L., Pe M.E. et al. (1992) Similarity of maize and *Sorghum* genomes as revealed by maize RFLP probes. *Theor. Appl. Genet.*, **84**, 10–16.

Bird, A.P. (1987) CpG islands as markers in the vertebrate nucleus. *Trends Genet.*, **3**, 342–347.

Blackburn, E.H. (1994) Telomeres: no end in sight. *Cell*, **77**, 621–623.

Bonierbale, M.W., Plaisted, R.L. and Tanksley, S.D. (1988) RFLP maps based on a common set of clones reveal modes of chromosomal evolution in potato and tomato. *Genetics*, **120**, 1095–1103.

Breyne, P., Van Montagu, M., Depicker, A. and Gheysen, G. (1992) Characterisation of a plant scaffold attachment region in a region that normalises transgene expression in tobacco. *Plant Cell*, **4**, 463–471.

Bureau, T.E. and Wessler, S.R. (1992) Tourist, a large range of small-inverted repeat elements frequently associated with maize genes. *Plant Cell*, **4**, 1284–1294.

Burr, B., Burr, F.A., Matz, E.C. and Romero-Severson, J. (1992) Pinning down loose ends: mapping telomeres and factors affecting their length. *Plant Cell*, **4**, 953–960.

Cheung, W.Y., Money, T.A., Abbo, S. et al. (1994) A family of related sequences associated with (TTTAGGG)n repeats are located in the interstitial regions of wheat chromosomes. *Mol. Gen. Genet.*, **245**, 349–354.

Civardi, L., Xia, Y., Edwards, K.J. et al. (1994) The relationship between the genetic and physical distances in the cloned $\alpha 1–sh2$ interval of the *Zea mays* L. genome. *Proc. Natl. Acad. Sci. USA*, **91**, 8268–8272.

Coen, E. S. and Carpenter, R. (1986) Transposable elements in *Antirrhinum majus*: generators of genetic diversity. *Trends Genet.*, **2**, 292–296.

Coen, E.S., Robbins, T.P., Almeida, J. et al. (1989) Consequences and mechanisms of transposition in *Antirrhinum majus*. In *Mobile DNA* (eds D.E. Berg and M.M. Howe), pp. 413–436. American Society for Microbiology, Washington.

Cross, S., Lindsey, J., Fantes, J. et al. (1990) The structure of a subterminal repeated sequence present on many human chromosomes. *Nucleic Acids Res.*, **18**, 6649–6657.

Dennis, E.S., Gerlach, W.L. and Peacock, W.J. (1980) Identical polypyrimidine–polypurine satellite DNAs in wheat and barley. *Heredity*, **44**, 349–366.

DePamphis, M.L. (1993) Eukaryotic DNA replication–anatomy of the origin. *Annu. Rev. Biochem.*, **62**, 29–63.

Deshpande, V.G. and Ranjekar, P.K. (1980) Repetitive DNA in the gramineae species with low DNA content. *Hoppe-Seylers's Z. Physiol. Chem.*, **361**, 1223–1233.

Deumling, B. and Greilhuber, J. (1982) Characterisation of heterochromatin in different species of the *Scilla siberica* group (liliaceae) by *in situ* hybridisation of satellite DNAs and fluorochrome banding. *Chromosoma (Berl.)*, **84**, 535–555.

Devos, K.M. and Gale, M.D. (1993a) Extended genetic maps of the homologous group 3 chromosomes of wheat, rye and barley. *Theor. Appl. Genet.*, **85**, 649–652.

Devos, K.M., and Gale, M.D. (1993b) The genetic maps of wheat and their potential in plant breeding. *Outlook on Agriculture*, **22**, 93–99.

Devos, K.M., Miller, T. and Gale, M.D. (1993) Comparative RFLP maps of homologous group 2-chromosomes of wheat, rye and barley. *Theor. Appl. Genet.*, **85**, 784–792.

Döring, H-P and Starlinger, P. (1986) Molecular genetics of transposable elements in plants. *Annu. Rev. Genet.*, **20**, 175–200.

Dover, G.A. (1986) Molecular drive in multigene families: how biological novelties arise, spread and are assimilated. *Trends Genet.*, **2**, 159–165.

Dunford, R.P., Kurata, N., Laurie, D.A. et al. (1995) Conservation of fine-scale DNA marker order in the genomes of rice and the Triticeae. *Nucleic Acids Res.*, **23**, 2774–2728.

Eggert, H. and Jack, R. (1991) An ecotopic copy of the Drosophila Hz associated SHR neither reorganises local chromatin structures nor hinders elution of a chromatin fragment from isolated nuclei. *EMBO J.*, **10**, 1237–1243.

Ellis, T.H.N., Lee, D., Thomas, C.M. et al. (1988) 5s rRNA genes in *Pisum*: sequence, long range and chromosomal organisation. *Mol. Gen. Genet.*, **214**, 333–342.

Fedoroff, N.V. (1989) Maize transposable elements. In *Mobile DNA* (eds D.E. Berg and M.M. Howe), pp. 375–412. American Society for Microbiology, Washington.

Flavell, R.B. (1980) The molecular characterisation and organisation of plant chromosomal DNA sequences. *Annu. Rev. Plant Physiol.*, **31**, 569–576.

Flavell, R.B. (1982) Amplification, deletion and rearrangement: major sources of variation during species divergence. In *Genome Evolution* (eds G.A. Dover and R.B. Flavell), pp. 301–324. Academic Press, London.

Flavell, R.B. (1986b) Repetitive DNA and chromosomal evolution in plants. *Phil. Trans R. Soc. Lond. B*, **312**, 227–242.

Flavell, R.B. (1986b) The structure and control of expression of ribosomal RNA genes. In *Oxford Surveys of Plant Molecular and Cell Biology* (ed. B.J. Miflin), Vol. 3, pp. 252–274. Oxford University Press, Oxford.

Flavell, R.B. (1994) Inactivation of gene expression in plants as a consequence of specific sequence duplication. *Proc. Natl. Acad. Sci. USA*, **91**, 3490–3496.

Flavell, R.B., Rimpau, J. and Smith, D.B. (1977) Repeated sequence DNA relationships in four cereal genomes. *Chromosoma*, **63**, 205–222.

Ganal, M.W., Young, N.D. and Tanksley, S.D. (1989) Pulsed field gel electrophoresis and physical mapping of large DNA fragments in the *Tm-2a* region of chromosome 9 in tomato. *Mol. Gen. Genet.*, **215**, 395–400.

Ganal, M.W., Lapitan, N.L.V. and Tanksley, S.P. (1991) Macrostructure of tomato telomeres. *Plant Cell*, **3**, 87–94.

Gebhardt, C., Ritter, E., Barone, A. et al. (1991) RFLP maps of potato and their alignment with the homologous tomato genome. *Theor. Appl. Genet.*, **83**, 49–57.

Gerats, A.G.M., Huits, H., Vrijlandt, E. et al. (1990) Molecular characterisation of a nonautonomous transposable element (*dTph1*) of petunia. *Plant Cell*, **2**, 1121–1128.

Gerlach, W.L. and Bedbrook, J.R. (1979) Cloning and characterisation of ribosomal RNA genes from wheat and barley. *Nucleic Acids Res.*, **1**, 1869–1885.

Gerlach, W.L. and Peacock, W.J. (1980) Chromosomal locations of highly repeated sequences in wheat. *Heredity*, **44**, 269–276.

Grandbastien, M.A. (1992) Retroelements in higher plants. *Trends Genet.*, **8**, 103–108.

Grellet, F., Delcasso, D., Panabieres, F. and Delseny, M. (1986) Organisation and evolution of a higher plant alphoid-like satellite DNA sequence. *J. Mol. Biol.*, **187**, 495–507.

Gruenbaum, Y., Naveh-Many, T., Cedar, H. and Razin, A. (1981) Sequence specificity of methylation in higher plant DNA. *Nature*, **292**, 860–862.

Gustafson, P., Butler, E. and McIntyre, C.L. (1990) Physical mapping of a low-copy DNA sequence in rye (*Secale cereale* L.). *Proc. Natl. Acad. Sci. USA*, **87**, 1899–1902.

Hall, G. Jr, Allen, G.C., Loer, D.S. et al. (1991) Nuclear scaffolds and scaffold-attachment regions in higher plants. *Proc. Natl. Acad. Sci. USA*, **88**, 9320–9324.

Helentjaris, T., Weber D. and Wright, S. (1988) Identification of the genomic locations of duplicate nucleotide sequences in maize by analysis of restriction fragment length polymorphisms. *Genetics*, **118**, 353–363.

Hernandez, P., Martin-Parras, L., Martinez-Robles, M.L. and Schrartzman, J.B. (1993) Conserved features in the mode of replication of eukaryotic ribosomal RNA genes. *EMBO J.*, **2**, 1475–1485.

Heslop-Harrison, J.S. and Bennett, M.D. (1990) Nuclear architecture in plants. *Trends Genet.*, **6**, 401–405.

Hillel, J., Schaap, T., Haberfeld, A. et al. (1990) DNA fingerprints applied to gene introgression in breeding programs. *Genetics*, **124**, 783–789.

Höfte, H., Desprez, T., Amselem, J. et al. (1993) An inventory of 1152 expressed sequence tags obtained by partial sequencing of cDNAs from *Arabidopsis thaliana*. *Plant J.*, **4**, 1051–1061.

Horowitz, H., Thornburn, P. and Haber, J.E. (1984) Rearrangements of highly polymorphic regions near telomeres of Saccharomyces cerevisiae. *Mol. Cell. Biol.*, **4**, 2509–2517.

Hutchinson, J. and Lonsdale, D. (1982) The chromosomal distribution of cloned highly repetitive sequences from hexaploid wheat. *Heredity*, **48**, 373–378.

Jackson, D.A., Dickinson, P. and Cook, P.R. (1990) The size of chromatin loops in Hela cells. *EMBO J.*, **4**, 567–571.

Jones, R.N. and Rees, H. (1982) *B Chromosomes*, Academic Press, London.

Kilian, A., Kudrra, D.A., Kleinhofs, A. et al. (1995) Rice–barley synteny and its application to saturation mapping of the barley Rpg1 region. *Nucleic Acids Res.*, **23**, 2729–2733.

Koop, B.F. and Hood, L. (1994) Striking sequence similarity over 100 kilobases of human and mouse T cell receptor DNA. *Nature Genet.*, **7**, 48–53.

Kowalski, S.P., Tien-Hung, L., Feldmann, K.A. and Paterson, A.H. (1994) Comparative mapping of *Arabidopsis thaliana* and *Brassica oleracea* chromosomes reveals islands of conserved organisation. *Genetics*, **138**, 499–510.

Kurata, N., Moore, G., Nagamura, H. et al. (1994) Conservation of genome structure between rice and wheat. *Bio/Technology*, **12**, 276–278.

Laemmli, U., Kas, E. and Poljak, L. (1992) Scaffold-associated regions as acting determinants of chromatin structural loops and functional domains. *Curr. Opin. Genes Dev.*, 275–285.

Lapitan, N.L.V., Ganal, M.W. and Tanksley, S.D. (1991) Organisation of the 5S ribosomal RNA genes in the genome of tomato. *Genome*, **34**, 509–514.

Leutwiler, L.S., Hough-Evans, B.R. and Meyerowitz, E.M. (1984) The DNA of *Arabidopsis thaliana*, *Mol. Gen. Genet.*, **194**, 15–23.

Lucas, H., Moore, G., Murphy, G. and Flavell, R.B. (1992) Inverted repeats in the long terminal repeats of the wheat retrotransposon Wis 2-1A. *Mol. Biol. Evol.*, **1**, 716–728.

Lydiate, D., Sharpe, A., Lagercrantz, U. and Parkin, I. (1993) Mapping the *Brassica* genome. *Outlook Agric.*, **22**, 85–92.

Maluszyska, J. and Heslop-Harrison, J.S. (1991) Localisation of tandemly repeated DNA sequences in *Arabidopsis thaliana*. *Plant J.*, **1**, 159–166.

Martinez-Zapater, J.M., Estelle, M.A. and Somerville, C.R. (1986) A highly repeated DNA sequence in *Arabidopsis thaliana*. *Mol. Gen. Genet.*, **204**, 417–423.

Matzke, M. and Matzke, A. (1995) How and why do plants inactivate homologous (trans)genes? *Plant Physiol.*, **107**, 679–685.

Menancio-Hautea, D., Fatokun, C.A., Kumar, L. et al. (1993) Comparative genome analysis of mungbean (*Vigna radiata* L. Wilczek) and cowpea (*V. unguiculata* L. Walpers) using RFLP mapping data. *Theor. Appl. Genet.*, **86**, 797–810.

Montero, L.M., Salinos, J., Matassic, G. and Bernardi, G. (1990) Gene distribution and isochore organisation in the nuclear genome of plants. *Nucleic Acids Res.*, **18**, 1859–1867.

Moore, G., Cheung, W., Schwarzacher, T. and Flavell, R.B. (1991a) *Bis-1* a major component of the cereal genome and a tool for studying genome organisation. *Genomics*, **10**, 469–476.

Moore, G., Lucas, H., Batty, N. and Flavell, R.B. (1991b) A family of retrotransposons and associated genomic variation in wheat. *Genomics*, **10**, 461–468.

Moore, G., Abbo, S., Cheung, W. et al. (1993a) Key features of cereal genome organisation as revealed by the use of cytosine methylation-sensitive restriction enzymes. *Genomics*, **15**, 472–482.

Moore, G., Foote, T., Helentjaris, T. et al. (1995a) Was there a single ancestral cereal chromosome? *Trends Genet.*, **11**, 81–83.

Morgante, M. and Olivieri, A.M. (1993) PCR-amplified microsatellites as markers in plant genetics. *Plant J.*, **3**, 175–182.

Newman, T., Retzel, E., Shoop, L. et al. (1994) Generation, analysis and dissemination of expressed sequence tags for Arabidopsis thaliana. In *Genomes and Chromosomes*, 20th EMBO Annual Symposium. EMBL, Heidelberg.

Richards, E.J. and Ausubel, F.M. (1988) Isolation of a higher eukaryotic telomere from *Arabidopsis thaliana*. *Cell*, **53**, 127–136.

Richards, E.J., Goodman, H.M. and Ausubel, F.M. (1991) The centromere region of *Arabidopsis thaliana*. *Nucleic Acids Res.*, **19**, 3351–3357.

Rimpau, J., Smith, D.B. and Flavell, R.B. (1978) Sequences organisation analysis of wheat and rye genomes by interspecies DNA/DNA hybridisation. *J. Mol. Biol.*, **123**, 327–359.

Roder, M.S., Lapitan, N.L.V., Sorrells, M.E. and Tanksley, S.D. (1993) Genetic and physical mapping of barley telomeres. *Mol. Gen. Genet.*, **238**, 294–303.

Sasaki, T. (1994) *Rice Genome, Newsletter for Rice Genome Analysis*, Vol. 312, Tsukuba, Japan. Rice Genome Research Program, NIAR/Staff 1994.

Schmidt, R. and Dean, C. (1992) Physical mapping of the *Arabidopsis thaliana* genome. In *Genome Analysis*, Vol. 4, *Strategies for Physical Mapping* (eds K.E. Davies and S.M. Tilghman), pp. 71–98. Cold Spring Harbor Laboratory Press, New York.

Schwarz-Sommer, Z., Giere, A., Klosgen, R. et al. (1985) The Spm (En) transposable 'element' controls the excision of a 2 kb DNA insert at the wx^{m-8} locus of *Zea mayo*. *EMBO J.*, **4**, 591–597.

Schweizer, G., Ganal, M., Ninnemann, H. and Hemleben, V. (1988) Species-specific DNA sequences for identification of somatic hybrids between *Lycopersicon esculentum* and *Solanum acaule*. *Theor. Appl. Genet.*, **75**, 679–684.

Segal, G., Sarfatti, M., Schaffer, M.A. et al. (1992) Correlation of genetic and physical structure in the region surrounding the I_2*Fusarium oxyspoprum* resistance locus in tomato. *Mol. Gen. Genet.*, **231**, 179–185.

Slatter, R.E. and Gray, J.C. (1991) Chromatin structure of plant genes. In *Oxford Surveys of Plant Molecular and Cell Biology* (ed. B.J. Miflin), Vol. 7, pp. 115–142. Oxford University Press.

Snape, J.W., Flavell, R.B., O'Dell, M. et al. (1985) Intrachromosomal mapping of the nucleolus organiser region relative to three marker loci on chromosome 1B of wheat (*Triticum aestivum*). *Theor. Appl. Genet.*, **69**, 263–270.

Springer, P.S., Edwards, K.J. and Bennetzen, J.L. (1994) DNA class organisation on maize *Adh* 1 yeast artificial chromosomes. *Proc. Natl. Acad. Sci. USA*, **91**, 863–867.

Steele Scott, N. and Timmis, J.N. (1984) Homologies between nuclear and plastid DNA in spinach. *Theor. Appl. Genet.*, **69**, 263–270.

Swartz, M.N., Trautner, T.A. and Kornberg, A. (1962) Enzymatic synthesis of deoxyribonucleic acid. *J. Biol. Chem.*, **237**, 1961–1967.

Tanksley, S.D., Bernatzky, R., Lapitan, N.L. and Prince, J.P. (1988) Conservation of gene repertoire but not gene order in pepper and tomato. *Proc. Natl. Acad. Sci. USA*, **85**, 6419–6423.

Thompson, W.F. and Murray, M.G. (1981) The nuclear genome structure and function. In *Biochemistry of Plants* (eds P.K. Stump and E.E. Conn), pp. 1–81. Academic Press, New York.

Thompson, W.F., Allen, G.C., Hall, G. Jr and Spiker, S. (1995) Matrix attachment regime and transgene expression in genomes. In *Genome*, 21st Stadler Symposium (eds. J.P. Gustafson and R.B. Flavell). Plenum Publishing Corporation, New York.

Tsay, Y-F., Frank, M.J., Page, T. et al. (1993) Identification of a mobil endogenous transposon in Arabidopsis thaliana. *Science*, **260**, 342–344.

Tyler-Smith, C., Oakey, R.J., Larin, Z. et al. (1993) Localisation of DNA sequences required for human centromere function through an analysis of re-arranged Y chromosomes. *Nature Genet.*, **5**, 368–375.

Van't Hof, J. and Lamm, S.S. (1992) Site of initiation of replication of the ribosomal genes of pea (*Pisum sativum*) detected by two dimensional gel electrophoresis. *Plant Mol. Biol.*, **20**, 377–382.

Weeden, N.F., Muehlbauer, F.J. and Ladizinsky, G. (1992) Extensive conservation of linkage relationships between pea and lentil genetic maps. *Heredity*, **83**, 123–129.

Werner, J.E., Endo, T.R. and Gill, B.S. (1992) Toward a cytogenetically based physical map of the wheat genome map of the wheat genome. *Proc. Natl. Acad. Sci. USA*, **89**, 11307–11311.

Whitkus, R., Doebly, J. and Lee, M. (1992) Comparative genome mapping of *Sorghum* and maize. *Genetics*, **132**, 1119–1130.

Wu, K.S. and Tanksley, S.D. (1993) PFGE analysis of the rice genome, estimation of fragment sizes organisation of repetitive sequences and relationships between genetic and physical distances. *Plant Mol. Biol.*, **23**, 243–254.

I Library Construction

2 cDNA Library Construction

PATRICK J. HUSSEY
Royal Holloway, University of London, Egham, Surrey, UK

JOHN HUNSPERGER
Haplogenetics, Gilroy, CA, USA

The ability to generate cDNA libraries from mRNA has had a considerable impact on our current understanding of genome organisation and gene expression. The generation of cDNA is the pivotal stage of many of the techniques described in this book. The purpose of this chapter is to provide an overview of methods for plant RNA isolation and the main strategies involved in cDNA library construction. Specific areas which will be covered by this chapter include:

(1) Isolation of RNA from plant tissue.
(2) Strategies for cDNA synthesis.
(3) Preferred vectors for cDNA cloning.
(4) *E. coli* strains.
(5) Transfer of DNA to *E. coli*.
(6) Availability of pre-made libraries.

1 INTRODUCTION

The successful isolation of intact mRNA is a critical step in many protocols in molecular biology, including cDNA library construction, northern analysis and *in vitro* translation. The methods for RNA isolation from plant tissues have had to address three main problems, each of which causes a reduction in RNA yield. These are: (1) breaking the cell wall of the majority, if not all, cells; (2) eliminating nuclease activity, which is relatively high in plant cells; and (3) removing contaminants such as DNA, starch and phenolics. The mRNA can be converted to double-stranded complementary DNA (cDNA) by a series of enzymatic procedures. The ability to generate cDNA from mRNA has had a considerable impact in leading to our current understanding of genome organisation (Chapter 1) and gene expression (see Chapters 5–8). Whether one chooses to isolate a cDNA clone or the respective gene would depend on the question being addressed, but in most cases the cDNA is cloned before the gene. This is because cloning genes directly from genomic DNA can be time-consuming, and the process is accelerated if the respective cDNA clones

Plant Gene Isolation: Principles and Practice. Edited by G. D. Foster and D. Twell.
© 1996 John Wiley & Sons Ltd.

are used as probes (Chapters 4 and 5). A cDNA library is a representation of the genes expressed in the tissue from which the mRNA was isolated. For this reason, cDNA libraries have been instrumental in the identification of tissue-specific and developmental-stage-specific genes (Chapter 6) and cDNA libraries have formed the basis for many of the current large-scale shotgun sequencing projects (Chapter 15). The purpose of this chapter is to present methods for plant RNA isolation and the main strategies involved in cDNA library construction.

2   RNA ISOLATION

The main RNA isolation protocol used by the authors is essentially the same as that described in Lefebvre et al. (1980), but modified for use with higher plants. Plant tissue is ground to a fine powder in liquid nitrogen, transferred to extraction buffer (50 mM Tris-HCl, pH 8.5, 300 mM NaCl, 5 mM EDTA, 2% (w/v) sodium dodecyl sulphate (SDS), 40 μg/ml proteinase K; 5 ml per gram fresh weight of tissue) and immediately mixed by vortexing for 1 min. The suspension is extracted twice with an equal volume of phenol/chloroform/isoamyl alcohol (25 : 24 : 1) and once with an equal volume of chloroform/isoamyl alcohol (24 : 1). The nucleic acid is precipitated from the aqueous phase by adding 0.1 volumes of 3 M sodium acetate and 2 volumes of filtered absolute ethanol and leaving overnight at $-20\,°C$. The nucleic acid is harvested by centrifugation (8000 rev/min for 10 min at 4 °C using the SS-34 rotor in a Sorvall RC-5B centrifuge), and the pellet is air-dried and then resuspended* (see below) in 1 M CsCl in TES buffer (1 M CsCl, 50 mM Tris-HCl, 50 mM NaCl, 5 mM EDTA, pH 8.0). The resuspended pellet is dispensed into 3 ml aliquots and loaded over 2 ml of 5.7 M CsCl in TES buffer in 5 ml Beckman polyallomer centrifuge tubes. After centrifugation (33 000 rev/min for 18 h at 22 °C using a SW50.1 rotor in a Beckman L8-M ultracentrifuge), the RNA pellets whilst the DNA forms a band at the 1 M CsCl/5 M CsCl interface and contaminants such as protein, starch and phenolics remain in suspension in the 1 M CsCl. The CsCl is carefully removed from the centrifuge tube with a Pasteur pipette, working from the top of the tube towards the base; the Pasteur pipette is kept just below the meniscus. The RNA pellet is resuspended in ice-cold TE buffer (10 mM Tris-HCl, 1 mM EDTA, pH 8.0) containing 1% (v/v) β-mercaptoethanol. The RNA is precipitated and harvested as described previously. The RNA pellet is resuspended in TE buffer and the concentration of an aliquot of the RNA is determined spectrophotometrically. The integrity of the RNA is assessed by running 1 μg of heat-denatured RNA on a 0.8% agarose gel. After staining the gel with ethidium bromide and visualisation of the RNA using a UV transilluminator, the two major ribosomal RNA bands should be of approximately equal intensity. The expected yield of total RNA for maize seedling material, for example, is between 0.5 and 0.7 mg per gram fresh weight of tissue. In this context it is important to note that at the stage marked with an asterisk above, not more than an estimated 0.75 mg of total RNA is loaded in each CsCl gradient. This will ensure that the final RNA preparations are not contaminated with DNA. The authors have used the procedure described above for isolating RNA from maize, tomato, *Arabidopsis*, French bean, tobacco and carrot tissues.

In any RNA isolation procedure, by far the greatest problem that has to be overcome is the removal of RNase activity arising from endogenous RNases in the tissue extract and from exogenous RNases in the extraction solutions, on glassware and on 'fingers'. All current protocols aim to reduce the level of endogenous RNase activity by including one or more of the following components in the extraction buffer: divalent ion chelators, e.g. EDTA; RNase inhibitors, e.g. vanadyl–ribonucleoside complexes and human placental RNase inhibitor; denaturants, e.g. SDS, β-mercaptoethanol, guanidine hydrochloride and guanidium isothiocyanate; and proteases, e.g. proteinase K. Levels of exogenous RNase activity can be reduced by taking the following precautions. All glassware should be soaked in 0.1% diethyl pyrocarbonate (DEPC) overnight and sterilised by baking for at least 2 h at 200 °C or by autoclaving (plasticware bought as sterile does not usually require any further treatment). All solutions should be prepared using DEPC-treated water (Sambrook et al., 1989) and sterilised by autoclaving. RNase contamination from fingers is minimised if gloves are worn at all times. Nucleic acids adhere to glass, and improved yields will result if all glassware is siliconised (Sambrook et al., 1989).

The purification of mRNA from total RNA is a usual step when preparing mRNA for use in the construction of cDNA libraries. The majority of eukaryotic mRNAs are post-transcriptionally modified by the addition of adenine nucleotides at the 3' end to produce a polyadenylate (poly(A)) tail. The traditional method for the purification of poly(A)$^+$ RNA is oligodeoxythymidylate (oligo(dT)) cellulose affinity column chromatography (Aviv & Leder, 1972). In this procedure poly(A)$^+$ RNA is bound to the oligo(dT)-cellulose by running total RNA in a high-salt buffer (0.5 M NaCl, 10 mM Tris-HCl, 1 mM EDTA, 0.1% (w/v) SDS, pH 7.6) through the column. The unbound RNA is poly(A)$^-$ (e.g. rRNA and tRNA). The poly(A)$^+$ RNA is eluted from the oligo(dT)-cellulose with low-salt buffer (10 mM Tris-HCl, 1 mM EDTA, pH 7.6). This procedure is standard and a detailed protocol can be found in Sambrook et al. (1989).

Several commercial groups now provide kits for the isolation of total RNA and poly(A)$^+$ RNA (e.g. Ambion, Amersham, Promega, Pharmacia). One of these kits, PolyATract^R 1000 (Promega), has been tested for the direct isolation of poly(A)$^+$ RNA from plant tissue homogenates, eliminating the need for total RNA isolation. This kit produces very good yields of plant poly(A)$^+$ RNA (Neil, 1993). The PolyATract system uses oligo(dT) probes coupled to biotin and takes advantage of fast solution hydridisation kinetics. In the procedure, biotinylated oligo(dT) and poly(A)$^+$ RNA are annealed in solution, after which paramagnetic streptavidin particles are added to bind the biotin moiety. A strong magnet then pulls the paramagnetic particle–RNA complexes out of solution. After a series of high-stringency washes, water is added to release the poly(A)$^+$ RNA into solution (Ekenberg et al., 1992).

Before poly(A)$^+$ RNA is used as a template for cDNA preparation, some cDNA synthesis protocols include a strong denaturation step (e.g. treatment with methyl mercury) to remove any secondary structure. This is to ensure that the polymerase enzymes are not prematurely halted by double-stranded regions.

3   cDNA SYNTHESIS

The first reports on the molecular cloning of cDNAs were those of globin cDNA (Rougeon & Mach, 1976; Efstratiadis et al., 1976) and involved two principal approaches that differed only in the strategy for the initiation of second-strand cDNA synthesis. Both methods required oligo(dT) (typically oligo(dT)12–18) to be annealed to the poly(A) tail, which could then act as a primer–initiator molecule for a RNA-dependent DNA polymerase (reverse transcriptase), such as that found in the RNA tumour virus, avian myeloblastosis virus (AMV). The product is primarily single-stranded cDNA. Second-strand cDNA synthesis and cloning of the resulting double-stranded cDNA was carried out by either of two methods:

(1)   The single-stranded cDNA alone was used as the primer–template for *E. coli* DNA polymerase (or AMV reverse transcriptase (RT), but this was not as efficient for all double-stranded cDNAs prepared). It was suggested that the AMV RT caused the elongation of the 3′ terminus of the cDNA to form a loop which could act as a primer for the *E. coli* DNA polymerase. After second-strand synthesis the single-stranded loop could be cut by S1 nuclease. Homopolymeric tails were added to the 3′ termini of the double-stranded DNA using terminal deoxytransferase. The tailed cDNA was subsequently ligated to cut pBR322 possessing complementary homopolymeric tails.

(2)   The single-stranded DNA was elongated with a homopolymeric tail using terminal deoxytransferase and was then replicated using a complementary oligonucleotide primer and *E. coli* DNA polymerase. Such double-stranded cDNA can be inserted into cut and appropriately tailed pBR322 (Rougeon et al., 1975).

There have now been many modifications to improve the procedure in order to obtain full-length cDNAs and a truly representative library of cDNA prepared from tissue mRNA. The modifications for first-strand synthesis include using improved methods of poly(A)$^+$ RNA isolation, removing secondary structure in the RNA and the use of more efficient reverse transcriptase enzymes. For second-strand synthesis, the formation and removal of the hairpin loop invariably resulted in loss of some cDNA sequence corresponding to the 5′ region of the mRNA. In addition, homopolymeric tailing of the single-stranded cDNA was difficult to control. As a result, improved strategies for second-strand synthesis have been developed. Also, methods for efficient

cloning of the cDNA into vectors have evolved, as have the vectors themselves
and the *E. coli* strains used for transformation or infection.

One of the protocols which has been successfully used by the authors is
based on the procedure devised by Okayama & Berg (1982) for constructing
unidirectional cDNA libraries (Fig. 1). The Okayama and Berg procedure
involved trapping poly(A)$^+$ RNA on oligo(dT)-tailed plasmid. The oligo(dT)
acts as the primer for the reverse transcriptase (AMV RT). In the next step,
oligo(dC) is added to the 3' end of the cDNA and also, as a consequence, the
vector. The oligo(dC) is removed from the vector terminus by cleavage at the
single *Hind*III site. The oligo(dC)-tailed cDNA is annealed to a short adapter
fragment containing a *Hind*III cohesive end and an oligo(dG) tail. The
vector–cDNA–mRNA is cyclised by *E. coli* DNA ligase. The mRNA is subsequently removed by the activity of *E. coli* RNase H, which introduces nicks in
the RNA strand, and *E. coli* DNA polymerase I, which performs nick translation in the presence of dNTPs to replace the RNA with DNA. Any gaps at the
3' termini of either strand would be filled in by the *E. coli* DNA polymerase I
activity. The resulting fragments are covalently joined using *E. coli* DNA ligase.
The recombinants are used to transform *E. coli* HB101 (section 5.1). This
procedure was applied to rabbit reticulocyte mRNA, producing 10^5 ampicillin-
resistant HB101 clones per microgram of starting mRNA. Analysis of globin
cDNA clones in this library revealed that 10% had the complete RNA sequence

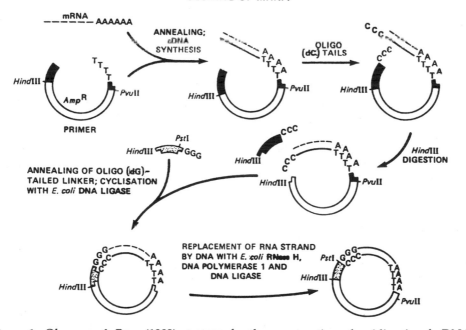

Figure 1. Okayama & Berg (1982) strategy for the construction of unidirectional cDNA
libraries. Reproduced by permission of The American Society for Microbiology

at the 5' end and 20–40% had all but two or three of the 5' nucleotides. This method has the advantage of high-efficiency cloning of full-length or nearly full-length cDNAs. The success of the procedure is attributed to the following:

(1)   The cDNA is made on the vector, reducing the loss of cDNAs through low cloning efficiencies.
(2)   Addition of homopolymeric tails to the 3' end favours blunt ends compared to 3' recessed ends and therefore more full-length cDNAs are likely to be obtained.
(3)   Replacement synthesis of the second strand is far more efficient than the method requiring DNA polymerase I to use the first-strand DNA as both primer and template.

The disadvantage of the Okayama and Berg procedure and the other earlier methods described above is that homopolymeric tailing is difficult to control and may have to be repeated several times. This would be fine for the vector tailing experiments but not for the oligo(dC) additions on the vector–cDNA–mRNA hybrid or double-stranded cDNA, where often only one attempt is possible.

Gubler & Hoffman (1983) reported a simpler protocol for generating cDNA libraries that combined the original oligo(dT)12–18 primer and the novel second-strand replacement synthesis devised by Okayama & Berg (1982). No S1 nuclease digestion step was included and the double-stranded cDNA was dG-tailed and ligated into EcoRV-cut dC-tailed pBR322. This procedure gave cloning efficiencies of 10^6 clones/μg mRNA.

The requirement for homopolymeric tailing was subsequently superseded by the use of synthetic DNA linkers or DNA adapters (Alexander et al., 1984; Helfman et al., 1983; Huynh et al., 1985). Linkers are synthetic double-stranded DNA molecules that possess one or more restriction sites. In the method described by Huynh et al. (1985) (Fig. 2), EcoRI linkers are ligated to blunt-end, methylated double-stranded cDNA using T4 DNA ligase. Blunt-end double-strand cDNA can be prepared by filling in or digesting the single-stranded ends using either E. coli DNA polymerase I or T4 DNA polymerase enzymes which have 5'–3' polymerase activity plus 3'–5' exonuclease activity, or Klenow fragment, which is the large fragment of DNA polymerase I having only the 5'–3' polymerase activity. The double-stranded cDNA is then treated with EcoRI methylase to methylate and therefore protect the internal EcoRI sites from the subsequent digestion with EcoRI. The excess linkers at the cDNA termini are removed by digestion with EcoRI, resulting in double-stranded cDNA with EcoRI-cohesive ends. Excess and digested linkers will interfere with the subsequent ligation of cDNAs to vector and are removed from the cDNA preparation by passage over a Bio-Gel A-50m column. The cDNA prepared can be ligated into any suitable EcoRI-cut vector, and Huynh et al. (1985) used λgt10 (see section 4). In order to ensure a high percentage of recombinants, it is preferred that the EcoRI vector be treated with phosphatase to remove the 5' phosphate groups and reduce the probability of obtaining false recombinants resulting from vector religation.

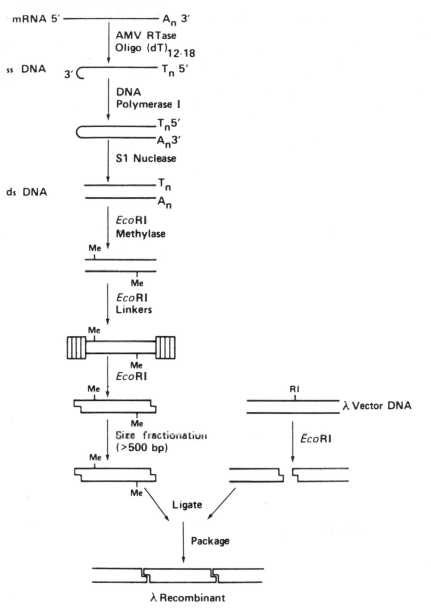

Figure 2. Scheme for double-stranded cDNA synthesis and cloning in λgt10 as described in Huynh et al. (1985). Reproduced from *DNA Cloning*, Vol. I (1985), ed. D.M. Glover, by permission of Oxford University Press

In contrast to linkers, adapters are short DNA molecules that possess a blunt end for ligation to the cDNA and a cohesive end for ligation into an appropriately prepared vector. The use of adapters that have enzyme half-sites eliminates the need to treat the double-stranded DNA with site-specific methylases.

The favoured protocol for the work carried out by the authors combined the vector-tailed primer and the replacement synthesis strategy for second-strand preparation devised by Okayama & Berg (1982) with a strategy for cyclising the vector–cDNA using synthetic DNA linkers (Hussey et al., 1990; Villemur et al., 1992; Rogers et al., 1993). A diagrammatic representation of the strategy is shown in Fig. 3. The plasmid Bluescribe M13$^+$ (see section 4) is cut with *Kpn*I and poly(dT) tracts (optimised for 100–150 Ts) are added to the 3′ termini using

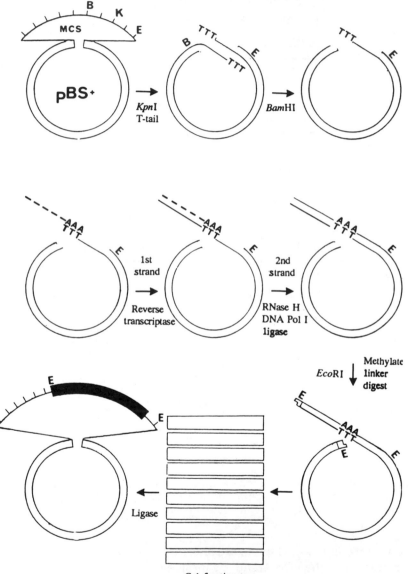

Figure 3. Flow diagram illustrating the main steps in the construction of directionally cloned cDNA libraries used by the authors (e.g. Hussey et al., 1990)

terminal deoxytransferase. One of the T-tracts is removed by cleavage at the single *Bam*HI site. Poly(A)$^+$ RNA is annealed to the T-tail vector, and first-strand synthesis is completed using AMV RT. Second-strand synthesis is initiated by treatment with RNase H followed by replacement synthesis using *E. coli* DNA polymerase I, and the fragments are joined using T4 DNA ligase. The vector–cDNA is treated with *Eco*RI methylase before ligation of *Eco*RI linkers to the ends of the vector–cDNA. Excess linkers at the ends of the vector–cDNA are removed by digestion with *Eco*RI. The vector–cDNA is separated on agarose gels and 10 consecutive size classes of the vector–cDNA are purified by cetyl triethylammonium bromide (CTAB) extraction (Langridge et al., 1980). Each vector–cDNA fraction is cyclised using T4 DNA ligase, and T4 DNA ploymerase is added to fill in any gaps. The recombinants are used to transform a high-efficiency-transformation strain of *E. coli*, DH5α F'IQ^{TM}, (Gibco BRL; see section 5.1) prepared as competent cells using the Hanahan (1985) protocol. The primary library is amplified by plating on solid media as opposed to suspension culture to ensure that 'slow growers' are represented in the library. The cloning efficiencies for libraries prepared using this method are routinely $< 10^6$ clones/μg mRNA. Commercial kits of the Okayama and Berg technique and vector-primer are now available from Pharmacia.

Modifications to the methods for cDNA preparation are continuing to develop and these modifications aim to match speed and efficiency of library preparation. The majority of these modifications are incorporated into cDNA synthesis kits that can be bought from commercial sources. In the majority of this kits, the basic protocol remains the same, i.e. that devised by Okayama & Berg (1982) as modified by Gubler & Hoffman (1983). The majority of the kits available boast quality full-length cDNA production and high cloning efficiencies. When deciding to use a particular protocol, several choices have to be made, as follows.

3.1 One primer, random primers or both

In the majority of library constructions, first-strand synthesis occurs from an oligo(dT) primer annealed to the poly(A) tail of mRNA. In certain cases this method may not be appropriate, i.e. when cDNA is required from mRNAs not possessing poly(A) tails or when there is strong RNA secondary structure that precludes the synthesis of cDNA representing the 5' regions of the RNA when primed by oligo(dT). In such cases the problems have been overcome by using random hexamer primers to prime first-strand synthesis (Birnstiel et al., 1985; Koike et al., 1987; Gruber et al., 1991; Kobs, 1991). As expected, the cDNA libraries constructed from random hexamer-primed and random hexamer plus oligo(dT)-primed mRNA have a smaller average insert size than oligo(dT)-primed mRNA (Gruber et al., 1991).

3.2 Type of reverse transcriptase

The first polymerase to be routinely used for cDNA synthesis was AMV RT (Efstratiadis et al., 1975). AMV RT is purified from AMV-infected chicks and consists of two polypeptides, one of which has primer-dependent 5'–3' polymerase activity that utilises RNA or DNA as template, and the other of which has a 3'–5' RNase H activity that degrades the RNA of DNA–RNA hybrids (Verma, 1975). Although AMV RT nucleolytic activity is an advantage when high levels of RNA secondary structure are present, it can have deleterious effects on the synthesis of full-length cDNA. This necessitated the search for a suitable replacement. Reverse transcriptase from Moloney murine leukaemia virus (M-MLV) is a single polypeptide and is a superior alternative, having a much reduced RNase H activity (Gerard & Grandgenett, 1975). However, M-MLV RT is less stable than AMV RT in standard mRNA reactions, and relatively large amounts are required to achieve high levels of cDNA synthesis (Gerard, 1985). Obtaining large amounts of M-MLV RT has been facilitated by the fact that the coding sequence for M-MLV RT has been cloned and the protein can be purified in an active form from *E. coli* expressing the gene (Kotewicz et al., 1985; Roth et al., 1985).

In order to achieve optimal conditions for full-length cDNA synthesis it would be preferable to remove or considerably reduce the RNase H activity in the reverse transcriptase enzyme. This has been done by producing genetically engineered mutant M-MLV RT genes that either have the RNase H sequence deleted, e.g. SuperScript™ RNase H⁻ (Gibco BRL) (Kotewicz et al., 1988), or have point mutations introduced into the RNase H active site, e.g. Super-Script™ II RNase H⁻ (Gibco BRL) (Gerard et al., 1992) and StrataScript™ RNase H⁻ (Stratagene) (Nielson et al., 1993). These genetically modified enzymes produce greater yields and a larger percentage of full-length first-strand cDNA than obtained with wild-type reverse transcriptases. Super-Script™ II RNase H⁻ (Gibco BRL) reverse transcriptase is superior to Super-Script™ RNase H⁻ (Gibco BRL) reverse transcriptase because wild-type polymerase activity is maintained.

3.3   Linkers, adapters or both

The choice of linkers or adapters depends on whether a randomly orientated (Huyhn et al., 1985) or a unidirectional cDNA library is required. In unidirectional cloning the vector and cDNA share an oriented pair of specific restriction sites or cohesive ends at their termini. Constructing unidirectional libraries has certain advantages, which include eliminating the requirement for treating the vector with phosphatase to prevent vector religation and also ensuring that DNA inserts could be cloned in the correct orientation with respect to vector-encoded expression signals.

Several methods for constructing unidirectional libraries are available and each depends on the novel design of either a primer–vector, a linker, an adapter or a primer–adapter. The Okayama & Berg (1982) procedure, which uses the primer–vector system, has been described above. In a procedure developed by Novagen from the method described in Meissner et al. (1987) (Fig. 4), a novel bifunctional linker is used that contains an *Eco*RI site that forms an additional *Hind*III site when ligated to the 3' ends of double-stranded

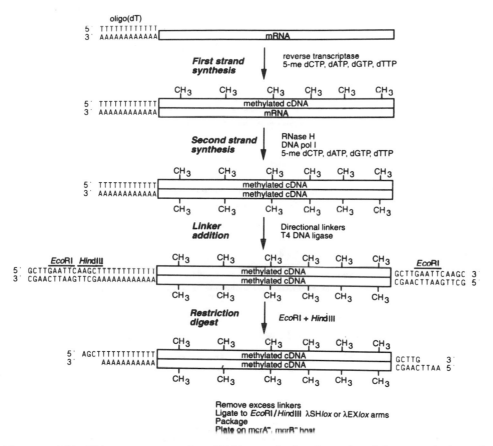

Figure 4. The Novagen system for cDNA synthesis. Reproduced by permission of Novagen Inc., Madison, WI

cDNA. Provided the cDNA is either treated with specific methylases or is synthesised incorporating 5-methyl-dCTP (see below) before linker ligation, digestion with *Hind*III and *Eco*RI would result in directional cDNA that could be cloned into an appropriately cut vector. A method using primer–adapters and adapters has been developed for Gibco BRL by D'Alessio et al. (1990). In this method (Fig. 5) a primer–adapter is used to prime first-strand synthesis and to provide a *Not*I restriction site which is needed to clone the cDNA directionally. After second-strand synthesis *Sal*I adapters are ligated to the cDNA termini. The cDNA is digested with *Not*I (a rare cutter), which produces cDNA with *Not*I- and *Sal*I-cohesive ends ready to be cloned into an appropriately cut vector.

3.4   Methylation, incorporation of 5-methyl-dCTP or neither

All cDNA preparations that use linkers or primer–adapters are subsequently digested with enzymes to generate cohesive ends for insertion of cDNA into

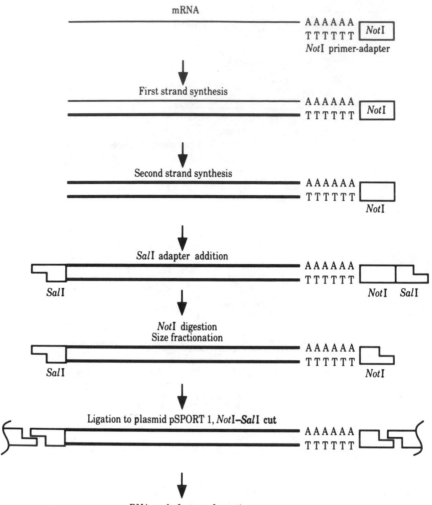

Figure 5. The SuperScript™ system for cDNA synthesis. Reproduced by permission of Life Technologies, Inc.

vectors. To prevent the enzymes from cutting at internal sites, one of two methods can be used. The first method involves methylating cytosine residues using restriction-site-specific methylases. The use of *Eco*RI methylase has already been described in the production of cDNA libraries by Huynh et al. (1985) (Fig. 2) and by the authors (Fig. 3). The second method involves incorporating an analogue of dCTP, 5-methyl-dCTP, into either the first-strand synthesis reaction (if primer–adapters are used) or both the first- and second-strand synthesis reactions (if linkers are used). Stratagene has tested the incorporation of 5-methyl-dCTP versus normal dCTP with Klenow fragment, *E. coli* DNA polymerase I, *E. coli* DNA polymerase III, AMV RT, M-MLV RT and StrataScript RNase H⁻ reverse transcriptase on various templates, and it proved to be an equally acceptable substrate for the enzymes listed. A flow chart for Stratagene's ZAP cDNA synthesis kit is shown in Fig. 6. A 50-base oligonucleotide primer–linker (or primer–adapter) is used to prime first-strand synthesis and provide an *XhoI* site. First-strand synthesis using StrataScript

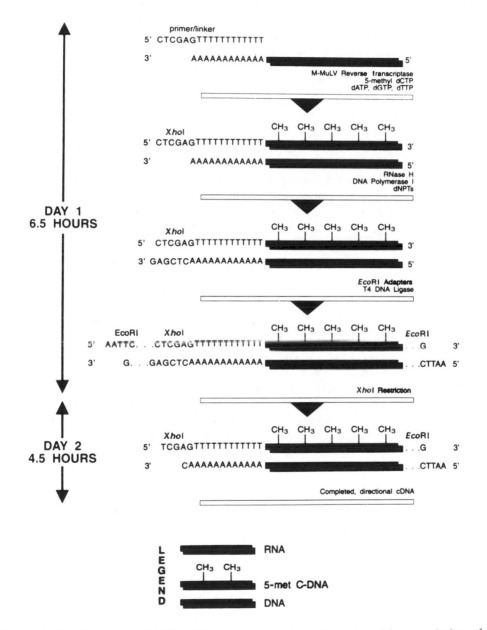

Figure 6. The Stratagene ZAP^R cDNA synthesis system. Reproduced by permission of Stratagene

RNase H$^-$ reverse transcriptase incorporates 5-methyl-dCTP in place of dCTP in the nucleotide mixture. In second-strand synthesis, dCTP is added to reduce the probability of 5-methyl-dCTP becoming incorporated in the second strand. This will ensure that the restriction site in the primer–linker is susceptible to enzyme digestion. EcoRI adapters are ligated to the blunt ends, and the resulting cDNA–adapters are cleaved with XhoI to generate the directional cDNA. In the Novagen protocol adapted from Meissner et al. (1987) (Fig. 4), the 5-methyl-dCTP is used in the synthesis of both strands and linkers are ligated to the blunt ends of the double-stranded cDNA.

When linkers or primer–adapters are used that provide rare cutter sites, e.g. NotI, neither treatment of the cDNA with methylases nor the incorporation of 5-methyl-dCTP is necessary. This is because the likely appearance of an eight-base cutter site, e.g. NotI, in a DNA sequence is only once in every 4^8 base pairs (i.e. 65 536 bp) of sequence. For comparison, the probability of finding a four-base cutter site in a cDNA sequence is one in every 254 bp, and of a six-base cutter site, one in every 4096 bp. The method of cDNA synthesis shown in Fig. 5 uses a primer–adapter giving a NotI restriction site.

The invention of the polymerase chain reaction (PCR) by Saiki et al. (1985) (also see Mullis & Faloona, 1987) has led to the development of new strategies of cDNA cloning (e.g. Frohman et al., 1988; Loh et al., 1989) and cDNA library construction (see Chapters 13 and 14). In the majority of cDNA library preparations at least 5 μg of poly(A)$^+$ RNA is required to obtain a representative library. It is not always possible to isolate so much mRNA, because of the lack of availability of certain cell types, and several methods have been developed that employ PCR to generate cDNA libraries from RNA isolated from small numbers of cells (Jepson et al., 1991; Gurr et al., 1991; Lambert & Williamson, 1993; Don et al., 1993; Dresselhaus et al., 1994). The method described by Jepson et al. (1991) (Fig. 7) uses < 50 mg of plant tissue. The total RNA purified from this tissue is used in a cDNA synthesis reaction (the Amersham cDNA synthesis kit). The cDNA ends are made flush by performing a Klenow fill-in reaction and then fractionated using a Bio-Gel A-150 column (Biorad). Two oligonucleotides, a 29-mer and a 12-mer, are annealed to produce an adapter with an EcoRI-cohesive end and a blunt end. The adapters are ligated to the blunt-end cDNA using T4 DNA ligase. The adapter–cDNA is heated to 73 °C in the presence of Taq polymerase and a nucleotide mixture. The 12-mer oligonucleotide melts off the DNA and allows the Taq polymerase to synthesise the complementary sequence to the 29-mer oligonucleotide using the 3' end of the cDNA as primer. The cDNA is then amplified by PCR using the 29-mer to prime the reaction. The cDNA is digested with EcoRI and ligated to dephosphorylated EcoRI-cut lambda ZAP II vector (section 4). This procedure has been used to prepare representative cDNA libraries from a number of tissue types from several species, including Pharbitis nil, Brassica napus, Beta vulgaris and Zea mays (Jepson et al., 1993). As protocols develop, fewer cells are required to prepare a cDNA library. In the protocol described by Dresselhaus et al. (1994), PCR cDNA libraries are prepared from poly(A)$^+$ RNA isolated from as few as 100 maize protoplasts.

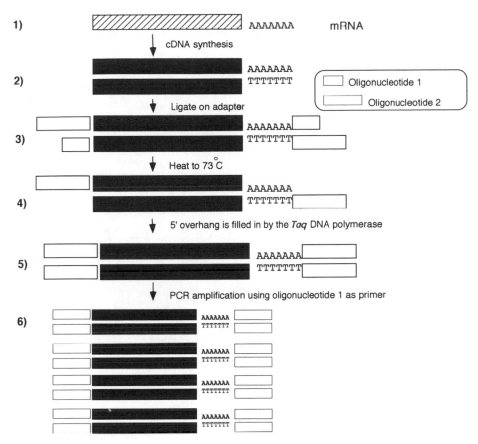

Figure 7. Flow diagram showing the main steps in the construction of a PCR cDNA library as described in Jepson et al. (1990). Reproduced by permission of Kluwer Academic Publishers

4 VECTORS FOR cDNA CLONING

The first vectors to be used for the molecular cloning of cDNAs were bacterial plasmids. The most popular plasmid cDNA cloning vectors are principally derived from the pUC series of plasmids designed by Messing and colleagues (Messing, 1983; Yanisch-Perron et al., 1985; Vieira & Messing, 1987), which essentially superseded the pBR322 vector (reviews: Balbas et al., 1986; Sambrook et al., 1989). The cDNA library constructions shown in Fig. 3 used the Bluescribe M13+ plasmid (Stratagene). The Bluescribe M13+/− (now called pBS+/−) vectors were synthesised by inserting T3 and T7 transcriptional promoters into the coding region of the *lacZ* gene of pUC19 and by inserting a 454-nucleotide *Rsa*I to *Dra*I intergenic region fragment from M13 into the *Nar*I site (see Stratagene protocols literature). The Bluescribe M13+/− vectors allow for histochemical selection of recombinants, *in vitro* RNA synthesis and rescue of single-stranded DNA, provided that an f1-sensitive host (e.g. JM101) is infected with a helper phage (e.g. M13K07). The later versions of pBS+/− are called pBluescript SK+/− (Fig. 8) and these have a more extensive polylinker containing 21 unique restriction sites.

In the early 1980s, Davies and colleagues (Young & Davies, 1983a,b; Huynh et al., 1985) developed two bacteriophage lambda vectors, λgt10 and λgt11, that

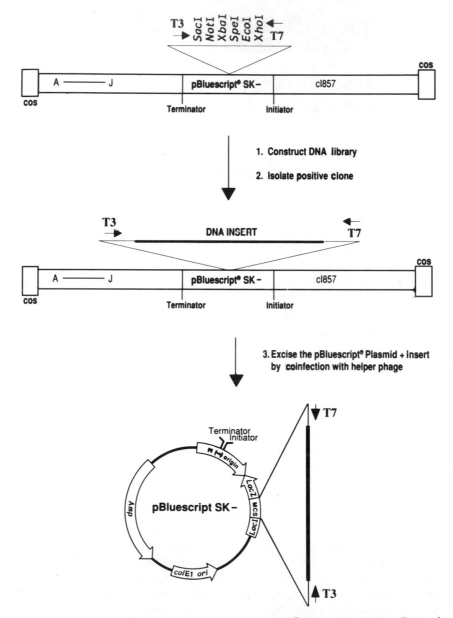

Figure 8. The map of Lambda ZAP[R] II and Lambda ZAP[R] II vector excision. Reproduced by permission of Stratagene

could be used for cDNA cloning. The λgt10 is an insertion vector having a single EcoRI target site for cloning up to 7.6 kb of foreign DNA. The EcoRI site is located in the cI repressor gene, and insertion of a DNA fragment generates a cI⁻ phenotype. The cI⁻ phage, in contrast to cI⁺ phage, form clear plaques when plated on E. coli strains carrying the high-frequency lysogeny mutation hflA150 (e.g. E. coli C600hfl strains). The λgt11 vector is an expression vector with a single EcoRI cloning site located within the lacZ gene and can accept foreign DNA up to 7.2 kb in length. The λgt11 produces blue plaques on lac⁻ hosts in the presence of isopropylthio-β-D-galactoside (IPTG) and 5-bromo-4-chloro-3-indoyl-β-D-galactoside (X-gal), whilst recombinant phage, due to the inactivation of β-galactosidase, produce colourless plaques. Libraries made in λgt10 and λgt11 can be screened with nucleic acid probes (Chapter 5). The λgt11 libraries can also be screened with antibody probes (Chapter 7). If library screening is only to be performed using nucleic acid probes, then a λgt10 vector is to be preferred because of the efficient recombinant selection technique and the fact that large recombinant phage plaques are produced which are of uniform size and shape (Huynh et al., 1985).

The lambda cDNA cloning vectors have the advantage of having a highly efficient and reproducible in vitro packaging system and subsequent infection of E. coli that results in far greater cloning efficiencies than could be achieved by bacterial plasmid transformation. In addition, lambda phage plaques can be plated and screened at a high density compared to plasmid bacterial colonies, allowing for rarer cDNAs to be identified. However, the subsequent purification and characterisation of specific clones is often time-consuming. To obtain pure clones can require two to three additional screens because of the diffusion of released phage across the plates. Also, restriction mapping and sequencing of the cDNA insert will usually require the fragments to be subcloned into a plasmid vector.

Several commercial groups have now developed vectors that combine the high-efficiency lambda vector with the versatility of the plasmid system. The vectors developed have a complete plasmid inserted into the lambda phage which can be excised in vivo. This excision process eliminates the need to subclone DNA inserts into a plasmid. Lambda ZAP^R and Lambda ZAP^R II (Stratagene) are examples of these novel vectors. The lambda ZAP vectors are insertion-type, expression cDNA cloning vectors. They were constructed as follows. A modified pBluescript SK− plasmid was constructed carrying separated initiator and terminator regions of the f1 phage origin of replication. This modified plasmid, pPreB, was inserted into a non-essential region of a lambda phage such that the plasmid sequences were flanked by the initiator and terminator domains. The lambda arms were constructed to contain few restriction sites in order to maintain six unique restriction sites within the polylinker region for cDNA cloning. In E. coli, excision of the plasmid sequence by f1 or M13 helper phage generates a phagemid which is identical to pBluescript SK−. A map of Lambda ZAP and the in vivo excision process is shown in Fig. 8. Lambda ZAP II differs from lambda ZAP in that it lacks the Sam100 mutation which allows for better growth of the bacteriophage (Short et al., 1988). Other examples of these novel vectors are λMOSS*lox* and λMOSE*lox* (Amersham), λEX*lox*™, λSH*lox*^R and λOCUS™ (Novagen), λExCell (Pharmacia), and λBlue-Mid™ (Clontech).

5   GENE TRANSFER TO *E. COLI*

5.1   Transformation and *E. coli* strains

Techniques for the transformation of *E. coli* have been reviewed by Hanahan (1985) and Sambrook et al. (1989). The choice of *E. coli* strain will dictate the transformation method to be used. *E. coli* strains that have been commonly used as hosts for plasmid cDNA libraries are HB101, DH1 and DH5 (Hanahan, 1985) and JM109 (Yanisch-Perron et al., 1985). Construction of expression cDNA libraries where cDNA inserts are cloned into sites within the *lacZ* gene require strains that provide α-complementation of the β-galactosidase gene, e.g. JM109, DH5α, DH1α, DH5α F'IQ™, DH5αMCR™ (Gibco BRL), XL1-Blue MRF' or SURE™ (Stratagene). If cDNA synthesis involves the incorporation of 5-methyl-dCTP, then *E. coli* strains lacking the McrA, McrBC and Mrr restriction systems which cleave C-methylated DNA are required, e.g. DI I5αMCR™ (Gibco BRL), XL1-Blue MRF' or SURE™ (Stratagene). The strains described above are all recombination deficient (*recA*), which ensures insert stability, and restriction deficient (*hsdS*, *hsdR* or *hsdRSM*), which allows cloning of DNA without cleavage by endogenous restriction endonucleases. All these strains are transformed efficiently using the procedure described in Hanahan (1985). Transformation efficiencies are in the range of $2-7 \times 10^8$ transformants/μg of plasmid DNA.

5.2   *In vitro* lambda packaging and *E. coli* infection

Protocols for the in vitro packaging of lambda phage essentially follow either the one-extract (Rosenberg, 1987) or two-extract (Hohn & Murray, 1977) packaging systems. The former system has the advantages of being free from endogenous phage and *Eco*K restriction. Modified and optimised packaging extracts can now be purchased from a number of commercial sources (e.g. Stratagene, Promega, Amersham, and Novagen) claiming packaging efficiencies of between 2×10^8 and 2×10^9 pfu/μg DNA. Stratagene's Gigapack^R II (Gold and Plus) packaging extracts (Kretz & Short, 1989) provide the highest efficiencies and these are all restriction free (*hsd*^−, *mcrA*^−, *mcrB*^−, *mrr*^−).

Several bacterial strains can be used for the growth of lambda phage clones and each may be suited to a particular vector or application. The *E. coli* strain LE392 is a permissive host allowing growth of parental and recombinant phage (Murray et al., 1977). This host is compatible with the majority of lambda cDNA cloning vectors for amplification or screening with nucleic acid probes.

The strains C600 and C600*hfl* (see above) are used to propagate λgt10 (Huyhn et al., 1985). The preferred host for screening expression libraries such as those produced in λgt11 is Y1090. This strain carries the Δ*lon* mutation, which decreases protease activity, and it also contains a plasmid, pMC9, which carries the *lac* repressor and antibiotic resistance (Huynh et al., 1985). The Y1090 strain may not be suitable for the propagation of lambda phage carrying plasmid sequences, e.g. Lambda ZAP^R, because the pMC9 plasmid may recombine with homologous sequences within the vector. The preferred host for Lambda ZAP^R vectors is the recombination-deficient XL1-Blue MRF' strain (Stratagene).

6 CONCLUDING REMARKS

In this chapter we have described protocols for plant RNA isolation and discussed several different strategies for cDNA library construction. As procedures and protocols continue to be developed, particularly by rival commercial laboratories, we will no doubt see new and improved methods that increase the speed and efficiency of mRNA isolation and cDNA cloning. For those research laboratories who wish to save the time and the expense of setting up such experiments, it may be more appropriate to purchase pre-made cDNA libraries or to have cDNA libraries custom-made. Several companies sell plant tissue cDNA libraries and, if they do not have an appropriate cDNA library, offer a custom library synthesis service (e.g. Clontech, Stratagene, Novagen). These companies often guarantee the cDNA library to have a minimum number of primary recombinants (around 10^6–10^7). As a final point, before initiating the type of methodology described in this chapter, the reader should be aware of the comprehensive practical molecular biology texts (Sambrook et al., 1989; Ausubel et al., 1987; Glover, 1985; Shaw, 1988).

Corresponding author: P. J. Hussey.

ESSENTIAL READING

Efstratiadis, A., Kafatos, F.C., Maxam, A.M. and Maniatis, T. (1976) Enzymatic *in vitro* synthesis of globin genes. *Cell*, **7**, 279–288.

Gubler, U. and Hoffman, B.J. (1983) A simple and very efficient method for generating cDNA libraries. *Gene*, **25**, 263–269.

Huynh, T.V., Young, R.A. and Davis, R.W. (1985) Constructing and screening cDNA libraries in λgt10 and λgt11. In *DNA Cloning, A Practical Approach*, Vol. 1 (ed. D.M. Glover), pp. 49–78. IRL Press, Oxford.

Okayama, H. and Berg, P. (1982) High-efficiency cloning of full-length cDNA. *Mol. Cell. Biol.*, **2**, 161–170.

Rougeon, F. and Mach, B. (1976) Stepwise biosynthesis *in vitro* of globin genes from globin mRNA by DNA polymerase of avian myeloblastosis virus. *Proc. Natl. Acad. Sci. USA*, **73**, 3418–3422.

REFERENCES

Alexander, D.C., McKnight, T.D. and Williams, B.G. (1984) A simplified and efficient vector-primer cDNA cloning system. *Gene*, **31**, 79–89.

Ausubel, F.M., Brent, R., Kingston, R.E. et al. (eds) (1995) *Current Protocols in Molecular Biology*, Wiley, NY.

Aviv, H. and Leder, P. (1972) Purification of biologically active globin messenger RNA by chromatography on oligo thymidylic acid-cellulose. *Proc. Natl. Acad. Sci. USA* **69**, 1408–1412.

Balbas, P., Soberon, X., Merino, E. et al. (1986) Plasmid vector pBR322 and its special purpose derivatives—a review. *Gene*, **50**, 3–40.

Birnstiel, M.L., Busslinger, M. and Strub, K. (1985) Transcription termination and 3′ processing: the end is in site. *Cell*, **41**, 349–359.

D'Alessio, J.M., Gruber, C.E., Cain, C. and Noon, M.C. (1990) Construction of directional cDNA libraries. *Focus*, **12**, 47–48.

Don, R.H., Cox, P.T. and Mattick, J.S. (1993) A 'one tube reaction' for synthesis and amplification of total cDNA from small numbers of cells. *Nucleic Acids Res.*, **21**, 783.

Dresselhaus, T., Lorz, H. and Kranz, E. (1994) Representative cDNA libraries from a few plant cells. *Plant J.*, **5**, 605–610.

Efstratiadis, A., Maniatis, T., Kafatos, F.C. et al. (1975) Full length and discrete partial reverse transcripts of globin and chorion mRNAs. *Cell*, **4**, 367–368.

Ekenberg, S., McCormick, M., Wu, L. and Smith, C. (1992) 30-minute poly(A)+ RNA isolation directly from tissue. *Promega Notes*, **39**, 7–11.

Frohman, M.A., Dush, M.K. and Martin, G.R. (1988) Rapid production of full-length cDNAs from rare transcripts: amplification using a single gene-specific oligonucleotide primer. *Proc. Natl. Acad. Sci. USA*, **85**, 8998–9002.

Gerard, G.F. (1985) Comparison of cDNA synthesis by avian and cloned murine reverse transcriptase. *Focus*, **7**, 1–3.

Gerard, G.F. and Grandgenett, D.P. (1975) Purification and characterisation of the DNA polymerase and RNase H activities in Moloney murine sarcoma-leukemia virus. *J. Virol.*, **15**, 785–797.

Gerard, G.F., Schmidt, B.J., Kotewicz, M.L. and Campbell, J.H. (1992) cDNA synthesis by moloney murine leukemia virus RNase H-minus reverse transcriptase possessing full DNA polymerase activity. *Focus*, **14**, 91–93.

Glover, D.M. (1985) *DNA Cloning, a Practical Approach*, Vols I, II and III. IRL Press, Oxford.

Gruber, C.E., Cain, C. and D'Alessio, J. (1991) Random hexamer-primed cDNA synthesis using the SUPERSCRIPT™ choice system. *Focus*, **13**, 88–90.

Gurr, S.J., McPherson, M.J., Scollan, C. et al. (1991) Gene expression in nematode-infected plant-roots. *Mol. Gen. Genet.*, **226**, 361–366.

Hanahan, D. (1985) Techniques for transformation of *E. coli*. In *DNA Cloning, A Practical Approach*, Vol. 1 (ed. D.M. Glover), pp. 109–135. IRL Press, Oxford.

Helfman, D.M., Feramisco, J.R., Fiddes, J.C. et al. (1983) Identification of clones that encode chicken tropomyosin by direct immunological screening of a cDNA expression library. *Proc. Natl. Acad. Sci. USA*, **80**, 31–35.

Hohn, B. and Murray, K. (1977) Packaging recombinant DNA molecules into bacteriophage particles *in vitro*. *Proc. Natl. Acad. Sci. USA*, **74**, 3259–3263.

Hussey, P.J., Haas, N., Hunsperger, J. et al. (1990) The β-tubulin gene family in *Zea mays*: two differentially expressed β-tubulin genes. *Plant Mol. Biol.*, **15**, 957–972.

Jepson, I., Bray, J., Jenkins, G. et al. (1991) A rapid procedure for the construction of PCR cDNA libraries from small amounts of plant tissue. *Plant Mol. Biol. Rep.*, **9**, 131–138.

Jepson, I., Greenland, A.J., Dickinson, H.G. et al. (1993) Construction of PCR cDNA libraries to search for genes involved in meiosis. *J. Exp. Bot.*, **44** (suppl.), P6.12.

Koike, S., Sakai, M. and Muramatsu, M. (1987) Molecular cloning and characterization of rat estrogen receptor cDNA. *Nucleic Acids Res.*, **15**, 2499–2513.

Kotewicz, M.L., D'Alessio, J.M., Driftmeir, K.M. et al. (1985) Cloning and overexpression of moloney murine leukemia-virus reverse transcriptase in *Escherichia coli. Gene*, **35**, 249–258.

Kotewicz, M.L., Sampson, C.M., D'Alessio, J.M. and Gerard, G.F. (1988) Isolation of cloned moloney murine leukemia-virus reverse transcriptase lacking ribonuclease-H activity. *Nucleic Acids Res.*, **16**, 265–277.

Kretz, P.L. and Short, J.M. (1989) Gigapack™: restriction free (*hsd⁻, mcrA⁻, mcrB⁻, mrr⁻*) lambda packaging extracts. *Strategies Mol. Biol.*, **2**, 25–26.

Lambert, K.N. and Williamson, V.M. (1993) cDNA library construction from small amounts of RNA using paramagnetic beads and PCR. *Nucleic Acids Res.*, **21**, 775–776.

Langridge, J., Langridge, P. and Bergquist, P.L. (1980) Extraction of nucleic acids from agarose gels. *Anal. Biochem.*, **103**, 264–271.

Lefebvre, P.A., Silflow, C.D., Wieben, E.D. and Rosenbaum, J.L. (1980) Increased levels of mRNAs for tubulin and other flagellar proteins after amputation or shortening of *Chlamydomonas* flagella. *Cell*, **20**, 467–477.

Loh, Y., Elliott, J.F., Cwirla, S. et al. (1989) Polymerase chain reaction with single sided specificity: analysis of T-cell receptor δ chain. *Science*, **243**, 217–220.

Meissner, P.S., Sisk, W.P. and Berman, M.L. (1987) Bacteriophage λ cloning system for the construction of directional cDNA libraries. *Proc. Natl. Acad. Sci. USA*, **84**, 4171–4175.

Messing, J. (1983) New M13 vectors for cloning. *Methods Enzymol.*, **101**, 20–78.

Mullis, K.B. and Faloona, F. (1987) Specific synthesis of DNA *in vitro* via a polymerase-catalysed chain reaction. *Methods Enzymol.*, **155**, 335–350.

Murray, N.E., Brammar, W.J. and Murray, K. (1977) Lambdoid phages that simplify the recovery of *in vitro* recombinants. *Mol. Gen. Genet.*, **150**, 53–61.

Neil, J.D. (1993) Isolation of mRNA from plant tissues using the PolyATract[R] System 1000. *Promega Notes*, **44**, 10–13.

Nielson, K., Sincox, T.G., Schoettlin, W. et al. (1993) Stratascript™ RNase H⁻ reverse transcriptase for larger yields of full-length cDNA transcripts. *Strategies Mol. Biol.*, **6**, 45–46.

Kobs, G. (1991) Generation of cDNA using the Riboclone[R] cDNA synthesis system (random primers). *Promega Notes*, **29**, 4.

Rogers, H.J., Greenland, A.G. and Hussey, P.J. (1993) Four members of the maize β-tubulin gene family are expressed in the male gametophyte. *Plant J.*, **4**, 875–882.

Rosenberg, S.M. (1987) Improved *in vitro* packaging of λ DNA. *Methods Enzymol.*, **153**, 95–103.

Roth, M.J., Tanese, N. and Goff, S.P. (1985) Purification and characterisation of murine retroviral reverse transcriptase expressed in *Escherichia coli. J. Biol. Chem.*, **260**, 9326.

Rougeon, F., Kourilsky, P. and Mach, B. (1975) Insertion of a rabbit β-globin gene sequence into an *E. coli* plasmid. *Nucleic Acids Res.*, **2**, 2365–2378.

Saiki, R.K., Scharf, S., Faloona, F. et al. (1985) Enzymatic amplification of β-globin genomic sequences and restriction site analysis for diagnosis of sickle cell anemia. *Science*, **230**, 1350–1354.

Sambrook, J., Fritsch, E.F. and Maniatis, T. (1989) *Molecular Cloning, a Laboratory Manual*, 2nd edn. Cold Spring Harbor Laboratory Press, Cold Spring Harbor, New York.

Shaw, C.H. (1988) *Plant Molecular Biology, a Practical Approach*. IRL Press, Oxford.

Short, J.M., Fernandez, J.M., Sorge, J. A. and Huse, W.D. (1988) λ ZAP: a bacteriophage λ expression vector with *in vivo* excision properties. *Nucleic Acids Res.*, **15**, 7583–7600.

Verma, I.M. (1975) Studies on reverse transcriptase of RNA tumor viruses: I. Localization of thermostable DNA polymerase and RNase H activities on one polypeptide. *J. Virol.*, **15**, 121–126.

Vieira, J. and Messing, J. (1987) Production of single-stranded plasmid DNA. *Methods Enzymol.*, **153**, 3–34.

Villemur, R., Joyce, C.M., Haas, N.A. et al. (1992) α-Tubulin gene family of maize (*Zea mays* L.): evidence for two ancient α-tubulin genes in plants. *J. Mol. Biol.*, **227**, 81–96.

Yanisch-Perron, C., Vieira, J. and Messing, J. (1985) Improved M13 phage cloning vectors and host strains: nucleotide sequences of M13mp18 and pUC19 vectors. *Gene*, **33**, 103–119.

Young, R.A. and Davies, R.W. (1983a) Efficient isolation of genes by using antibody probes. *Proc. Natl. Acad. Sci. USA*, **80**, 1194–1198.

Young, R.A. and Davies, R.W. (1983b) Yeast RNA polymerase genes: isolation with antibody probes. *Science*, **222**, 778–782.

3 Genomic DNA Isolation and Lambda Library Construction

SIMON A. J. WARNER
University of Leicester, Leicester, UK

Lambda provides a versatile and easily manipulated cloning vector that is able to accept DNA fragments up to 23 kb in size. This chapter examines some of the salient points in the design of lambda vectors and the protocols required to produce libraries that are representative of the entire genome and/or enriched for specific restriction fragments. Lambda still has an important role to play as a tool for plant gene isolation, as it offers a rapid and accurate route to interesting plant genomic sequences.

Topics discussed are as follows:

(1) DNA extraction protocols
(2) Theory of genomic library construction
(3) Practical approaches and protocols of library construction
(4) Design and function of lambda vectors
(5) Availability of pre-made libraries

1 INTRODUCTION

Before the 1980s, the technology available to isolate plant genes was limited in comparison to the many varied techniques such as the polymerase chain reaction (PCR) or map-based cloning that are available to scientists now. Conventionally, the starting point in gene isolation was a cDNA or at least a purified protein of interest. DNA probes would be synthesised and genomic libraries would be screened for clones hybridising with the DNA probe. Although in many cases there are quicker methods to isolate genomic sequences corresponding to genes or sequences of interest, the conventional route of screening genomic libraries is a powerful method for isolating moderately large portions of genomic sequence. In comparison to PCR approaches, direct genomic cloning offers the advantage that larger fragments of DNA may be isolated, giving a greater probability of obtaining the whole gene in one clone, containing far fewer sequence errors than a fragment amplified via PCR. However, not all sequences are stable and some are toxic in host cells and therefore may be difficult to isolate using conventional cloning strategies.

Plant Gene Isolation: Principles and Practice. Edited by G. D. Foster and D. Twell
© 1996 John Wiley & Sons Ltd.

Other problems may also arise, depending on the organisation and structure of the genomic DNA in particular cases. This chapter will examine some of these possibilities and discuss approaches currently available to facilitate the isolation of plant gene sequences of interest in lambda vectors. Commonly used protocols will be outlined, together with the more general protocols useful for the construction of random overlapping genomic libraries in lambda replacement vectors.

2   GENOME ORGANISATION
(review: Lapitan, 1992)

Chapter 1 discusses the plant genome structure in detail, so only information relevant to the construction of genomic libraries will be discussed in the following subsections.

2.1   Size of the nuclear genome

Direct chemical analysis or microdensitometry of feulgen-stained nuclei allows estimates of nuclear DNA content to be made (Bennet & Smith, 1976; Bennet et al., 1982). The range of genome sizes for higher plants is approximately 1×10^8 bp to 1×10^{11} bp, with *Arabidopsis thaliana* at about 2×10^8 bp and wheat at about 2×10^{10} bp being commonly studied representatives of plants with small and large genomes respectively. It is estimated that the human genome is approximately 6×10^9 bp, and the bacterium *E. coli* has a genome of 4×10^6 bp, putting the representative average plant genome in a size range similar to that of humans, e.g. tobacco (*Nicotiana tabacum* 4×10^9 bp).

2.2   Complexity of the genome

DNA reassociation studies, isopycnic centrifugation and other molecular techniques have shown that the majority of plant genomes are not composed of sequences that directly code for proteins. Instead, much of the genome consists of repetitive, satellite or non-coding DNA. Other elements exist in the genome which may be mobile, such as retroviral DNA, or transposons, which are discussed in Chapter 12. It is believed that this satellite or non-coding DNA may be important in defining chromosome structure.

2.3   Methylation of nuclear DNA

Chemical and restriction analysis of nuclear DNA from plants has revealed that approximately one quarter of all cytosine residues are methylated to 5-methyl-cytosine. This does not occur randomly but in specific positions that consist of either the CG dinucleotide or the CXG trinucleotide (Shapiro, 1976; Gardiner-garden et al., 1992). The role of methylation in gene expression is being studied, and results suggest that when DNA is hypermethylated in a particular region, gene expression may be switched off, possibly via changes in chromatin structure, or via reduced accessibility of transcription factors to promoter elements (Inamdar et al., 1991). Most restriction enzymes are sensitive to methylated DNA and will not cleave at their normal recognition sequences if these sequences contain 5-methylcytosine instead of cytosine. This is an important limitation of restriction enzymes that must be considered when manipulating plant genomic DNA.

2.4   Organellar DNA organisation
(reviews: Kuroiwa, 1991; Sugiura, 1989, Salganik et al., 1991)

The evolution of the organelles is believed to be via the symbiosis of prokaryotes with early eukaryotic cells to perform specialised functions. This theory is supported in the structure, transcription and translation of genes and corresponding mRNA, which are more akin to the mechanisms found in prokaryotes than to those in eukaryotes. The two organelles containing their own genomes are the plastids and the mitochondria. Their roles in the cell are varied: primarily they are responsible for the provision of free energy via photosynthesis and oxidative phosphorylation, but they also have many other important biosynthetic and physiological roles.

The chloroplast (the photosynthetic plastid) genome consists of a double-stranded circular DNA molecule of about 150 kb. The DNA is not associated with nucleosomes and is not as heavily methylated as nuclear DNA, although DNA methylation does play a role in gene regulation in plastids (Ngernprasirt-siri et al., 1988; Ohta et al., 1991). The full DNA sequence of the chloroplast genome has been obtained for several plant species, including rice, tobacco and liverwort (Hiratsuka et al., 1989; Olmstead et al., 1993; Shinozaki et al., 1986; Ohyama et al., 1986).

Plant mitochondrial genomes (reviews: Hanson & Folkerts, 1992; Gray et al., 1992; Mackenzie et al., 1994; Schuster & Brennicke, 1994) are generally double-stranded circular molecules of varying sizes. Recombination between repeat regions on the mitochondrial genome during evolution has caused rearrangement and led to new and different forms of DNA that comprise the genome. Multiple circular DNAs and linear DNAs of varying sizes may make up the genome. Introns are also present in some mitochondrial genes and some of the genome may have been mobile, switching from other organelles during evolution via processes possibly directly or indirectly involving retroelements. The complete sequence of the liverwort mitochondrial genome, a double-stranded circular molecule of 184 kb, has been elucidated (Oda et al., 1992), as has the 55-kb mitochondrial genome of the alga *Prototheca wickerhamii* (Wolff et al., 1994) and mitochondrial plasmid-like DNAs from male-sterile rice (Shikanai et al., 1989).

3 LAMBDA AS A CLONING VECTOR
(reviews: Kaiser & Murray, 1985; Sambrook et al., 1989)

Genetic studies and the creation of mutants of wild-type lambda made it possible to exploit this bacteriophage as a vector. During the latter half of the 1970s lambda molecules were produced that contained convenient restriction enzyme sites that could replace the 40% of the genome not required for lytic growth in *E. coli* with foreign DNA. It was also possible to assemble infectious lambda phage particles in vitro by adding cell extracts to lambda DNA. The development of two types of vector was achieved: the replacement vector, where restriction digestion of the vector leads to the removal of the non-essential DNA contained between two restriction sites which may be replaced by foreign DNA, and insertional vectors, where there is a unique restriction site that is used to insert the foreign DNA. Because lambda is not viable if it exceeds 105% or is smaller than 78% of the size of the wild-type genome, replacement vectors can accept foreigh DNA between approximately 9 and 22 kb, and insertional vectors can accept between 0 and 10 kb of foreign DNA (maps of some vectors showing unique restriction sites are shown in Fig. 1).

3.1 Selection procedures

Libraries containing a large proportion of nonrecombinant clones can create problems, since greater numbers of clones must be screened. Also, nonrecombinant clones often grow faster than the recombinant clones during replication and may dilute out the recombinant clones during library amplification. Therefore, it is advisable to minimise the number of nonrecombinant clones present in a library.

3.2 Replacement vectors

Nonrecombinants may be prevented in replacement vector libraries by a combination of three strategies:

(1) Physically separating the stuffer fragment from the arms.
(2) Preventing the stuffer fragment ligating to the arms by designing the vector so that following double restriction digestion the arms and stuffer have non-compatible termini, or by partially filling in the termini, rendering the arms and stuffer incompatible.
(3) Biological selection that relies on genes that are normally present on the stuffer being absent, allowing efficient replication in a particular host strain of *E. coli* of only recombinant clones.

3.3 Insertional vectors

The selection for insertional vector recombinants is harder, as there is no stuffer fragment to remove, allowing biological selection. Lambda gt10 has a

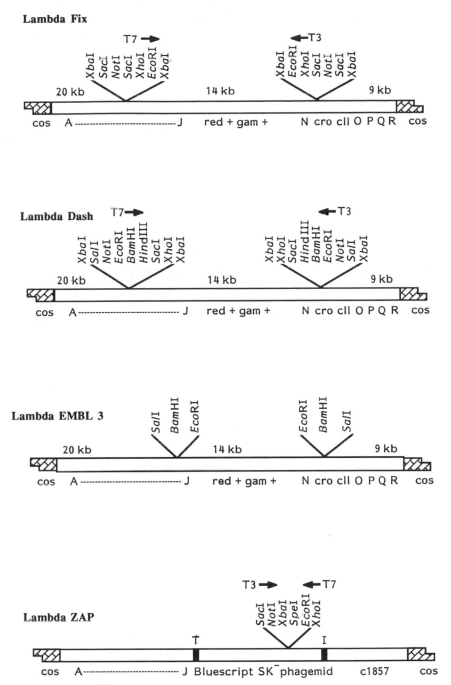

Figure 1. Maps of commonly used lambda vectors showing restriction sites present in the vectors' polylinkers. Lambda FIX, lambda DASH and lambda ZAP are supplied by Stratagene. Lambda EMBL 3 is supplied by many molecular biology companies. Lambda ZAP is the only insertional vector shown.

biological selection whereby the *cI* repressor is inactivated when an insert is present in the vector. If nonrecombinants are grown on *hfl* mutants of *E. coli* that allow high levels of *cI* repressor to accumulate, the bacteriophage becomes lysogenic and does not replicate as rapidly as in the lytic cycle. Similarly, lambda NM1149 (Murray, 1983) is rendered non-lysogenic when foreign DNA is inserted into its cloning site and it is grown on the appropriate host strain.

Other vectors rely on efficient dephosphorylation of the restriction termini, possibly coupled with unidirectional cloning utilising two restriction enzymes yielding non-compatible termini. Visual assessments of the proportion of nonrecombinants in the library are available for some insertional vectors that contain the β-galactosidase gene if 5-bromo-4-chloro-3-indoyl-β-D-galactoside (X-gal) and isopropylthio-β-D-galactoside (IPTG) are included in the media.

4   DNA ISOLATION

Most current protocols for DNA extraction involve the efficient disruption of the plant cell wall in an aqueous buffer, with a compromise being made between the extent of disruption and the yield of high molecular weight DNA. Often, plant nucleic acid extracts are contaminated with polysaccharides, polyphenolics and pigments which render the DNA recalcitrant to manipulation via DNA modification and restriction enzymes. Caesium chloride gradient centrifugation removes proteins and other contaminants, yielding a large amount of pure DNA. Alternatively, there are many commercial kits available that incorporate column chromatography as a purification process (e.g. Promega and Flowgen). For the purpose of cloning into lambda replacement vectors, DNA greater than 100 kb as judged by electrophoresis against lambda concatomer markers is required. For more accurate size estimates of purified DNA samples, pulsed field gel electrophoresis (PFGE) against appropriate markers is recommended (described in Chapter 4).

4.1   Total nucleic acid isolation

Total nucleic acids may be extracted from freeze dried, frozen or fresh plant material by a variey of standard techniques that have proved successful for many plant species. Many basic protocols are variations of the CTAB nucleic acid protocol, which exploits the ability of the detergent CTAB (cetyl triethyl-ammonium bromide) to form soluble stable complexes with nucleic acids at high salt concentrations which may then be precipitated by lowering the salt concentration (Murray & Thompson, 1980). Other techniques incorporate the use of the strong anionic detergent sodium dodecyl sulphate (SDS) in conjunction with proteinase K and other components such as antioxidants and chelating agents, and are described elsewhere (Draper et al., 1988).

Protocol 1 described below facilitates rapid isolation of nucleic acid of sufficient purity to be used in PCR techniques (Chapter 6) or be used to carry out Southern blotting analysis without further purification on caesium chloride gradients. Both these techniques are important to check that the gene of interest is present in the DNA sample and also to provide information regarding the distribution of restriction sites within the sequence of interest (section 4.3).

PROTOCOL 1: CTAB DNA EXTRACTION OF PLANT DNA

This method has been used with success at Leicester University for a wide variety of different plants, including tobacco, asparagus, tomato, potato, maize, barley, wheat, lilly, Japanese knotweed and many other plant species. This technique allows the purification of nucleic acid from plants containing high concentrations of problematic carbohydrates and polyphenols that are characteristically responsible for 'brown' DNA, and does not require large volumes of phenol for protein extraction.

Solutions

- 2 × CTAB extraction buffer: 100 mM Tris-HCl, pH 8.0; 0.7 M NaCl; 20 mM EDTA; 2% CTAB w/v; 1.0% polyvinylpyrolidone (PVP-360) w/v; 0.2% 2-mercaptoethanol
- 2 × CTAB precipitation buffer: 100 mM Tris-HCl, pH 8.0; 20 mM EDTA; 2% CTAB w/v

Method

(1) Preheat the extraction buffer to 60 °C.
(2) Pulverise plant material under liquid nitrogen in a chilled pestle and mortar. Add 5 ml/g fresh weight material of 2 × CTAB extraction buffer at 60 °C and incubate with occasional gentle swirling at 60 °C for 20 min.
(3) Add an equal volume of chloroform/isoamyl alcohol (24 : 1) mix by inverting the tube several times and spin at 1500g for 5 min at room temperature. Remove the top aqueous phase, using a serological disposable pipette, to a new tube.
(4) Reduce the NaCl concentration to below 0.4 M by adding an equal volume of 2 × CTAB precipitation buffer and leave the CTAB–nucleic acid complex to precipitate for 5 min to 1 h at room temperature.
(5) Collect the CTAB–nucleic acid complex by centrifugation at 2000g at room temperature for 5 min. Pour off the supernatant, removing the remainder with a pipette.
(6) Fully resuspend the pellet in the minimum volume of 1 M NaCl and transfer to a microcentrifuge tube. Add 2.5 volumes of ethanol, mix and centrifuge at room temperature for 5 min. Pipette off the supernatant and wash the pellet with 70% ethanol, recentrifuging for 2 min to recover the pellet.
(7) Remove the dregs of the 70% ethanol and allow the pellet to dry on the bench. Resuspend the DNA in sterile distilled water and analyse via agarose gel electrophoresis.

The total nucleic acid will contain a large proportion of RNA and at this point it may be worthwhile purifying the DNA further on a CsCl gradient. However, if the DNA digests efficiently with restriction endonucleases and is of sufficient molecular weight, the contaminating RNA will not hinder any of the processes required to make a library and the total nucleic acid may be used directly.

PROTOCOL 2: CsCl/EtBr PURIFICATION OF PLANT DNA

(This protocol is modified from Draper et al., 1988.)

CsCl/EtBr-banded DNA is pure and cleaves well with restriction enzymes. Although this process is time-consuming, it is not labour-intensive, and the extra time invested eases subsequent quantitation and manipulation of the DNA. The volumes and spin times of this protocol can be adjusted according to tube size, rotor and ultracentrifuge available.

Method

(1) Add TE (10 mM Tris-HCl, pH 8.0; 1 mM EDTA) to the DNA to a final volume of 10 ml.

(2) For each millilitre of DNA solution add 1 g of finely ground CsCl and warm to 30 °C with gentle mixing to aid dissolution of the CsCl until completely dissolved. Note the change in volume of the DNA/CsCl solution and add 0.8 ml of 10 mg/ml ethidium bromide solution for every 10 ml of DNA/CsCl solution and gently mix. This should give a solution with a final density of 1.55 g/ml and concentration of ethidium bromide of 740 μg/ml.

(3) Transfer the solution to a heat-sealable polyallomer centrifuge tube (e.g. Beckman Quick-Seal 13.5-ml tube No. 342413) using a Pasteur pipette. Seal and centrifuge in a suitable rotor (e.g. Beckman Ti80) at 48 000 rev/min, 15 °C, for two days. If vertical rotors are available, spin times can be reduced to overnight.

(4) Visualise the DNA band using a hand-held long-wave (320 nm) UV source. Puncture the top of the tube with a 21 gauge needle to allow air into the tube. Remove the DNA band using a 2-ml syringe fitted with a 19 gauge needle by penetrating the tube bevelled side up approximately 0.5 cm below the DNA band and slowly withdrawing the DNA into the syringe.

(5) Remove the ethidium bromide by partitioning with an equal volume of CsCl/water-saturated isopropanol four to six times until the organic phase is no longer pink after mixing gently with the aqueous phase by slow inversion.

(6) Dilute the CsCl/DNA solution three-fold with water, precipitate the DNA with two volumes of ethanol at room temperature and collect the DNA by centrifugation at 10 000 rev/min at room temperature for 10 min. The DNA pellet should be white in colour with no traces of pink colouring. If the DNA is pink, then resuspend in water and reprecipitate by adding 1/20

volume 4 M sodium acetate, pH 6, and two volumes of ethanol to remove traces of ethidium bromide present.
(7) Wash the DNA pellet with 70% ethanol, recollect the pellet by centrifugation, remove all the supernatant and allow the pellet to dry on the bench. Resuspend the DNA in 0.5–1.0 ml of sterile double-distilled water.
(8) Scan the DNA solution by UV spectroscopy between 200 and 300 nm in a quartz cuvette. A good preparation will show a minimum at 230 mm and a clear maximum at 260 nm.

The OD_{280}/OD_{260} ratio should be about 1.8, and the concentration can be estimated from the OD_{260} reading as 1 OD_{260} unit = 50 $\mu g/ml$. Aliquot the DNA and store at −20 °C.

4.2   Organellar DNA extractions

To isolate nuclear DNA it is important first to isolate pure intact nuclei. This is best achieved starting from fresh plant material. The material may be pre-treated with ether to enhance cell disruption, and then homogenised in the presence of membrane-stabilising agents such as hexylene glycol, sucrose or glycerol and chelating agents such as EDTA. The lysate is then filtered and nuclei may be differentially pelleted from the cell debris following percoll step gradient centrifugation or differential sedimentation. High-salt-concentration buffers are used to burst the nuclei and DNA may be further purified on CsCl gradients.

Purification of chloroplastic and mitochondrial DNA is performed similarly, except the nuclei are removed by centrifugation and the organellar solution is treated with DNase to remove any contaminating DNA from the preparation. Differential or step sucrose gradient centrifugation are employed to enrich for either mitochrondria or chloroplasts. The organelles are then disrupted by osmotic shock and treatment with sarkosyl and the DNA purified on CsCl gradients.

PROTOCOL 3: NUCLEAR DNA PREPARATION

(This protocol is modified from Watson & Thompson, 1986.)

Solutions

- Nuclei extraction buffer: 1.0 M hexylene glycol; 10 mM Pipes-KOH, pH 7.0; 10 mM $MgCl_2$; 5 mM 2-mercaptoethanol
- Nuclei wash buffer: as nuclei extraction buffer but contains, in addition, 0.5% triton X-100
- Percoll gradient buffers: 35% and 80% percoll made in nuclei extraction buffer
- Lysis buffer: 10 mM Tris-HCl, pH 8.0; 10 mM EDTA; 0.5% triton X-100
- 2 × CTAB extraction buffer as in Protocol 1.

Method

(1) Collect 50–200 g of fresh leaf material and submerse in ice-cold ether in a fume hood for 1–2 min. Wash the leaves in ice-cold nuclei extraction buffer. Remove midveins or any other fibrous material and cut into small pieces using scissors, add 3 ml/g fresh weight of ice-cold nuclei extraction buffer and homogenise in a Waring blender with a number of 5-s bursts until the plant material is completely disrupted, stopping if the medium starts to froth.

(2) Squeeze the homogenate through four layers of muslin into a beaker and then sieve through a 300-μm nylon mesh into polypropylene GSA bottles and centrifuge at 500g for 10 min at 4 °C. Resuspend the pellet in 1 ml/g fresh weight of nuclei wash buffer by gentle swirling or by stirring with a soft paintbrush and centrifuge at 500g for 10 min at 4 °C.

(3) Resuspend the pellet in approximately 7 ml of nuclei isolation buffer and carefully layer over a step percoll gradient consisting of a 7-ml layer of 80% percoll and a 7-ml layer of 30% percoll in nuclei isolation buffer. Spin at 200g for 10 min at 4 °C in a swing-out rotor with gentle acceleration and deceleration.

(4) Using a serological pipette, remove the upper zones and keep the greyish-white band at the 30–80% interface, which should contain the nuclei, and place into a new tube in a volume of 1 or 2 ml. If during the next stage little DNA is released following lysis, the nuclei may not have banded at the interface but pelleted at the bottom of the tube. In this case it may be necessary to increase the concentration of percoll in the bottom gradient to 90% to obtain efficient banding.

(5) Add nuclei lysis buffer up to a final volume of 10 ml and mix by inversion. Then add 10 ml of 2 × CTAB extraction buffer at 60 °C (the high salt concentration should be sufficient to cause the nuclei to burst), mix by gentle inversion and proceed as described in Protocol 1, Step 2. The solution should become viscous during this stage, and this is a good indicator of efficient purification of nuclei on the step gradient.

PROTOCOL 4: ISOLATION OF CHLOROPLAST AND MITOCHONDRIAL
            DNA

(This protocol is modified from Palmer, 1986.)

This method does not yield ultrapure organellar preparations, as no step density gradients are used. However, the organelles are sufficiently enriched and free of nuclear DNA to use for DNA isolation to construct libraries or carry out Southern analysis.

Solutions

- Isolation buffer: 0.35 M sorbitol; 50 mM Tris-HCl, pH 8.0; 5 mM EDTA; 0.1% w/v bovine serum albumin (BSA); 0.1% w/v 2-mercaptoethanol
- Wash buffer: 0.35 M sorbitol; 50 mM Tris-HCl, pH 8.0; 25 mM EDTA

- Lysis buffer: 5% w/v sodium sarcosinite; 50 mM Tris-HCl, pH 8.0; 25 mM EDTA
- DNase I (2500 Kunitz units/mg; Sigma)
- DNase I buffer: 0.35 M sorbitol; 50 mM Tris-HCl, pH 8.0; 15 mM $MgCl_2$

Method

(1) Cut 100 g of plant material into small pieces using scissors and add 300 ml of ice-cold isolation buffer. Homogenise in a Waring blender with 5-s bursts until the plant material is completely disrupted. Filter the homogenate through four layers of muslin into a beaker and then sieve through a 300-μm nylon mesh into polypropylene GSA bottles. Centrifuge at 500g for 10 min at 4 °C.

(2) Transfer the supernatant to a new centrifuge tube and spin down the crude chloroplast pellet at 1000g, 4 °C, for 10 min. Recentrifuge the supernatant at 12000g, 4 °C, for 10 min to collect the mitochondria.

(3) Resuspend the organelle pellets in 10 ml of DNase I buffer by gentle swirling or stirring with a soft paintbrush and add 3 mg of DNase I freshly prepared in DNase I buffer. Incubate on ice with gentle swirling for 1 h.

(4) Terminate the DNase I digest by adding three volumes of wash buffer and collecting the mitochondria at 12000g and the chloroplasts at 1000g at 4 °C for 10 min. Resuspend the pellets in 10 ml of wash buffer and repeat the washing processes twice more.

(5) Following the final wash, resuspend each organelle pellet in 1 ml of wash buffer and add 0.2 ml of lysis buffer to each. Leave the tubes at room temperature for 10 min with occasional inverting.

(6) Bring the final volume of the organellar nucleic acid preparations up to 10 ml with wash buffer and purify the DNA on CsCl/EtBr gradients as described in Protocol 2.

4.3 Determination of DNA quantity and quality

If the DNA preparation has been purified by CsCl gradient it is sufficient to use standard spectrophotometric measurements to calculate the concentration. The integrity of the genomic DNA may be analysed on 0.3% agarose gels run with appropriate markers (e.g. lambda concatomers; Sigma). Alternatively, PFGE may be used (Chapter 4).

At this point, before a library is constructed it is a good idea to carry out Southern blot analysis on the genomic DNA to reveal the organisation and number of cross-hybridising target sequences in the plant genome. The stringency of the washes may be varied to give optimal signal/noise ratio, and it is likely that the same washing conditions will be optimal in the library screening process. A large number of cross-hybridising restriction fragments could mean several things: the gene could be part of a multigene family, there could be a number of pseudogenes present in the genome or there may be a repetitive element in the probe sequence. If there are many hybridising sequences it may be better to design a different, more specific, probe or at least beware of the consequences of using the probe for library screening. Fig. 2A shows a representative restriction digest and Southern blot of total plant DNA probed with a cDNA probe encoding S-adenosyl homocysteine hydrolase.

5   LAMBDA VECTORS AND A CLONING STRATEGY

5.1   Which lambda vector?

There are many lambda vectors commercially available that are predigested and ready for direct cloning. The best choice of vector depends on the cloning strategy employed.

First, it must be decided whether to opt for a genomic library that is representative of the whole genome or a library that contains particular-sized restriction fragments of the genome. This decision can be based on the results of Southern analyses.

5.2   Representative libraries

In order to construct a library representative of the genome with as few clones as possible, the DNA should be fragmented as randomly as possible. Given this, the number of clones required to be representative of the genome, N, is given by

$$N = \ln(1 - P)/\ln(1 - f)$$

where P is the probability of a particular unique DNA sequence being represented and f is the average size of the DNA fragments cloned as a fraction of the total genome (Clark & Carbon, 1976). (This equation is also described in Chapter 5.) For example, if the *Arabidopsis* genome is approximately 2×10^8 bp and the clones in the library have an insert size of 9 kb, then for a 95% probability of obtaining a particular unique sequence a library of 67 000 clones must be screened, which represents three genome equivalents. For a 99% probability, over 100 000 clones must be screened, representing five genome equivalents. If a library was constructed in a cosmid vector allowing a 50-kb insert size, only 12 000 clones would need to be screened for a 95% probability. For wheat, nearly 7 000 000 clones with 9-kb inserts must be screened for a 95% chance of obtaining a unique sequence.

To ease the screening process it is best to screen as few clones as possible. As the equation shows, the number of clones to be screened depends on the size of the inserts in the library and the genome size. Therefore, replacement vectors should be used in preference to insertional vectors, as they can accept far larger fragments of foreign DNA. In the example give for wheat, screening 7×10^6 clones is a formidable task, and other vector systems allowing much larger inserts such as cosmids, BACS and YACS (discussed in Chapter 4) should be considered.

Some of the many commonly used replacement vectors commercially available are lambda 2001 (Stratagene), DASH (Stratagene), FIX (Stratagene) and EMBL (Stratagene, Clontech, Pharmacia and other manufacturers). Fig. 1 shows the restriction maps of some of these vectors.

5.3 Cloning specific restriction fragments

Southern blotting analysis may show that the required gene is single copy and may be totally contained within a small single restriction fragment. A specific library containing the desired restriction fragment may be constructed. The number of clones to be screened to give the desired probability of obtaining a positive can be estimated using the equation of Clark and Carbon described in the previous section. f is now the size of the unique DNA sequence cloned as a fraction of the total genome divided by the proportion of size-selected DNA recovered from the total DNA digest. If the restriction fragment isolated represents one-tenth of the total DNA digested, then the number of clones to be screened is reduced by 10-fold from the number necessary if a random representative library were to be screened. In practice the enrichment is better than 10-fold.

The choice of vector used for this purpose will depend on the size of the restriction fragment required; if it is between 0 and 9 kb, then an insertional vector must be used. Commonly used insertional vectors include the λgt series (Stratagene, Clontech, Promega, Pharmacia and other manufacturers) and ZAP (Stratagene). Lambda ZAP offers the added bonus of an automatic excision of the desired clone as a plasmid (restriction map shown in Fig. 1).

6 METHODOLOGY OF LIBRARY CONSTRUCTION

Plant genomic libraries may be constructed from total DNA or nuclear DNA. If total DNA is used, it must be remembered that plastid DNA may represent up to 20% of the DNA isolated, and this should be taken into consideration when calculating the number of clones required for screening. Plant nuclear DNA is highly methylated, which can be problematic when cleaving with restriction enzymes that show sensitivity to methylated DNA. Information regarding restriction enzyme sensitivity to methylation can be obtained from manufacturers' catalogues.

6.1 Restriction digestion of genomic DNA—a total digest

If a small specific library of restriction fragments is required, total DNA digestion is necessary to prepare the desired restriction fragments between 0 and 10 kb for ligation into insertional vectors.

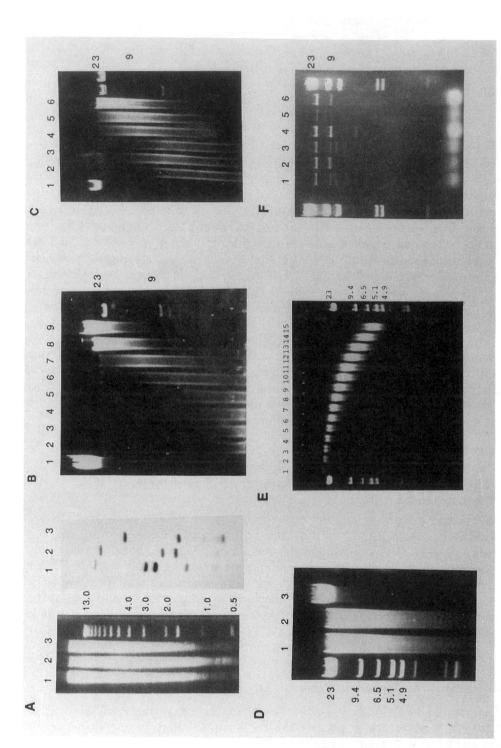

Figure 2. Typical stages of DNA preparation in constructing a representative plant genomic library in a lambda replacement vector. (A) Plant genomic DNA digested with EcoRI (lane 1), HindIII (lane 2) and HindIII/XbaI (lane 3) with a Southern blot of the same gel probed with a cDNA clone. (B) MboI partial digest of plant genomic DNA shown with lambda HindIII marker: lane 1, 4 μg undigested DNA; lane 2, 4 μg DNA digested with 4 U MboI/μg DNA for 1 h; lane 3, 4 μg DNA digested with 2 U MboI/μg DNA for 1 h; lane 4, 4 μg DNA digested with 1 U MboI/μg DNA for 1 h; lane 5, 4 μg DNA digested with 0.5 U MboI/μg DNA for 1 h; lane 6, 4 μg DNA digested with 0.25 U MboI/μg DNA for 1 h; lane 7, 4 μg DNA digested with 0.125 U MboI/μg DNA for 1 h; lane 8, 4 μg DNA digested with 0.0625 U MboI/μg DNA for 1 h; lane 9, 4 μg DNA digested with 0.03125 U MboI/μg DNA for 1 h. (C) Accurate partial digests: lane 1, 4 μg undigested DNA; lane 2, 4 μg DNA digested with 0.125 U MboI/μg DNA for 1 h; lane 3, 4 μg DNA digested with 0.112 U MboI/μg DNA for 1 h; lane 4, 4 μg DNA digested with 0.100 U MboI/μg DNA for 1 h; lane 5, 4 μg DNA digested with 0.0875 U MboI/μg DNA for 1 h; lane 6, 4 μg DNA digested with 0.0625 U MboI/μg DNA for 1 h; lane 7, 4 μg DNA digested with 0.0375 U MboI/μg DNA for 1 h. (D) Scaled-up partial digests: lane 1, 4 μg undigested DNA; lane 2, 4 μg DNA digested with 0.0343 U MboI/μg DNA for 1 h; lane 3, 4 μg undigested DNA. (E) Sucrose gradient fractions of partially digested DNA. Fractions 6–8 contain DNA of the correct size for cloning into replacement vectors. (F) Mini phage preparations of recombinants in EMBL 3 digested with SalI.

In general, digestions are carried out overnight in restriction buffers supplied by the manufacturers, together with a 10-fold excess of each restriction enzyme which is added to ensure complete digestion. Multiple digestions are carried out either in buffers that are compatible for both enzymes, or by digesting first with one enzyme, which is then heat-inactivated, followed by the optimisation of the salt concentration for the second enzyme. As a result of inhibition or alteration of enzyme activity at high glycerol concentrations, it is not possible to add a volume of restriction enzyme in excess of 10% of the total volume. The result of this is that generally large reaction volumes are required. DNA may, however, be concentrated following digestion by ethanol precipitation and analysed by gel electrophoresis.

6.2   Random fragmentation of genomic DNA

Random overlapping DNA fragments between 9 and 23 kb in size with cohesive compatible termini are required to construct a representative library in lambda replacement vectors. This may be achieved by partial restriction digestion or by random shearing as described below.

6.3   Partial digestion with restriction enzymes

When attempting to construct a representative library in a lambda replacement vector, the most common way to generate a set of random, overlapping DNA fragments between 9 and 23 kb in size is by partial restriction digestion with *Mbo*I. This restriction enzyme cleaves at the tetrameric sequence GATC and generates cohesive termini that are compatible with *Bam*HI. This strategy is used because, statistically, the tetrameric sequence should frequently occur in the DNA, allowing random overlapping fragments to be generated on partial digestion. It should be noted that *Mbo*I is not methylation sensitive, and so should be used in preference to its isoschizomer, *Sau*3A, when dealing with plant DNA, to avoid biasing the library to hypomethylated DNA sequences.

The starting size of the DNA is important. The greater the DNA size, the greater the proportion of DNA fragments with two cohesive ends generated by the partial digest. Molecules with one sheared end will directly compete with the double-cohesive-ended molecules for the vector arms and reduce cloning efficiency. For practical purposes, DNA greater than 100 kb, as judged by electrophoresis against lambda concatamers, is useful for cloning.

To get a good partial digest a time-course may be carried out or a dilution series of the restriction enzyme may be used to determine optimum conditions empirically for each DNA preparation. Digests are carried out in the buffer systems recommended by the manufacturer, but clearly the amount of enzyme used is limiting. Products are analysed on agarose gels and conditions are best that give a good distribution of DNA between 9 and 23 kb. Scaled-up digestions are performed identically to the pilot reactions (see Fig. 2B, C).

PROTOCOL 5: PARTIAL DIGESTION OF PLANT GENOMIC DNA WITH *Mbo*I

Solutions

- *Mbo*I and supplied 10× buffer

Method

(1) In a microcentrifuge tube mix 15 *μ*l of 10× *Mbo*I buffer with 40 *μ*g of genomic DNA and make up to 150 *μ*l with sterile water. Label nine tubes and aliquot 30 *μ*l into tube 1 and 15 *μ*l into tubes 2–9. Preheat all tubes to 37 °C.

(2) Add 32 U of *Mbo*I to the first tube mix and quickly transfer 15 *μ*l to tube 2 with a clean pipette tip. Repeat the two-fold serial dilution with subsequent tubes but add nothing to tube 9 and discard 15 *μ*l from tube 8 so that each tube contains identical volumes.

(3) Incubate at 37 °C for 1 h, and then at 65 °C for 20 min. Add a one-fifth volume of DNA loading buffer to each sample and analyse on a 0.5% agarose gel against lambda *Hind*III markers.

(4) Assess the enzyme concentration that gives a good size distribution between 9 and 23 kb and carry out a further series of pilot digests that vary from half to double the optimum enzyme concentration from the preliminary digest. Analyse the products on an agarose gel.

(5) Once the optimum concentration has been obtained for the pilot reaction experiment, the reaction can be scaled up to digest 10 times the quantity of DNA, i.e. 400 *μ*g. In order to keep the concentration of DNA identical to that in the pilot reaction, this means that the digestion must be carried out in a 1.5-ml volume which is 100× the volume of the succcessful pilot reaction. It has been shown that the optimum enzyme concentration is half that of the optimum for the 15 *μ*l pilot reaction (Seed et al., 1982). An aliquot can be analysed on an agarose gel to check for a successful size distribution (Fig. 2D).

(6) Gently extract the partial digest with phenol/chloroform, ethanol precipitate the DNA and resuspend in TE in a 500 *μ*l volume.

6.4   Random shearing

Random shearing offers the advantage that DNA fragments really are random and overlapping, whereas restriction digestion will not cleave all parts of the genome equally. Unfortunately it is much harder to clone sheared DNA fragments, and the conditions required to generate a good size distribution are not as controllable as enzymatic digestion. DNA may be sheared through vigorous manipulation, such as passing rapidly through a narrow-bore needle, or via sonication with a horn sonicator. The effect of the manipulation may be assessed by agarose gel electrophoresis. The sheared ends must then be treated with T4 DNA polymerase to blunt the termini, and then adapters, in excess concentration, are ligated onto the repaired fragments, which will then enable ligation to the vector. This technique is described in the next chapter but is not recommended in the construction of lambda libraries, as partial digestion by restriction endonucleases results in a larger proportion of DNA fragments that are readily ligatable.

6.5   Size fractionation of insert DNA

Size fractionation of the insert DNA is important if a specific restriction fragment is required or if a representative genomic library is being constructed. If smaller or larger inserts are present, cloning in replacement vectors will be inefficient due to the large proportion of unsuitably sized inserts. For a representative genomic library in lambda, the required size of the random overlapping DNA fragments must be between 9 and 23 kb, whilst following partial restriction digestion, the fragmented DNA will be between 1 and 50 kb.

   DNA may be size fractionated via sucrose density gradients (Sambrook et al., 1989) or directly following agarose gel electrophoresis from the agarose gel (Fig. 2E).

PROTOCOL 6: SUCROSE GRADIENT FRACTIONATION OF PARTIALLY
DIGESTED DNA

A continuous gradient between 10% and 40% is required. This can be formed using a gradient former or by setting up a series of step gradients and allowing them to diffuse overnight at 4 °C.

Solutions

- 40%, 35%, 30%, 25%, 20%, 15% and 10% sucrose in 20 mM Tris-HCl, pH 8.0; 20 mM EDTA.

Method

(1)   Using a long wide-bore needle and 10-ml syringe, load 5.4 ml of the 10% sucrose solution into the bottom of a 38 ml Beckman tube. Fill the syringe with 15% sucrose and, placing the end of the needle at the bottom of the centrifuge tube, carefully and slowly underlay the 10% sucrose solution with 5.4 ml of 15% sucrose solution. Continue with other solutions. Place at 4 °C overnight (16 h).

(2)   Add the partially digested DNA in 500 μl of TE to the top of the gradient

and spin in a swing-out rotor (Beckman SW28 or equivalent) at 26 000 rev/min at 15 °C for 24 h.

(3) Collect 1-ml fractions from the bottom of the tube using a fraction collector attached to a peristaltic pump; a good flow rate is between 0.5 and 1 ml per minute. Run out approximately 25 μl of each fraction against markers diluted with an equal volume of 40% sucrose solution.

(4) Precipitate relevant DNA fractions by adding a 1/20 volume of 4 M sodium acetate, and an equal of volume of isopropanol, and centrifuging in a high-speed microcentrifuge for 10 min at room temperature. Remove all the supernatant and resuspend the pellet in 50 μl of water. Ethanol precipitate once more to ensure that no sucrose remains in the DNA fractions and resuspend in 50 μl of water. Run an aliquot against markers to check size distribution and concentration of fractionated DNA.

6.6 Dephosphorylation of the insert DNA

Once fractions of the desired size have been obtained, the insert DNA may be treated with alkaline phosphatase to prevent self-ligation (Sambrook et al. 1989), which were it to occur would reduce the cloning efficiency. If the fractionated DNA contains fragments below 12 kb in size, the possibility exists of having scrambled clones containing two inserts. When lambda FIX is being used this is not necessary, as the inserts are prevented from ligating to one another following partial fill-in of the restriction overhang. When using an insertional vector to clone a specific restriction fragment, it may not be possible to dephosphorylate the insert if the vector requires dephosphorylation to prevent self-ligation.

6.7 Ligation into lambda, packaging and plating

Most lambda vectors are sold ready to clone by the manufacturers together with recommended host strains. As many strains and packaging extracts are available that maximise the cloning of methylated DNA in lambda systems, it is advantageous to use these when constructing a plant genomic library. As an example, Gigapack gold II from Stratagene offers high-efficiency packaging of lambda clones containing methylated DNA.

6.8 Analysis of library size and diversity

An aliquot of the library is plated out and the size of the library determined. The range of insert size may be determined by performing mini phage preparations on a few clones packed at random (Fig. 2F). If insertional vectors are being used and the insert sizes are relatively small, PCR can be used to check for a random size distribution of insert sizes. An alternative method for determining library size and diversity is described in Chapter 5.

PROTOCOL 7: MINI LAMBDA PHAGE PREPARATIONS

(1)   Touch each individual plaque with a sterile pipette tip and agitate in 200 μl of plating cells in a 10-ml tube. Add 5 ml of NZY broth to each tube and shake at 37 °C overnight.

(2)   The following day add 100 μl of chloroform to each tube and shake at 37 °C for a further 15 min. The cultures should be completely lysed. Spin at 10 000g at 4 °C for 10 min to remove cellular debris.

(3)   Add 5 μg of unboiled RNase I to each tube (the DNase activity in the unboiled RNase is required to digest the *E. coli* DNA) and incubate at 37 °C for 30 min.

(4)   Place the tube on ice and mix in an equal volume of 20% polytheylene glycol (PEG) 8000, 2 M NaCl and 10 mM MgSO$_4$, and incubate on ice for 20 min.

(5)   Spin down precipitated phage at 10 000g at 4 °C for 10 min. Remove all the supernatant and resuspend the pellets in 0.5 ml of 10 mM MgSO$_4$.

(6)   Spin down any insoluble matter in a microcentrifuge for 2 min and transfer the clear supernatant to a new tube. To each tube add 5 μl of 10% SDS and 5 μl of 0.5 M EDTA, pH 8.0, and heat to 65 °C for 15 min.

(7)   Perform two phenol/chloroform extractions and ethanol precipitate the DNA. Collect by centrifugation in a microcentrifuge at 4 °C for 10 min. Wash the pellet with 70% ethanol, air-dry and resuspend in 10–50 μl of water. Carry out restriction analysis in the presence of 0.1 mg/ml boiled RNase I.

Analyse the digestions on an agarose gel. Check that each clone has a different insert within the expected size range for the particular vector used.

6.9   Libraries from alternative sources

For many of the model plants, libraries are available from research programmes, databases or commercial sources. Using the Mosaic NCSA system, information regarding many plant research programmes can be found on the worldwide web (WWW) on the Internet under the subject of agriculture and the subsection of plant biology. For example, *Arabidopsis* has a very comprehensive research programme, and information regarding libraries may be obtained from the ABRC at Ohio State University from Keith Davis, e-mail address Davis.68@osu.edu. Information is also available regarding the genomes of sorgum, tomato, rice, maize and soybean in the plant biology subsection. Table 1 gives some examples of commercially available sources advertised in manufacturer's catalogues.

   Other commercially available libraries prepared for Stratagene or Clontech include barley, wheat, tobacco, potato, pea, oilseed rape and soybean. When using commercially supplied libraries it must be remembered that these have been amplified at least once and sometimes twice, probably resulting in the relative dilution of some clones. Some commercial libraries do not contain

Table 1. Commercially available libraries

Library source	Vector	No. of clones	Insert size range (kb)
Alfalfa (18-day post-emergence seed-lings grown under ambient light—Clontech)	EMBL 3	2.0×10^6	8–20
Arabidopsis (4.5-week-old leaves and stem from columbia—Clontech)	EMBL 3	1.3×10^6	8–21
Arabidopsis (whole plant from columbia—Stratagene)	ZAP II	$> 1.5 \times 10^6$	4–10
Corn (Zea mays var. B73 whole seed-ling two-leaf stage—Clontech)	EMBL 3	1.7×10^6	8–21
Corn (Zea mays var. Missouri 17 inbred line etiolated seed-lings—Stratagene)	FIX II	$> 1.5 \times 10^6$	9–23

sufficient independent clones to comprise a library that is representative of the plant's genome.

7   CONCLUDING REMARKS

Once the library has been constructed and preliminary analysis suggests that the library is representative of the genome, screening is possible without prior amplification. The biggest obstacle in creating a representative library is gener-ating enough clones. It may be necessary to pool the clones from several packaging reactions to generate a library of sufficient size to be representative of the plant genome. There will many sequences that will be under-repre-sented in the library due to hypermethylation, scarcity of MboI sites or potential toxicity of expressed sequences in E. coli. Ultimately, the true test of the library is whether it contains your gene of interest and, in practice, often many more clones must be screened than would be predicted from the equation of Clark and Carbon (Clark & Carbon, 1976). A variety of screening strategies may be employed and some of these are discussed in Chapters 5 and 6. Other cloning strategies are described elsewhere in this book that rely on different vectors and host cell systems, using powerful DNA amplification technology or other methods. These techniques may provide a good alternative to genomic cloning in lambda vectors if screening lambda libraries proves unsuccessful, but these techniques are more error prone than conventional cloning and this should be an important consideration when embarking on a cloning strategy.

ACKNOWLEDGMENTS

Thanks go to Mark Skipsey, Liz Walker, Mike Roberts and Eleanor Warner for their contributions and help in the preparation of this chapter. Thanks also go to all at Leicester, especially John Draper for providing me with a happy and productive work environment.

ESSENTIAL READING

Bennet, M.D. and Smith, J.B. (1976) Nuclear DNA amounts in Angiosperms. *Phil. Trans. R. Soc. Lond. Ser. B.*, 227–274.
Draper, J., Scott, R., Armitage, P. and Walden, R. (1988) *Plant Genetic Transformation and Gene Expression*. Blackwell Scientific Publications. Oxford.
Murray, N.E. (1983) Phage lambda and molecular cloning. In *Lambda II* (eds R.W. Hendrix, J.W. Roberts, F.W. Stahl and R. Weissberg), pp. 395–432. Cold Spring Harbor Press, Cold Spring Harbor, New York.
Sambrook, J., Fritsh, E.F. and Maniatis, T. (eds) (1989) In *Molecular Cloning: A Laboratory Manual*, 2nd edn. Cold Spring Harbor Press, New York.

REFERENCES

Bennet, M.D., Smith, J.B. and Heslop-Harrison, J.S. (1982) Nuclear DNA amounts in Angiosperms. *Proc. R. Soc. Lond. Ser. B.*, 179–192.
Clark, L. and Carbon, J. (1976) A colony bank containing synthetic Col E1 hybrid plasmids representative of the entire *E. coli* genome. *Cell*, **9**, 91.
Gardinergarden, M., Sued, J.A. and Frommer, M. (1992) Methylation sites in Angiosperm genes. *J. Mol. Evol.*, **34**, 219–230.
Gray, M.W., Hanicjoyce, P.J. and Covello, P.S. (1992) Transcription, processing and editing in plant mitochondria. *Annu. Rev. Plant. Physiol. Plant. Mol. Biol.*, **43**, 145–175.
Hanson, M.R. and Folkerts, O. (1992) Structure and function of the higher plant mitochondrial genome. *Int. Rev. Cytol.*, **141**, 129–172.
Hiratsuka, J., Shimada, H., Whittier, R. et al. (1989) The complete sequence of the rice (*Oryza sativa*) chloroplast genome: intermolecular recombination between distinct TrNA genes accounts for a major plastid DNA interconversion during the evolution of the cereals. *Mol. Gen. Genet.*, **217**, 184–194.
Inamdar, N.M., Ehrlich, K.C. and Ehrlich, M. (1991) CpG methylation inhibits binding of several sequence-specific DNA-binding proteins from pea, wheat soybean and cauliflower. *Plant Mol. Biol.*, **17**, 113–123.
Kaiser, K. and Murray, N.E. (1985) The use of phage lambda replacement vectors in the construction of representative genomic DNA libraries. In *DNA Cloning: A Practical Approach* (ed. D.M. Glover), pp. 1–20. IRL Press, Oxford.
Kuroiwa, T. (1991) The replication, differentiation and inheritance of plastids with emphasis on the concept of organelle nuclei. *Int. Rev. Cytol.*, **128**, 1.
Lapitan, N.L.V. (1992) Organization and evolution of higher plant nuclear genomes. *Genome*, **35**, 171–181.
Mackenzie, S., He, S. and Lyznik, A. (1994) The elusive plant mitochondrion as a genetic system. *Plant Physiol.*, **105**, 775–780.
Murray, M.G. and Thompson, W.F. (1980) Rapid isolation of high molecular weight plant DNA. *Nucleic Acids Res.*, **8**, 4321–4325.
Ngernprasirtsiri, J., Kobayashi, H. and Akazawa, T. (1988) DNA methylation as a

mechanism of transcription in nonphotosynthetic plastids in plant cells. *Proc. Natl. Acad. Sci. USA*, **13**, 4750–4754.

Oda, K., Yamato, K., Ohta, E. et al. (1992) Gene organization deduced from the complete sequence of the liverwort *Marchantia polymorpha* mitochondrial DNA. A primitive plant mitochondrial genome. *J. Mol. Biol.*, **223**, 1–7.

Ohta, N., Sato, N., Kawano, S. and Kuroiwa, T. (1991) Methylation of DNA in chloroplasts and amyloplasts of pea, *Pisum sativum*. *Plant Sci.*, **78**, 33–42.

Ohyama, K., Fukezawa, H., Kohchi, T. et al. (1986) Chloroplast gene organisation deduced from the complete sequence of liverwort, *Marchantia polymorpha* chloroplast DNA. *Nature*, **322**, 572–574.

Olmstead, R.G., Sweeere, J.A. and Wolfe, K.H. (1993) Ninty extra nucleotides in the ndh F gene of tobacco chloroplast DNA: a summary of revisions to the 1986 genome sequence. *Plant Mol. Biol.*, **22**, 1191–1193.

Palmer, J.D. (1986) Isolation and structural analysis of chloroplast DNA. *Methods Enzymol.*, **118**, 167–186.

Salganik, R.I., Dudareva, N.A. and Kiseleva, E.V. (1991) Structural organisation and transcription of plant mitochrondrial and chloroplast genomes. *Elect. Microsc. Rev.*, **4**, 221–247.

Schuster, W. and Brennicke, A. (1994) The plant mitochondrial genome—physical structure, information-content, RNA editing and gene migration to the nucleus. *Annu. Rev. Plant Physiol. Plant Mol. Biol.*, **45**, 61–78.

Seed, B., Parker, R.C. and Davidson, N. (1982) Representation of DNA-sequences in recombinant DNA libraries prepared by restriction enzyme partial digestion. *Gene*, **19**, 201–209.

Shapiro, H.S. (1976) Distribution of purines and pyrimidines in deoxyribonucleic acids. In *Handbook of Biochemistry and Molecular Biology* (ed. G.D. Fasman), Vol. 2, p. 282. CRC Press, New York.

Shikanai, T., Yang, Z.Q. and Yamada, Y. (1989) Nucleotide sequence and molecular characterisation of plasmid-like DNAs from the mitochondria of cytoplasmic male-sterile rice. *Curr. Genet.*, **15**, 349–354.

Shinozaki, K., Ohme, M., Tanaka, M. et al. (1986) The complete nucleotide sequence of the tobacco chloroplast genome: its gene organisation and expression. *EMBO J.*, **5**, 2043–2049.

Sugiura, M. (1989) The chloroplast chromosomes in land plants. *Annu. Rev. Cell Biol.*, **5**, 51–70.

Watson, J.C. and Thompson, W.F. (1986) Purification and restriction of plant nuclear DNA. *Methods Enzymol.*, **118**, 57–75.

Wolff, G., Plante, I., Lang, B.F. et al. (1994) Complete sequence of the mitochondrial DNA of the chlorophyte alga Prototheca wickerhammi: gene content and genome organisation. *J. Mol. Biol.*, **237**, 75–86.

4 BAC, YAC and Cosmid Library Construction

HONG-BIN ZHANG
SUNG-SICK WOO
ROD A. WING
Texas A&M University, Texas, USA

The advent of pulse field gel electrophoresis (PFGE) and cosmid, yeast artificial chromosome (YAC), and bacterial artificial chromosome (BAC) cloning of large genomic DNA fragments has fuelled an explosion of genome research leading to the physical characterisation of complex genomes and isolation of many important genes for which only a phenotype and map position were known. This chapter will discuss: (1) the isolation of megabase-size DNA from protoplasts and nuclei suitable for physical mapping and construction of large-insert DNA libraries; (2) the construction of cosmid, YAC and BAC libraries; and (3) the application of these large-insert DNA libraries to construct large overlapping contigs for physical mapping of plant genomes and the isolation of agriculturally significant genes.

1 INTRODUCTION

Large-insert genomic DNA libraries have served a pivotal role in the isolation and characterisation of important genomic regions and genes. This chapter will discuss the construction of three types of large-insert DNA libraries—cosmids (Collins & Hohn, 1978), yeast artificial chromosomes (YACs) (Burke et al., 1987), and bacterial artificial chromosomes (BACs) (Shizuya et al., 1992) and the isolation of DNA that is suitable for cloning in all three of these vector systems. Cosmids have been used since the earliest beginnings of plant molecular biology and remain valuable as general vectors for cloning DNA fragments up to approximately 40 kb. Cosmids are basically plasmid vectors; however, they contain lambda cos sites which allow DNA to be packaged *in vitro*. This feature allows for the efficient transfection of in vitro packaged DNA into *E. coli*. Cosmid vectors have also been modified into plant transformation vectors and thus can be mobilised efficiently into plants using *Agrobacterium tumefaciens*. YAC vectors have been used quite extensively since 1987, especially for physical mapping, contig construction, and map-based gene cloning. The YAC vector system is designed to maintain foreign DNA fragments in yeast as linear chromosomes. Thus the vector system contains all of the elements of a eukaryotic chromosome, including telomeres, a centromere and a

Plant Gene Isolation: Principles and Practice. Edited by G. D. Foster and D. Twell.
© 1996 John Wiley & Sons Ltd.

yeast origin of DNA replication, which are required for chromosome maintenance and stability. The cloning capacity in YACs is limitless and depends
primarily on the quality and size of the DNA to be cloned. BAC vectors are
relatively new and have yet to be fully tested, although recent results from
several laboratories indicate that the BAC system will be useful for many plant
genomic applications. BAC vectors are derived from the E. coli F factor
plasmid, which contains genes for strict copy number control and unidirectional DNA replication. Both features promote plasmid maintenance and
stability. The upper limit for cloning capacity in BACs at present is about
400 kb (Shizuya et al., 1992).

Large-insert DNA libraries are used for a variety of applications from
physical mapping to map-based gene cloning (see Chapter 10). The construction of a representative genomic library is very important. Table 1 shows the
number of clones that are needed to have a probability of 99% that a library
contains a desired clone based on an average insert size of 40, 150 and 500 kb
for a cosmid, BAC or YAC library, respectively. Here we have listed 21

Table 1. The number of clones (N) required to have a 99% probability of a particular
clone represented in a library having an average insert size of 40 kb, 150 kb, or 500 kb
for selected crop plants.

Scientific name	Common name	Genome size (Mbp/1C)	Cosmids (40 kb)	BACs (150 kb)	YACs (500 kb)
Allium cepa	Onion	15 290	1.76×10^6	4.7×10^5	1.4×10^5
Arabidopsis thaliana	Arabidopsis	145	1.7×10^4	4.5×10^3	1.3×10^3
Avena sativa	Oat	11 315	1.3×10^6	3.5×10^5	1.0×10^5
Beta vulgaris ssp. esculenta	Sugar beet	758	8.7×10^4	2.3×10^4	7.0×10^3
Brassica napus	Canola	1182	1.36×10^5	3.2×10^4	1.1×10^4
Capsicum annuum	Pepper	2702	3.5×10^5	9.4×10^4	3.5×10^3
Citrus sinensis	Orange	382	4.4×10^4	1.2×10^4	3.5×10^3
Glycine max	Soybean	1115	1.3×10^5	3.4×10^4	1.0×10^4
Gossypium hirsutum	Cotton	2246	1.8×10^5	6.9×10^4	2.1×10^4
Hordeum vulgare	Barley	4873	5.6×10^5	1.5×10^5	4.5×10^4
Lactuca sativa	Lettuce	2639	3.0×10^5	8.1×10^4	2.4×10^4
Lycopersicon esculentum	Tomato	953	1.1×10^5	2.9×10^4	8.8×10^3
Malus x domestica	Apple	769	8.9×10^4	2.4×10^4	7.1×10^3
Manihot esculenta	Cassava	760	8.7×10^4	2.3×10^4	7.0×10^3
Musa sp.	Banana	873	1.0×10^5	2.7×10^4	8.0×10^3
Nicotiana tabacum	Tobacco	4434	5.1×10^5	1.4×10^5	4.1×10^4
Oryza sativa sspp. indica & japonica	Rice	431	5.0×10^4	1.3×10^4	4.0×10^3
Phaseolus vulgaris	Common bean	637	7.4×10^4	2.0×10^4	5.9×10^3
Saccharum sp.	Sugarcane	3000	3.5×10^5	9.2×10^4	2.8×10^4
Sorghum bicolor	Sorghum	760	8.7×10^4	2.3×10^4	7.0×10^3
Triticum aestivum	Wheat	15 966	1.8×10^6	4.9×10^5	1.5×10^5
Zea mays	Maize	2504	2.9×10^5	7.7×10^4	2.3×10^4

$N = \ln(1 - P)/\ln(1 - I/GS)$ (Clarke & Carbon, 1976), where N = number of clones, P = probability, I = insert
size and GS = genome size (genome sizes are from Arumuganathan & Earle, 1991).

important crop plants and *Arabidopsis thaliana*. Theoretically the number of clones required for each type of library is attainable; however, if the objective of the experiment is to perform a walk or assemble a 1-Mb overlapping contig then the larger the insert, the fewer steps are required, and the probability of completing a contig or walk is increased. To illustrate the power of chromosome walking with cosmids, BACs and YACs, we have plotted the number of clones required for a 1000-kb walk, assuming a 50% overlap for each clone, versus the probability of completing the walk (Fig. 1). If a cosmid library is used with a 50-kb insert size, 39 steps would be required and the probability of completing the walk is nil. However, if a BAC library was used (e.g. 150-kb average insert size), it would require about 13 steps and would have a 90% probability of completing the walk. For the YAC library (e.g. 500 kb average insert size), about three steps would be required to cross the same 1000-kb distance and would have a 99% chance of success.

2 MEGABASE-SIZE DNA ISOLATION FROM PLANTS

To construct large-insert DNA libraries in BAC and YAC vectors, methods must be developed to isolate very high molecular weight (HMW) DNA—megabase-size DNA—from plants. Fig. 2A shows a general scheme for the isolation

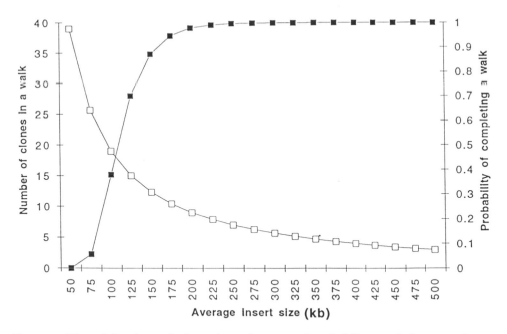

Figure 1. Plot of the theoretical number of steps and probability needed to complete a 1000 kb walk. Vector insert size versus number of clones needed to complete a walk (light boxes) and probability of completing a chromosome walk (dark boxes). Calculations are based on a library with 30 000 clones, a genome size of 1000 Mb, a walking distance of 1000 kb, and an average of 50% overlap between adjacent clones

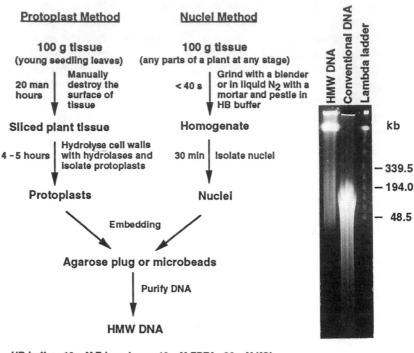

A. Preparation of HMW DNA from Plants   B. HMW vs conventional DNA

Protoplast Method

100 g tissue
(young seedling leaves)

20 man hours | Manually destroy the surface of tissue

Sliced plant tissue

4 - 5 hours | Hydrolyse cell walls with hydrolases and isolate protoplasts

Protoplasts

Nuclei Method

100 g tissue
(any parts of a plant at any stage)

< 40 s | Grind with a blender or in liquid N₂ with a mortar and pestle in HB buffer

Homogenate

30 min | Isolate nuclei

Nuclei

Embedding

Agarose plug or microbeads

Purify DNA

HMW DNA

kb

— 339.5

— 194.0

— 48.5

HB buffer: 10 mM Trizma base, 10 mM EDTA, 80 mM KCl, 500 mM sucrose, 1 mM spermine, 1 mM spermidine, 0.15% beta-mETOH, and 0.5% triton X-100, pH = 9.40.

Figure 2. (A) A general scheme for preparation of HMW DNA from plants. (B) PFGE analysis of wheat DNA prepared by a HMW method (Zhang et al., 1995) and by a conventional method. The DNA was subjected to PFGE on a 1% agarose gel in 0.5× TBE (0.045 M Tris-borate, 0.001 M EDTA, pH 8.0) for 20 h at 150 V, 35 s pulse, at 11 °C

of megabase-size DNA from plants. To isolate such DNA, protoplasts or nuclei must first be embedded in agarose plugs or microbeads. The agarose acts as a solid yet porous matrix which allows for the diffusion of various reagents for DNA purification and subsequent manipulations while preventing the DNA from being sheared (Schwartz & Cantor, 1984). Microbeads are preferred over plugs because the use of beads increases the surface area surrounding the tissue sample by approximately 1000-fold, thereby allowing for more efficient and rapid diffusion of chemicals and enzymes into and out of the agarose beads (Cook, 1984; Overhauser & Radic, 1987; Wing et al., 1993). Once embedded, the protoplasts or nuclei are lysed and proteins degraded in the presence of 0.5 M EDTA, 1% sarcosyl and 0.1–1.0 mg/ml of proteinase K at 50 °C (Schwartz & Cantor, 1984). After cell lysis and protein degradation, the remaining DNA is suitable for enzymatic modifications. Fig. 2B shows a pulsed-field gel that contains unrestricted wheat DNA isolated: (1) from nuclei embedded in agarose or (2) by a conventional method. As can be seen, the majority of wheat DNA prepared from agarose-embedded nuclei is retained in the well of the gel and at the compression zone, which is greater than 1 Mb in size. For comparison, the majority of DNA prepared by a conventional method is between 100 and 150 kb in size (see Chapter 3).

Most protocols for the isolation of megabase-size DNA from plants utilise the protoplast method (Cheung & Gale, 1990; Ganal et al., 1989; Honeycutt et al., 1992; Sobral et al., 1990; van Daelen et al., 1989; Wing et al., 1993; Woo et al., 1995). Although the protoplast method yields megabase-size DNA of high quality, the process is costly and labour-intensive. For example, to prepare protoplasts from tomato, young leaves are manually feathered with a razor blade before being incubated for 4–5 h with cell-wall-degrading enzymes (Ganal & Tanksley, 1989). With sorghum, Woo et al. (1995) found that the best way to generate high yields of protoplasts for megabase-size DNA isolation is to rub carborundum on both sides of the leaves with a paintbrush, 50 strokes/side, before a 4–5 h incubation with cellulysin. Thus the amount of time before embedding in agarose can be between 7 and 9 h, depending on the amount of leaf material being processed. Furthermore, since each plant species requires a different set of conditions to generate protoplasts, the method will only work if a high-yielding protoplast method has been developed for a given plant species.

Some investigators have tried with varying degrees of success to prepare megabase-size DNA from nuclei (Guidet & Langridge, 1992; Hatano et al., 1992). Zhang et al. (1995) have recently developed a universal nuclei method that works well for several divergent plant taxa. Fresh or frozen tissue is homogenised with a blender or mortar and pestle, respectively. Nuclei are then isolated and embedded as above. The quality of the DNA is as good as that of DNA prepared from protoplasts; the DNA is often more concentrated, and was shown to contain lower amounts of chloroplast DNA. The primary advantage of the method is that it is economical and not as labour-intensive as the protoplast method. To isolate nuclei and embed in agarose routinely takes less than 2 h.

3 GENERAL INTRODUCTION TO PROCEDURES FOR THE CONSTRUCTION OF COSMID, YAC AND BAC LIBRARIES

The construction of large-insert DNA libraries generally follows the same protocols and differs primarily in two important areas: (1) the size of the DNA fragments to be cloned; and (2) the method of transformation which is particular to the vector and host system being used.

3.1 Generation of large DNA fragments from megabase-size DNA

Once HMW DNA has been prepared, it must somehow be fragmented and DNA in the desired size range isolated. In general, as the desired DNA fragment size increases, fewer manipulations of the DNA can be tolerated. DNA fragmentation utilises two general approaches: (1) physical shearing; and

(2) partial digestion with a restriction enzyme that cuts relatively frequently within a genome. Since physical shearing is not dependent upon the frequency and distribution of particular restriction enzyme sites, this method should yield the most random distribution of DNA fragments (Ward & Jen, 1990). However, the ends of the sheared DNA fragments must be repaired and cloned directly, or restriction enzyme sites added by the addition of synthetic linkers. These subsequent steps may damage the HMW DNA and lead to lower yields of clonable DNA. Because of the subsequent steps required to clone DNA fragmented by shearing, most researchers fragment DNA by partial restriction enzyme digestion. The advantage of partial restriction enzyme digestions is that no further enzymatic modifications of the ends of the restriction fragments are necessary. Four common techniques can be used to achieve reproducible partial digestions of HMW DNA: (1) varying the concentration of the restriction enzyme (Burke & Olson, 1991); (2) varying the time of incubation with the restriction enzyme (Anand et al., 1989); (3) varying the concentration of an enzyme cofactor (e.g. Mg^{2+}) (Albertsen et al., 1989); and (4) varying the ratio of endonuclease to methylase (Larin et al., 1991).

3.2   Size selection of HMW DNA

After partial digestion, the DNA must be size-selected to remove the smaller DNA fragments that can compete more effectively than the larger DNA fragments for vector ends and which can also be more efficiently transformed. The two most common size-selection techniques are sucrose gradient centrifugation (Burke & Olson, 1991) and PFGE (Albertsen et al., 1990; Birren & Lai, 1993). Most cosmid libraries have been constructed with DNA selected on sucrose gradients. All but one of the plant YAC and BAC libraries have used PFGE for size selection. The notable exception is the *A. thaliana* YAC library constructed by Ward & Jen (1990), who used sucrose gradients. The reason why PFGE is primarily used for size selection is that the DNA usually does not get as diluted as DNA on sucrose gradients; therefore, it is easier to concentrate DNA from agarose gels without substantial loss in yield and size. Additionally, 5–20% gradients do not efficiently resolve DNA above 100 kb. Cosmid libraries can be constructed with DNA size-selected on sucrose gradients because the yields are good in the size range of 25–40 kb.

3.3   Vector preparation

All of the large DNA cloning vectors can be propagated in *E. coli*, so most standard large-scale preparations for plasmid DNA work well (Ausubel et al., 1992; Sambrook et al., 1989; Silhavy et al., 1984). Special care is required for BAC vectors because they are single copy in *E. coli* (Shizuya et al., 1992; Woo et al., 1994). Once isolated, the vector DNA is digested with the appropriate enzyme(s) and the resulting ends are usually dephosphorylated to prevent self-ligation, which can increase the efficiency of ligating vector DNA to insert DNA. Several phosphatases are available and include calf intestinal alkaline phosphatase (CIAP) and heat-killed (HK) phosphatase (Epicenter, USA). HK phosphatase is preferred because it can easily been inactivated with a short incubation at 65 °C.

A clever alternative to dephosphorylation is termed the 'half-site fill-in' method (Zabarovsky & Allimets, 1986). To illustrate, if the vector cloning site is *Xho*I, which produces a 4-bp 5' overhang, and if the first two bases only are filled in, the resulting 2-bp overhang is no longer complementary with the other end of the vector and thus the vector can no longer self-ligate. Next, if genomic DNA is digested with *Sau*3AI, which also creates a 4-bp 5' overhang, and its first two base pairs are filled in, the *Sau*3AI ends are no longer complementary to each other. In this example, neither the vector nor the genomic DNA can self-ligate. However, the genomic DNA and vector DNA now have complementary ends that are 'ligatable'. The 'half-site fill-in' technique, in principle, allows the ligation of inserts into the cloning site of the vector, prevents the formation of chimeric inserts and blocks the self-ligation of the vector. Zhang et al. (1994a) used the 'half-site fill-in' method to construct a 11 280-clone cosmid library of *Chlamydomonas reinhardtii* DNA. The drawback to this technique is that there are only a limited number of enzyme combinations which can be used (Zabarovsky & Allimets, 1986).

3.4 Transfer of recombinant DNA molecules into host cells

Transfer of recombinant DNA into host cells depends on the vector system used. Cosmids are introduced by *in vitro* packaging a ligation mixture followed by transfecting *E. coli*. BACs can be introduced by standard *E. coli* transformation protocols that utilise CaCl$_2$ (Hanahan, 1983) or electroporation (Shizuya et al., 1992; Woo et al., 1994). YACs are introduced by generating yeast spheroplasts and incubating the ligation mixture with the yeast in the presence of polyethylene glycol (Burgess & Percivil, 1987). The *E. coli* vectors commonly use antibiotic selection, whereas the YAC vectors use auxotrophic complementation for selection.

3.5 Clone storage

Libraries can be stored as cells or DNA. Large-insert DNA libraries are most commonly stored as individual clones in either 96- or 384-well microtitre dishes as glycerol stocks. Usually, multiple copies of libraries are made and stored in separate −80 °C freezers to guard against unforeseen disasters such as freezer breakdowns.

4 CONSTRUCTION OF COSMID, BAC AND YAC LIBRARIES

4.1 Cosmid libraries

Cosmid cloning vectors originally introduced in 1978 by Collins and Hohn were the principal vectors for cloning DNA greater than 20 kb but less than

50 kb until the introduction of YACs in 1987 (Burke et al., 1987). Cosmids are basically plasmid vectors; however, they contain one or two lambda cos sites which allow DNA to be packaged into lambda phage particles *in vitro* (see Chapter 3). This feature allows for the efficient transfer of encapsulated DNA into *E. coli* by transfection. Fig. 3 shows a diagram outlining the construction of cosmid libraries with a vector that contains one or two cos sites. In Fig. 3A (e.g. Ish-Horowicz & Burke, 1981), a single vector is used and digested to produce two 'arms'. First, the vector is digested with either *Hind*III or *Sma*I and then both arms are digested with *Bam*HI. A large and a small fragment are isolated. One fragment contains a single cos site plus an origin of DNA replication and an antibiotic-selectable marker (ampicillin). The other fragment only contain a cos site. The two fragments are added together with size-selected genomic DNA that has compatible ends (*Bam*HI, *Sau*3AI, or *Mbo*I), ligated and then in vitro packaged into lambda phage particles. Finally, the packaged DNA is used to transfect *E. coli* cells. *E. coli* containing cosmid DNA is selected for by growth on media containing antibiotics. In Fig. 3B (e.g. Bates & Swift, 1983) the plasmid contains two cos sites, so only a single set of digestions is needed. In this example the vector is digested with *Bam*HI and *Sma*I to produce two 'arms'. Each arm has a cos site and a selectable marker. Additionally, one arm has an origin of DNA replication.

Because there are so many cosmid cloning vectors available, we will list references for those being used in several plant-related applications instead of listing specific cosmids and libraries. For single cos vectors see Collins & Hohn (1978), Cross & Little (1986), Hohn & Collins (1980), Ish-Horowicz & Burke (1991), Lazo et al. (1991), Ma et al. (1992), Meyerowitz (1992), Olsewski et al. (1988), Simoens et al. (1986) and Wahl et al. (1987). For double cos vectors see Bates & Swift (1983) and Lerner et al. (1992).

The cosmid system has been used for numerous applications in plants. The first plant cosmid libraries were constructed primarily for library screening and physical mapping. One of the first plant cosmid libraries was developed for physical mapping and map-based cloning in *A. thaliana* (Hauge et al., 1991). Briefly, *A. thaliana* DNA was sheared and DNA of approximately 40 kb inserted into a derivative of the Lorist B cosmid vector (Cross & Little, 1986), which contains SP6 and T4 bacteriophage promoters for rapidly generating end-specific RNA probes useful for contig building and chromosome walking (see Chapter 10) and the lambda origin of replication. Using this library and a fingerprinting strategy developed by Coulson et al. (1986), Hauge et al. (1991) fingerprinted approximately 17 000 *A. thaliana* cosmid clones. The fingerprints were compared with all other fingerprints to assemble a set of 800 contigs that represent approximately 90–95% of the *A. thaliana* genome. These contigs were and are currently being linked together by using YACs (Hwang et al., 1991). Additionally, this library was used to isolate a set of overlapping clones in the *ABI3* genomic region which led to the first successful demonstration of map-based gene cloning in plants (Giraudat et al., 1992).

Several laboratories have engineered various cosmid vectors that were suitable for *Agrobacterium*-mediated transformation (Lazo et al., 1991; Ma et al.,

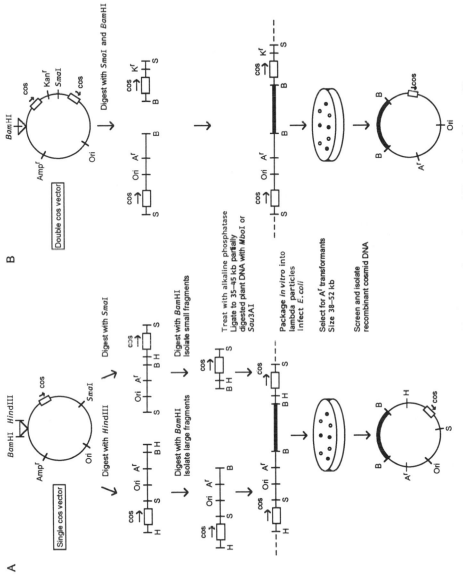

Figure 3. A general scheme for the construction of cosmid libraries with single cos (A) and double cos (B) vector

1992; Meyerowitz, 1992; Olszewski et al., 1988; Simoens et al., 1986). The vectors were modified by the addition of T-DNA borders, plant-selectable markers, and broad-range origins of DNA replication (see Chapter 11). These vectors can be utilised to clone large DNA fragments and then be directly transformed into plants via *Agrobacterium*-mediated transformation. Complete plant cosmid libraries in binary vectors have been constructed in *E. coli* and *Agrobacterium*. Lazo et al. (1991) found that if the *E. coli lamB* gene was expressed in *Agrobacterium*, the lambda infection receptor, *in vitro*-packaged DNA could be directly transfected into *Agrobacterium* without having to shuttle through *E. coli* first. This led to the construction of an ordered 20 000-clone *A. thaliana* cosmid library in *Agrobacterium*.

One potentially powerful application of binary cosmid vectors would be to shotgun-clone plant genes by transforming a complete cosmid library into a mutant line (Chapter 8). Olszewski et al. (1988) constructed a specialised binary cosmid vector pOCA18 which contains the *cis*-signals necessary for the efficient transfer of DNA into plants, left and right T-DNA borders, the overdrive element, NPTII and RK2 origin of replication. Olszewski argued that if an *A. thaliana* library contained 10^7 clones, then 10^4 transformed plant cells would be required to have a 95% probability of obtaining a transgenic plant with the desired construct. With present technology, the generation of 10^4 transformed plants would only be feasible in *A. thaliana*. This vector has been used to construct several *A. thaliana* libraries and used to isolate at least two plant genes by molecular complementation of *A. thaliana* mutants: *GA1* (Sun et al., 1992) and *AXR1* (Leyser et al., 1993).

Cosmid libraries have been used to assemble contigs in several targeted regions of the *A. thaliana* genome and have led to the successful isolation of at least three genes by positional cloning: *ABI3* (Giraudat et al., 1992), *RPS2* (Mindrinos et al., 1994, Bent et al., 1994), and *GA1* (Sun et al., 1992).

All of the vectors and strains may be obtained from the authors of the papers cited. Alternatively, most of the plasmids and strains may be obtained through the American Type Culture Collection (ATCC) (USA and Canada 1-800-638-6597; outside USA 1-301-881-2600). Several biotechnology companies sell cosmid cloning vectors and kits. Gibco BRL and Boehringer Mannheim sell pHC79 (Hohn & Collins, 1980). Stratagene (USA) sells Super CosI and pWE15 (Wahl et al., 1987) cosmid vector kits. Clontech, USA also sells the pWE15 (Wahl et al., 1987) cosmid vector. *In vitro* packaging kits can be obtained from most biotechnology suppliers, including Stratagene, Gibco BRL, Clontech, Invitrogen and Promega.

4.2 YAC libraries

YACs were originally introduced in 1987 by Burke, Carl and Olson. YACs have revolutionised the study of large genomes because they provided a vector that could essentially be used to clone DNA of any size. The primary YAC vector (pYAC4; Fig. 4) and *S. cerevisiae* host (AB1380: *MATα ura3 trp1 ade2-1 can1-100 lys2-1 his5*) have been used for the construction of the majority of plant YAC

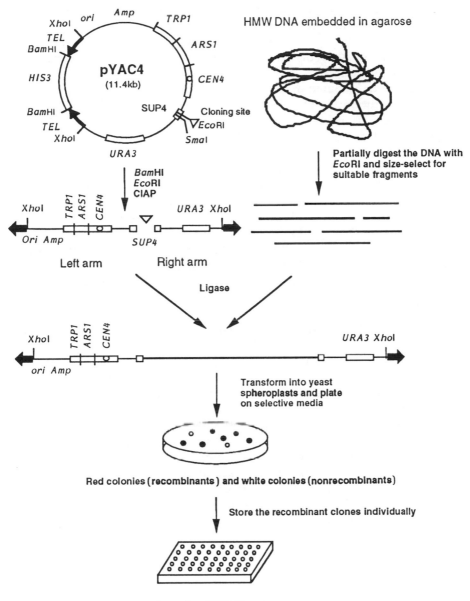

Figure 4. A general scheme for the construction of a YAC library

libraries (Burke et al., 1987). The basic YAC vector system consists of two arms which provide all of the elements of a eukaryotic chromosome: two telomeres, one centromere (*CEN4*), and one origin of DNA replication (*ARS1*). In addition, each arm contains a selectable marker that can complement a corresponding auxotrophic mutation in the yeast host, namely the *TRP1* and *URA3* genes.

Finally the cloning site is located within a tRNA suppressor gene, *SUP4*. When the *SUP4* tRNA is intact it can suppress the *ade1* mutation, which produces yeast colonies which are white in colour. However, when the *SUP4* gene is disrupted by the insertion of a genomic DNA fragment, the *ade1* mutation is no longer suppressed and the yeast colonies appear red. DNAs cloned into the YAC vector are maintained as yeast chromosomes and are thus low copy and segregate correctly during mitotic and meiotic division.

YAC libraries have been constructed for a number of plant species, including *Arabidopsis* (Grill & Somerville, 1991; Ward & Jen, 1990; Ecker, 1990), maize (Edwards et al., 1992), tomato (Martin et al., 1992), rice (Umehara et al., 1995), barley (Kleine et al., 1987), and sugarbeet (Eyers et al., 1992). The principal cloning vector for all of these libraries has been pYAC4 or a derivative (Burke et al., 1987). Other vectors have been constructed: pYAC-RC (rare cutter) (Marchuk & Collins, 1988), which contain several rare restriction enzyme sites for cloning; pYAC41 and pYAC45 (Grill & Somerville, 1991), which contain transcriptional promoters on either side of the cloning sites and can be used for rapid generation of YAC end probes; and pJS97 and pJS98 (Shero et al., 1991), which were designed as discrete left- and right-arm plasmids that can be used for the efficient isolation of both YAC ends by plasmid rescue for chromosome walking.

Fig. 4 shows a diagram depicting the construction of a plant YAC library in pYAC4. First, the YAC vector is digested with *Bam*HI and *Eco*RI and then dephosphorylated. The genomic DNA is partially digested with *Eco*RI, size-selected once or twice, and finally ligated to the cloning vector. The ligation reaction is then transformed into *S. cerevisiae* spheroplasts using the method of Burgess & Percivil (1987). This transformation method is one of the best and usually yields transformation frequencies of between 10^6 and 10^7 colony-forming units per microgram of control vector DNA. Transformants are first selected for uracil prototrophy and red colour. Red clones are then transferred to plates lacking both tryptophan and uracil to verify the presence of both YAC arms. DNA from recombinant clones is prepared by embedding yeast cells in agarose plugs or beads, and enzymatically removing the cell wall, followed by cell lysis and a thorough washing (Burke et al., 1987). The DNA in agarose is then analysed by contour-clamped homogeneous electric field (CHEF) electrophoresis (Chu et al., 1986). Fig. 5 shows the analysis of seven YAC clones containing tomato DNA using CHEF electrophoresis (Zhang et al., 1994b). Here the common bands in all of the lanes correspond to some of the endogenous yeast chromosomes and the additional band or intensity in each lane corresponds to a YAC.

Plant YAC libraries have been extensively used for physical mapping and positional cloning (Chapter 10). As mentioned for the *A. thaliana* genome, a cosmid library of 17 000 clones could be divided into 800 contigs. These contigs have been further divided by using YACs. Hwang et al. (1991) published the isolation of YACs that hybridise with 125 *A. thaliana* RFLPs that comprise one-third of the genome. Matallana et al. (1993) are using a similar approach with an *A. thaliana* YAC library that has an average insert size of 250 kb.

YACs can be used to isolate more markers in a genomic region of interest.

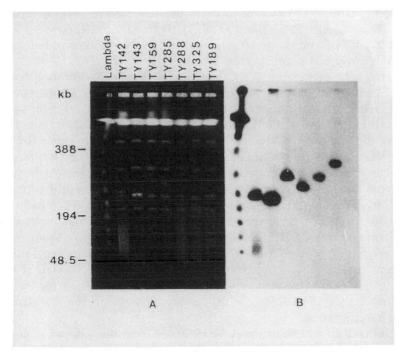

Figure 5. PFGE analysis of YACs containing tomato DNA homologous isolated by hybridization to RFLP marker TG523. (A) The YACs on the ethidium bromide-stained PFGE gel. The bands common to all lanes are from *Saccharomyces cerevisiae* chromosomes, and the unique, or higher-density, band if a YAC comigrates with a yeast chromosome, in each lane, indicates a YAC. (B) The YACs on the PFGE blot from the gel shown in (A) hybridised with TG523. Note that the TY189 did not hybridise with TG523

For example, Zhang et al,. (1994b) used such a strategy for the analysis of the *jointless* locus in tomato, which completely suppresses the formation of pedicel abscission zones (Butler, 1936). Six YACs were isolated using a single RFLP marker tightly linked to *jointless*. The ends of the YACs were isolated and mapped. This resulted in the isolation of seven new RFLP markers, two of which were found to completely cosegregate with the *jointless* phenotype. Similar approaches have been used during the course of map-based gene cloning of several genes from plants.

Recently, YACs have been used to isolate plant genes by map-based cloning. The first plant gene isolated using YACs was the *FAD3* locus from *A. thaliana* Arondel et al. (1992) used two RFLPs linked to *FAD3* to build a four-YAC contig that encompassed the *FAD3* locus. The first use of YACs to isolate a gene from a crop plant was demonstrated by Martin et al. (1993), who cloned the tomato *Pto* disease resistance gene. Martin et al. (1993) used RFLP markers tightly linked to *Pto* and isolated a set of YACs that encompassed the locus. In this case no chromosome walking was necessary; hence it is termed chromosome 'landing'. YACs have been used to isolate several other genes from *A. thaliana*, e.g. *RPS2* (Bent et al., 1994) and *ABI1* (Leung et al., 1994; Meyer et al., 1994), and many target genes are in the process of being isolated using YACs in various plant species.

One final note concerning YAC libraries is that YAC libraries have been found to contain significant amounts of chimeric clones (Anderson, 1993). Chimeric clones can be problematic because a part of the insert comes from one region of the genome while another part comes from a different, unlinked region of the genome. Chimeric clones are thought to occur by *in vivo* recombination between homologous regions and/or co-cloning. *In vivo* recombination can occur because the yeast host AB1380 does not have any mutations in its recombination pathways. Chartier et al. (1992) recently reported the construction of a mouse YAC library in a new yeast strain, 3a, which carries a mutation in *RAD52*, a gene that has been shown to mediate recombination between cotransformed plasmids in host cells (Larionov et al., 1994). Although the level of chimeric clones in the mouse library was not assessed, a plant YAC library constructed in 3a could potentially eliminate the formation of some chimeric clones. Unfortunately, the transformation frequency for 3a is about 10-fold lower that of AB1380, which is significant for the construction of complete YAC libraries.

All of the vectors and strains may be obtained from the authors of the papers cited. Alternatively, most of the plasmids and strains may be obtained through the ATCC. pJS97 and pJS98 (Shero et al., 1991) can be purchased from Gibco BRL. Yeast strain 3a, which contains the *RAD52* mutant, can be obtained from Chartier et al. (1992).

4.3   BAC libraries

BACs (Shizuya et al., 1992) and P1-derived artificial chromosomes (PACs) (Oannou et al., 1994) are relatively new types of large DNA fragment cloning vectors available to plant molecular biologists. These vectors permit the cloning of DNA of at least 350 kb in *E. coli*. Fig. 6 shows a diagram of a popular BAC vector pBeloBAC11 (Shizuya & Simon, unpublished). The basic structure of the BAC/PAC vectors is derived from the endogenous plasmid F. The F backbone contains four essential regions that function in plasmid stability and copy number (review: Willetts & Skurry, 1987). Both *ParA* and *ParB* are required for partitioning and plasmid stability. Additionally, *ParB* is also required for incompatibility with other F factors. *OriS* is the origin of DNA replication, which is unidirectional. *RepE* encodes protein E, which is essential for replication from *OriS* and for copy number control. A chloramphenicol-resistance gene was incorporated for antibiotic selection of transformants. The original BAC vector pBAC108L did not incorporate the *LacZ* gene for colour selection (Shizuya et al., 1992). pBeloBAC11 contains the *LacZ* gene, and thus the identification of recombinant DNA clones is simplified (Shizuya & Simon, unpublished). The PAC vector (Oannou et al., 1994) has most of the features of the BAC system; however, the vector contains the *SacB* gene, which provides a

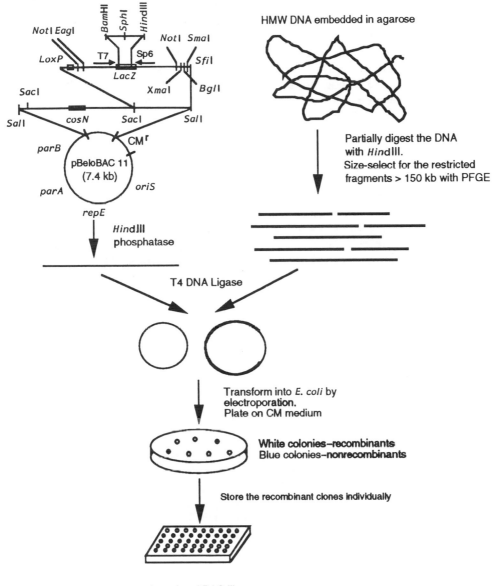

Figure 6. A general scheme for the construction of a BAC library

positive selection for recombinant clones during library construction. *SacB* encodes sucrose synthase. When cells are grown in the presence of saccharose, sucrose synthase will degrade saccharose into levan, which is highly toxic to *E. coli*. The *Bam*HI cloning site is within the *SacB* gene, and thus disruption of the *SacB* gene by inserting a large DNA fragment allows for growth of the cell on media containing saccharose. Additionally, the vector has a 'pUC19 link', containing a high-copy-number origin of DNA replication, which is used for convenient vector propagation and is later removed during vector preparation for library construction.

The most widely used E. coli strain for BAC cloning is DH10B (Hanahan et al., 1991). Key features of this strain include mutations that block: (1) restriction of foreign DNA by endogenous restriction endonucleases (hsd/RMS); (2) restriction of DNA containing methylated DNA (5'-methylcytosine or methyladenine residues, and 5'-hydroxymethylcytosine) (mcrA, mcrB, mcrC and mrr); and (3) recombination (recA1).

The construction of BAC libraries is outlined in Fig. 6 using pBeloBAC11. First, the vector is digested with HindIII and then dephosphorylated to prevent self-ligation. Next, HMW DNA is partially digested with HindIII and DNA > 150 kb is size-selected on a CHEF gel. Woo et al. (1994) have found that it is essential to perform two such size selections to increase the average insert size of BACs, due to trapping of small DNA fragments after the first size selection. Finally, the vector and genomic DNA are ligated and then electroporated into E. coli. Recombinant transformants are selected on media containing chloramphenicol, 5-bromo-4-chloro-3-indoyl-β-D-galactaside (X-gal) and isopropylthio-β-D-galactaside (IPTG). After recombinant transformants are detected, their sizes are assayed by simple DNA minipreparations followed by digestion with NotI to free the DNA insert from the vector, and CHEF electrophoresis. Fig. 7A shows the analysis of 18 randomly selected rice BAC clones from a library that we have recently constructed (Zhang et al., unpublished). As can be seen, all of the BAC clones have inserts and the DNA yield is satisfactory. Fig. 7B shows a Southern filter of this gel hybridised with total rice genomic DNA to determine the relative amount of low- and high-copy-number DNA sequences present in these clones.

Since the BAC and PAC systems are relatively new, only a few plant libraries have been constructed. Woo et al. (1994) constructed a 13 750-clone BAC library from Sorghum bicolor having an average insert size of 157 kb. This insert size is comparable with most of the published plant YAC libraries. This BAC library contains approximately 90% of the sorghum genome, and 14% is contaminated with chloroplast-derived sequences. To evaluate the sorghum BAC library, the library was tested for clone representation, clone stability and chimerism. The library was screened with nine random RFLP probes, six from sorghum and three from maize. In all but one case, at least one BAC clone was found for each probe. To evaluate clone stability, four BAC clones having inserts of between 280 and 315 kb were grown over 100 generations and restriction digestion patterns were compared. No rearrangements were detected, which suggests that large plant DNA sequences can be stably maintained as BACs in E. coli. Finally, because chimerism has been an important problem with large-insert DNA libraries, especially with YACs, Woo et al. (1994) estimated the level of chimerism in the library by fluorescence in situ hybridisation (FISH). Eleven of the largest BAC clones, 250–350 kb, were labelled and hybridised in situ to both metaphase and interphase sorghum nuclei. For each BAC clone, no chimerism was detected, which suggests that the level of

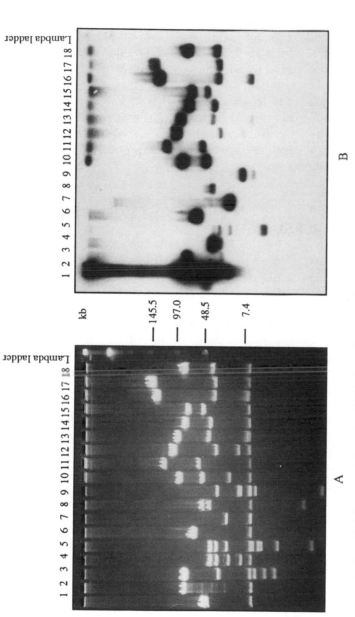

Figure 7. PFGE analysis of BACs containing rice DNA from Teqing (*Oryza sativa* subsp. *indica*) (lanes 1–9) and Lemont (*Oryza sativa* subsp. *japonica*) (lanes 10–18). (A) The *Not*I digested BAC DNAs on the ethidium bromide-stained PFGE gel. Note that the common band across all the BAC lanes is the 7.4-kb BAC vector pBeloBAC11. The BAC DNA in lane 2 was partially digested. (B) The BACs on the PFGE blot from the gel shown in (A) hybridised with Teqing and Lemont total DNA. Note that the vector did not hybridise with the rice DNA. Some bands did not hybridise or weakly hybridised with the probe, indicating that those DNA fragments consist of single or low-copy DNA sequences. The remaining bands hybridised strongly with the probe, indicating that these fragments contain highly repetitive DNA sequences

chimerism in the sorghum BAC library is very low. Additionally, our group has constructed two complete rice BAC libraries from *japonica* and *indica* types (Zhang et al., unpublished). Evaluation of these libraries has given similar results to those obtained with the sorghum system.

The initial results with the sorghum and rice BAC libraries suggest that the BAC system will be useful for many plant genomic applications. The construction and evaluation of additional BAC libraries should substantiate these results, which will probably lead to wider usage of the BAC vector system for plant molecular biology.

To obtain pBeloBAC11, write to Dr H. Shizuya and Dr M. Simon, Department of Biology, California Institute of Technology, Pasadena, CA 91125, FAX: 808-796-7066. High-efficiency electrocompetent *E. coli* cells can be obtained from most biotechnology suppliers, including Stratagene, Gibco BRL, Clontech, Invitrogen and Promega.

5   EVALUATION OF LARGE DNA FRAGMENT CLONING SYSTEMS: BACs VERSUS YACs

When evaluating the BAC cloning system, a comparison must be made with the proven YAC system.

The primary similarity between the two systems is that they can, in principle, handle any size of DNA that is cloned into a cloning site. Additionally, both systems are maintained as low-copy clones in the host cell.

The differences between the two system are broad and can be attributed mainly to the host systems—yeast versus bacteria. *E. coli* divides faster, and is easier to isolate DNA from and to transform. These factors contribute to the speed and efficiency for the construction and analysis of BAC libraries and individual BAC clones.

For large-insert DNA library construction the transformation efficiency of the host cell is critical. For *S. cerevisiae*, Burgess & Percivil (1987) reported a system that reproducibly yields competent yeast spheroplasts with transformation efficiencies of 10^7 transformants/μg. However, in our hands it is very hard to reproducibly make competent yeast cells that are this efficiently transformed. In addition, the amount of time required to conduct a yeast transformation, which includes growing the cells to the optimal cell density and incubating the cells with the optimal amount of spheroplasting enzyme, is very substantial. In contrast, the sorghum and rice BAC libraries we constructed (Woo et al., 1994; Zhang et al., unpublished) used electroporation to introduce the ligated DNA into *E. coli*. Homemade competent cells yield transformation efficiencies of 10^9–10^{10} transformants/μg, and commercially available cells guarantee at least 10^{10} transformants/μg. Under optimal conditions, a 1000-fold difference in transformation efficiencies between yeast and bacteria (10^7 versus 10^{10}) can make a significant difference when the objective is to construct a complete library. Because of the higher transformation frequencies for *E. coli*, the amount of size-selected DNA required to make a complete BAC library should be less than that needed for a YAC library. Furthermore, BACs may be more efficiently introduced than YACs, because the BAC is circular and probably more stable than linear YAC DNA.

We should note that we have not recovered recombinant BACs larger than 350 kb. This is in contrast to the YAC system, where human and mouse libraries have been constructed with average insert sizes of greater than 470 kb (Chumakov et al., 1992; Larin et al., 1991). Since E. coli can replicate its own genome of about 4 Mb, it is unlikely that E. coli can not replicate a 1 Mb BAC. One possible reason why we and others have not detected larger BAC clones is that there may be a limit to the size of a molecule that can be delivered by electroporation. It therefore might be possible to create larger-insert BAC libraries using conventional $CaCl_2$ transformation techniques (Hanahan, 1983) or biolistics (Klein et al., 1987).

Once large-insert DNA libraries are constructed, the ability to screen for and analyse individual clones efficiently becomes important (Section 3). Whole YACs (e.g. Driesen et al., 1991) and BACs (Oannou et al., 1994; Shizuya et al., 1992; Woo et al., 1994) have been used as probes for physical mapping by FISH and to isolate genome-specific cDNAs (Arondel et al., 1992; Lovett et al., 1991; Martin et al., 1993). Normally the isolation of YAC probes requires standard megabase DNA isolation in agarose plugs or beads followed by CHEF gel electrophoresis and gel isolation. This process can take 3–5 days to complete. Such isolations do not always guarantee that the YAC can be separated from the 17 endogenous yeast chromosomes, and YAC DNA yields are often low. In contrast, BAC DNA isolations use standard plasmid DNA isolation techniques which take advantage of the separation of supercoiled plasmid DNA from bacterial genomic DNA. Fig. 7A shows BAC DNA isolated from standard 5-ml minipreparations (Silhavy et al., 1984). The DNA preparation takes 3 h and the whole process, including CHEF electrophoresis and insert isolation, takes 36–48 h.

In short, the BAC system is more 'user-friendly' and will probably lead to more rapid progress in cloning biologically significant genes.

6   BAC PLANT TRANSFORMATION VECTORS?

The last step in map-based gene cloning of plant genes usually involves the conversion of a line that does not contain a functional gene to a line that contains the functional gene (e.g. genetic complementation). This is usually accomplished by subcloning a YAC or BAC that contains the target gene into a plant transformation vector and then transforming overlapping subclones into the test plant. This process is labour-intensive, and requires several subclones and multiple tranformants to account for position effects. A possible alternative to the above scenario would be to transform the entire large YAC/BAC DNA fragment into the tester line and thus avoid having to obtain overlapping subclones and multiple transformants. Such a system could, in principle, streamline the gene identification step and provide a yes/no answer more quickly than with existing procedures. In mammalian systems some groups have demonstrated that whole YACs could be integrated into mammalian cells (Huxley & Gnirke, 1991; Pachnis et al., 1990; Pavan et al., 1990). Such a system has not been demonstrated for plants.

Because of the ease with which plant BAC libraries can be constructed, maintained and used, several groups are now actively pursuing the idea of converting the present BAC vectors into plant transformation vectors in the hope of streamlining the gene identification step of map-based cloning. At present two transformation methods seem the most feasible. First, the BAC vector could be converted into a binary vector and DNA transferred into plants via standard *Agrobacterium*-mediated methods. Recent evidence from Miranda et al. (1992) demonstrated that *Agrobacterium* is capable of transferring approximately 96% of the Ti plasmid (170 kb) into plants. Second, it might be possible to utilise one of several site-specific recombination systems to integrate a BAC into a specific site within a plant chromosome: Cre-*lox* (Fukushige & Sauer, 1992; Sauer & Henderson, 1990) or FLP-*FRT* (Ogorman et al., 1991). Ow and coworkers have demonstrated that the Cre-*lox* site-specific recombination system works efficiently in tobacco (Bayley et al., 1992; Dale & Ow, 1990, 1991; Qin et al., 1994). Recently, by using a mutation(s) in the *lox* site and expressing the Cre recombinase transiently, Albert et al. (1995) have been able to drive the forward reaction of genomic integration instead of plasmid resolution. Since the pBeloBAC11 vector already contains a *lox* recombination site, all that is necessary to convert the BAC vector into a plant transformation vector is to add a plant selectable marker (e.g. 35S-neomycinphosphotransferase, NPTII). The advantage of using a site-specific recombination system versus *Agrobacterium* is that site-specific recombination will result in integration of a BAC clone into a specific site on the chromosome instead of randomly throughout the genome as with *Agrobacterium*. Therefore potential position effects could be reduced by using the Cre-*lox* system.

7 CONCLUDING REMARKS

The field of plant genome research is expanding at an exponential rate and is leading to fundamental insights into genome organisation, evolution and biology. The methods described in this chapter should provide a summary of the current techniques used to isolate and clone relatively large plant DNA fragments. The primary starting materials are megabase-size plant DNA and cloning vectors with large-insert cloning capacity. The routine isolation of megabase-size plant DNA has been a significant hurdle. The recent paper by Zhang et al. (1995), demonstrating a simple nuclei isolation method adaptable to a wide range of plant taxa, should provide a more convenient and economical way to isolate DNA for both physical mapping and YAC and BAC cloning. The three classes of cloning vectors discussed, cosmids, YACs and BACs, are all valuable and will continue to serve as important tools for plant genomics.

Although the BAC vector system is relatively new, we feel that this system may replace several applications where either a cosmid or YAC would have previously been used. An obvious example would be for the construction of complete plant genomic libraries of a moderate size of 150 kb (Woo et al., 1994). Finally, if the BAC vector system can be converted into a plant transformation system, either by *Agrobacterium*-mediated methods or by site-specific recombination, then it may be possible to routinely transfer 100–150 kb inserts into plants, which could dramatically streamline the final map-based gene cloning step of genetic complementation.

ACKNOWLEDGMENTS

The authors would like to thank S.D. Choi for providing unpublished data, and our collaborators B. Gill, J. Jiang, A. Paterson and X. Zhao. Additionally, RAW would like to thank D. Ogrydziak, B. Landry, S. McCormick, R.W. Michelmore, S.D. Tanksley and Texas A&M University for providing the opportunity, training and support to pursue research in plant genomics. This chapter was funded in part by grants from the following sources: USDA-NRICGP#91 and 94-37300, 93-00918; The Rockefeller Foundation; Cotton Incorporated #94-138; and the Texas Agriculture Experiment Station Research Enhancement Program.

Author for correspondence: R. A. Wing.

ESSENTIAL READING

Birren, B. and Lai, E. (1993) *Pulsed Field Gel Electrophoresis, A Practical Guide.* Academic Press, Inc., New York.

Burke, D.T. and Olson, M.V. (1991) Preparation of clone libraries in yeast artificial chromosome vectors. *Methods Enzymol.*, **194**, 251–270.

Shizuya, H., Birren, B., Kim, U.-J. et al. (1992) Cloning and stable maintenance of 300-kilobase-pair fragments of human DNA in *Escherichia coli* using an F-factor-based vector. *Proc. Natl. Acad. Sci. USA*, **89**, 8794–8797.

Woo, S.-S., Jiang, J., Gill, B.S. et al. (1994) Construction and characterization of a bacterial artificial chromosome library of *Sorghum bicolor. Nucleic Acids Res.*, **22**, 4922–4931.

Zhang, H.-B., Zhao, X.P., Ding, X.D. et al. (1995) Preparation of megabase-size DNA from plant nuclei. *Plant J.*, **7**, 175–184.

REFERENCES

Albersten, H.M., Abderrahim, H., Cann, H.M. et al. (1990) Construction and characterization of a yeast artificial chromosome library containing seven haploid human genome equivalents. *Proc. Natl. Acad. Sci. USA*, **87**, 4256–4260.

Albert, H., Dale, E.C., Lee, E. and Ow, D.W. (1995) Site-specific integration of DNA into wild-type and mutant lox sites placed in the plant genome. *Plant J.*, **7**, 649–659.

Albertsen, H.M., Paslier, D.L., Abderrahim, H. et al. (1989) Improved control of partial DNA restriction enzyme digest in agarose using limiting concentration of Mg. *Nucleic Acids Res.*, **17**, 808.

Anand, R., Villasante, A. and Tyler-Smith, C. (1989) Construction of yeast artificial chromosome libraries with large inserts using fractionation by pulsed-field gel electrophoresis. *Nucleic Acids Res.*, **17**, 3425–3433.

Anderson, C. (1993) Genome shortcut leads to problems. *Science*, **259**, 1684–1687.

Arondel, V., Lemieux, B., Hwang, I. et al. (1992) Map-based cloning of a gene controlling omega-3 fatty acid desaturation in *Arabidopsis*. *Science*, **258**, 1353–1355.

Arumuganathan, K. and Earle, E.D. (1991) Nuclear DNA content of some important plant species. *Plant Mol. Biol. Rep.*, **9**, 208–219.

Ausubel, F.M., Brent, R., Kingston, R.E. et al. (1992) *Short Protocols in Molecular Biology*, 2nd edn. John Wiley & Sons, New York.

Bates, P.F. and Swift, R.A. (1983) Double cos site vectors: simplified cosmid cloning. *Gene*, **26**, 137–146.

Bayley, C.C., Morgan, M., Dale, E.C. and Ow, D.W. (1992) Exchange of gene activity in transgenic plants catalyzed by the Cre-*lox* site-specific recombination system. *Plant Mol. Biol.*, **18**, 353–362.

Bent, A.F., Kunkel, B.N., Dahlbeck, D. et al. (1994) RPS2 of *Arabidopsis thaliana*: A leucine-rich repeat class of plant disease resistance genes. *Science*, **265**, 1856–1860.

Burgess, P.M. and Percivil, K.J. (1987) Transformation of yeast spheroplasts without cell fusion. *Anal. Biochem.*, **163**, 391–397.

Burke, D.T., Carle, G.F. and Olson, M.V. (1987) Cloning of large segments of exogenous DNA into yeast by means of artificial chromosome vectors. *Science*, **236**, 806–811.

Butler, L. (1936) Inherited characters in the tomato. *J. Hered.*, **27**, 25–26.

Chartier, F.L., Keer, J.T., Sutcliffe, M.J. et al. (1992) Construction of a mouse yeast artificial chromosome library in a recombination-deficient strain of yeast. *Nature Genet.* **1**, 132–136.

Cheung, W.Y. and Gale, M.D. (1990) The isolation of high molecular weight DNA from wheat, barley and rye for analysis by pulse-field gel electrophoresis. *Plant Mol. Biol.*, **14**, 881–888.

Chu, G., Vollrath, D. and Davis, R.W. (1986) Separation of large DNA molecules by contour-clamped homogeneous electric fields. *Science*, **234**, 1582–1585.

Chumakov, I., Rigault, P., Guillou, S. et al. (1992) Continuum of overlapping clones spanning the entire human chromosome 21q. *Nature*, **359**, 380–387.

Clarke, L. and Carbon, J. (1976) A colony bank containing synthetic ColE1 hybrid plasmids representative of the entire E. coli genome. *Cell* **9**, 91–100.

Collins, J. and Hohn, B. (1978) Cosmids: a type of plasmid gene-cloning vector that is packageable in vitro in bacteriophage lambda heads. *Proc. Natl. Acad. Sci. USA*, **75**, 4242.

Cook, P.R. (1984) A general method for preparing intact nuclear DNA. *EMBO J.*, **3**, 1837–1842.

Coulson, A., Sulston, J., Brenner, S. and Karn, J. (1986) Toward a physical map of the nematode *Caenorhabditis elegans*. *Proc. Natl. Acad. Sci. USA*, **83**, 7821–7825.

Cross, S.H. and Little, P.F.R. (1986) A cosmid vector for systematic chromosome walking. *Gene*, **49**, 9–22.

Dale, E.C. and Ow, D.W. (1990) Intra- and intermolecular site-specific recombination in plant cells mediated by bacteriophage P1 recombinase. *Gene*, **91**, 79–85.

Dale, E.C. and Ow, D.W. (1991) Gene transfer with subsequent removal of the selection gene from the host genome. *Proc. Natl. Acad. Sci. USA*, **88**, 10558–10562.

Driesen, M.S., Dauwerse, J.G., Wapenaar, M.C. et al. (1991) Generation and fluorescent *in situ* hybridization mapping of yeast artificial chromosomes of 1p, 17p, 17q, and 19q from a hybrid cell line by high-density screening of an amplified library. *Genomics*, **11**, 1079–1087.

Ecker, J.R. (1990) PFGE and YAC analysis of the *Arabidopsis* genome. *Methods*, **1**, 186–194.

Edwards, K.J., Thompson, H., Edwards, D. et al. (1992) Construction and characterization of a yeast artificial chromosome library containing three haploid maize genome equivalents. *Plant Mol. Biol.*, **19**, 299–308.

Eyers, M., Edwards, K. and Schuch, W. (1992) Construction and characterization of a yeast artificial chromosome library containing two haploid *Beta vulgaris* L. genome equivalents. *Gene*, **121**, 195–201.

Fukushige, S. and Sauer, B. (1992) Genome targeting with a positive-selection lox integration vector allows highly reproducible gene expression in mammalian cells. *Proc. Natl. Acad. Sci. USA*, **89**, 7905–7909.

Ganal, M.W. and Tanksley, S.D. (1989) Analysis of tomato DNA by pulsed-field gel electrophoresis. *Plant Mol. Biol. Rep.*, **7**, 17–41.

Ganal, M.W., Young, N.D. and Tanksley, S.D. (1989) Pulsed-field gel electrophoresis and physical mapping of large DNA fragments in the *Tm-2a* region of chromosome 9 in tomato. *Mol. Gen. Genet.*, **215**, 395–400.

Giraudat, J., Hauge, B.M., Valon, C. et al. (1992) Isolation of the *Arabidopsis ABI3* gene by positional cloning. *Plant Cell*, **4**, 1251–1261.

Grill, E. and Somerville, C. (1991) Construction and characterization of a yeast artificial chromosome library of *Arabidopsis* which is suitable for chromosome walking. *Mol. Gen. Genet.*, **226**, 484–490.

Guidet, F. and Langridge, P. (1992) Megabase DNA preparation from plant tissue. *Methods Enzymol.* **216**, 3–12.

Hanahan, D. (1983) Studies on transformation of *Escherichia coli* with plasmids. *J. Mol. Biol.*, **166**, 557–580.

Hanahan, D., Jersee, J. and Bloom, F.R. (1991) Plasmid transformation of *Escherichia coli* and other bacteria. *Methods Enzymol.*, **204**, 53–113.

Hatano, S., Yamaguchi, J. and Hirai, A. (1992) The preparation of high-molecular weight DNA from rice and its analysis by pulsed-field gel electrophoresis. *Plant Sci.*, **83**, 55–64.

Hauge, B.M., Giraudat, J., Hanley, S. et al. (1991) Physical mapping of the *Arabidopsis* genome and its applications. In *Plant Molecular Biology 2* (eds R.G. Herrmann and B. Larkins) pp. 239–248, Plenum Press, New York.

Hohn, B. and Collins, J. (1980) A small cosmid for efficient cloning of large DNA fragments. *Gene*, **11**, 291–298.

Honeycutt, R.J., Sorbral, B.W.S., McClelland, M. and Atherly, A.G. (1992) Analysis of large DNA from soybean (*Glycine max* L. Merr.) by pulsed-field gel electrophoresis. *Plant J.*, **2**, 133–135.

Huxley, C. and Gnirke, A. (1991) Transfer of yeast artificial chromosomes from yeast to mammalian cells. *BioEssays*, **13**, 545–550.

Hwang, I., Kohchi, T., Hauge, B.M. and Goodman, H.M. (1991) Identification and map position of YAC clones comprising one-third of the *Arabidopsis* genome. *Plant J.*, **1**, 367–374.

Ish-Horowicz, D. and Burke, J.F. (1981) Rapid and efficient cosmid cloning. *Nucleic Acids Res.*, **9**, 2989–2998.

Klein, T.M., Wolf, E.D., Wu, R. and Sanford, J.C. (1987) High-velocity microprojectiles for delivering nucleic acids into living cells. *Nature*, **327**, 70–73.

Larin, Z., Monaco, A.P. and Lehrach, H. (1991) Yeast artificial chromosome libraries containing large inserts from mouse and human DNA. *Proc. Natl. Acad. Sci. USA*, **88**, 4123–4127.

Larionov, V., Graves, J., Kouprina, N. and Resnick M.A. (1994) The role of recombination and *RAD52* in mutation of chromosomal DNA transformed into yeast. *Nucleic Acids Res.*, **22**, 4234–4241.

Lazo, G.R., Stein, P.A. and Ludwig, R.A. (1991) A DNA transformation-competent *Arabidopsis* library in *Agrobacterium*. *Bio/Technology*, **9**, 963–967.

Lerner, T., Wright, G., Leverone, B. et al. (1992) Molecular analysis of human chromosome 16 cosmid clones containing *Not*I sites. *Mammal. Genome*, **3**, 92–100.

Leung, J., Bouvier-Durand, M., Morris, P.C. et al. (1994) *Arabidopsis* ABA response gene *ABI*1: features of a calcium-modulated protein phosphatase. *Science*, **264**, 1448–1451.

Leyser, H.M.O., Lincoln, C.A., Timpte, C. et al. (1993) *Arabidopsis* auxin-resistance gene *AXR*1 encodes a protein related to ubiquitin-activating enzyme E1. *Nature*, **364**, 161–164.

Lovett, M., Kere, J. and Hinton, L.M. (1991) Direct selection: a method for the isolation of cDNAs encoded by large genomic regions. *Proc. Natl. Acad. Sci. USA*, **88**, 9628–9631.

Ma, H., Yanofsky, M.F. Klee, H.J. et al. (1992) Vectors for plant transformation and cosmid libraries. *Gene*, **117**, 161–167.

Marchuk, D. and Collins, F.S. (1988) pYAC-RC, a yeast artificial chromosome vector for cloning DNA cut with infrequently cutting restriction endonucleases. *Nucleic Acids Res.*, **16**, 7743.

Martin, G.B., Ganal, M.W. and Tanksley, S.D. (1992) Construction of a yeast artificial chromosome library of tomato and identification of cloned segments linked to two disease resistance loci. *Mol. Gen. Genet.*, **233**, 25–32.

Martin, G., Brommonschenkel, S., Chunwongse, J. et al. (1993) Map-based cloning of a protein kinase gene conferring disease resistance in tomato. *Science*, **262**, 1432–1463.

Matallana, E., Bell, C.J., Dunn, P.J. et al. (1993) Genetic and physical linkage of the *Arabidopsis* genome: methods for anchoring yeast artificial chromosomes. In *Methods in Arabidopsis Research* (eds C. Koncz, N.H. Chua and J. Schell), pp. 144–169. World Scientific, Singapore, River Edge, NJ.

Meyer, K., Leube, M.P. and Grill, E. (1994) A protein phosphatase 2C involved in ABA signal transduction in *Arabidopsis thaliana*. *Science*, **264**, 1452–1455.

Meyerowitz, E. (1992) Vectors for plant transformation and cosmid libraries. *Gene*, **117**, 161–167.

Mindrinos, M., Katagiri, F., Yu, G.L. and Ausubel, F.M. (1994) The *A. thaliana* disease resistance gene *RPS2* encodes a protein containing a nucleotide-binding site and leucine-rich repeats. *Cell*, **78**, 1089–1099.

Miranda, A., Janssen, G., Hodges L. et al. (1992) *Agrobacterium tumefaciens* transfers extremely long T-DNAs by a unidirectional mechanism. *J. Bacteriol.*, **174**, 2288–2297.

Oannou, P.A., Amemiya, C.T., Grames, J. et al. (1994) A new bacteriophage P1-derived vector for the propagation of large human DNA fragments. *Nature Genet.*, **6**, 84–90.

Ogorman, S., Fox, D.T. and Wahl, G.M. (1991) Recombinase-mediated gene activation and site-specific integration in mammalian cells. *Science*, **251**, 1351–1355.

Olszewski, N.E., Martin, F.B. and Ausubel, F.M. (1988) Specialized binary vector for plant transformation: expression of the *Arabidopsis thaliana AHAS* gene in *Nicotiana tabacum*. *Nucleic Acids Res.*, **16**, 10765–10782.

Overhauser, J. and Radic, M.Z. (1987) Encapsulation of cells in agarose beads for use with pulsed-field gel electrophoresis. *Focus*, **9**, 8–9.

Pachnis, V., Pevny, L., Rothstein, R. and Costantini, F. (1990) Transfer of yeast artificial chromosome carrying human DNA from *Saccharomyces cerevisiae* into mammalian cells. *Proc. Natl. Acad. Sci. USA*, **87**, 5109–5113.

Pavan, W.J., Hieter, P. and Reeves, R.H. (1990) Modification and transfer into an embryonal carcinoma cell line of a 360-kilobase human-derived yeast artificial chromosome. *Mol. Cell. Biol.*, **10**, 4163–4169.

Qin, M., Bayley, C., Stockton, T. and Ow, D.W. (1994) Cre recombinase mediated site-specific recombination between plant chromosomes. *Proc. Natl. Acad. Sci. USA*, **91**, 1706–1710.

Sambrook, J., Fritsch, E.F. and Maniatis, T. (1989) *Molecular cloning, A Laboratory Manual*, 2nd edn. Cold Spring Harbor Laboratory Press, Cold Spring Harbor, New York.

Sauer, B. and Henderson, N. (1990) Targeted insertion of exogenous DNA into the eukaryotic genome by Cre recombinase. *New Biol.*, **4**, 441–449.

Schwartz, D.C. and Cantor, C.R. (1984) Separation of yeast chromosome-sized DNAs by pulsed field gradient gel electrophoresis. *Cell*, **37**, 67–75.

Shero, J., McCormick, M., Antenarakis, S. and Hieter, P. (1991) Yeast artificial chromosome vectors for efficient clone manipulation and mapping. *Genomics*, **10**, 505–580.

Silhavy, T.J., Berman, M.L. and Enquist, L.W. (1984) *Experiments with Gene Fusions*. Cold Spring Harbor Laboratory Press, Cold Spring Harbor, New York.

Simoens, C., Alliotte, T., Mendel, R. et al. (1986) A binary vector for transferring genomic libraries to plants. *Nucleic Acids Res.*, **14**, 8073–8090.

Sobral, B.W.S., Honeycutt, R.J., Atherly, A.G. and McClelland, M. (1990) Analysis of rice (*Oryza sativa* L.) genome using pulsed-field electrophoresis and rare cutting restriction endonucleases. *Plant Mol. Biol.Rep.*, **8**, 253–275.

Sun, T.-P., Goodman, H.M. and Ausubel, F.M. (1992) Cloning the *Arabidopsis GA1* locus by genomic subtraction. *Plant Cell*, **4**, 119–128.

Umehara, Y., Inagaki, A., Tanoue, H. et al. (1995) Construction and characterization of a rice YAC library for physical mapping. *Mol. Breeding*, **1**, 79–89.

van Daelen, R.A.J., Jonkers, J.J. and Zabel, P. (1989) Preparation of megabase-sized tomato DNA and separation of large restriction fragments by field inversion gel electrophoresis (FIGE). *Plant Mol. Biol.*, **12**, 341–342.

Wahl, G.M., Lewis, K.A., Ruiz, J.C. et al. (1987) Cosmid vectors for rapid genomic walking, restriction mapping, and gene transfer. *Proc. Natl. Acad. Sci. USA*, **84**, 2160–2164.

Ward, E.R., and Jen, G.C. (1990) Isolation of single-copy-sequence clones form a yeast artificial chromosome library of randomly-sheared *Arabidopsis thaliana* DNA. *Plant Mol. Biol.*, **14**, 561–568.

Willetts, N. and Skurry, R. (1987) Structure and function of the F factor and mechanism of conjugation. In *Escherichia coli and Salmonella typhimurium: Cellular and Molecular Biology* (ed. F.C. Neidhardt), Vol. 2, pp. 1110–1133. American Society for Microbiology, Washington, DC.

Wing, R.A., Zhang, H.-B. and Tanksley, S.D. (1994) Map-based cloning in crop plants: tomato as a model system: I. Genetic and physical mapping of *jointless*. *Mol. Gen. Genet.*, **242**, 681–688.

Wing, R.A., Rastogi, V.K., Zhang, H.-B. et al. (1993) An improved method of plant megabase DNA isolation in agarose microbeads suitable for physical mapping and YAC cloning. *Plant J.*, **4**, 893–898.

Woo, S.-S., Rastogi, V.K., Zhang, H.-B. et al. (1995) Isolation of sorghum megabase-size DNA and application for physical mapping and bacterial and yeast artificial chromosome library construction. *Plant Mol. Biol. Rep.*, **13**, 82–94.

Zabarovsky, E.R. and Allimets, R.L. (1986) An improved technique for the efficient construction of genomic libraries by partial filling-in of cohesive ends. *Gene*, **42**, 119–123.

Zhang, H., Herman, P.L. and Weeks, D.P. (1994a) Gene isolation through genomic complementation using an indexed library of *Chlamydomonas reinhardtii* DNA. *Plant Mol. Biol.*, **24**, 663–672.

Zhang, H.-B., Martin, G.B., Tanksley, S.D. and Wing, R.A. (1994b) Map-based cloning in crop plants. Tomato as a model system: II. Isolation and characterization of a set of overlapping yeast artificial chromosomes encompassing the *jointless* locus. *Mol. Gen. Genet.*, **244**, 613–621.

II Library Screening

5 Heterologous and Homologous Gene Probes

NIGEL G. HALFORD
IACR Long Ashton Research Station, Bristol, UK

(1) Plating of lambda and plasmid libraries; transfer and immobilisation of DNA from plaques and colonies onto hybridisation membranes.
(2) Strategies available for library screening and clone identification; practical considerations.
(3) Probe selection; generation of probes using cloned cDNAs, genomic sequences and PCR products as templates; design and use of oligonucleotide probes.
(4) Use of DNA probes to isolate homologous and heterologous sequences; screening cDNA and genomic libraries; analysing gene families; isolating sequences from distantly related species.

1 INTRODUCTION

A genomic or cDNA library, whether in a plasmid or in a lambda vector, contains a large number of different recombinants. There are a number of methods for screening libraries for specific sequences and this chapter deals with the use of nucleic acid probes. The basis of this procedure is the ability of two single-stranded DNA molecules of complementary or nearly complementary sequence to form a duplex (Britten & Davidson, 1985). A radiolabelled single-stranded DNA molecule can therefore be used as a 'probe' to screen DNA from individual recombinants in the library. In this context, identical sequences are termed homologous probes, while similar but not identical sequences are called heterologous probes. The ability of heterologous sequences to form duplexes depends on the length of matching sequence, its base composition, the number of 'mismatches', the reaction (or hybridisation) conditions and the stringency of the conditions in which the probe is washed off after hybridisation.

2 TYPES OF PROBE

The probe may be a sequence which has already been cloned, such as a cDNA, polymerase chain reaction (PCR) product (Chapter 13) or genomic fragment

---

Plant Gene Isolation: Principles and Practice. Edited by G. D. Foster and D. Twell.

from a different species, an incomplete part of the gene or cDNA to be isolated, or a sequence from a different gene from the same species, such as a different member of a gene family. Alternatively, it may be a synthetic oligonucleotide. An oligonucleotide probe may be preferable if no closely related cDNA or gene clones are available to use, or if the only sequence information available is derived by back-translating from the amino acid sequence of a protein. The use of a third type of probe, mixed sets of sequences derived from particular mRNA populations (differential screening) is described in Chapter 6.

3   PROCEDURE

The construction of cDNA and genomic libraries is described in Chapters 2 and 3 respectively, and BAC, YAC and cosmid libraries are discussed in Chapter 4. To screen a library with a nucleic acid probe, plates are prepared with either bacterial colonies containing a plasmid library, or with plaques from a phage library growing on a bacterial lawn. In the case of the former, the colonies are replicated on a nitrocellulose or other hybridisation filter and grown further, whereas plaques are 'lifted' onto a filter and processed immediately.

The filters are processed to lyse the bacterial cells or to break open (disrupt) the phage protein coat and denature and fix the DNA *in situ* (Nygaard & Hall, 1963). They are then probed with a radiolabelled DNA fragment which will hybridise to the sequence of interest (Denhardt, 1966; Gillespie & Spiegelman, 1965). The position of a hybridisation signal is determined by autoradiography and used to identify the colony or plaque containing that sequence.

The colony or plaque is 'picked' from the master plate and replated at a density to give individual colonies or plaques and screened again. This is repeated until a pure population is obtained. Autoradiographs of plaque lifts made in the direction and purification of a potato cDNA clone are shown in Fig. 1.

4   APPLICATIONS

(1)   Isolation of homologous gene sequences from the same species, e.g. use of a cDNA probe to isolate a genomic clone, or vice versa, use of a PCR product to isolate cDNA or genomic clones, or use of a partial gene or cDNA sequence to isolate a full-length sequence.

(2)   Identification of closely related genes in a gene family.

(3)   Isolation of related genes from other species. cDNA or gene probes may cross-hybridise with sequences from closely related species. Otherwise, degenerate oligonucleotide probes which allow for a degree of sequence divergence may be used.

(4)   Isolation of genes encoding proteins which have been completely or partially sequenced. The protein sequence is back-translated to give a DNA sequence, and this is used to design an oligonucleotide probe.

5 THE USE OF cDNA, PCR PRODUCT AND GENE PROBES TO ISOLATE HETEROLOGOUS AND HOMOLOGOUS SEQUENCES

There are now many examples of the applications of this technique, and a selection of recent publications in which it features is given in Table 1. One of the best examples is the analysis of the seed storage protein genes of the small-grain cereals. It illustrates the use of a probe from one species to identify gene sequences from another and shows how the isolation of one cDNA sequence can lead to the characterisation of a whole gene family. It is also a good example of the huge amount of information which can be derived from this sort of analysis.

The largest fraction of the seed proteins of wheat, barley and rye comprises the prolamins (review: Shewry & Tatham, 1990). These are monomeric and polymeric proteins which are characterised by their solubility in aqueous alcohols, though some of the polymeric proteins have to be reduced to individual subunits first. The structure and composition of the prolamins is a major determinant of the nutritional and processing properties of the grain and this has resulted in a great deal of effort being put into their characterisation. cDNA and gene isolation and characterisation have played an important role in this because the number and similarity of the individual polypeptides have made them difficult to study at the protein level.

There are around 50 individual polypeptides in the prolamin fraction of bread wheat seed proteins (Miflin et al., 1983). The monomeric components are usually referred to as gliadins, while the polymeric components are called glutenins. Allelic variation in the polypeptide composition of one small group of glutenins, called the high molecular weight subunits of glutenin (HMW subunits), has been correlated with the breadmaking quality of European wheat cultivars, and molecular studies of the genes encoding this group of proteins have provided one of the most complete pictures of the structure, organisation and evolution of any plant gene family (review: Shewry et al., 1989).

Bread wheat is a hexaploid, with three genomes called A, B and D. There are two HMW subunit genes, designated x and y, located on chromosome 1 of each genome, making a total of six. The first HMW subunit DNA sequences to be isolated were two cDNAs identified by screening a library using a barley D hordein sequence as a probe (Forde et al., 1983). Barley D hordeins are homologous to the wheat HMW subunits. The cDNAs were then used to screen a genomic library from the cultivar Cheyenne, and all six of the HMW subunit genes from that cultivar have been isolated and sequenced (Forde et al., 1985; Halford et al., 1987; Anderson et al., 1989; Anderson & Greene, 1989). Genes from other cultivars have been isolated in a similar manner (Flavell et al., 1990; Halford et al., 1992; Sugiyama et al., 1985). The sequence data confirmed and extended the knowledge gained from the analysis of the proteins themselves. It revealed that the proteins have a central domain made up of blocks of degenerate nine, six and three amino acid repeats, flanked by N- and C-terminal non-repetitive domains. It revealed differences between

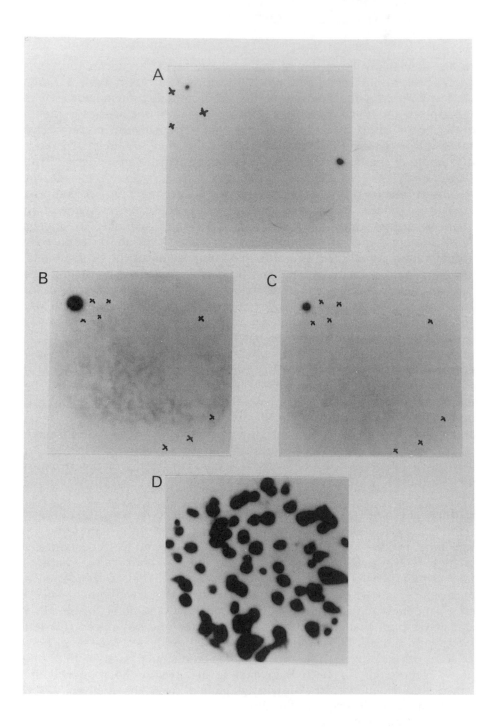

allelic forms, particularly the distribution of cysteine residues, which could be responsible for differences in breadmaking quality between cultivars. It showed that, in some cases, quantitative rather than qualitative differences in HMW subunit composition were responsible for differences in breadmaking quality between some cultivars. It provided further evidence for the formation of a β-spiral tertiary structure which had been predicted from the physical data of the proteins. It supported a hypothesis for a particular evolutionary pathway for the gene family. Finally, it led to further experiments to determine the regulatory elements in the gene promoters (Halford et al., 1989; Thomas & Flavell, 1991).

6 PRACTICAL CONSIDERATIONS FOR THE USE OF cDNA, PCR PRODUCT AND GENE PROBES

The HMW subunit family of proteins and the genes which encode them are particularly suitable for this technique. They are highly conserved within a species and amongst closely related species and readily cross-hybridise. Things are not always so easy and a number of practical considerations have to be taken into account before beginning a similar investigation. Some of the strategies which might be employed are described in Fig. 2.

Perhaps the most important decision to make is which species to attempt to isolate a gene or cDNA from. In the above case, where the properties of specific wheat proteins were being investigated, there was no choice, and other constraints, such as the availability of libraries, may apply. However, bread wheat is a hexaploid and it is common for the evolution of polyploidy to lead to the silencing of some genes which are surplus to requirements. In some wheat cultivars, three of the six HMW subunit genes are silent. This is not important if a cDNA library is to be screened, but it may be impossible to distinguish between silent and expressed genes in a genomic library. Barley is a close relative of wheat but is a diploid and has fewer silent genes; in some cases it may be a more preferable cereal species to study. However, the identification of clones of single- or low-copy genes is still difficult because of its relatively large genome, and it would not be a suitable species to study unless it were

---

Figure 1. Autoradiographs of plaque lifts made in the purification of a potato tuber cDNA clone, probed with a 500-bp potato PCR product. (A) shows two hybridisation signals on a lift taken from a 22 × 22 cm plate with approximately 200 000 plaques. Only one corner (approximately 10% of the total area) is shown. The crosses show the positions of three of the stab-marks made in the filter and the plate so that the position of hybridising plaques on the plate could be determined. (B) and (C) show replica lifts made in the first purification step of one of the plaques. This time a 90-mm plate with approximately 200 plaques was used. The crosses show the positions of the asymmetric pattern of stab-marks made in identical positions on the two filters. (D) shows a plaque lift made in the final stage of plaque purification. The increased signal results from the plaques being plated at a lower density and being allowed to grow larger. (Autoradiographs kindly provided by Angela Man)

Table 1. Some recent publications describing the isolation of genomic and cDNA clones by library screening with DNA probes

Cloned sequence	Type of probe	Reference
Arabidopsis nitrilase cDNA	cDNA of second member of gene family	Bartling et al. (1994)
Potato basic chitinase and 1,3-beta-glucanase cDNAs	Heterologous cDNAs from tobacco	Beerhues & Kombrink (1994)
Vigna radiata calmodulin genes	Heterologous cDNA from *Arabidopsis*	Botella & Areca (1994)
Coriander endosperm oleoyl–acyl carrier protein (ACP) thioesterase cDNA	Heterologous cDNA from safflower	Dormann et al. (1994)
Alfalfa aspartate aminotransferase genes	Homologous cDNA and PCR products	Gregerson et al. (1994)
Arabidopsis plastid omega-6 fatty acid desaturase cDNA	Oligonucleotides derived from conserved amino acid sequences	Hitz et al. (1994)
Arabidopsis plasma membrane H$^+$-ATPase gene	Heterologous cDNA from *Nicotiana plumbaginifolia*	Houlne & Boutry (1994)
Rice cDNAs encoding putative peroxidases	Homologous PCR product	Ito et al. (1994)
Tomato chloroplastic Cu/Zn superoxide dismutase gene	Homologous cDNA	Kardish et al. (1994)
Arabidopsis soluble epoxide hydrolase cDNAs and gene	Homologous partial cDNA isolated from subtractive library	Kiyosue et al. (1994)
Castor bean casbene synthase cDNA	Partial homologous cDNA	Mau & West (1994)
Cucumber ascorbate oxidase gene	Homologous cDNA ard partial genomic sequence	Ohkawa et al. (1994)
Tomato GACA-hybridising DNA repeat	GACA (4) oligonucleotide	Phillips et al. (1994)
Maize cyclin cDNAs	Homologous PCR products	Renaudin et al. (1994)
Isolectin cDNAs from *Clivia miniata*	*Amaryllis* lectin cDNA	Vandamme et al. (1994)
Rice lipid transfer protein (LTP) gene	LTP cDNA from maize	Vignols et al. (1994)
Bean malic enzyme gene	Homologous cDNA	Walter et al. (1994)
Soybean leginsulin cDNA	Oligonucleotides based on determined amino acid sequence	Watanabe et al. (1994)
Arabidopsis chloroplast ribosomal GTPase centre protein L12 gene	Heterologous cDNA probe from spinach	Weglohner & Subramanian (1994)
Potato mitochondrial NAD$^+$-dependent malic enzyme cDNA from leaf cDNA library	Potato mitochondrial NAD$^+$-dependent malic enzyme cDNA from tuber cDNA library	Winning et al. (1994)
Maize acidic class I chitinase cDNAs	Homologous PCR product	Wu et al. (1994)
Overlapping YACS encompassing the *jointless* locus of *Arabidopsis*	Molecular marker linked to *jointless*	Zhang et al. (1994a)
Arabidopsis protein kinase genes	Homologous PCR products	Zhang et al. (1994b)
Soybean brassino-steroid-regulated gene: full-length cDNA	Oligonucleotide with sequence corresponding to 5′ end of partial cDNA	Zurek & Clouse (1994)

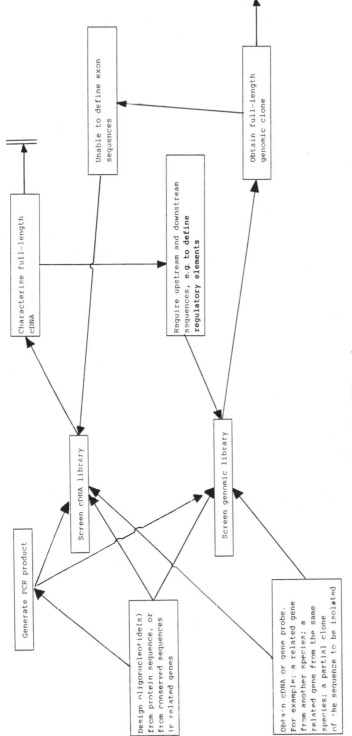

Figure 2. The strategies available for library screening using DNA probes

necessary to work on a cereal. In many cases, *Arabidopsis* has been selected for gene isolation because it has an unusually small genome for a higher plant.

The number of clones which have to be screened in order to identify a clone of a single-copy gene is given by the following equation (Clarke & Carbon, 1976):

$$N = \ln(1 - P)/\ln[1 - (I/G)]$$

where

$$N = \text{number of plaques to be screened}$$

$$P = \text{probability of obtaining a single-copy clone}$$

$$I = \text{average insert size of library}$$

$$G = \text{genome size}$$

This equation can also be used to give the number of independent clones which have to be present in a library for it to be representative (see Chapter 3). An approximation can be obtained from the simpler equations:

$$N = 5 \times (G/I)$$

for a 99% probability of obtaining a single-copy clone.

$$N = 3 \times (G/I)$$

for a 95% probability of obtaining a single-copy clone.

The barley genome is 5.2×10^9 bp in size (Grierson & Covey, 1988), so if the average insert size of a library is 1.5×10^4 bp, $N = 1.73 \times 10^6$ where $P = 99\%$, and 1.1×10^6 where $P = 95\%$.

In contrast, the *Arabidopsis* genome is only 7×10^7 bp in size (Grierson & Covey, 1988). This means that $N = 2.3 \times 10^4$ where $P = 99\%$, and 1.4×10^4 where $P = 95\%$. Clearly, generating and screening *Arabidopsis* genomic libraries is a much simpler proposition.

There are dangers in considering *Arabidopsis* as a perfect model for plant development. Its gene families are generally simpler than those in many other plant species. This may be an advantage, as there are fewer silent genes (genes which are not expressed; also called pseudogenes), for example, but it may mean that complexities of differential regulation of gene expression within a gene family may be missing. In addition, some aspects of its physiology and development are very different from those of distantly related plant species. Its seed development, structure and composition are very different from those of the cereals, for example. Despite these drawbacks, however, it has proved to be extremely useful in the isolation of plant genes, and an analysis of Table 1 indicates that it is the species of choice in a high proportion of studies of this kind.

I have already mentioned the problem of the presence of silent genes in genomic libraries. This is not the only disadvantage associated with screening genomic libraries. Most eukaryotic genes contain introns, and the exact position of splicing sites can be difficult to predict from bare sequence data

(Breathnach & Chambon, 1981; Goodall et al., 1989). This is particularly true in plant genes, because the consensus sequences associated with intron–exon boundaries are relatively poorly conserved. These problems can be avoided by screening a cDNA library, as cDNAs must originate from active genes and do not contain introns. The number of cDNA clones which should be screened depends on the number of independent clones in the library and the abundance of the transcript of interest in the mRNA population from which the cDNA was synthesised. Fully representative libraries contain 10^6 or more independent clones, and an exhaustive screen of such a libray is a labour-intensive and time-consuming process.

7   LIMITATIONS OF USING cDNA, PCR PRODUCT AND GENE PROBES

Screening libraries with cloned sequences relies on the availability of a sequence of sufficient similarity to cross-hybridise with a clone containing an insert of interest. The degree of conservation of proteins and the genes which encode them between different species varies a great deal. The wheat HMW subunit genes will cross-hybridise with homologous sequences in barley and rye, but there are no cross-hybridising sequences in less closely related species. In contrast, a rye *SNF1*-related protein kinase gene was used successfully to isolate an *Arabidopsis* homologue in a low-stringency screen (Le Guen et al., 1992). Some plant proteins have even retained a considerable degree of sequence conservation with homologues from animals and yeast. The protein kinases are classic examples. They are a large family of proteins, all eukaryotic examples of which have a number of characteristic, highly conserved regions within a large catalytic domain (Hanks et al., 1988). Perhaps the most intensively studied of these are the cdc2 protein kinases, which play a fundamental role in the regulation of the eukaryotic cell cycle (review: Francis & Halford, 1995). These are highly conserved at the amino acid level; for example, the sequence of the *Arabidopsis* cdc2 protein is 64.5% identical to that of the fission yeast (*Saccharomyces pombe*) homologue, and the recognition of a plant protein by an antibody raised to a conserved cdc2-specific motif (Glu-Gly-Pro-Ser-Thr-Ala-Ile-Arg-Glu-Ile-Ser-Leu-Leu-Lys-Glu) was the first evidence that plants contained a cdc2-like protein (John et al., 1989). The similarity of the DNA sequences, however, is not sufficient for cross-hybridisation between a plant cdc2 DNA sequence and a non-plant homologue, and PCR, which requires oligonucleotides designed from short conserved sequences, was eventually used to clone plant cdc2 sequences.

8   THE USE OF OLIGONUCLEOTIDE PROBES

Cloning by PCR is described in Chapter 13. It requires two synthetic oligonucleotide primers which will anneal to the sequence to be amplified to form a template for the polymerase. Oligonucleotides can also be used as probes; indeed, this technique predates PCR and may still be preferable if two sequences are not available from which to design primers, or if the primers anneal to sequences too far apart to allow efficient amplification of the intervening sequence. The oligonucleotide is radiolabelled by the addition of a [^{32}P]phosphate group to the 5' end using T4 polynucleotide kinase. [γ-^{32}P]ATP is the phosphate donor.

The oligonucleotide sequence is derived from a sequence in a related DNA sequence or sequences, or by back-translating from a protein sequence. The use of oligonucleotides as opposed to cloned sequences as probes has a number of advantages. First, the sequence from which it is derived need only be very short. Theoretically, a 12-mer should hybridise to a complementary sequence, although the shortest oligonucleotides used in practice are 14-mers. Second, a degree of degeneracy can be incorporated into the nucleotide sequence. This may be necessary if the oligonucleotide sequence is based on two or more sequences which are not absolutely identical, or if it is derived by back-translating from a protein sequence, in which case allowance has to be made for the degeneracy of the genetic code. However, it should be remembered that each degenerate base in the sequence increases the number of different oligonucleotides in the probe population by a factor of two, and the number of degenerate bases should therefore be kept to a minimum. This may have a bearing on the selection of the amino acid sequence from which to derive the oligonucleotide sequence. For example, the amino acids methionine and tryptophan are encoded by only one codon, whereas there are six possible codons for each of leucine, arginine and serine.

9 EXAMPLES OF THE USE OF OLIGONUCLEOTIDE PROBES

The first plant protein kinase transcripts to be cloned were isolated using oligonucleotide screening (Lawton et al., 1989). A number of conserved sequences had been identified in protein kinases from yeasts and animals, and two of these were back-translated to give the sequence for two oligonucleotide probes (Fig. 3A). These probes had been used to isolate protein kinase sequences from a HeLa cDNA library (Hanks, 1987), and in some positions, such as the second codon of the first oligonucleotide, which encodes a leucine residue, some of the possible codons had not been used. This selection was made on the basis of human codon preference, but the same principle can be applied for designing oligonucleotides specifically for screening plant libraries using plant codon usage tables (Campbell & Gowri, 1990). Despite the tailoring of these oligonucleotides for screening a human library, they were used successfully to isolate protein kinase cDNAs from bean and rice.

A similar example is the cloning of cDNA clones for plant mitotic cyclins reported by Hata et al. (1991). Mitotic cyclins interact with the cdc2 protein (see above) to form maturation promotion factor. As with cdc2, it was apparent that these proteins have been highly conserved through evolution and it was reasonable to expect that plant mitotic cyclins would retain some similarities to the yeast and animal homologues which had already been cloned. Two amino acid sequences which were highly conserved in yeast and animal cyclins were back-translated to give oligonucleotide probe sequences. One of the oligo-nucleotide sequences is notable for its incorporation of deoxyinosine at positions of ambiguity to reduce degeneracy (Fig. 3B). Deoxyinosine will 'pair' with any of the four natural bases with or without hydrogen bonding (Ohtsuka et al., 1985). These oligonucleotides identified cyclin homologues in cDNA libraries from carrot and soybean.

The commonest problem encountered with the use of oligonucleotide probes is the isolation of false positives, i.e. the hybridisation of the probe to a sequence unrelated to the one being searched for. Oligonucleotide sequences are short, and short sequences may be repeated at random within a genome; this is compounded by the use of mixed oligonucleotide populations in a probe. The longer the oligonucleotide sequence, the more likely it is to be unique among the sequences being screened, but the advantage of using a longer probe may be offset by the requirement to introduce more degeneracy into the sequence. In both the examples described above, two oligonucleotide probes were used. Clones which hybridise to two independent probes are likely to be genuine positives. However, this may not be possible if the probe sequence is derived from N-terminal sequencing of a protein and no other sequence information is available.

The risk of picking up false positives can be kept to a minimum by ensuring that the probe is washed off the filters at a high enough stringency. Oligo-nucleotide hybridisation to a complementary sequence is very specific and will not tolerate any mismatches under certain conditions. It is imperative that these conditions are applied in the experiment. Duplex stability is a function of the length of the oligonucleotide and its base composition, guanine–cytosine pairing being stronger than adenine–thymine pairing. For oligonucleotides up to 20 bases long, the temperature at which half of the duplex is dissociated (T_d) at 1M Na^+ is given by the following equation (Wallace et al., 1979):

$$T_d = 2(A + T) + 4(G + C)$$

where A, C, G and T are the number of adenines, cytosines, guanines and thymines, respectively. In our experience, a strong signal is still obtained from oligonucleotide probing after washing at $T_d + 2\,°C$. Sequential washings at increasing temperature between $T_d - 5\,°C$ and $T_d + 2\,°C$ should discriminate between hybridisation to an absolutely complementary sequence and one where there is a mismatch.

Another problem to be taken into account in the design of oligonucleotide probes is self-complementarity. This is the ability of the molecule to fold into a hairpin loop or stem–loop structure if it contains complementary sequences. This affects the efficiency of the labelling reaction and the hybridisation to other sequences.

Despite these problems, the use of oligonucleotide probes is a powerful technique. The successful use of cDNA or gene probes from yeast or animals to

114

A

```
                          1                                    2

Consensus         Asp Leu Lys Pro Glu Asn        Gly Thr Pro Glu Tyr Leu Ala Pro Glu
sequences

Back-             5'-GAC CUN AAA CCN GAA AAC      5'-GGN ACN CCN GAA UAC CUN GCN CCN GAA
translation            U U       G       U                       G       U   U           G

Selected          5'-GAC CUC AAA CCN GAA AA       5'-GGC ACC CCC GAG UAC CUC GCA CCN GA
codons                 U  G       G      G                        U       U   G       C
                                                                                       U

Probes            3'-CTAGACTTCGGNCTCTT            3'-CCGTGGGACTCATAGACCGAGNCT
                         G  G    T   T                     G      G   G   G
                                                                          T
```

B

Consensus   Ile Leu Val Asp Trp Leu Val Glu Val
sequence   Gln

Back-   5'-AUA CUN GUN GAC UGG CUN GUN CAA GTN
translation   C   U   U   U   G G
                              U

Probe   5'-ATICTIGTIGATTGGCTIGTIGTICAAGT
                              T   T   G G

Figure 3. (A) Consensus sequences of two highly conserved regions in eukaryotic protein kinases, the DNA sequences derived by back-translation from the amino acid sequences, selected codons, and oligonucleotide probe sequences used to isolate protein kinase cDNAS from bean and rice (Lawton et al., 1989). (B) Consensus sequence identified in cyclin proteins, the DNA sequence derived by back-translation from the amino acid sequence, and the sequence of an oligonucleotide probe incorporating deoxyinosines to reduce degeneracy and used to isolate cyclin sequences from carrot and soybean (Hata et al., 1991)

isolate plant protein kinase or cyclin sequences has not been reported, and these are excellent examples of how oligonucleotide screening can be used to make the jump between homologous sequences from evolutionarily distant organisms.

10   TECHNICAL HINTS

Detailed descriptions of methods used in library screening with DNA probes are given in a number of texts, of which Berger & Kimmel (1987), Sambrook et al. (1989) and Hames & Higgins (1985) are particularly recommended. Researchers new to the field may find themselves with a bewildering number of techniques to choose from, so I have described below some procedures which we have used routinely and successfully in our laboratory in order to isolate plant genes.

10.1   Preparation of *Escherischia coli* cells for plating

A variety of different *E. coli* strains are available, though the supplier of the lambda vector should recommend a strain to use. It is important to check that the host is considered a disabled one, i.e. it has biological limitations which mean that it is unlikely to survive in the gut. Use of a non-disabled host may mean that a greater degree of containment will have to be used when plating out the library. Disabled strains include most K-12 derivatives (refer to HSE guidance notes on genetic modification). The cells are prepared for plating in the following way (Sambrook et al., 1989).

Inoculate 50 ml of liquid 2× YT medium (Sambrook et al., 1989) containing 0.2% w/v maltose in a 250-ml conical flask. Incubate overnight and then pellet the cells by centrifugation at 4000g for 5 min at 4 °C. Drain the pellet thoroughly and resuspend in 12 ml of cold 10 mM $MgSO_4$. Keep at 4 °C.

10.2   Plating the phage

Plate the phage on plates containing solidified 2× YT medium containing 0.2% $MgSO_4$ and 0.2% maltose (both added after autoclaving). For round 90-mm-diameter Petri dishes, mix up to 10 μl of the phage with 100 μl of plating bacteria and incubate at 37 °C for 20 min, and then add to 3 ml top (0.7%) 2× YT agar at 42 °C and pour onto the plate. Allow to cool, and then incubate at 37 °C overnight. Accurately determine the titre of the phage by plating out serial dilutions of the library.

The first screening of the library is best done on large plates (e.g. Nunc, 210 × 210 mm square dishes). Obviously, the more plaques per plate, the fewer plates required to screen the library. However, a high plaque density reduces the chance of recovering a clone of interest. We recommend a plaque density of 5×10^4 per plate. Mix 5×10^4 plaque-forming units (PFU) (< 100 μl) with 1 ml of plating bacteria and proceed as before, using 30 ml of top agarose to pour onto the plates (agarose is used in place of agar because it sets harder). Allow to cool, and then incubate the plates at 37 °C for approximately 8 h. This time will vary, but it is important that the plaques are distinct, but are not allowed to grow together. The size of the plaques can also be reduced by increasing the density of the plating cells. After incubation, place the plates at 4 °C for at least 2 h (or overnight) to cool. This allows the agarose to harden further.

10.3 Plaque lifts
(Benton & Davis, 1977)

This procedure transfers phage particles to a filter, disrupts them, denatures the phage DNA and irreversibly fixes it to the membrane. Nitrocellulose filters can be used, but nylon-based alternatives (such as Hybond-N from Amersham) are stronger and quicker to process. It is important that two lifts are taken from each plate so that the results of hybridisation with a probe can be verified.

Cut the filter slightly smaller than the area of the surface of the plate. Label it with a soft pencil and place it on the surface of the plate for 1 min. During this time record its position on the plate by stabbing a needle dipped in permanent ink through the filter into the surface of the agar several times to make an asymmetric pattern (Fig. 1). Transfer the filter DNA side up to a tray containing several layers of Whatmann 3MM filter paper soaked in denaturing solution (0.5 M NaOH, 1.5 M NaCl). After 1 min, transfer it to a second tray containing filter paper soaked in neutralising solution (3 M NaCl, 1 M Tris-HCl, pH 6). Leave the filter in this tray for 5 min, and then remove it and leave it to air-dry before fixing the DNA by UV crosslinking or baking, according to the manufacturer's instructions. The second lift is performed in the same way, except that the filter is left on the place for an additional minute.

10.4 Probe preparation

There are now many ways of synthesising a radiolabelled probe. Oligonucleotides can be labelled using T4 polynucleotide kinase and $[\gamma\text{-}^{32}P]ATP$. Longer DNA molecules are used to make a template for a DNA polymerase. The polymerase reaction is then performed using three 'cold' nucleotides and a radiolabelled nucleotide. One of the simplest methods of generating the template is to anneal random primers to denatured double-stranded DNA or to single-stranded DNA (Feinberg & Vogelstein, 1983; Feinberg, 1984). The Klenow fragment of DNA polymerase 1 is the enzyme usually used for the reaction, and $[\alpha\text{-}^{32}P]dATP$ or dCTP is incorporated into the newly synthesised DNA strand. Denature the probe by placing it in a boiling water bath before adding it to the hybridisation solution.

Non-radioactive methods for labelling DNA are now available, the most popular of which is the incorporation of biotinylated nucleotides. The biotinylated DNA can be detected colorimetrically. This is described in more detail in Chapters 13 and 14.

10.5 Hybridisation and washing

Place the filters in a bag, box or hybridisation bottle containing 4 ml of prehybridisation buffer (6× standard saline citrate (SSC) (Sambrook et al., 1989), 0.2% non-fat dried milk) per 100 cm² of filter. Incubate at 65 °C in a shaking water bath or hybridisation oven for at least 2 h. Remove the buffer and add 2 ml of hybridisation buffer (6× SSC, 0.2% non-fat dried milk, 1% nonidet P-40) per 100 cm² of filter, plus the radiolabelled probe. Hybridise overnight at 65 °C. Remove the buffer containing the probe and rinse the filters with 2× SSC/0.1% sodium dodecyl sulphate (SDS). The stability of DNA hybrids and, therefore, the determination of temperature and salt concentration at which non-specifically bound probe will wash off a hybridisation filter is determined by complex equations which I will not describe here (Anderson & Young, 1985). As a rough guide, for high stringency, wash the filters in 0.2× SSC/0.1% SDS for 20 min, and then repeat. If a less stringent wash is required, either increase the salt concentration (e.g. up to 0.5× SSC), or decrease the temperature (e.g. to 50 °C), or both. A degree of trial and error may be required if the exact similarity between the probe and target sequences is not known. Finally, wrap the filters in Cling Film and autoradiograph.

10.6 Plaque purification

Develop the autoradiograph and mark it at the positions of the asymmetric holes in the filters. Line up the marks on the autoradiographs from the two filters from one plate and look for hybridisation spots which align on the two autoradiographs. The presence of a spot in the same place on both autoradiographs almost certainly indicates a hybridising plaque. Place the plate on top of the autoradiograph on a light-box and align the stab-marks in the plate with the corresponding marks on the autoradiograph. Remove a plug of agar over the hybridisation spot with the wide end of a sterile Pasteur pipette and transfer it to 1 ml of SM buffer (Sambrook et al., 1989). Add 10 μl of chloroform and leave at room temperature for 1 h for the phage to diffuse into the buffer. This phage suspension can be stored at 4 °C for several months. Check its titre by plating out serial dilutions, and then plate out approximately 100 PFU on a 90-mm Petri dish. Take duplicate plaque lifts and hybridise with the probe as before. This time it should be possible to take a smaller plug containing a single plaque using the narrow end of a Pasteur pipette. Repeat the plating, lifting and hybridising until a phage suspension derived from a single hybridising plaque is obtained.

10.7 Phage DNA isolation

A comprehensive description of the different lambda vectors which are available is given by Sambrook et al. (1989). A number of them contain plasmid or phagemid sequences which can be excised with helper phage to generate viable vectors containing the cloned DNA sequence. However, if this is not possible, phage DNA has to be isolated so that the cloned fragment can be removed and inserted into a more versatile vector. Phage DNA isolation is generally considered to be difficult, but the following is a rapid method for generating several micrograms of phage DNA which is clean enough to act as a substrate for PCR, or to be cut with restriction enzymes.

Plate out 10^5 PFU from the pure phage suspension on a 90-mm Petri dish and incubate overnight. This will give confluent lysis. Scrape the top agar into a 30-ml centrifuge tube and add 5 ml of phage buffer plus 0.1 ml of chloroform. Add crude DNase and RNase to 1 $\mu g/ml$ and incubate at 37 °C for 30 min. Spin at 4000g for 5 min at 4 °C to pellet the agar and cell debris and transfer the supernatant to a Corex tube. Add 1 ml of 20% polyethylene glycol (PEG) 6000, 2.5 M NaCl, mix gently, and stand at room temperature for 1 h. Spin at 6000g for 10 min at 4 °C. An off-white precipitate should be clearly visible. Drain the pellet thoroughly and resuspend in 0.5 ml of phage buffer. Extract with an equal volume of chloroform to remove the PEG, add EDTA to 30 mM, protein-ase K to 0.5 mg/ml, and SDS to 0.5%, and incubate for 30 min at 37 °C. This disrupts and digests away the protein coat. Extract with an equal volume of phenol/chloroform (1 : 1), and then with an equal volume of chloroform, and then precipitate with ethanol.

Phage DNA can also be isolated from liquid cultures using a method described in Chapter 3.

10.8 Screening plasmid libraries by colony hybridisation
(Grunstein & Hogness, 1975; Sambrook et al., 1989)

Bacterial colonies can be transferred from a plate to a nitrocellulose filter simply by placing the filter on top of a plate on which the colonies are growing, marking the filter and plate in a pattern with permanent ink, and then placing the filter colony side up on another plate. Replica filters are made by laying a duplicate filter on top of the first filter and incubating the 'sandwich' at 37 °C to allow the replica colonies to grow. The 'master' plate should be incubated at 37 °C after the lift has been taken so that the colonies regrow to a reasonable size. The filters are kept as a sandwich during denaturation and neutralisation, the procedure for which is the same as described above for the plaque lifts.

10.9 Hybridisation with oligonucleotide probes
(Wallace & Miyada, 1987)

Prewet the filters in 6× SSC for 5 min, and then prehybridise in 10× Den-hardt's solution (Denhardt, 1966), 0.2% SDS, at 67 °C for 5 min. Hybridise in 6× SSC at room temperature for 1 h, wash three times in 6× SSC at room temperature, 1 min per wash, and autoradiograph. Wash three times at $T_d - 5$ °C and autoradiograph again. A comparison between the two autoradio-graphs should identify positive clones, but it is worth repeating the process at T_d and even $T_d + 2$ to be sure.

11   CONCLUDING REMARKS

The contribution that this technique has made to the study of plant genes and gene families is huge. Table 1 lists some of the examples of its use which have been published in recent months. Clearly, a comprehensive list of the genes and cDNAs which have been isolated by library screening with DNA probes would fill many pages on its own. I have described in detail some examples of the use of different types of probe, and the application of the technique in the analysis of closely related genes and gene families, and less closely related genes, such as plant homologues of yeast and mammalian genes. Although alternative approaches, such as PCR, have been developed, library screening with homologous and heterologous gene probes will continue to be an important part of many gene isolation projects.

ESSENTIAL READING

Hames, B.D. and Higgins, S.J. (eds) (1985) *Nucleic Acid Hybridisation: A Practical Approach*. IRL Press, Oxford.

Hata, S., Kouchi, H., Suzuka, I. and Ishii, T. (1991) Isolation and characterisation of cDNA clones for plant cyclins. *EMBO J.*, **10**, 2681–2688.

Lawton, M.A., Yamamoto, R.T., Hanks, S.K. and Lamb, C.J. (1989) Molecular cloning of plant transcripts encoding protein kinase homologs, *Proc. Natl. Acad. Sci. USA*, **86**, 3140–3144.

Sambrook, J., Fritsch, E.F. and Maniatis, T. (eds) (1989) *Molecular Cloning, a Laboratory Manual*. Cold Spring Harbor Laboratory Press, New York.

Shewry, P.R., Halford, N.G. and Tatham, A.S. (1989) The high molecular weight subunits of wheat, barley and rye: genetics, molecular biology, chemistry and role in wheat gluten structure and functionality. *Oxford Surveys Plant Mol. Cell Biol.*, **6**, 163–219.

REFERENCES

Anderson, M.L.M. and Young, B.D. (1985) Quantitative filter hybridisation. In *Nucleic Acid Hybridisation, A Practical Approach* (eds B.D. Hames and S.J. Higgins), pp. 73–112. IRL Press, Oxford.

Anderson, O.D. and Greene, F.C. (1989) The characterisation and comparative analysis of high molecular weight glutenin genes from genomes A and B of hexaploid bread wheat. *Theor. Appl. Genet.*, **77**, 689–700.

Anderson, O.D., Greene, F.C., Yip, R.E. et al. (1989) Nucleotide sequences of the two high molecular weight glutenin genes from the D-genome of a hexaploid bread wheat, *Triticum aestivum* L. cv. Cheyenne. *Nucleic Acids Research*, **17**, 461–462.

Bartling, D., Seedorf, M., Schmidt, R.C. and Weiller, E.W. (1994) Molecular characterisation of two cloned nitrilases from *Arabidopsis thaliana*—key enzymes in biosynthesis of the plant hormone indole-3-acetic acid. *Proc. Natl. Acad. Sci. USA*, **91**, 6021–6025.

Beerhues, L. and Kombrink, E. (1994) Primary structure and expression of messenger RNAs encoding basic chitinase and 1,3-beta glucanase in potato. *Plant Mol. Biol.*, **24**, 353–367.

Benton, W.D. and Davis, R.W. (1977) Screening lambda gt recombinant clones by hybridisation to single plaques *in vitro*. *Science*, **196**, 180–182.

Berger, S.L. and Kimmel, A.R. (eds) (1987) *Methods Enzymol.*, **152**. *Guide to Molecular Cloning Techniques*. Academic Press, Inc., San Diego.

Botella, J.R. and Areca, R.N. (1994) Differential expression of two calmodulin genes in response to physical and chemical stimuli. *Plant Mol. Biol.*, **24**, 757–766.

Breathnach, R. and Chambon, P. (1981) Organisation and expression of eukaryotic split genes coding for proteins. *Annu. Rev. Biochem.*, **50**, 349–383.

Britten, R.J. and Davidson, E.H. (1985) Hybridisation strategy. In *Nucleic Acid Hybridisation, A Practical Approach* (eds B.D. Hames and S.J. Higgins), pp. 3–16. IRL Press, Oxford.

Campbell, W.H. and Gowri, G. (1990) Codon usage in higher plants, green algae and cyanobacteria. *Plant Physiol.*, **92**, 1–11.

Clarke, L. and Carbon, J. (1976) A colony bank containing synthetic Col E1 hybrid plasmids representative of the entire *E. coli* genome. *Cell*, **9**, 91–99.

Denhardt, D.T. (1966) A membrane filter technique for the detection of complementary DNA. *Biochem. Biophys. Res. Commun.*, **23**, 641–646.

Dormann, P., Kridl, J.C. and Ohlrogge, J.B. (1994) Cloning and expression in *Escherischia coli* of a cDNA coding for the oleoyl-acyl carrier protein thioesterase from coriander. *Biochim. Biophys. Acta—Lipids and Metabolism*, **1212**, 134–136.

Feinberg, A.P. (1984) Addition. *Anal. Biochem.*, **137**, 266–267.

Feinberg, A.P. and Vogelstein, B. (1983) A technique for radiolabelling DNA restriction endonuclease fragments to high specific activity. *Anal. Biochem.*, **132**, 6–13.

Flavell, R.B., Goldsbrough, A.P., Robert, L.S. et al. (1990) Genetic variation in wheat HMW glutenin subunits and the molecular basis of breadmaking quality. *Bio-technology*, **7**, 1281–1285.

Forde, J., Forde, B.G., Fry, R.P. et al. (1983) Identification of barley and wheat cDNA clones related to the high M_r polypeptides of wheat gluten. *FEBS Lett.*, **162**, 360–366.

Forde, J., Malpica, J.-M., Halford, N.G. et al. (1985) The nucleotide sequence of a HMW subunit gene located on chromosome 1A of wheat (*Triticum aestivum* L.). *Nucleic Acids Res.*, **13**, 6817–6832.

Francis, D. and Halford, N.G. (1995) The plant cell cycle—minireview. *Physiol. Plant.*, **93**, 365–374.

Gillespie, D. and Spiegelman, S. (1965) A quantitative assay for DNA–RNA hybrids with DNA immobilised on a membrane. *J. Mol. Biol.*, **12**, 829–842.

Goodall, G., Wiebauer, K. and Filipowicz, W. (1989) Intron recognition in plants. In *Molecular Biology of RNA* (ed. T.R. Cech), pp. 155–163. Alan R. Liss, Inc., New York.

Gregerson, R.G., Miller, S.S., Petrowski, M. et al. (1994) Genomic structure, expression and evolution of the Alfalfa aspartate aminotransferase genes. *Plant Mol. Biol.*, **25**, 287–399.

Grierson, D. and Covey, S.N. (1988) *Plant Molecular Biology*, 2nd edn. Blackie and Son Ltd, Glasgow.

Grunstein, M. and Hogness, D.S. (1975) Colony hybridisation: a method for the isolation of cloned DNAs that contain a specific gene. *Proc. Natl. Acad. Sci. USA*, **72**, 3961–3965.

Halford, N.G., Forde, J., Anderson, O.D. et al. (1987) The nucleotide and deduced amino acid sequences of an HMW glutenin subunit gene from chromosome 1B of bread wheat (*Triticum aestivum* L.) and comparison with those of genes from chromosomes 1A and 1D. *Theor. Appl. Genet.*, **75**, 117–126.

Halford, N.G., Forde, J., Shewry, P.R. and Kreis, M. (1989) Functional analysis of the upstream regions of a silent and an expressed member of a family of wheat seed protein genes in transgenic tobacco. *Plant Sci.*, **62**, 207–216.

Halford, N.G., Field, J.M., Blair, H. et al. (1992) Analysis of HMW glutenin subunits encoded by chromosome 1A of bread wheat (*Triticum aestivum* L.) indicates quantitative effects on grain quality. *Theor. Appl. Genet.*, **83**, 373–378.

Hanks, S.K. (1987) Homology probing—identification of cDNA clones encoding members of the protein-serine kinase family. *Proc. Natl. Acad. Sci. USA*, **84**, 388–392.

Hanks, S.K., Quinn, A.M. and Hunter, T. (1988) The protein kinase family: conserved features and deduced phylogeny of the catalytic domains. *Science*, **241**, 42–52.

Hitz, W.D., Carlson, T.J., Booth, J.R. et al. (1994) Cloning of a higher plant plastid omega-6 fatty acid desaturase cDNA and its expression in a cyanobacterium. *Plant Physiol.*, **105**, 635–641.

Houlne, G. and Boutry, M. (1994) Identification of an *Arabidopsis thaliana* gene encoding a plasma membrane H$^+$-ATPase whose expression is restricted to anther tissues. *Plant J.*, **5**, 311–317.

Ito, H., Kimizuka, F., Ohbayashi, A. et al. (1994) Molecular cloning and characterisation of two complementary cDNAs encoding putative peroxidases from rice (*Oryza sativa* L.) shoots. *Plant Cell Rep.*, **13**, 261–366.

John, P.C.L., Sek, F.J. and Lee, M.G. (1989) A homolog of the cell cycle control protein p34cdc2 participates in the division cycle of *Chlamydomonas*, and a similar protein is detectable in higher plants and remote taxa. *Plant Cell*, **1**, 1185–1193.

Kardish, N., Magal, N., Aviv, D. and Galun, E. (1994) The tomato gene for the chloroplastic Cu, Zn superoxide-dismutase—regulation of expression imposed in transgenic tobacco plants by a short promoter. *Plant Mol. Biol.*, **25**, 887–897.

Kiyosue, T., Beetham, J.K., Pinot, F. et al. (1994) Characterisation of an *Arabidopsis* cDNA for a soluble epoxide hydrolase gene that is inducible by auxin and water-stress. *Plant J.*, **6**, 259–269.

Le Guen, L., Thomas, M., Bianchi, M. et al. (1992) Structure and expression of a gene from *Arabidopsis thaliana* encoding a protein related to SNF1 protein kinase. *Gene*, **120**, 249–254.

Mau, C.J.D. and West, C.A. (1994) Cloning of casbene synthase cDNA—evidence for conserved structural features among terpenoid cyclases in plants. *Proc. Natl. Acad. Sci. USA*, **91**, 8497–8501.

Miflin, B.J., Field, J.M. and Shewry, P.R. (1983) Cereal storage proteins and their effect on technological properties. In *Seed Proteins*. (eds J. Daussant, J. Mosse and J. Vaughan), pp. 255–319. Academic Press, London.

Nygaard, A.P. and Hall, B.D. (1963) A method for the detection of RNA–DNA complexes. *Biochem. Biophys. Res. Commun.*, **12**, 98–104.

Ohkawa, J., Ohya, T., Ito, T. et al. (1994) Structure of the genomic DNA encoding cucumber ascorbate oxidase and its expression in transgenic plants. *Plant Cell Rep.*, **13**, 481–488.

Ohtsuka, E., Matsuki, S., Ikehara, M. et al. (1985) An alternative approach to deoxy-oligonucleotides as hybridisation probes by insertion of deoxyinosine at ambiguous codon positions. *J. Biol. Chem.*, **85**, 2605–2608.

Phillips, W.J., Chapman, C.G.D. and Jack, P.L. (1994) Molecular cloning and analysis of one member of a polymorphic family of GACA-hybridising DNA repeats in tomato. *Theor. Appl. Genet.*, **88**, 845–851.

Renaudin, J.P., Colasanti, J., Rime, H. et al. (1994) Cloning of four cyclins from maize indicates that higher plants have three structurally distinct groups of mitotic cyclins. *Proc. Natl. Acad. Sci. USA*, **91**, 7375–7379.

Shewry, P.R. and Tatham, A.S. (1990) The prolamin storage proteins of cereal seeds: structure and evolution. *Biochem. J.*, **267**, 1–12.

Sugiyama, T., Rafalski, A., Peterson, D. and Soll, D. (1985) A wheat HMW glutenin subunit gene reveals a highly repeated structure. *Nucleic Acids Res.*, **13**, 8729–8739.

Thomas, M.S. and Flavell, R.B. (1991) Identification of an enhancer element for the endosperm-specific expression of high-molecular-weight glutenin. *Plant Cell*, **2**, 1171–1180.

Vandamme, E.J.M., Smeets, K., Vanleuven, F. and Peumans, W.J. (1994) Molecular cloning of mannose-binding lectins from *Clivia miniata*. *Plant Mol. Biol.*, **24**, 825–830.

Vignols, F., Lund, G., Pammi, S. et al. (1994) Characterisation of a rice gene coding for a lipid transfer protein. *Gene*, **142**, 265–270.

Wallace, R.B. and Miyada, C.G. (1987) Oligonucleotide probes for the screening of recombinant DNA libraries. In *Guide to Molecular Cloning Techniques* (eds S.L. Berger and A.R. Kimmel), *Methods Enzymol.*, **152**, pp. 432–443. Academic Press, Inc., San Diego.

Wallace, R.B., Schaffer, J., Murphy, R.F. et al. (1979) Hybridization of synthetic oligodeoxyribonucleotides to φx174 DNA: the effect of single base pair mismatch. *Nucleic Acids Res.*, **6**, 3543–3656.

Walter, M.H., Grimapettenati, J. and Feuillet, C. (1994) Characterisation of a bean (*Phaseolus vulgaris* L.) malic enzyme gene. *Eur. J. Biochem.*, **224**, 999–1009.

Watanabe, Y., Barbashov, S.F., Komatsu, S. et al. (1994) A peptide that stimulates phosphorylation of the plant insulin-binding protein—isolation, primary structure and cDNA cloning. *Eur. J. Biochem.*, **224**, 167–172.

Weglohner, W. and Subramanian, A.R. (1994) Multicopy CTPase centre protein L12 of *Arabidopsis* chloroplast ribosome is encoded by a clustered nuclear gene family with the expressed members closely linked to tRNA (Pro) genes. *J. Biol. Chem.*, **269**, 7330–7336.

Winning, B.M., Bourgignon, J. and Leaver, C.J. (1994) Plant mitochondrial NAD^+-dependent malic enzyme—cDNA cloning, deduced primary structure of the 59kDa and 62kDa subunits, import, gene complexity and expression analysis. *J. Biol. Chem.*, **269**, 4780–4786.

Wu, S.C., Kriz, A.L. and Widholm, J.M. (1994) Molecular analysis of 2 cDNA clones encoding acidic class-1 chitinase in maize. *Plant Physiol.*, **105**, 1097–1105.

Zhang, H.B., Martin, G.B., Tanksley, S.D. and Wing, R.A. (1994a) Map-based cloning in crop plants—tomato as a model system. 2. Isolation and characterisation of a set of overlapping yeast artificial chromosomes encompassing the *jointless* locus. *Mol. Gen. Genet.*, **244**, 613–621.

Zhang, S.H., Lawton, M.A., Hunter, T. and Lamb, C.J. (1994b) ATPK1, a novel ribosomal protein kinase gene from *Arabidopsis*. 1. Isolation, characterisation and expression. *J. Biol. Chem.*, **269**, 17586–17592.

Zurek, D.M. and Clouse, S.D. (1994) Molecular cloning and characterisation of a brassinosteroid-regulated gene from elongating soybean (*Glycine max* L.) epicotyls. *Plant Physiol.*, **104**, 161–170.

6 Differential Screening

PAOLO A. SABELLI

University of Bristol, Bristol, UK

The isolation of plant genes by differential screening has had a major impact in plant molecular biology studies, often resulting not only in the cloning of differentially regulated genes, but also in the identification of the regulating *cis*-elements and *trans*-acting factors. The following issues are addressed in this chapter:

(1) Importance of the regulation of gene transcription in plant biology.
(2) The description and discussion of a standard differential screening method.
(3) An evaluation of improved procedures for higher sensitivity and practicality.
(4) The description of recent alternative approaches to differential screening.
(5) The characterisation of differentially regulated genes.
(6) An overview of plant molecular biology studies in which cloning by differential screening has played a key role.

1 INTRODUCTION

The isolation of nucleic acid sequences from a library by means of differential hybridisation with complex nucleic acid probes (differential screening) is a powerful technique which has been extensively used in both plant and non-plant molecular biology studies. The principle is based on differences in the concentration of nucleic acid species between two or more samples. Under given conditions, the intensity of hybridisation signal between a labelled probe and the complementary target sequences depends on their relative abundancies. Differential screening uses this property to identify nucleic acid sequences which are present at different concentrations in different biological sources. As, generally, messenger RNAs are the nucleic acid sequences studied by this technique (as cDNAs), differential screening aims at isolating differentially transcribed mRNAs. Differential screening represents, therefore, a means to isolate nucleic acid sequences on the basis of a common regulatory mechanism rather than their identities or functions. The advantage of such an approach is that no information on specific nucleic acid sequences is required, as their isolation is based solely on their different concentrations in the two (or more) samples.

In the vast majority of cases, differential screening has been used to identify

Present Address: *National University of Singapore, Singapore.*

Plant Gene Isolation: Principles and Practice. Edited by G. D. Foster and D. Twell.
© 1996 John Wiley & Sons Ltd.

cDNA clones that reflect mRNAs present at different levels in different cell types (Sargent, 1987). The amount of a given mRNA in the cell is controlled at various levels, such as transcription, RNA processing and transport, and mRNA stability. These, together with translational and post-translational controls, contribute to the overall control over gene expression which is reflected in the phenotypic expression brought about by the synthesised proteins. Of the controls which regulate the levels of most mRNAs in a cell, the control at the transcriptional level is paramount. Thus, by comparing the levels of mRNA concentrations in two samples we are mainly looking at levels of gene transcription. Throughout this chapter, differentially transcribed mRNAs will be referred to as the subjects of cloning by differential screening. However, it should be noted that, from a rigorous point of view, this is incorrect. Indeed, differential screening can only detect general variations in the steady-state levels of mRNAs, which are the result of different controls operating at different levels (see above).

Control over gene expression is a dominant theme in biology. As virtually all of an organism's cells possess the same genetic content, it is by the differential modulation of gene expression that individual cells or groups of cells differentiate, specialise and become devoted to perform a specific set of functions. Differentiation is a generally irreversible process in animals but plant cells can, under opportune conditions, dedifferentiate, multiply and differentiate again into a new individual's components. In addition, cells react to external and internal stimuli by activating or repressing particular sets of genes. This may be even more true for higher plants, which are sessile organisms and, therefore, must respond and adapt quickly to the surrounding environmental conditions. All these remarkable modifications in cell structure and function are, to a large extent, brought about by different sets of proteins being synthesised, which, in turn, are the result of different sets of genes being transcribed. Thus, there is no doubt that regulation of gene expression, and in particular gene transcription, plays a central role in development, differentiation and response to stimuli in eukaryotic cells.

For example, in tobacco, each organ (leaf, root, stem, petal, ovary and anther) has 24 000–27 000 different mRNAs. Of these, 6000–11 000 (25–40%) are organ-specific and about 8000 (30–33%) are common to all organs (Kalamay & Goldberg, 1980, 1984). It is obvious that among the organ-specific genes there are genes encoding proteins which may play important roles in development and organogenesis (this without considering the 5000–13 000 mRNAs shared by two or more organs). Differential screening is a technique that allows the analysis of differentially regulated gene transcription and the cloning of mRNAs which are differentially transcribed between two or more samples. These samples may be different cell types, tissues, organs, the same cells which have been subjected to different treatments, or cells which have a different metabolic or physiological status. Before the advent of recombinant DNA technology, patterns of mRNA complexity were studied by RNA-excess/single-copy DNA hybridisation techniques (Kalamay & Goldberg, 1980, 1984), and by polyacrylamide gel electrophoresis analysis of in vitro translated mRNA populations. Even when coupled to two-dimensional gel separations, not more than the most abundant $1-2 \times 10^3$ polypeptides could be resolved and it was rarely possible to recover enough proteins to carry out further analyses and to isolate the corresponding mRNAs and genes. Thus, these techniques were characterised by a low sensitivity and were mainly descriptive. Molecular cloning and differential screening have made the analysis and isolation of differentially transcribed genes possible.

Generally, in differential screening experiments, cellular RNA is extracted from two samples (Fig. 1). Poly(A)$^+$ RNAs are then purified from the total RNAs and the poly(A)$^+$ RNA fraction obtained from one sample (+) is used as a template for the synthesis of the corresponding cDNA, which is then cloned into a plasmid or bacteriophage vector. The cDNA library is then plated at a relatively low density to facilitate subsequent identification of individual clones by colony (Grunstein & Hogness, 1975) or plaque (Villarreal & Berg, 1977) hybridisation. Two replica filters are taken from this master plate and hybridised independently to ^{32}P-labelled first-strand cDNA probes obtained by reverse transcription of the mRNA fraction used to synthesise the library (+), and the mRNA fraction purified from the other sample (−). After autoradiography, clones which show different intensities of hybridisation signal with the two probes identify mRNAs for differentially transcribed genes (Fig. 2). These clones are then selected from the master plate and usually subjected to a second round of differential hybridisation to confirm the results obtained in the first round and eliminate artefacts. Finally, the selected clones are characterised.

Differential screening was first used to identify DNA sequences which are induced when yeast (*Saccharomyces cerevisiae*) cells are grown on galactose-containing medium (St John & Davis, 1979). Shortly after, this technique was implemented in plant biology, and a cDNA, encoding the precursor of the small subunit of chloroplast ribulose 1,5-biphosphate carboxylase, was isolated from pea (Bedbrook et al., 1980). Since its introduction, differential screening has been extensively used and has proved to be particularly useful in tackling novel biological problems on which little or no molecular information was available.

Differential screening has also been used to directly identify genomic clones in yeast (Clancy et al., 1983), with the added benefit of the isolation of important regulatory sequences involved in the differential regulation of gene transcription. In this, a genomic library is screened with two different cDNA pool probes. Differentially hybridising clones identify genes (and not cDNAs as in conventional differential screening experiments) which are differentially transcribed. In plants, this approach has had only limited use because of their large genome rich in non-coding sequences; therefore, an impractically high number of clones should be screened compared to the screening of cDNA clones. However, there are cases in which relatively abundant transcripts have been isolated as genomic clones from the weed *Arabidopsis thaliana* (Simoens et al., 1988), which is characterised by a very small genome, or when the analysis of differentially transcribed mRNAs could not allow the identification of

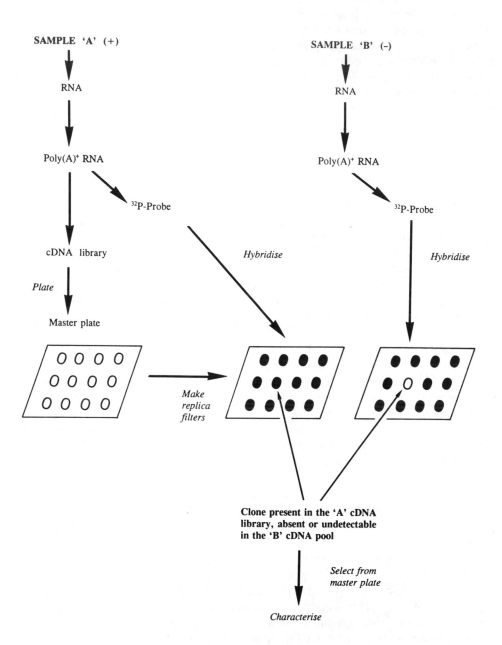

species-specific clones, such as in the case of pathogen genes induced *in planta* following infection (Pieterse et al., 1993a). Also, a cluster of three members of a pollen-specific gene family has been isolated from *Brassica napus* by differential screening of a genomic library (Albani et al., 1990). In addition, differential hybridisation has been used as a complementary technique for the cloning of genetically characterised mutations from maize (McLaughlin & Walbot, 1987). In mammals, differential screening using genomic probes has been shown to be a valid approach for cloning DNA sequences which are amplified in the genome and may be involved in phenomena such as drug resistance, transformation and differentiation (Brison et al., 1982).

In this chapter, a standard method of cDNA library screening by differential hybridisation is outlined and the advantages and limitations of the technique are discussed together with improved protocols for sensitivity and practicality. Some of the applications in studies of plant molecular biology are reviewed and examples given based on our studies on the molecular dissection of root meristem organisation in maize.

2   METHOD

A schematic diagram of the usual steps followed for the isolation of nucleic acid sequences by differential screening of a cDNA library is given in Fig. 1. The description of the procedures for cDNA library construction and handling will not be given, as it is the subject of Chapter 2.

2.1   Preparing replica filters

Currently, the majority of cDNA libraries are constructed in bacteriophage lambda DNA-derived vectors. These vectors offer several advantages over plasmid vectors, including a higher level of representation of the nucleic acid sequence complexity present in the sample from which the library is made (Chapter 2).

Figure 1. Diagrammatic representation of procedure for the isolation of transcripts by differential screening of a cDNA library. Two samples are used for the extraction of cellular RNA; a sample designated 'A', in which one or more genes of interest are transcribed (indicated by (+)), and a sample designated 'B', which differs from 'A' by the lack of these mRNAs (indicated by (−)). Usually, poly(A)+ RNAs are purified from both samples. The poly(A)+ RNA from 'A' is then used to construct a cDNA library which is plated at a relatively low density onto one or more master plates. Two replica filters are then prepared from the master plate(s). ^{32}P-labelled single-stranded cDNA probes are prepared from both 'A' and 'B' poly(A)+ RNA pools. These probes are hybridised to the two replica filters independently. Following autoradiography, the clones detected by the 'A' probe, but not by the 'B' probe, are candidates for being specifically (or preferentially) transcribed in sample 'A'. They can be recovered from the master plate(s) and characterised. Often, a second round of differential hybridisation is needed to confirm the results obtained after the first round and to discard false positives

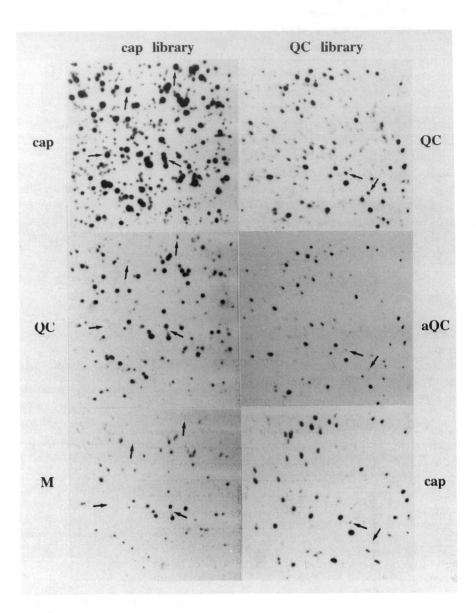

Figure 2. Example of identification of putative cDNAs specifically (or preferentially) transcribed in the cap or quiescent centre (QC) of the maize root apex by differential plaque hybridisation. Aliquots of PCR-cDNA libraries from the root cap (left) and the quiescent centre (right) were plated at low density and hybridised to cap and QC PCR-cDNA probes (labelled by random priming), respectively, as positive controls. Hybridisation of the same (or replica) filters with PCR-cDNA probes from the meristem (M) or activated quiescent centre (aQC), as indicated at the sides, allows the identification of mRNAs preferentially or specifically transcribed in the cap and QC. From Sabelli *et al.* (1993). Reprinted by permission of Kluwer Academic Publishers

(1)   Once the cDNA library is constructed, it is necessary to estimate its titre. This can be conveniently accomplished by plating serial dilutions of the library and by calculating the average number of plaque-forming units (PFUs) obtained per unit of volume of the original library. Then, an appropriate dilution of the library titre is made which will give a relatively low density of plaques ($3-7\ PFU/cm^2$) (approximately 1500–4000 PFU for a square 23-cm Nunc plate) in order to facilitate the identification of individual hybridising clones. Because each plate contains relatively widely spaced plaques, it is advisable to let the plaques grow for longer than would normally be done for library screening. This can be accomplished by letting the plaques grow overnight at 37 °C. Large, discrete plaques result in a larger than normal amount of target DNA being available on the filter for hybridisation with the probe and, therefore, in an increased autoradiographic signal. This is particularly important when using heterogeneous probes, as in the case of differential screening, for the actual concentration of the hybridising probe sequences is low due to the dilution of sequences in the pool.

(2)   From each plate, two or more (up to 5–6) filters are taken to allow parallel hybridisation of the plated library aliquot to two (or more) different probes. In order to facilitate this DNA blotting step, it is important to harden the top agar/agarose layer by incubating the plates at 4 °C for several hours. This prevents the top layer from sticking to the filter (which could result in high background) and disturbing the bacterial lawn, causing loss of plaques or cross-contamination of clones. As the identification of positive clones relies entirely on differences in signal intensities with the different probes, it is crucial to ensure that equal or nearly equal amounts of DNA from each plaque are transferred to the replica filters. This can be achieved by taking particular care both in placing the filters on the plate and in peeling them off after the time allowed for DNA transfer to occur, and by allowing approximately 1 min longer for each subsequent lift. Thus, if for the first lift one allows 1 min, for the second lift allow 2 min and so on. Some investigators have observed considerable changes in hybridisation intensities between the first and subsequent lifts, while there is little variation among subsequent lifts. This has been ascribed to the considerable differences in the amounts of DNA transferred between the first lift and the others and, therefore, to minimise this problem, the first lift could be discarded and only the subsequent lifts used.

   Another way to minimise artefacts due to differences in DNA transfer is to take only one lift and to hybridise it to the two different probes in two successive rounds of hybridisation/signal detection/probe stripping. This procedure has the disadvantage of taking twice as long to get to the final stage of identifying interesting clones; in the case of screening with several different probes, it would make it difficult to maintain the master plate free from contaminating fungi, which could even result in the impossibility of selecting positive clones from the plate for further analyses. For this reason, in the case of multiple rounds of hybridisation of a set of filters or if low-activity probes or low-performance hybridisation protocols are used, resulting in long autoradiographic exposure times between each round, it is sensible to maintain the master plate as free as possible from microbiological contamination when taking the lifts. This can be conveniently achieved by using autoclaved filters and forceps and working in a sterile flow-cabinet.

(3) It is important to make reference marks on the filter for its orientation and the interpretation of the autoradiographic patterns obtained when overlapping it with the master plate. One simple method is to make three asymmetric holes in each filter with a needle (before autoclaving) and to mark a specific code on each filter with a pencil (e.g. 1A, 1B, 2A, 2B, etc., where 1 and 2 are the plates' numbers and A and B represent the two probes used). During the lift, marks corresponding to the positions of the holes on the filters can be made on the bottom of the plate with a marker pen (different colours can be used for each replica filter taken from the same plate). After hybridisation and autoradiography, the holes usually give unmistakable, sharp dot-shaped signals which can be used to orient the autoradiograph with respect to the different replica filters taken and finally to the coloured marks on the plates in order to identify and isolate clones of interest. In differential screening experiments, the orientation and matching of the plate and autoradiograph are facilitated by the relatively simple hybridisation patterns due to the low-density plating of the library.

(4) Once the filter has been carefully lifted from the plate, DNA should be denatured with alkali, neutralised and irreversibly bound to it. This can be done by sequentially incubating the filter for 5–7 min each time (DNA-side up) on filter paper soaked with 0.5 M NaOH, 1.5 M NaCl, then 0.5 M Tris-HCl, pH 7.5, 1.5 M NaCl, and finally 2× standard saline citrate (SSC) buffer (0.3 M NaCl, 0.03 M sodium citrate, pH 7.0). Irreversible binding of DNA is achieved in the case of nitrocellulose filters by baking the filter at 80 °C for 2 h under vacuum, and for nylon filters (e.g. Hybond or equivalent) by wrapping them in Cling Film and irradiating on a long-wave transilluminator DNA-side down for 2–3 min (we also then bake the filters for 1 h without a vacuum). The master plate should be stored at 4 °C until needed for the selection of positive clones at the end of the screening procedure.

2.2 Hybridisation and post-hybridisation

Probe preparation, hybridisation, washing and signal detection are virtually as in other filter hybridisation experiments, such as Southern hybridisation. The major difference is that, in differential hybridisation, probes consist of heterogeneous pools of sequences, resulting in reduced sensitivity. Emphasis, therefore, must be given to labelling techniques which result in increased specific activity and sensitivity. For this reason, isotopic labelling and detection procedures are preferred.

Early experiments included ^{32}P-5'-end-labelling of RNA probes (Bedbrook et al., 1980). This had the disadvantage of being a low-specific-activity labelling procedure insufficient for the detection of non-abundant mRNAs. Also, nuclear run-off RNAs have been used as hybridisation probes (Somssich et al.,

1989). However, the vast majority of cloning experiments by differential screening have used first-strand cDNA probes reverse-transcribed from poly(A)$^+$ RNA in the presence of ^{32}P-labelled dNTPs. cDNAs synthesised in this way faithfully reflect the sequence complexity of the original mRNA pools from which they are synthesised and, therefore, are valuable probes to identify differentially expressed mRNAs. Several modifications which have recently been introduced, particularly to increase the sensitivity of the method, include random-priming labelling of double-stranded cDNAs, the use of vector sequences as probes to amplify the signal, and the amplification in vivo or in vitro of cDNA probes; these will be discussed later. The conditions, parameters and protocols for nucleic acid hybridisation have been described and discussed in detail elsewhere (Sambrook et al., 1989; Sabelli & Shewry, 1993).

After differentially hybridising clones have been identified, they are isolated from the master plate. An easy way is to collect an agar plug, containing the plaque of interest, with the wide end of a pipette tip and to transfer it to a 1.5-ml Eppendorf tube containing 0.5 ml of sterile SM buffer (100 mM NaCl, 15 mM MgSO$_4$, 50 mM Tris-HCl, pH 7.5, 0.01% gelatin) and a drop (20–50 μl) of chloroform (to kill bacteria).

Normally, not all clones which appear to be differentially transcribed after one round of screening are indeed so. A major reason for obtaining false positives resides in the differences in the efficiency of DNA transfer from the master plate to the replica filters. Thus, a second round of screening is usually necessary to narrow the range of the clones to be analysed. Although this may sound cumbersome, it is definitely less labour-intensive than having to characterise a large number of clones which are not, in fact, differentially regulated. Thus, each resuspended plaque is plated at low density on a 9-cm Petri dish, replica filters prepared and screening carried out as for the first round. Before plating, the PFU titre of each resuspended plaque should be estimated. As a rough guide, plating 10–50 μl of a 10^{-4} dilution of the resuspended plaque gives the low-density plating required for the screening. If a large number of clones has to be processed and the possibility of missing out some of them is not of concern, several resuspended plaques could be pooled and an aliquot of the pool plated and screened. Alternatively, a highly accurate and fast way to carry out the secondary screening is to screen by Southern blotting inserts amplified by the polymerase chain reaction (PCR) (see below).

3   IMPROVED PROCEDURES

As mentioned in Section 1, the main limitation to gene cloning by differential screening is its low sensitivity. This is due to the heterogeneous pool of probe sequences resulting in very low actual concentrations of hybridising sequences. A typical higher eukaryotic cell contains 10000–30000 different mRNAs (Alberts et al., 1994). It is generally thought that the detection of cDNA clones corresponding to rare mRNA species is beyond the sensitivity of this technique. There are about 5–20 molecules of each rare mRNA in the cell, equivalent to less than 0.1% of total mRNAs, but about one-third of all different mRNAs belongs to this class of abundance. Thus, rare mRNAs represent about 30% of all the different mRNA sequences. The relatively low sensitivity of the technique causes two major problems. First, a large portion of genes (those encoding rare mRNAs) appears to be outside its reach. Second, multiple identical clones are often isolated which correspond to the most abundant mRNAs. Another drawback is that this technique is demanding in the amounts of labour and time required. In this section, several procedures will be discussed that can be followed to optimise the technique, thereby increasing sensitivity and reliability, and to minimise the work input.

3.1   Choice of samples

The choice of plant material is important. Samples should be carefully chosen to maximise the probability of isolating differentially transcribed mRNAs which are relevant to the biological problem studied. Techniques are available to analyse differential gene transcription between small samples, such as two discrete populations of cells. PCR amplification can be used to construct representative cDNA libraries from RNA extracted from a few cells (Gurr & McPherson, 1991; Sabelli et al., 1993; Bertioli et al., 1994). This is made possible by quantitative or semiquantitative amplification by PCR (Chapter 13). PCR-amplified cDNA libraries can, in fact, closely reflect the mRNA complexity in vivo. Thus, by using PCR, cDNA libraries can be constructed from two small sets of cells differing only in a few characteristics, and there is a good chance of identifying relevant differentially regulated clones. On the other hand, differential screening of very different tissues may provide a large number of cDNAs (which may be useful, for example, as developmental markers), but it could be difficult to gain further clues as to their functions.

The choice of the RNA template for constructing the library and/or to synthesise probes could be important for optimising the procedure. Only the mRNA polysomal fraction is translated into proteins and should be the best template for cloning differentially expressed mRNAs. However, most mRNAs are regulated at the transcriptional level, and the poly(A)$^+$ RNA fraction is conveniently used in most differential screening experiments.

3.2   Hybridisation efficiency and signal intensity

Low sensitivity is the major limitation to cloning by differential hybridisation. How can the hybridisation efficiency be increased? There are three major factors that can be controlled by the investigator.

First, the amount of DNA transferred to the filter could be increased. Although the amount of filter-bound DNA is not usually a limiting factor in differential screening experiments, its increase is beneficial, especially in terms of reproducibility of the results and when increased concentrations of radiolabelled probes are used. Thus, besides plating at a low density and growing plaques for an extended time (see Section 2), the plaque size can also be increased by using thick agar plates (e.g. 25–40 ml of bottom agar/9-cm plate) (Ginn & Rapp, 1991).

Screening libraries by differential dot-blot hybridisation has been reported as a valuable approach to increase reliability and sensitivity. Conkling et al. (1990) identified rare root-specific transcripts by dot-blotting a small aliquot of station-ary overnight cultures of randomly picked recombinants from a root cDNA library of tobacco. Although the analysis of random clones can be tedious, this procedure did not require any plasmid purification and the overall speed of the method was acceptable. Other approaches to increase the amount of filter-bound DNA and reproducibility, based on the differential hybridisation of cDNAs on a single lift, have been developed which also allow the detection of marginal differences in gene transcription (Lemke et al., 1993).

Second, in filter hybridisation experiments, the hybridisation rate (and therefore the signal intensity) depends mainly on the probe concentration, its specific activity and the hybridisation conditions (Cochran et al., 1983). Thus, hybridisation conditions can be chosen to shift the equilibrium towards the formation of stable duplex molecules (Sabelli & Shewry, 1993). It is particularly useful to add high molecular weight polymers, such as polyethylene glycol (PEG), to the hybridisation buffer. These occupy a large proportion of the volume, resulting in a considerable increase in actual probe concentration. Also, the concentration of radioactivity could be increased (as compared to normal hybridisation experiments), but we usually do not exceed 10^6 counts/min per ml in PEG-containing hybridisation buffer in order to minimise background signal.

Considerations regarding stringency conditions are less important in dif-ferential than in normal library screening, as signal differences reflect different concentrations of nucleic acid rather than sequence similarity. Hence, strin-gency conditions should be chosen mainly to maximise specific hybridisation versus background. However, under low-stringency conditions, related but not identical sequences may cross-hybridise (e.g. cDNAs encoded by multigene families) and it could become impossible to identify members which are differentially transcribed. In our experiments of differential screening of PCR-cDNA libraries from the maize root apex, high-stringency conditions were found to give the best results (Figs 2 and 3).

Third, both probe specific activity and hybridisation performance could be maximised. Because sensitivity is such an important factor in differential screening, ^{32}P-labelled probes are to be preferred, although non-radioactive labelling methods have also been successfully used (Etscheid et al., 1993). Most commonly, first-strand cDNA probes are obtained by reverse transcription of poly(A)$^+$ RNA. Usually this is limited by the low amounts of mRNA generally available and by the amount of radioactive precursors which have to be used in order to achieve a satisfactory high probe concentration during hybridisation, as these probes can hybridise to only one DNA strand of the target clone. Labelling double-stranded cDNAs by random priming yields more efficient

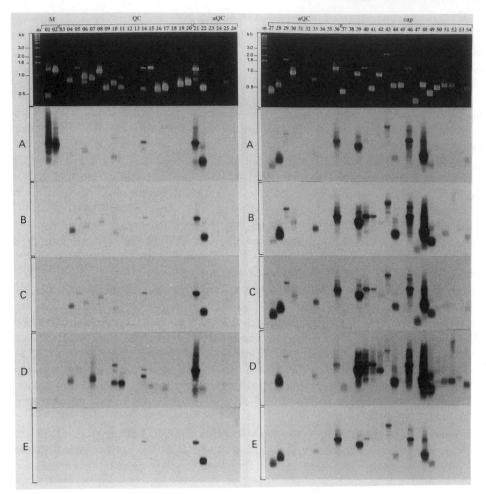

Figure 3. Southern hybridisation analysis of 54 putative cell population-specific or preferentially transcribed cDNA clones in the maize root apex. Two putative meristem-specific (M) clones, 18 quiescent centre-specific (QC) clones, 16 activated quiescent centre-specific (aQC) clones, and 18 cap-specific (cap) clones obtained by a first round of differential screening of PCR-cDNA libraries (Fig. 2) were PCR-amplified using primers flanking the vector's multicloning site. They were then size-fractionated by agarose gel electrophoresis, stained with ethidium bromide (top panel), blotted and hybridised to the different ^{32}P-labelled PCR-cDNA pool probes from the meristem (A), quiescent centre (B), activated quiescent centre (C), cap (D) and detaching cap cells (E)

probes which hybridise to both DNA strands with approximately double the sensitivity (Feinberg & Vogelstein, 1983) (Fig. 2).

Edwards et al., (1985) and Parfett et al., (1989), while studying changes in gene expression during the mammalian cell cycle, used in vivo amplified cDNA libraries as templates for obtaining unlimited quantities of probes. Using high concentrations of probes, they were able to show that rare nucleic acids (corresponding to as little as 0.003% of the total probe) could be detected.

A dual-labelling method using ^{35}S- and ^{32}P-labelled probes has been used to identify cDNAs corresponding to mRNAs that are potentially gibberellin regulated in the flowers of tomato plants (Olszewski et al., 1989). In this procedure, clones that hybridise to RNAs that are differentially transcribed are identified using differential autoradiography/fluorography to discriminate between the ^{32}P and ^{35}S isotopes. The main advantage of this method is that it does not require the production of duplicate replica filters and, therefore, is not subject to the screening artefacts due to uneven transfer of DNA to the different replica filters.

Vector sequences can also be used as probes to identify cDNAs representing rare mRNA species. Boll et al. (1986) developed a two-step 'sandwich' hybridisation procedure using ^{35}S-labelled vector DNA as a probe to amplify the signal. In the first step, filter-bound DNA is hybridised to saturation with non-radioactive cDNA probes cloned in a single-stranded vector. Then, the filter is hybridised with a ^{35}S-labelled probe specific to the vector DNA. In order to visualise differentially hybridising clones, the procedure is carried out in parallel using non-radioactive cDNAs from both samples. In this way, the signal is highly amplified and sharpness is ensured by using ^{35}S-probes. mRNAs as rare as 1 mRNA in 15 000 were isolated (corresponding to a level of about 35 mRNA molecules per cell), which could not be detected by means of conventional differential hybridisation. Obviously, when using probes containing vector sequences, it is crucial that such sequences do not hybridise to the cloning vector used to construct the cDNA library, which is present in every colony or plaque. Hybridisation specificity should be due only to cDNA sequences, labelled vector sequences being a means of enhancing an otherwise very weak or undetectable signal. This implies that, in the above experiments, cDNAs should be cloned into two heterologous vectors, one for plating the library and preparing lifts, and the other to generate radioactive probes. This can complicate the construction and maintenance of cDNA libraries.

If the amplified libraries are constructed in transcription vectors, then single-stranded RNA probes can be used to replace first-strand cDNA probes. Lu et al. (1990) showed that cDNA libraries prepared from cytoplasmic RNA not enriched in poly(A)$^+$ RNA are suitable for differential screening experiments. This can be particularly useful if only small amounts of the starting material are available.

3.3   Selecting non-hybridising clones

In differential screening, clones which do not hybridise to either probe are usually discarded. However, such clones could represent important low-abundance mRNAs. 'Cold-plaque' screening is a simple method which has been used to identify cDNAs from rare mRNAs which exhibit anther-specific transcription in oilseed rape (Hodge et al., 1992). This procedure, which could be applied to most differential screening studies, consists of isolating non-hybridising ('cold') clones as candidates for low-level-transcribed genes. Transcription specificity is then analysed by using the selected cDNAs to probe diagnostic northern dot-blots containing RNAs from different samples. This method, which is made relatively fast by PCR-amplifying the selected cDNAs and by purifying them on a low-gelling-temperature agarose gel before labelling, represents a clever adaptation of differential screening to the isolation of scarce mRNAs.

3.4 PCR-assisted screening

PCR could also be used to synthesise large quantities of sequences as templates for producing labelled probes. The sequence complexity of cellular mRNAs can be maintained after careful PCR amplification of corresponding cDNAs. PCR-amplified complex probes have been used to detect differentially transcribed mRNAs in the root apex of maize (Sabelli et al., 1993) and in *Chenopodium quinoa* plants infected with arabis mosaic virus (Bertiolo et al., 1994).

PCR technology can also considerably improve both accuracy and speed of the second round of differential screening. Selected cDNAs can be amplified by PCR, gel separated and analysed by replica Southern blots using the same probes as for the first round of screening (Mutchler et al., 1992). We have used this approach to further select the number of putative cell population-specific clones isolated during the first screening of maize root apex PCR-cDNA libraries (Sabelli et al., 1993). As shown in Fig. 3, 54 clones were isolated by a first round of differential screening as cell population specific. After Southern hybridisation of PCR-amplified inserts with the different probes, a dozen cDNAs were further selected based on transcription specificities and sizes. For example, clones No. 1 and 2 appear to be meristem-specific, and clone No. 11 cap-specific. Several cDNAs could not be visualised by PCR amplification (Nos 3, 12, 23, 24, 31, 32, 34 and 38), suggesting that they were nonrecombinant (or recombinant with a small insert), while others gave two bands, indicating cross-contamination (or PCR-mispriming artefacts) (Nos 6, 10, 11, 14, 16 and 41). Nine clones were selected and hybridised to corresponding root tip mRNAs (Fig. 4). Comparison of the hybridisation intensities in relation to the different stringency conditions and autoradiography exposure times suggested that their levels varied by 2–3 orders of magnitude and that both abundant and relatively non-abundant root tip mRNAs have been isolated by differential screning.

Other instances of this application have been reported by Luo et al. (1994) and Thomas et al. (1994). This method is indeed highly recommended. Its advantages are speed (one or several rounds of differential screening can be replaced by a differential Southern blot hybridisation), and accuracy (colony/plaque hybridisation is replaced by a highly reproducible PCR-DNA Southern hybridisation, where cross-contaminating clones can be distinguished, and the sizes of the cDNAs estimated directly).

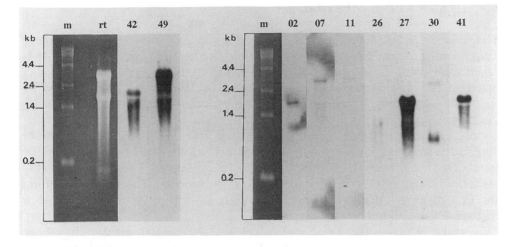

Figure 4. Northern blot analysis of cDNA clones selected for their specificity of gene transcription (Figs 2 and 3). Both relatively abundant and rare mRNAs expressed in the maize root apex could be detected among selected cell population-specific clones. Nine positive ^{32}P-labelled cDNA clones (indicated by numbers as in Fig. 3) were used as probes with root tip RNA (rt) replicates (stained with ethidium bromide on the left). Membranes were washed and exposed as follows: 02, 07, 11, 26, and 30, 1 × SSC at 65 °C for three days; 27, 41, 42, and 49, 0.5 × SSC at 65 °C, overnight. Lane m is the RNA ladder size marker (Gibco). Reproduced in part from Sabelli & Shewry (1995) by permission of Humana Press Inc.

Differential library screening by Southern blotting of PCR-amplified cDNA clones is so fast and accurate that some investigators have successfully used it in place of a conventional first round of screening. Tagu et al. (1993) have isolated symbiosis-related cDNAs from eucalypt mycorrhiza by Southern analysis of randomly picked clones which have been amplified by PCR. In this case, the labour-intensive and time-consuming first step of analysing a relatively large number of random clones has been compensated for by the speed and reliability of the following PCR-mediated screening steps.

3.5 Signal evaluation

Traditionally, the evaluation of hybridisation signal specificity is done by visual inspection of autoradiographs. Although it is generally straightforward to select clones based on the presence or absence of signal, it becomes difficult to distinguish between subtle differences, often resulting in the selection of a large number of putatively interesting clones whose specificity of transcription is artefactual. Advanced digital image processing systems have been shown to be highly accurate and sensitive for the evaluation of marginal differences in gene expression (Lu et al., 1990). The use of phosphor imaging technology could also result in a significant increase in the speed of the detection of the hybridisation signal.

4 ALTERNATIVE METHODS

Besides differential screening, other techniques are available for the isolation of differentially transcribed genes. Subtraction cloning (Hedrick et al., 1984) is based on cloning from normal or enriched cDNA or genomic libraries using enriched cDNA or genomic probes (Chapter 14). Enrichment is achieved by means of eliminating from the screening procedure sequences common to two samples. Usually, the final step of subtraction cloning is by differential hybridisation, but because the sequence complexity is greatly reduced in the probe or the library screened, or both (and, most importantly, limited to sample-specific sequences), the overall sensitivity of the technique is very high and the isolation of rare nucleic acid sequences can be routine. The original method and improved variations are being increasingly used in plant molecular biology and are discussed in detail in Chapter 14.

Recently, a novel technique called 'mRNA differential display' has been developed for the isolation of scarce mRNAs which are differentially regulated between two or more systems (Liang & Pardee, 1992). The method is based on PCR synthesis, in the presence of radiolabelled dNTPs, of cDNAs from subsets of the cellular mRNA population, using one oligo(T) 3' primer annealed to the polyadenylated tail of a subset of mRNAs and another short oligonucleotide of arbitrary sequence (Fig. 5). By the use of different sets of primers, virtually all the mRNAs expressed in a sample can be analysed. The PCR products from different samples are then separated side by side on a sequencing gel and the patterns compared. Unique autoradiographic bands identify specifically expressed cDNAs that can be easily eluted from the gel, and either characterised or used to probe a cDNA or genomic library.

The advantages of this method are its sensitivity (cDNAs corresponding to at least as low as 30 mRNA copies per cell can be isolated) and speed (the entire procedure can be completed in a few days). In addition, unlike differential and subtractive hybridisation, mRNA differential display allows the simultaneous detection of differentially transcribed genes in both samples. Clone redundancy and false positives should also be minimised. Thus, this technique has the potential to become a valuable alternative to differential screening and subtraction hybridisation in studies of differential gene transcription, and commercial kits are becoming available. Improved versions of the original method have already been applied to plants to isolate senescence-related genes from *Arabidopsis thaliana* cell suspension cultures (Callard et al., 1994).

5 CHARACTERISATION OF ISOLATED CLONES

Generally, differential screening experiments result in the isolation of a relatively large number of clones. How can these clones be characterised? In this section, some methods will be briefly referred to which could be followed to gain further information about the clones isolated. Although every cDNA may encode different and important proteins, for reasons of practicality it is important to further select for clones representing unrelated sequences. Similar cDNAs could be characterised subsequently.

In an initial step of characterisation, the cDNAs should be cross-hybridised and grouped into different hybridisation classes. This can be conveniently done by dot-blot hybridisation experiments (Clancy et al., 1983; Marco et al., 1990; Etscheid et al., 1993). The length of each cDNA could be estimated either by PCR amplification of the insert (Fig. 3) or by restriction analysis of individual recombinants, and the largest clones from each hybridisation class further selected.

The next step is to sequence the selected clones. A partial sequence (200–300 bp) can be sufficient to search nucleic acid and protein databases for related clones. This can be considerably simplified if the cDNA libraries have been constructed by directional cloning. It would be possible, therefore, to predict which cDNA strand is the coding strand, and the search for open reading frames will be restricted to three reading frames rather than six.

Southern blotting analysis would provide information on whether the sequences correspond to single genes or to a gene family (Conkling et al., 1990; Marco et al., 1990). Northern blot experiments with RNAs extracted from different tissues/organs or, for example, from differentially treated cells would give information about the sizes of the corresponding mRNAs, and their patterns of temporal and/or spatial transcription in the plant (Theologis et al., 1985; van der Zaal et al., 1987; Marty et al., 1993). In situ hybridisation with in vitro transcribed RNA probes would give a finer characterisation of the transcription patterns (Koltunow et al., 1990; Drews et al., 1992; Cheung et al., 1993; Heintzen et al., 1994). By cloning the cDNAs into an expression vector, it is possible to obtain considerable quantities of the encoded proteins synthesised in vivo in a heterologous system. Thus, specific antibodies can be obtained which can be used in immunocytochemistry experiments to study the spatial and temporal patterns of protein accumulation (Mundy & Chua, 1988) and their subcellular locations (Rothbarth et al., 1993).

Information obtained at the protein level by one- or two-dimensional gel separations of in vivo or in vitro translation products can be complemented by the analysis of RNA hybrid-selected translation products corresponding to the cDNAs isolated by differential screening (Bedbrook et al., 1980; Fuller et al., 1983; Czarnecka et al., 1984; Slater et al., 1985; Galau et al., 1986). A transgenic approach can also be undertaken. Antisense experiments can downregulate the steady-state levels of specific mRNAs in vivo in plant cells (Gray et al., 1992). In addition, using constructs in which the cDNAs have been put under the control of constitutive, tissue-specific or inducible promoters, it should be possible to express them and to study the induced phenotypes.

Because cDNAs obtained by differential screening are characterised by a common regulatory mechanism, most likely at the transcriptional level, their isolation provides a means to isolate corresponding *cis*-regulatory sequences on the chromosome, which can be further characterised in transgenic experiments (Koltunow et al., 1990; Medford et al., 1991), and eventually the *trans*-acting factors (Chapter 7). Thus, the importance of clones isolated by differential

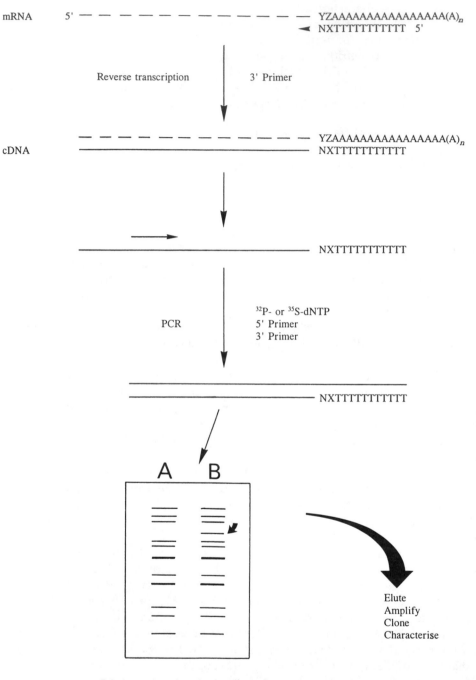

Gel electrophoresis and autoradiography

screening is extended beyond their identities and functions, and sequence motifs and enzymes which play crucial roles in the regulation of gene transcription in a given system could be identified. Both basic research and biotechnological applications could greatly benefit from the identification of regulatory factors, rather than the isolation of the regulated products.

6   DIFFERENTIAL SCREENING APPLIED TO PLANTS

Gene cloning by differential screening has been extensively used in plant molecular biology and can be considered a technique which has substantially contributed to our knowledge of plant gene transcription. Virtually every aspect of plant biology has benefited from cloning approaches based on the analysis of differential gene transcription. Thus, important new insights into plant development, reproduction, and responsiveness to phytohormones and to both biotic and abiotic external factors have been obtained by studying genes which have been isolated by differential screening. Some of the applications of this technique are listed in Table 1.

Differential screening was used originally for the isolation of a mRNA from pea encoding the small subunit of ribulose, 1,5-biphosphate carboxylase, one of the most abundant plant proteins (Bedbrook et al., 1980). Subsequent refinements of the technique and its application to a wide range of biological problems, particularly where it had been difficult to obtain molecular information, have resulted in a high number of reports in the last decade.

In the early and mid-1980s the first results on the isolation of cDNAs which are preferentially or specifically transcribed at the meiotic prophase of *Lilium* microsporocytes (Appels et al., 1982), or responsive to auxin in elongating soybean hypocotyls (Walker & Key, 1982) and pea epicotyls (Theologis et al., 1985), were published.

Stress-responsive genes were also identified. Czarnecka et al. (1984) reported the isolation of several soybean stress-related cDNAs which were induced either by heat shock or by other physical or chemical stress, such as water and salt stress, anaerobiosis and high concentrations of phytohormones.

Figure 5. Schematic flow chart of a typical mRNA differential display experiment. mRNAs are purified from two samples (A and B) and reverse-transcribed into first-strand cDNAs using a 3' primer which anneals to the polyadenylated tails of mRNAs. Because the primer has two specific bases at its 3' end (N and X), only a subset of the total mRNAs is reverse-transcribed. Double-stranded ^{32}P-labelled cDNAs are then synthesised and amplified by PCR using the same 3' primer, and a decamer of arbitrary sequence (5' primer, indicated by the horizontal arrow). cDNAs from the two samples are then separated on a sequencing gel and visualised by autoradiography. Bands present in only one sample represent specifically transcribed genes (indicated by the bold arrow). These PCR fragments can be eluted from the gel, reamplified, cloned and characterised. By using different combinations of 5' and 3' primers, virtually all mRNAs in a sample can be analysed. Abbreviations: N = A, C, G or T; X = A, C or G; Y and Z = nucleotides complementary to N and X, respectively. Modified from Callard et al. (1994) by permission of Eaton Publishing Co.

Table 1. Examples of the application of differential screening to the isolation of plant genes and cDNAs

Field of research	Plant species	Reference
Embryogenesis	*Brassica napus*	Harada et al. (1989)
	Daucus carota	Aleith & Richter (1990)
		Thomas & Wilde (1985)
		Wilde et al. (1988)
	Gossypium hirsutum	Galau et al. (1986)
	Helianthus annuus	Almoguera & Jordano (1992)
	Hordeum vulgare	Aalen et al. (1994)
		Bartels et al. (1991)
		Chen et al. (1994)
	Zea mays	Goday et al. (1988)
Hormonal regulation	*Cicer arietinum*	Colorado et al. (1994)
	Fragaria ananassa	Reddy et al. (1990)
		Reddy & Poovaiah (1990)
	Glycine max	Hagen et al. (1984)
		Walker & Key (1982)
	Hordeum vulgare	Hong et al. (1988)
	Lycopersicon esculentum	Lincoln et al. (1987)
		Olszewski et al. (1989)
	Nicotiana plumbaginifolia	Dominov et al. (1992)
	Nicotiana tabacum	Memelink et al. (1987)
		Takahashi et al. (1989)
		van der Zaal et al. (1987)
	Oryza sativa	Mundy & Chua (1988)
	Spirodela polyrrhiza	Smart & Fleming (1993)
	Pisum sativum	Theologis et al. (1985)
	Zea mays	Gómez et al. (1988)
Germination	*Brassica napus*	Harada et al. (1988)
Cell proliferation	*Catharanthus roseus*	Kodama et al. (1991)
	Nicotiana tabacum	Tatahashi et al. (1989)
Flower development	*Antirrhinum*	Baldwin et al. (1992)
	Brassica napus	Hodge et al. (1992)
		Roberts et al. (1991)
		Scott et al. (1991)
	Brassica oleracea	Nasrallah et al. (1985)
	Helianthus annuus	Herdenberger et al. (1990)
	Lilium	Kim et al. (1993)
	Lycopersicon esculentum	Gasser et al. (1989)
		Smith et al. (1990)
		Twell et al. (1989)
	Nicotiana alata	Anderson et al. (1986)

Table 1. *(cont.)*

Field of research	Plant species	Reference
	Nicotiana tabacum	Cheung et al. (1993) Drews et al. (1992) Goldberg (1988) Koltunow et al. (1990)
	Petunia hybrida	Brugliera et al. (1994)
	Zea mays	Larkin (1994) Wright et al. (1993)
Pollen development	*Brassica napus*	Albani et al. (1990)
	Lilium	Appels et al. (1982)
	Nicotiana tabacum	Weterings et al. (1992)
	Oenothera organensis	Brown & Crouch (1990)
	Petunia inflata	Mu et al. (1994)
	Tradescantia paludosa	Stinson et al. (1987)
	Zea mays	Rogers et al. (1991) Stinson et al. (1987)
Fruit development	*Lycopersicon esculentum*	Lincoln et al. (1987) Mansson et al. (1985) Pear et al. (1989) Slater et al. (1985)
Meristem development	*Brassica oleracea*	Medford et al. (1991)
	Sinapis alba	Staiger & Apel (1993)
	Zea mays	Sabelli et al. (1993)
Senescence	*Brassica napus*	Buchanan-Wollaston (1994)
	Dianthus caryophyllus	Lawton et al. (1989) Wang et al. (1993)
Root-specific transcription	*Nicotiana tabacum*	Conkling et al. (1990)
	Zea mays	Collazo et al. (1992) Montoliu et al. (1989)
Epidermis-specific transcription	*Pachyphytum*	Clark et al. (1992)
Tuber-specific transcription	*Solanum tuberosum*	Stiekema et al. (1988) Taylor et al. (1992)
Seed-specific transcription	*Helianthus annuus*	Almoguera & Jordano (1992)
	Hordeum vulgare	Aalen et al. (1994) Jakobsen et al. (1989)
Coleoptile-specific transcription	*Zea mays*	Stiefel et al. (1988)
Tissue-culture-specific transcription	*Arabidopsis thaliana*	Simoens et al. (1988)
	Catharanthus roseus	Kodama et al. (1991)

continued overleaf

Table 1. (*cont.*)

Field of research	Plant species	Reference
	Nicotiana tabacum	Marty et al. (1993) Takahashi et al. (1989)
	Zinnia elegans	Demura & Fukuda (1993)
Constitutive transcription	*Arabidopsis thaliana*	Alliotte et al. (1989)
Biotic-factor-specific transcription	*Arabidopsis thaliana*	Aufsatz & Grimm (1994)
	Eucalyptus globulus	Tagu et al. (1993)
	Glycine max	Franssen et al. (1987) Fuller et al. (1983)
	Hordeum vulgare	Brandt et al. (1992) Gregersen et al. (1993) Titarenko et al. (1993)
	Lycopersicon esculentum	Van der Eycken et al. (1993)
	Nicotiana tabacum	Hooft van Huijsduijnen et al. (1986) Marco et al. (1990)
	Oryza sativa	Talbot et al. (1993)
	Petroselinum crisoum	Somssich et al. (1989)
	Pisum sativum	Riggleman et al. (1985)
	Solanum tuberosum	Gurr et al. (1991) Harper et al. (1993) Marineau et al. (1987) Pieterse et al. (1991) Pieterse et al. (1993a) Pieterse et al. (1993b) Taylor et al. (1990)
Abiotic-factor-specific transcription		
Light-regulated transcription	*Arabidopsis thaliana*	Simoens et al. (1988)
	Lemna gibba	Okubara & Tobin (1991)
	Pisum sativum	Bedbrook et al. (1980)
	Sinapis alba	Heintzen et al. (1994)
Water-stress-induced transcription	*Arabidopsis thaliana*	Yamaguchi-Shinozaki et al. (1992)
	Craterostigma plantagineum	Bartels et al. (1990)
	Oryza sativa	Mundy & Chua (1988)
	Pisum sativum	Guerrero & Mullet (1988)
Cold-induced transcription	*Hordeum vulgare*	Cattivelli & Bartels (1990)
	Lycopersicon esculentum	Schaffer & Fischer (1988)
	Medicago sativa	Mohapatra et al. (1988)

Table 1. (*cont.*)

Field of research	Plant species	Reference
Heat-induced transcription	*Glycine max*	Czarnecka et al. (1984)
Salt-stress-induced transcription	*Lycopersicon esculentum*	Espartero et al. (1994) Godoy et al. (1990)
Wound-induced transcription	*Populus*	Parsons et al. (1989)
	Solanum tuberosum	Logemann et al. (1988)
Chemical/pollutant-induced transcription	*Picea abies*	Etscheid et al. (1993)
	Triticum aestivum	Snowden & Gardner (1993)
	Zea mays	Didierjean et al. (1993)

Among the initial fields of research in which this technique was successfully used was embryogenesis. Thomas & Wilde (1985) identified mRNAs regulated during carrot somatic embryo development, and Galau et al. (1986) isolated several late embryogenesis abundant (*Lea*) mRNAs from cotton.

Plant–microbe interactions were also studied by differential screening. Fuller et al. (1983) isolated five unique soybean sequences (leghemoglobin, *NodA*, *NodB*, *NodC* and *NodD*) which are specific to root nodules induced by *Rhizobium*. cDNAs corresponding to a range of mRNAs induced by *Fusarium* in pea (Riggleman et al., 1985) and by tobacco mosaic virus in tobacco (Hooft van Huijsduijnen et al., 1986) were also isolated. These and subsequent results of cloning by differential hybridisation (Table 1) represent a solid basis for the development of an increasingly important branch of biology which studies plant–pathogen interaction and signalling.

Early differential screening experiments have also triggered the understanding of two systems with high potential for important biotechnological applications, flower development and the biology of sexual reproduction, and the physiology of fruit development and ripening. Self-incompatible flowering plants are prevented from inbreeding by pollen–stigma interaction resulting in the synthesis of style glycoproteins which inhibit pollen tube growth. cDNAs encoding such proteins were first isolated by differential screening (Nasrallah et al., 1985; Anderson et al., 1986). Hybrid genes containing toxin-encoding sequences under the control of the promoter from style glycoprotein genes have been used to target cell ablation to the carpel in transgenic plants (review: Nasrallah et al., 1991). In an analogous approach, male-sterile transgenic plants have been obtained by using a tapetal-specific promoter (identified by differential screening) (Goldberg, 1988), fused to a ribonuclease gene. Both the industry of ornamental crops and the breeding of hybrid cultivars could greatly benefit from these advances.

In addition, the isolation of genes which are transcribed at different stages of tomato fruit development (Mansson et al., 1985) and ripening-related cDNAs

(Slater et al., 1985), together with other studies, has resulted in the eventual engineering of tomato plants producing fruits with a prolonged shelf-life (review: Leemans, 1993).

7   CONCLUDING REMARKS

The isolation of genes and cDNAs by differential screening has represented the first purpose-designed strategy for isolating nucleic acids which are under regulated control of gene transcription. This has triggered an explosion of molecular information on a variety of biological systems and has stimulated further research and new insights in the biology of yeast, animals and plants. Although differential screening is characterised by a generally high chance of success, its drawbacks are thought to be low sensitivity and a relatively high work input. In this chapter, I have tried to review current methods and applications of differential screening in plant molecular biology, describing improved techniques for high sensitivity, speed and practicality. I have also briefly mentioned some of the experiments that could be carried out for the characterisation of differentially transcribed mRNAs and genes, which I hope will be particularly useful to the novice.

Although more sensitive (subtraction cloning) and convenient (mRNA differential display) techniques have been recently developed, improved protocols for differential screening can be both sensitive and practical, and the basic strategy is still widely used some 15 years after its introduction. Specific genes can now be targeted by more sophisticated molecular cloning and molecular genetics techniques; however, differential screening is likely to remain a valuable tool in studies of differential gene transcription in any molecular biology laboratory.

ACKNOWLEDGMENTS

I thank Shirley Burgess and Peter Holloway for comments on the manuscript. Part of the work was supported by the Plant Molecular Biology Programme of the AFRC (grant No. PG 206/517).

ESSENTIAL READING

Bedbrook, J.R., Smith, S.M. and Ellis, R.J. (1980) Molecular cloning and sequencing of cDNA encoding the precursor to the small subunit of chloroplast ribulose-1,5-biphosphate carboxylase. *Nature*, **287**, 692–697.

Kalamay, J.C. and Goldberg, R.B. (1980) Regulation of structural gene expression in tobacco. *Cell*, **19**, 935–946.

Kalamay, J.C. and Goldberg, R.B. (1984) Organ-specific nuclear RNAs in tobacco. *Proc. Natl. Acad. Sci. USA*, **81**, 2801–2805.

Liang, P. and Pardee, A.B. (1992) Differential display of eukaryotic messenger RNA by means of the polymerase chain reaction. *Science*, **257**, 967–971.

St. John, T.P. and Davis, R.W. (1979) Isolation of galactose-inducible DNA sequences from *Saccharomyces cerevisiae* by differential plaque filter hybridization. *Cell*, **16**, 443–452.

REFERENCES

Aalen, R.B., Opsahl-Ferstad, H.-G., Linnestad, C. and Olsen, O.-A. (1994) Transcripts encoding an oleosin and a dormancy-related protein are present in both the aleurone layer and the embryo of developing barley (*Hordeum vulgare* L.) seeds. *Plant J.*, **5**, 385–396.

Albani, D., Robert, L.S., Donaldson, P.A. et al. (1990) Characterization of a pollen-specific gene family from *Brassica napus* which is activated during early microspore development. *Plant Mol. Biol.*, **15**, 605–622.

Alberts, A., Bray, D., Lewis, J. et al. (1994) *Molecular Biology of the Cell*. Garland Publishing Inc., New York.

Aleith, F. and Richter, G. (1990) Gene expression during induction of somatic embryogenesis in carrot cell suspensions. *Planta*, **183**, 17–24.

Alliotte, T., Tiré, C., Engler, G. et al. (1989) An auxin-regulated gene of *Arabidopsis thaliana* encodes a DNA-binding protein. *Plant Physiol.*, **89**, 743–752.

Almoguera, C. and Jordano, J. (1992) Developmental and environmental concurrent expression of sunflower dry-seed-stored low-molecular-weight heat-shock protein and Lea mRNAs. *Plant Mol. Biol.*, **19**, 781–792.

Anderson, M.A., Cornish, E.C., Mau, S.-L. et al. (1986) Cloning of a cDNA for a stylar glycoprotein associated with expression of self-incompatibility in *Nicotiana alata*. *Nature*, **321**, 38–44.

Appels, R., Bouchard, R.A. and Stern, H. (1982) cDNA clones from meiotic-specific poly(A)⁺ RNA in *Lilium*: homology with sequences in wheat, rye, and maize. *Chromosoma*, **85**, 591–602.

Aufsatz, W. and Grimm, C. (1994) A new, pathogen-inducible gene of *Arabidopsis* is expressed in an ecotype specific manner. *Plant Mol. Biol.*, **25**, 229–239.

Baldwin, T.C., Coen, E.S. and Dickinson, H.G. (1992) The *ptl1* gene expressed in the transmitting tissue of *Antirrhinum* encodes an extensin-like protein. *Plant J.*, **2**, 733–739.

Bartels, D., Schneider, K., Terstappen, G. et al. (1990) Molecular cloning of abscisic acid-modulated genes which are induced during desiccation of the resurrection plant *Craterostigma plantagineum*. *Planta*, **181**, 27–34.

Bartels, D., Engelhardt, K., Roncarati, R. et al. (1991) An ABA and GA modulated gene expressed in the barley embryo encodes an aldose reductase related protein. *EMBO J.*, **10**, 1037–1043.

Bertioli, D.J., Smoker, M., Brown, A.C.P. et al. (1994) A method based on PCR for the construction of cDNA libraries and probes from small amounts of tissue. *Biotechniques*, **16**, 1054–1058.

Boll, W., Fujisawa, J.-C., Niemi, J. and Weissmann, C. (1986) A new approach to high sensitivity differential hybridization. *Gene*, **50**, 41–53.

Brandt, J., Thordal-Christensen, H., Vad, K. et al. (1992) A pathogen-induced gene of barley encodes a protein showing high similarity to a protein kinase regulator. *Plant J.*, **2**, 815–820.

Brison, O., Ardeshir, F. and Stark, G.R. (1982) General method for cloning amplified DNA by differential screening with genomic probes. *Mol. Cell. Biol.*, **2**, 578–587.

Brown, S.M. and Crouch, M. (1990) Characterization of a gene family abundantly expressed in *Oenothera organensis* pollen that shows sequence similarity to polygalacturonase. *Plant Cell*, **2**, 263–274.

Brugliera, F., Holton, T.A., Stevenson, T.W. et al. (1994) Isolation and characterization of a cDNA clone corresponding to the *Rt* locus of *Petunia hybrida*. *Plant J.*, **5**, 81–92.

Buchanan-Wollaston, V. (1994) Isolation of cDNA clones for genes that are expressed during leaf senescence in *Brassica napus*. *Plant Physiol.*, **105**, 839–846.

Callard, D., Lescure, B. and Mazzolini, L. (1994) A method for the elimination of false positives generated by the mRNA differential display technique. *BioTechniques*, **16**, 1096–1103.

Cattivelli, L. and Bartels, D. (1990) Molecular cloning and characterization of cold-regulated genes in barley. *Plant Physiol.*, **93**, 1504–1510.

Chen, F., Hayes, P.M., Mulrooney, D.M. and Pan, A. (1994) Identification and characterization of cDNA clones encoding plant calreticulin in barley. *Plant Cell*, **6**, 835–843.

Cheung, A.Y., May, B., Kawata, E.E. et al. (1993) Characterization of cDNAs for stylar transmitting tissue-specific proline-rich proteins in tobacco. *Plant J.*, **3**, 151–160.

Clancy, M.J., Buten-Magee, B., Straight, D.J. et al. (1983) Isolation of genes expressed preferentially during sporulation in the yeast *Saccharomyces cerevisiae*. *Proc. Natl. Acad. Sci. USA*, **80**, 3000–3004.

Clark, A.M., Verbeke, J.A. and Bonhert, H.J. (1992) Epidermis-specific gene expression in *Pachyphytum*. *Plant Cell*, **4**, 1189–1198.

Cochran, B.H., Reffel, A.C. and Stiles C.D. (1983) Molecular cloning of gene sequences regulated by platelet-derived growth factor. *Cell*, **33**, 939–947.

Collazo, P., Montoliu, L., Puigdomènech, P. and Rigau, J. (1992) Structure and expression of the lignin O-methyltransferase gene from *Zea mays* L. *Plant Mol. Biol.*, **20**, 857–867.

Colorado, P., Rodríguez, A., Nicolás, G. and Rodríguez, D. (1994) Abscisic acid and stress regulate gene expression during germination of chick-pea seeds. Possible role of calcium. *Physiol. Plant.*, **91**, 461–467.

Conkling, M.A., Cheng, C., Yamamoto, Y.T. and Goodman, H.M. (1990) Isolation of transcriptionally regulated root-specific genes from tobacco. *Plant Physiol.*, **93**, 1203–1211.

Czarnecka, E., Edelman, L., Schöffl, F. and Key, J.L. (1984) Comparative analysis of physical stress responses in soybean seedlings using cloned heat shock cDNAs. *Plant Mol. Biol.*, **3**, 45–58.

Demura, T. and Fukuda, H. (1993) Molecular cloning and characterization of cDNAs associated with tracheary element differentiation in cultured *Zinnia* cells. *Plant Physiol.*, **103**, 815–821.

Didierjean, L., Frendo, P., Nasser, W. et al. (1993) Plant genes induced by chemicals and pollutants. In *Mechanisms of Plant Defense Responses* (eds B. Fritig and M. Legrand), pp. 276–285. Kluwer Academic Publishers, The Netherlands.

Dominov, J.A., Stenzler, L., Lee, S. et al. (1992) Cytokinins and auxins control the expression of a gene in *Nicotiana plumbaginifolia* cells by feedback regulation. *Plant Cell*, **4**, 451–461.

Drews, G.N., Beals, T.P., Bui, A.Q. and Goldberg, R.B. (1992) Regional and cell-specific gene expression patterns during petal development. *Plant Cell*, **4**, 1383–1404.

Edwards, D.R., Parfett, C.L.J. and Denhardt, D.T. (1985) A pBR322-derived vector for cloning blunt-ended cDNA: its use to detect molecular clones of low-abundance mRNAs. *DNA*, **4**, 401–408.

Espartero, J., Pintor-Toro, J.A. and Pardo, J.M. (1994) Differential accumulation of S-adenosylmethionine synthetase transcripts in response to salt stress. *Plant Mol. Biol.*, **25**, 217–227.

Etscheid, M., Buschmann, K., Köhler, R. et al. (1993) Differential screening in a cDNA-library from spruce for clones associated with forest decline reveals accumulation of ribulose-1,5-biphosphate carboxylase small subunit mRNA. *J. Phytopathol.*, **137**, 317–343.

Feinberg, A.P. and Vogelstein, B. (1983) A technique for radiolabelling DNA restriction endonuclease fragments to high specific activity. *Anal. Biochem.*, **132**, 6–13.

Franssen, H.J., Nap, J.-P., Gloudemans, T. et al. (1987) Characterization of cDNA for nodulin-75 of soybean: a gene product involved in early stages of root nodule development. *Proc. Natl. Acad. Sci. USA*, **84**, 4495–4499.

Fuller, F., Künstner, P.W., Nguyen, T. and Verma, P.S. (1983) Soybean nodulin genes: analysis of cDNA clones reveals several major tissue-specific sequences in nitrogen-fixing root nodules. *Proc. Natl. Acad. Sci. USA*, **80**, 2594–2598.

Galau, G.A., Hughes, D.W. and Dure III, L. (1986) Abscisic acid induction of cloned cotton late embryogenesis-abundant (*Lea*) mRNAs. *Plant Mol. Biol.*, **7**, 155–170.

Gasser, C.S., Budelier, K.A., Smith, A.G. et al. (1989) Isolation of tissue-specific cDNAs from tomato pistils. *Plant Cell*, **1**, 15–24.

Ginn, D.I. and Rapp, J.P. (1991) Larger lambda plaques using thick agar plates. *BioTechniques*, **10**, 612–614.

Goday, A., Sánchez-Martínez, D., Gómez, J. et al. (1988) Gene expression in developing *Zea mays* embryos: regulation by abscisic acid of a highly phosphorylated 23- to 25-kD group of proteins. *Plant Physiol.*, **88**, 564–569.

Godoy, J.A., Pardo, J.M. and Pintor-Toro, J.A. (1990) A tomato cDNA inducible by salt stress and abscisic acid: nucleotide sequence and expression pattern. *Plant Mol. Biol.*, **15**, 695–705.

Goldberg, R.B. (1988) Plants: novel developmental processes. *Science*, **240**, 1460–1467.

Gómez, J., Sánchez-Martínez, D., Stiefel, V. et al. (1988) A gene induced by the plant hormone abscisic acid in response to water stress encodes a glycine-rich protein. *Nature*, **334**, 262–264.

Gray, J., Picton, S., Shabbeer, J. et al. (1992) Molecular biology of fruit ripening and its manipulation with antisense genes. *Plant Mol. Biol.*, **19**, 69–87.

Gregersen, P.L., Brandt, J., Thordal-Christensen, H. and Collinge, D.B. (1993) cDNA cloning and characterization of mRNAs induced in barley by the fungal pathogen, *Erysiphe graminis*. In *Mechanisms of Plant Defense Responses* (eds B. Fritig and M. Legrand), pp. 304–307. Kluwer Academic Publishers, The Netherlands.

Grunstein, M. and Hogness, D.S. (1975) Colony hybridization: a method for the isolation of cloned DNAs that contain a specific gene. *Proc. Natl. Acad. Sci. USA*, **72**, 3961–3965.

Guerrero, F.D. and Mullet, J.E. (1988) Reduction of turgor induces rapid changes in leaf translatable RNA. *Plant. Physiol.*, **88**, 401–408.

Gurr, S.J. and McPherson, M.J. (1991) PCR-directed cDNA libraries. In *PCR, A Practical Approach* (eds M.J. McPherson, P. Quirke and G.R. Taylor), pp. 147–170. Oxford University Press, New York.

Gurr, S.J., McPherson, M.J., Scollan, C. et al. (1991) Gene expression in nematode-infected plant roots. *Mol. Gen. Genet.*, **226**, 361–366.

Hagen, G., Kleinschmidt, A. and Guilfoyle, T. (1984) Auxin-regulated gene expression in intact soybean hypocotyl and excised hypocotyl sections. *Planta*, **162**, 147–153.

Harada, J.J., Baden, C.S. and Comai, L. (1988) Spatially regulated genes expressed during seed germination and postgerminative development are activated during embryogeny. *Mol. Gen. Genet.*, **212**, 466–473.

Harada, J.J., DeLisle, A., Baden, C.S. and Crouch, M.L. (1989) Unusual sequence of an abscisic acid-inducible mRNA which accumulates late in *Brassica napus* seed development. *Plant Mol. Biol.*, **12**, 395–401.

Harper, G., Gurr, S.J., Scollan, C. et al. (1993) Gene expression during a plant–nematode interaction. In *Mechanisms of Plant Defense Responses* (eds B. Fritig and M. Legrand), pp. 340–343. Kluwer Academic Publishers, The Netherlands.

Hedrick, S.M., Cohen, D.I., Nielsen, E.A. and Davis, M.M. (1984) Isolation of cDNA clones encoding T cell-specific membrane-associated proteins. *Nature*, **308**, 149–153.

Heintzen, C., Melzer, S., Fischer, R. et al. (1994) A light- and temperature-entrained circadian clock controls expression of transcripts encoding nuclear proteins with homology to RNA-binding proteins in meristematic tissue. *Plant J.*, **5**, 799–813.

Herdenberger, F., Evrard, J.-L., Kuntz, M. et al. (1990) Isolation of flower-specific cDNA clones from sunflower. *Plant Sci.*, **69**, 111–122.

Hodge, R., Wyatt, P., Draper, J. and Scott, R. (1992) Cold-plaque screening: a simple technique for the isolation of low abundance, differentially expressed transcripts from conventional cDNA libraries. *Plant J.*, **2**, 257–260.

Hong, B., Uknes, S.J. and Ho, T.H.D. (1988) Cloning and characterization of a cDNA encoding a mRNA rapidly-induced by ABA in barley aleurone layers. *Plant Mol. Biol.*, **11**, 495–506.

Hooft van Huijsduijnen, R.A.M., van Loon, L.C. and Bol, J.F. (1986) cDNA cloning of six mRNAs induced by TMV infection of tobacco and characterization of their translation products. *EMBO J.*, **5**, 2057–2061.

Jakobsen, K., Sletner Klemsdal, S., Aalen, R.B. et al. (1989) Barley aleurone cell development: molecular cloning of aleurone-specific cDNAs from immature grains: *Plant Mol. Biol.*, **12**, 285–293.

Kim, S.-R., Kim, Y. and An, G. (1993) Molecular cloning and characterization of anther-preferential cDNA encoding a putative actin-depolymerizing factor. *Plant Mol. Biol.*, **21**, 39–45.

Kodama, H., Ito, M., Hattori, T. et al. (1991) Isolation of genes that are preferentially expressed at the G1/S boundary during the cell cycle in synchronized cultures of *Catharanthus roseus* cells. *Plant Physiol.*, **95**, 406–411.

Koltunow, A.M., Truettner, J., Cox, K.H. et al. (1990) Different temporal and spatial gene expression patterns occur during anther development. *Plant Cell*, **2**, 1201–1224.

Larkin, J.C. (1994) Isolation of a cytochrome P450 homologue preferentially expressed in developing inflorescences of *Zea mays*. *Plant Mol. Biol.*, **25**, 343–353.

Lawton, K.A., Huang, B., Goldsbrough, P.B. and Woodson, W.R. (1989) Molecular cloning and characterization of senescence-related genes from carnation flower petals. *Plant Physiol.*, **90**, 690–696.

Leemans, J. (1993) Ti to tomato, tomato to market. *Bio/Technology*, **11**, 522–526.

Lemke, S.J., Burke, T., Boder, G.B. and Moore, R.E. (1993) SPOT: an improved differential screening protocol that allows the detection of marginally induced mRNAs. *BioTechniques*, **14**, 415–419.

Lincoln, J.E., Cordes, S., Read, E. and Fischer, R.L. (1987) Regulation of gene expression by ethylene during *Lycopersicon esculentum* (tomato) fruit development. *Proc. Natl. Acad. Sci. USA*, **84**, 2793–2797.

Logemann, J., Mayer, J.E., Schell, J. and Willmitzer, L. (1988) Differential expression of genes in potato tubers after wounding. *Proc. Natl. Acad. Sci. USA*, **85**, 1136–1140.

Lu, X., Dengler, J., Rothbarth, K. and Werner, D. (1990) Differential screening of murine ascites cDNA libraries by means of *in vitro* transcripts of cell-cycle-phase-specific cDNA and digital image processing. *Gene*, **86**, 185–192.

Luo, G., An, G. and Wu, R. (1994) A PCR differential screening method for rapid isolation of clones from a cDNA library. *BioTechniques*, **16**, 672–675.

Mansson, P.-E., Hsu, D. and Stalker, D. (1985) Characterization of fruit specific cDNAs from tomato. *Mol. Gen. Genet.*, **200**, 356–361.

Marco, Y.J., Ragueh, F., Godiard, L. and Froissard, D. (1990) Transcriptional activation of 2 classes of genes during the hypersensitive reaction of tobacco leaves infiltrated with an incompatible isolate of the phytopathogenic bacterium *Pseudomonas solanacearum*. *Plant Mol. Biol.*, **15**, 145–154.

Marineau, C., Matton, D.P. and Brisson, N. (1987) Differential accumulation of potato tuber mRNAs during the hypersensitive response induced by arachidonic acid elicitor. *Plant Mol. Biol.*, **9**, 335–342.

Marty, I., Brugidou, C., Chartier, Y. and Meyer, Y. (1993) Growth-related gene expression in *Nicotiana tabacum* mesophyll protoplasts. *Plant J.*, **4**, 265–278.

McLaughlin, M. and Walbot, V. (1987) Cloning of a mutable *bz2* allele of maize by transposon tagging and differential hybridization. *Genetics*, **117**, 771–776.

Medford, J.I., Elmer, J.S. and Klee, H.J. (1991) Molecular cloning and characterization of genes expressed in shoot apical meristems. *Plant Cell*, **3**, 359–370.

Memelink, J., Hoge, J.H.C. and Schilperoort, R.A. (1987) Cytokinin stress changes the developmental regulation of several defence-related genes in tobacco. *EMBO J.*, **6**, 3579–3583.

Mohapatra, S.S., Poole, R.J. and Dhindsa, R.S. (1988) Abscisic acid-regulated gene expression in relation to freezing tolerance in alfalfa. *Plant Physiol.*, **87**, 468–473.

Montoliu, L., Rigau, J. and Puigdomènech, P. (1989) A tandem of α-tubulin genes preferentially expressed in radicular tissues from *Zea mays*. *Plant Mol. Biol.*, **14**, 1–15.

Mu, J.-H., Stains, J.P. and Kao, T. (1994) Characterization of a pollen-expressed gene encoding a putative pectin esterase of *Petunia inflata*. *Plant Mol. Biol.*, **25**, 539–544.

Mundy, J. and Chua, N.-H. (1988) Abscisic acid and water-stress induce the expression of a novel rice gene. *EMBO J.*, **7**, 2279–2286.

Mutchler, K.J., Klemish, S.W. and Russo, A.F. (1992) A rapid PCR protocol for identification of differentially expressed genes from a cDNA library. *PCR Methods Applic.*, **1**, 195–198.

Nasrallah, J.B., Kao, T.-H., Goldberg, M.L. and Nasrallah, M.E. (1985) A cDNA clone encoding an S-locus-specific glycoprotein from *Brassica oleracea*. *Nature*, **318**, 263–267.

Nasrallah, J.B., Nishio, T. and Nashrallah, M.E. (1991) The self-incompatibility genes of *Brassica*: expression and use in genetic ablation of floral tissues. *Annu. Rev. Plant Physiol. Plant Mol. Biol.*, **42**, 393–422.

Okubara, P.A. and Tobin, E.M. (1991) Isolation and characterization of three genes negatively regulated by phytochrome action in *Lemna gibba*. *Plant Physiol.*, **96**, 1237–1245.

Olszewski, N.E., Gast, R.T. and Ausubel, F.M. (1989) A dual-labeling method for identifying differentially expressed genes: use in the identification of cDNA clones that hybridize to RNAs whose abundance in tomato flowers is potentially regulated by gibberellins. *Gene*, **77**, 155–162.

Parfett, C.L.J., Hofbauer, R., Brudzynski, K. et al. (1989) Differential screening of a cDNA library with cDNA probes amplified in a heterologous host: isolation of murine GRP78 (BiP) and other serum-regulated low-abundance mRNAs. *Gene*, **82**, 291–303.

Parsons, T.J., Bradshaw, H.D. and Gordon, M.P. (1989) Systemic accumulation of specific mRNAs in response to wounding in poplar trees. *Proc. Natl. Acad. Sci. USA*, **86**, 7895–7899.

Pear, J.R., Ridge, N., Rasmusson, R. et al. (1989) Isolation and characterization of a fruit-specific cDNA and the corresponding genomic clone from tomato. *Plant Mol. Biol.*, **13**, 639–651.

Pieterse, C.M.J., Risseeuw, E.P. and Davidse, L.C. (1991) An *in planta* induced gene of *Phytophthora infestans* codes for ubiquitin. *Plant Mol. Biol.*, **17**, 799–811.

Pieterse, C.M.J., Riach, M.B.R., Bleker, T. et al. (1993a) Isolation of putative pathogenicity genes of the potato late blight fungus *Phytophthora infestans* by differential hybridization of a genomic library. *Physiol. Mol. Plant Pathol.*, **43**, 69–79.

Pieterse, C.M.J., Verbakel, H.M., Hoek Spaans, J. et al. (1993b) Increased expression of the calmodulin gene of the late blight fungus *Phytophthora infestans* during pathogenesis on potato. *Mol. Plant-Microbe Interact.*, **6**, 164–172.

Reddy, A.S.N. and Poovaiah, B.W. (1990) Molecular cloning and sequencing of a cDNA for an auxin-repressed mRNA: correlation between fruit growth and repression of the auxin-regulated gene. *Plant Mol. Biol.*, **14**, 127–136.

Reddy, A.S.N., Jena, P.K., Mukherjee, S.K. and Poovaiah, B.W. (1990) Molecular cloning of cDNAs for auxin-induced mRNAs and developmental expression of the auxin-inducible genes. *Plant Mol. Biol.*, **14**, 643–653.

Riggleman, R.C., Fristensky, B. and Hadwiger, L.A. (1985) The disease resistance response in pea is associated with increased levels of specific mRNAs. *Plant Mol. Biol.*, **4**, 81–86.

Roberts, M.R., Robson, F., Foster, G.D. et al. (1991) A *Brassica napus* mRNA expressed specifically in developing microspores. *Plant Mol. Biol.*, **17**, 295–299.

Rogers, H.J., Allen, R.L., Hamilton, W.D.O. and Lonsdale, D.M. (1991) Pollen specific cDNA clones from *Zea mays*. *Biochim. Biophys. Acta*, **1089**, 411–413.

Rothbarth, K., Petzelt, C., Lu, X. et al. (1993) cDNA-derived molecular characteristics and antibodies to a new centrosome-associated and G2/M phase-prevalent protein. *J. Cell Sci.*, **104**, 19–30.

Sabelli, P.A. and Shewry, P.R. (1993) Nucleic acid blotting and hybridization. In *Methods in Plant Biochemistry*, Vol. 10 (ed. J. Bryant), pp. 79–100. Academic Press, London.

Sabelli, P.A. and Shewry, P.R. (1995) Northern analysis and nucleic acid probes. In *Methods in Molecular Biology, Plant Gene Transfer and Expression Protocols*, Vol. 49 (ed. H. Jones), pp. 213–228. Humana Press Inc., Totowa, NJ.

Sabelli, P.A., Burgess, S.R., Carbajosa, J.V. et al. (1993) Molecular characterization of cell populations in the maize root apex. In *Molecular and Cell Biology of the Plant Cell Cycle* (eds J.C. Ormrod and D. Francis), pp. 97–109. Kluwer Academic Publishers, The Netherlands.

Sambrook, J., Fritsch, E.F. and Maniatis, T. (1989) *Molecular Cloning. A Laboratory Manual*. Cold Spring Harbor Laboratory Press, Cold Spring Harbor, New York.

Sargent, T.D. (1987) Isolation of differentially expressed genes. *Methods Enzymol.*, **152**, 423–432.

Schaffer, M.A. and Fischer, R.L. (1988) Analysis of mRNAs that accumulate in response to low temperature identifies a thiol protease gene in tomato. *Plant Physiol.*, **87**, 431–436.

Scott, R., Dagless, E., Hodge, R. et al. (1991) Patterns of gene expression in developing anthers of *Brassica napus*. *Plant Mol. Biol.*, **17**, 195–207.

Simoens, C.R., Peleman, J., Valvekens, D. et al. (1988) Isolation of genes expressed in specific tissues of *Arabidopsis thaliana* by differential screening of a genomic library. *Gene*, **67**, 1–11.

Slater, A., Maunders, M.J., Edwards, K. et al. (1985) Isolation and characterisation of cDNA clones for tomato polygalacturonase and other ripening-related proteins. *Plant Mol. Biol.*, **5**, 137–147.

Smart, C.C. and Fleming, A.J. (1993) A plant gene with homology to D-*myo*-inositol-3-phosphate synthase is rapidly and spatially up-regulated during an abscisic-acid-induced morphogenic response in *Sprirodela polyrrhiza*. *Plant J.*, **4**, 279–293.

Smith, A.G., Gasser, C.S., Budelier, K.A. and Fraley, R.T. (1990) Identification and characterization of stamen- and tapetum-specific genes from tomato. *Mol. Gen. Genet.*, **222**, 9–16.

Snowden, K.C. and Gardner, R.C. (1993) Five genes induced by aluminum in wheat (*Triticum aestivum* L.) roots. *Plant Physiol.*, **103**, 855–861.

Somssich, I.E., Bollmann, J., Hahlbrock, K. et al. (1989) Differential early activation of defense-related genes in elicitor-treated parsley cells. *Plant. Mol. Biol.*, **12**, 227–234.

Staiger, D. and Apel, K. (1993) Molecular characterization of two cDNAs from *Sinapis alba* L. expressed specifically at an early stage of tapetum development. *Plant J.*, **4**, 697–703.

Stiefel, V., Pérez-Grau, L., Albericio, F. et al. (1988) Molecular cloning of cDNAs encoding a putative cell wall protein from *Zea mays* and immunological identification of related polypeptides. *Plant Mol. Biol.*, **11**, 483–493.

Stiekema, W.J., Heidekamp, F., Dirkse, W.G. et al. (1988) Molecular cloning and analysis of four potato tuber mRNAs. *Plant Mol. Biol.*, **11**, 255–269.

Stinson, J.R., Eisenberg, A.J., Willing, R.P. et al. (1987) Genes expressed in the male gametophyte of flowering plants and their isolation. *Plant Physiol.*, **83**, 442–447.

Tagu, D., Python, M., Crétin, C. and Martin, F. (1993) Cloning symbiosis-related cDNAs from eucalypt ectomycorrhiza by PCR-assisted differential screening. *New Phytol.*, **125**, 339–343.

Takahashi, Y., Kuroda, H., Tanaka, T. et al. (1989) Isolation of an auxin-regulated gene cDNA expressed during the transition from G0 to S phase in tobacco mesophyll protoplasts. *Proc. Natl. Acad. Sci. USA*, **86**, 9279–9283.

Talbot, N.J., Ebbole, D.J. and Hamer, J.E. (1993) Identification and characterization of *MPG1*, a gene involved in pathogenicity from the rice blast fungus *Magnaporthe grisea*.

Plant Cell, **5**, 1575–1590.

Taylor, J.L., Fritzemeier, K.-H., Häuser, I. et al. (1990) Structural analysis and activation by fungal infection of a gene encoding a pathogenesis-related protein in potato. *Mol. Plant-Microbe Interact.*, **3**, 72–77.

Taylor, M.A., Mad Arif, S.A., Kumar, A. et al. (1992) Expression and sequence analysis of cDNAs induced during the early stages of tuberisation in different organs of the potato plant (*Solanum tuberosum* L.). *Plant Mol. Biol.*, **20**, 641–651.

Theologis, A., Huynh, T.V. and Davis, R.W. (1985) Rapid induction of specific mRNAs by auxin in pea epicotyl tissue. *J. Mol. Biol.*, **183**, 53–68.

Thomas, M.G., Hesse, S.A., Al-Mahdawi, S. et al. (1994) A procedure for second-round differential screening of cDNA libraries. *BioTechniques*, **16**, 229–232.

Thomas, T. and Wilde, D. (1985) Analysis of gene expression in carrot somatic embryos. In *Somatic Embryogenesis* (eds M. Terzi, L. Pitto and Z.R. Sung), pp. 77–85. IPRA, Rome.

Titarenko, E., Hargreaves, J., Keon, J. and Gurr, S.J. (1993) Defence-related gene expression in barley coleoptile cells following infection by *Septoria nodorum*. In *Mechanisms of Plant Defense Responses* (eds B. Fritig and M. Legrand), pp. 308–311. Kluwer Academic Press, The Netherlands.

Twell, D., Wing, R., Yamaguchi, J. and McCormick, S. (1989) Isolation and expression of an anther-specific gene from tomato. *Mol. Gen. Genet.*, **217**, 240–245.

Van der Eycken, W., Inzé, D., Van Montagu, M. and Gheysen, G. (1993) Study of tomato genes affected by infection with the nematode *Meloidogyne incognita*. In *Mechanisms of Plant Defense Responses* (eds B. Fritig and M. Legrand), pp. 336–339. Kluwer Academic Publishers, The Netherlands.

van der Zaal, E.J., Memelimk, A.M., Mennes, A.M. et al. (1987) Auxin-induced mRNA species in tobacco cell culture. *Plant Mol. Biol.*, **10**, 145–157.

Villarreal, L.P. and Berg, P. (1977) Hybridization *in situ* of SV40 plaques: detection of recombinant SV40 virus carrying specific sequences of nonviral DNA. *Science*, **196**, 183–185.

Walker, J.C. and Key, J.L. (1982) Isolation of cloned cDNAs to auxin-responsive poly(A)$^+$ RNAs of elongating soybean hypocotyl. *Proc. Natl. Acad. Sci. USA*, **79**, 7185–7189.

Wang, H., Brandt, A.S. and Woodson, W.R. (1993) A flower senescence-related mRNA from carnation encodes a novel protein related to enzymes involved in phosphonate biosynthesis. *Plant Mol. Biol.*, **22**, 719–724.

Weterings, K., Reijnen, W., van Aarssen, R. et al. (1992) Characterization of a pollen-specific cDNA clone from *Nicotiana tabacum* expressed during microgametogenesis and germination. *Plant Mol. Biol.*, **18**, 1101–1111.

Wilde, H.D., Nelson, W.S., Booij, H. et al. (1988) Gene-expression programs in embryogenic and non-embryogenic carrot cultures. *Planta*, **176**, 205–211.

Wright, S.Y., Suner, M.-M., Bell, P.J. et al. (1993) Isolation and characterization of male flower cDNAs from maize. *Plant J.*, **3**, 41–49.

Yamaguchi-Shinozaki, K., Koizumi, M., Urao, S. and Shinozaki, K. (1992) Molecular cloning and characterization of 9 cDNAs for genes that are responsive to desiccation in *Arabidopsis thaliana*: sequence analysis of one cDNA clone that encodes a putative transmembrane channel protein. *Plant Cell Physiol.*, **33**, 217–224.

7 Expression Library Screening

IMRE E. SOMSSICH
BERND WEIßHAAR

Max-Planck-Institut für Züchtungsforschung, Köln, Germany

In this chapter we describe the screening of expression libraries in order to obtain specific cDNAs and genes. Two procedures are described: (1) use of polyclonal antiserum or mixtures of monoclonal antibodies; and (2) employment of oligonucleotide probes comprising specific DNA sequences (recognition site DNA) which are bound by protein factors. In both cases, the cloning strategy depends on the functional expression in *E. coli* of relatively high levels of recombinant protein. Immunoscreening requires an appropriate antibody but no previous knowledge of the protein sequence encoded by the gene of interest. Successful recognition site DNA probing demands that the DNA-binding activity of the expressed product is maintained. Both methods are highly sensitive, generally applicable, and have led to the isolation of numerous DNA clones encoding diverse proteins, including several transcription factors.

1 INTRODUCTION

Ever since its inception in the mid-1970s (Rougeon et al., 1975; Efstratiadis et al., 1976; Higuchi et al., 1976; Rabbitts, 1976), cDNA cloning has become an extremely valuable and indispensable tool in molecular biology. Improvements in the screening (Grunstein & Hogness, 1975; Benton & Davis, 1977; Hanahan & Meselson, 1980) and construction (Chapter 2; Okayama & Berg, 1982; Gubler & Hoffman, 1983) of cDNA libraries using poly(A)$^+$ RNA from various sources, from distinct tissues, organs or cell types, has led to the isolation of hundreds of gene representatives from a large number of prokaryotic and eukaryotic organisms. The availability of these molecular probes for studies related to gene structure, expression and regulation has greatly facilitated our knowledge concerning the role and interplay of specific genes in diverse biological processes such as embryogenesis, morphogenesis or tumorigenesis, to name just a few. Currently, attempts are being made to completely sequence representative cDNA libraries to yield information about 'expressed sequence tags' (ESTs) (Chapter 15; Höfte et al., 1993).

In the majority of cases, the respective cDNA clones were isolated on the basis of hybridisation to denatured (single-stranded) DNA or RNA probes with

Plant Gene Isolation: Principles and Practice. Edited by G. D. Foster and D. Twell.
© 1996 John Wiley & Sons Ltd.

related nucleotide sequences (Chapter 5). These probes often represented regions of genes known to be highly conserved in nucleotide sequence, or they were prepared from synthetic oligonucleotides corresponding to amino acid sequences obtained by analysis of the purified protein of interest. In some cases the probes comprised complete mixtures of cDNAs representing differentially expressed mRNA transcripts (differential screening; Chapter 6), or simply full-length cDNA or genomic clones originating from heterologous sources.

The establishment of an efficient method to rapidly screen large cDNA libraries expressing foreign proteins in E. coli using antibody probes (Young and Davis, 1983a, b) represented yet another important advance. This technique enabled us to isolate genes expressing certain proteins against which antibodies could be raised, but which were not present in sufficient amounts for detailed analysis. The widespread adoption of phage vectors for the construction of cDNA expression libraries has led to a broad application of the immunological screening technique resulting in the cloning of over 200 gene representatives from various organisms, including plants (Table 1). Expression libraries have also been successfully screened with ligands, including small proteins, which resulted in the cloning of cDNAs encoding the respective target proteins. For example, radiolabelled calmodulin has been used as a probe to detect a calmodulin-dependent kinase II-producing clone in a murine brain cDNA expression library as well as a glutamate decarboxylase encoding cDNA from a petunia library (Sikela & Hahn, 1987; Sambrook et al., 1989; Fromm & Chua, 1992; Baum et al., 1993).

A still rather new development has been the use of double-stranded DNA (recognition site DNA) to successfully screen cDNA expression libraries in search of DNA-binding proteins (Singh et al., 1988; Vinson et al., 1988). In this protocol, DNA corresponding in sequence to important DNA elements present in the regulatory regions of genes is used to detect cDNA clones encoding proteins which recognise the regulatory element under investigation. By this screening method, genes encoding sequence-specific DNA-binding proteins have been isolated first from a number of mammalian species but later also from other organisms, including plants (Singh et al., 1988, 1989; Hai et al., 1989; Maekawa et al., 1989; Katagiri & Chua, 1992; Table 2). Characterisation of the various DNA-binding proteins cloned by this screening procedure revealed them to contain structural motifs common to different classes of transcription factors. DNA-binding proteins containing zinc-finger motifs, helix-turn-helix motifs, bZIP domains or homeodomain regions have all been isolated, indicating that the procedure shows no strong bias towards a certain subclass of DNA-binding domain (Singh et al., 1989; Katagiri & Chua, 1992; Pabo & Sauer, 1992).

In this chapter, we will attempt to summarise our own experience in constructing and employing plant cDNA expression libraries, focusing mainly on the use of recognition site DNA probes but also including information relevant to antibody screenings. It is our intention to avoid including detailed protocols for methods that have now become standard in most molecular biology laboratories and for which extensive step-by-step manuals are easily

Table 1. Plant cDNA clones isolated by immunological screening of expression libraries

Plant species	Protein product	Library	Reference
Tomato	Endoploygalacturonase	λCharon 16	DellaPenna et al. (1986)
		λgt11	Sheehy, 1987
Bean	Chitinase	λgt11	Broglie et al. (1986)
	Chalcone isomerase (CHI)	λgt11	Mehdy & Lamb (1987)
	Chitinase	λgt11	Hedrick et al. (1988)
Potato	4-Coumarate: CoA ligase (4CL)	λgt11	Fritzemeier et al. (1987)
Parsley	Bergaptol O-methyltransferase (BMT)	λgt11	Hauffe (1988)
Petunia	Chalcone isomerase (CHI)	λgt11	van Tunen et al. (1988)
Pea	Cu/Zn superoxide dismutase (SOD)	λgt11	Scioli & Zilinskas (1988)
	NADPH: protochlorophyllide oxidoreductase	λgt11	Spano et al. (1992)
Zucchini	ACC (1-aminocyclopropane-1-carboxylate) synthase	λgt11	Sato & Theologis (1989)
Periwinkle	Tryptophan decarboxylase	lambdaZAP	DeLuca et al. (1989)
	Strictosidine synthase (SSS)	lambdaZAP	Pasquali et al. (1992)
White birch	Betvl	λgt11	Breiteneder et al. (1989)
	Profilin	λgt11	Valenta et al. (1991)
Castor bean	Casebene synthetase	λgt11	Lois & West (1990)
Peanut	Cationic peroxidase	λgt11	Buffard et al. (1990)
Soybean	β-1,3-Glucanase	λgt11	Takeuchi et al. (1990)
Winter squash	ACC synthase	pTTQ18	Nakajima et al. (1990)
Alfalfa	Phenylalanine ammonia lyase (PAL)	λgt11	Gowri et al. (1991)
Tobacco	β-1,3-Glucanase (PR2)	λgt11	Côté et al. (1991)
Flaveria	Flavanol sulphotransferase (ST)	lambdaZAP	Varin et al. (1992)
Maize	GF14(bZIP)	λgt11	de Vetten et al. (1992)
Artichoke	Cinnamate 4-hydroxylase (4CH)	λgt11	Teutsch et al. (1993)

available. In these cases we will merely refer the interested reader to the relevant publications.

2 CONSIDERATIONS

DNA expression libraries can be constructed in plasmid or in bacteriophage lambda expression vectors (for a comparison see Chapter 2 and Sambrook et al.

Table 2. Plant cDNA clones isolated by DNA recognition site probes from expression libraries

Plant species	Protein product	Library	Reference
Tobacco	TGA1a/TGA1b (bZIP)	λgt11	Katagiri et al. (1989)
	TAF-1 (bZIP)	lambdaZAP	Oeda et al. (1991)
	GT-1a	lambdaZAP	Gilmartin et al. (1992)
	B2F, or 'GT-1a'	λgt11	Perisic & Lam (1992)
	ATBP-1	lambdaZAP	Tjaden & Coruzzi (1994)
	TFHP-1	lambdaZAP	Kawaoka et al. (1994)
Wheat	HBP-1a(17) (bZIP)	λgt11	Tabata et al. (1989)
	EmBP-1 (bZIP)	λgt11	Guiltinan et al. (1990)
	HBP-1b(c38) (bZIP)	λgt11	Tabata et al. (1991)
	HBP-1a(1)/1a(c14)/1b(c1) (bZIP)	λgt11	Mikami et al. (1994)
Tomato	HSF	λgt11	Scharf et al. (1990)
	GBF4/9/12 (bZIP)	lambdaZAP	Meier & Gruissem (1994)
Maize	OCSBF-1 (bZIP)	λgt11	Singh et al. (1990)
	Zmhox1a (HD protein)	λgt11	Bellmann & Werr (1992)
	OBF3.1/2 (bZIP)	lambdaZAP	Foley et al. (1993)
	IBP1	λgt11	Lugert & Werr (1994)
Rice	GT-2	λgt11	Dehesh et al. (1990)
Parsley	CPRF1 (bZIP)	lambdaZAP	Weißhaar et al. (1991)
	CPRF2/3 (bZIP)	λgt11	Weißhaar et al. (1991)
	BPF-1	λgt11/ lambdaZAP	da Costa e Silva et al. (1993)
	PRHP (HD protein)	lambdaZAP	Korfhage et al. (1994)
Petunia	EPF1 (zinc finger)	λgt11	Takatsuji et al. (1992)
Bean	VBP1 (bZIP)	lambdaZAP	Ehrlich et al. (1992)
Arabidopsis	GBF1 (bZIP)	lambdaZAP	Schindler et al. (1992)
	HAT3.1 (HD protein)	lambdaZAP	Schindler et al. (1993)
	PRHA (HD protein)	lambdaZAP	Korfhage et al. (1994)
Rice	PF1	λgt11	Nieto-Sotelo et al. (1994)
Potato	MBF1/2/3 (bZIP)	λgt11	Feltkamp et al. (1994)

(1989). One major disadvantage of plasmid libraries is that generation of the replicas required for screening on the basis of protein expression is rather labour-intensive. Therefore, when screening large libraries, as will mostly be the case when employing representative plant cDNA libraries, a bacteriophage lambda-derived vector will be the cloning vehicle of choice. A recent improvement in this field is the development of cloning systems which allow unidirectional cloning of the cDNA, thereby increasing the number of in-frame fusions by a factor of 2. Convenient lambda vectors and corresponding cDNA modification techniques allowing directional cloning can now be purchased from several companies (e.g. lambdaZAP, Stratagene; λgt11 Sfi-Not, Promega). An additional advantage of these vectors is that two different restriction sites must be used at the two arms of the lambda vector, thus reducing the frequency of 'empty' vector phages in the library by preventing religation of the left and right vector arms. Furthermore, in the case of the vector lambdaZAP, direct excision of the cDNA from the phage (a procedure termed 'zapping') can circumvent tedious subcloning steps (Short et al., 1988).

A number of factors can influence the successful outcome of screening cDNA expression libraries. First of all, it is obvious that the protein one is looking for must be expressed in the tissue which is used to extract the RNA needed for the construction of the cDNA library. Therefore, the selection of the correct starting material is an important point. The stability or toxicity of the foreign gene product in E. coli will also influence the outcome of any screening attempt. When bacteriophage λgt11 is used as the vector, the majority of the foreign proteins will be expressed as fusions to the E. coli protein β-galactosidase. Such fusions can significantly increase the stability of the respective protein (Stanley, 1983). Also, the ability to induce expression of the foreign gene product (in most cases by isopropylthio-β-D-galactoside (IPTG)) has proven beneficial in cloning recombinant phages expressing proteins toxic to E. coli cells. Nevertheless, some phages will still not multiply well for several reasons and are therefore under-represented in amplified libraries. For this reason, it is advisable to screen unamplified libraries. In some cases, as with the Stratagene cDNA synthesis kit for directional cloning in lambdaZAP, the plating of unamplified libraries requires, due to the hemimethylated nature of the cDNA, special E. coli host strains (e.g. SURE or PLK-F'). These strains sometimes cause problems in expression screenings, since their slow growth often results in poor plaque formation. In lambdaZAP the fusion protein only contains a small region derived from β-galactosidase and from the polylinker at the N-terminus. For this reason, λgt11 and derivatives might still be a better choice than lambdaZAP or related phages, especially for expression screenings, even if the subsequent subcloning is somewhat more laborious. In this case, subcloning can be facilitated by the use of the polymerase chain reaction (PCR) technique (Chapter 13), employing a combination of λgt11 primers adjacent to the cDNA insert cloning site (e.g. primers #1218 and #1222 from New England Biolabs). However, one must bear in mind that random errors in nucleotide sequence may be produced during PCR. Another factor which may have an influence on the outcome is the type of cDNA prepared: either oligo(dT) or randomly primed. Depending on the domain structure of the target protein, the epitopes or DNA-binding domains may be located more at the C- or the N-terminus. If possible, a mixed library containing cDNAs originally primed by either form of oligomer should be screened. However, if one likes to take advantage of directional cloning, one will have to choose oligo(dT) priming.

When employing immunological probes, the most obvious factor is the quality of the antibody itself. Both monoclonal and polyclonal antibodies have been successfully used although polyclonal antibodies are better suited, as they can recognise numerous epitopes on the protein (Young & Davis, 1983b; Sambrook et al., 1989). When using monoclonals, which by definition will detect only one epitope, several antibodies should be used, since more than one protein may contain the same epitope, thereby potentially resulting in the cloning of numerous 'false' positives. A problem with polyclonal antisera is that most rabbit sera contain considerable amounts of anti-*E. coli* antibodies. We therefore suggest removing them by immunoabsorption (Snyder et al., 1987; Sambrook et al., 1989) prior to screening.

In the cases where double-stranded DNA probes (recognition site probes) are used, parameters influencing the equilibrium association and kinetic constants of a protein–DNA interaction must be taken into consideration, as has been nicely outlined by Singh et al. (1989). Parameters such as pH, temperature, presence or absence of detergents, quality and concentration of competitor DNA, and ionic strength of the binding and wash buffers, should therefore be optimised to detect the relevant recombinant protein. Such test experiments are usually performed by electrophoretic mobility shift assays (EMSAs). If the nature of the DNA-binding protein under investigation allows, a pretest to optimise the screening conditions can be performed by dot-blotting (Shao & Pratt, 1994), i.e. under conditions more closely resembling the actual screening. With regard to the DNA probe, the use of concatenated oligonucleotides representing multiple recognition sites has also proven to be of great advantage (Vinson et al., 1988; Sambrook et al., 1989). Still, the technique has its intrinsic limitations. For certain eukaryotic proteins the bacterial environment may not allow for proper folding, thereby leading to conformations not able to recognise the DNA probe. In this case, denaturation followed by slow renaturation of such bacteria-produced foreign proteins bound to nitrocellulose replica filters may prove successful (Vinson et al., 1988). Also, it will be impossible to detect DNA-binding proteins which require in vivo post-translational modifications, e.g. phosphorylation or heteromeric subunits for binding activity.

3 METHODOLOGY

3.1 Construction of cDNA expression libraries

Obviously, the quality of any library can only be as good as the RNA used for its construction. Therefore, great effort should be made to check for the integrity of the RNA isolated from the starting material; see Chapter 2 and Sambrook et al., (1989). Although numerous methods are available to obtain good RNA from plant tissue, we have had good success with parsley and *Arabidopsis* cell cultures as well as with parsley plant material using a modified version of the procedure described by Dunsmuir et al., (1988).

PROTOCOL 1: RNA EXTRACTION AND POLY(A)$^+$ RNA PREPARATION

(1) Grind 5–40 g fresh weight cells to a fine powder under liquid nitrogen.
(2) Thaw the powder in lysis buffer (2 ml/g; 100 mM Tris-HCl, pH 8.5,

100 mM NaCl, 20 mM EGTA, 2% sodium dodecyl sulphate (SDS)). Note that all buffers are made up using diethylpyrocarbonate (DEPC)-treated H_2O.

(3) Add 0.5 volume of Tris-HCl-saturated phenol and 100 μl of β-mercapto-ethanol and thoroughly mix for 5 min.

(4) Add 0.5 volume of chloroform and mix again for 5 min.

(5) Centrifuge the mixture briefly.

(6) Re-extract the aqueous phase twice (or until the interphase is greatly reduced) with phenol/chloroform (1:1) and subsequently precipitate with isopropanol.

(7) Following centrifugation, resuspend the pellet in 10 ml of H_2O, add 5 ml of 4 M LiCl and incubate the solution for 3 h on ice to precipitate the RNA.

(8) Following centrifugation, resuspend the RNA pellet in 1.8 ml of H_2O and precipitate with ethanol.

(9) Following centrifugation, resuspend the RNA pellet in H_2O and determine the RNA content photometrically.

(10) Poly(A)$^+$ RNA is obtained by two rounds of purification on oligo(dT) spun columns (Pharmacia) according to the instructions supplied by the manufacturer. It is advisable to save aliquots of the intermediate RNA fractions and to include them in the final RNA hybridisation experiment.

(11) The success of the poly(A)$^+$ purification procedure and the quality of the final poly(A)$^+$ RNA should be tested by RNA hybridisation experiments (Chapter 2) using known probes or probes detecting highly conserved genes (e.g. actin, ubiquitin).

For the construction of expression libraries the reader is referred to the extensive protocols given in Young & Davis (1983a), Huynh et al. (1985), Snyder et al. (1987) and Sambrook et al. (1989). Additionally, detailed manuals are also provided by various companies now offering superb cDNA library construction kits and high-grade molecular biology reagents. In fact, cDNA expression libraries from various sources can also be purchased from several companies. We have had good experience making complex cDNA libraries (both random and oligo(dT) primed) containing a high portion of full-length cDNA clones starting from RNA extracted from parsley and *Arabidopsis* cell cultures, using λgt11 or lambdaZAP expression vectors and construction kits purchased from Pharmacia and Stratagene. In contrast to some of the protocols provided by the suppliers, we always radiolabelled the cDNA used for library construction, rather than performing a parallel preparation. This allows one to follow the cDNA in column runs and during precipitation steps. The last step of the cDNA synthesis and modification protocol should be chromatography on Sephacryl 400 columns to get rid of small (< 400–500 bp) cDNA fragments. This column can be equilibrated with ligation buffer. Once the cDNA is ready for ligation, one should test the ratio of the amount of cDNA to vector DNA empirically by performing test ligations with, for example, 0.5 μg of vector DNA and varying amounts of cDNA. The best ratio can then be determined after in vitro packaging and phage plating. To reduce the need for the expensive packaging extracts, we found that we can successfully package 1 µl of test ligations using only one-third of a commercially available packaging extract (Gigapack gold, Stratagene).

3.2   Screening of expression libraries using antibodies

A detailed discussion on the use of antibody probes, methods to detect the antibodies bound to its antigen on nitrocellulose filters and the subsequent methods required to verify the cloning of the isolated gene representative are presented in Huynh et al. (1985), Snyder et al. (1987) and Sambrook et al. (1989). For immunological screenings it is best to keep the density of the plaques low. That is, the borders of individual plaques should still be recognised and should not make direct contact with the borders of neighbouring plaques (e.g. not denser than 400–500 PFU/cm^2).

PROTOCOL 2: IMMUNOLOGICAL SCREENING

Probe preparation

Detection of the antigen–antibody complexes can be achieved either using an enzyme-conjugated second antibody followed by a colour reaction or with ^{125}I-labelled protein A or anti-IgG followed by autoradiography.

Antibodies coupled to horse-radish peroxidase or alkaline phosphatase that react with species-specific determinants on primary antibodies, and chromogenic substrates such as 4-chloro-1-naphtol, and 5-bromo-4-chloro-3-indolyl phosphate (BCIP) in combination with nitroblue tetrazolium (NBT), for detection of the complexes, are available from numerous commercial sources which also provide detailed step-by-step manuals (e.g. Amersham, Boehringer Mannheim, Pharmacia, Promega, Stratagene).

Radiolabelled protein A is also available from commercial sources. Preparation of radioiodinated second antibody is given in Sambrook et al. (1989). However, whenever possible we advise the use of non-radioactive probes.

Screening procedure

(1) Plate out the library using an appropriate *E. coli* strain (e.g. Y1090r⁻ for λgt11; XL1-Blue or BB4 for lambdaZAP) on 2–3-day-old LB plates. Incubate plates for 3–4 h at 42 °C.

(2) Soak numbered nitrocellulose filters (S&S nitrocellulose BA85) in 10 mM IPTG solution for a few minutes, remove filters from the solution, and allow them to dry at room temperature.

(3) Transfer the plates to a 37 °C incubator and quickly overlay each plate with the dry previously saturated IPTG filters. Incubate the plates for at least 4 h. If a duplicate filter is required, remove the first filter and overlay the plate with the second one and continue to incubate the plates for an additional 4–6 h.

(4) Following incubation, mark the position of the filter on the plates and remove them carefully and immediately immerse them in TNT buffer (10 mM Tris-HCl, pH 8.0, 150 mM NaCl, 0.05% Tween-20). Perform all of the following washing and incubation steps at room temperature with gentle shaking.

(5) Store all plates at 4 °C until the results of the immunological screening are obtained.

(6) After all filters are removed and rinsed, transfer them to a fresh batch of TNT buffer. Agitate filters for another 30 min at room temperature. (If necessary, the filters can be removed from the buffer at this stage and stored in Saran Wrap for up to 24 h at 4 °C.)

(7) Transfer the filters individually to a glass tray or Petri dish containing blocking buffer (TNT + 20% fetal bovine serum; 5–7.5 ml/82-mm filter; 10–15 ml/138-mm filter) and agitate the buffer slowly for 15–30 min on a rotary platform.

(8) Transfer the filters to fresh glass trays or Petri dishes containing the primary antibody in blocking buffer (use the same amount of volume per filter as in Step 7). Use the highest dilution of antibody that gives acceptable background levels but can still detect 50–100 pg of denatured protein (e.g. analysed by western blotting). Agitate the solution containing the submerged filters gently on a rotary platform for 1–4 h at room temperature. (This solution can be stored at 4 °C and reused several times!)

(9) Wash the filters for 5–10 min in each of the following buffers (volume per filter as in Step 7). Always transfer the filters individually from one buffer to the next: (a) TNT + 0.1% bovine serum albumin; (b) TNT + 0.1% bovine serum albumin + 0.1% nonidet P-40; (c) TNT + 0.1% bovine serum albumin.

(10) Dry the filters and detect the antigen–antibody complexes with the radiochemical or chromogenic reagent of your choice.

(11) Identify the locations of plaques on the plates giving positive signals. Remove a plug of agar from this area using the large end of a Pasteur pipette. Transfer the plug to an Eppendorf tube containing 1 ml of SM buffer (50 mM Tris-HCl, pH 7.5, 10 mM $MgCl_2$, 100 mM NaCl, 0.02% gelatine) and then add two drops of chloroform. Allow the bacteriophage particles to elute from the agar plug for several hours at 4 °C.

(12) Determine the titre of the bacteriophages in the eluate. Replate them at approximately 2000–3000 PFU/90-cm plate and repeat the screening procedure. Repeat screenings until all plaques on a plate show positive signals.

Most primary antibodies can be used in the 1 : 1000 dilution range. As mentioned above, the blocking buffer containing the primary antibody (Step 8) can be reused several times. In fact, our experience is that reuse will also reduce unspecific background. Additionally, the relatively high background obtained with some antibodies can be markedly reduced by performing a preliminary screening on plated wild-type bacteriophages.

Other possible blocking agents that have been used in different laboratories include 1% gelatine, 3% bovine serum albumin, casein and 5% non-fat dried milk. We recommend carrying out trial experiments to determine which works best in your particular case.

3.3   Screening of expression libraries using recognition site probes

This screening approach was first published by Singh et al. (1988), who used a recognition site probe to clone a human cDNA encoding an enhancer element binding protein (H2TF1). Several high-quality protocols of this 'southwestern' technique are now available (Sambrook et al., 1989; Singh et al., 1989; Cowell & Hurst, 1993; Garabedian et al., 1993; Werr et al., 1994). Here we will list the procedure which works best in our hands.

The protocol described below was used by us to clone cDNAs encoding DNA-binding proteins, including plant bZIP (Weißhaar et al., 1991) and homeodomain-type proteins (Korfhage et al., 1994). For first-round screenings, we always prepare two sets of filter lifts from each plate. The protein bound to one set of filter replicas undergoes a subsequent denaturation/renaturation treatment, whereas the other set remains untreated ('native'). In our case, most of the positive clones were identified after the denaturation/renaturation procedure; however, some clones were only detected without the denaturation/renaturation step.

One important point concerns the composition of the binding buffer. As mentioned above, this buffer should be optimised for the DNA–protein interaction under investigation. In our case the optimal binding buffer was 25 mM HEPES/KOH, pH 7.4, 4 mM KCl, 5 mM $MgCl_2$, 1 mM EDTA, 7% glycerol, and 0.5 mM DTT (added just prior to use). The binding reaction also contained 5 μg/ml denatured salmon sperm DNA. In some cases it has been claimed that use of sonicated double-stranded salmon sperm DNA (1 μg/ng probe) instead of denatured DNA can be critical and can reduce background signals.

PROTOCOL 3: SOUTHWESTERN SCREENING

Probe preparation

(1) Separately phosphorylate two complementary oligonucleotides (5 μg each), on average 30 nucleotides long, representing the binding site of interest and building compatible ends, with T4 polynucleotide kinase, using buffer conditions (50 μl volume) as described in Sambrook et al., 1989, for 30 min *on ice*.

(2) Combine the two reactions, incubate the tube containing the reaction mixture for 5 min at 70 °C in a water bath, and allow the oligonucleotides to slowly anneal by turning off the water bath and letting it to cool to room temperature.

(3) Add 4 units of T4 DNA ligase (Boehringer) and fresh ATP to the reaction mixture to ligate the double-stranded oligonucleotides via their sticky ends

to multimers of between 6-mers and 35-mers. Incubate for 12 h at 16 °C. Check 1/20 of the reaction (0.5 μg DNA) on a 2.5% agarose gel. A ladder of the proper size range should be visible. Store the concatenated oligonucleotides at −20 °C. (Depending on the actual results, the ligation time or the amount of ligase added to the reaction must be modified to get products of the correct length. Additionally, gel-purified oligonucleotides often increase the length of the multimeric forms.)

(4) Label 1 μg of the multimerized binding site oligonucleotide with 100 μCi each of the four NTPs (3000 Ci/mmol). We use the Boehringer nick-translation kit and the protocol supplied by the manufacturer except that the reaction volume was increased to 60 μl. Following nick translation, add 25 mM EDTA to increase the total volume to 100 μl, and separate the labelled DNA from unincorporated nucleotides by two Sephadex G25 spun column centrifugations. Bring the volume to 1 ml with TE buffer, mix well and count 1 μl. The activity should be greater than 1×10^8 counts/min (Cerenkow counts) per 1 μg of DNA. The labelled oligonucleotide multimers should be analysed on a polyacrylamide gel (using the 1 μl set aside for counting). A ladder must be visible after autoradiography of the gel. For best results, the probe should be used within 24 h after preparation. However, for second- and third-round screenings, the probe can be reused provided the work progresses speedily.

Alternatively, cloned multimers of about 100 bp in length (e.g. 5-mers of a 20-bp recognition site DNA) have also been successfully used as probes. This can have a number of advantages. One is dealing with a better defined probe that can be labelled using the Klenow fragment. Labelling in such a manner results in less nicks or single-stranded DNA ends which may decrease background signals.

Screening procedure

One aspect which turned out to be crucial during our screenings was the plating density. We plate $1.5-2.0 \times 10^4$ phages per 145-mm-diameter Petri dish. Higher densities resulted in plaques too small for efficient protein expression in infected bacteria.

Since the described protocol encompasses about two days of work, including the night in between (at least for the first-round screening), it is important, when planning your experiment, to realise that there will be only one major break of about 6 h (between Steps 3 and 4).

(1) Plate out the library at a density of about $15-20 \times 10^3$ PFU per 145-mm-diameter prewarmed NZY-containing (1% casein hydrolysate, 0.5% yeast extract, 0.5% NaCl, 0.2% $MgSO_4 \cdot 7H_2O$, 1.5% agar) agar plate. Use 1.2 ml of a freshly grown culture of appropriate plating cells (e.g. Y1090r⁻, XL1-Blue or SURE) and 9 ml of top agarose (0.8% agarose in SM) per plate. Incubate the numbered plates at 42 °C until tiny little plaques are visible

(takes about 4 h). This step should be ready by about 8.00 p.m. We screen 12 large plates in a typical first round.

(2) While the plates are incubating, soak two nitrocellulose filters (S&S nitrocellulose BA85, 132 mm diameter) per large plate in 10 mM IPTG and allow them to dry on a plastic surface. In the case of a lambdaZAP library, the IPTG concentration was increased to 20 mM. Also prepare the binding buffer (you will need about 5 l) as well as 500 ml of 6 M guanidinium-HCl (GuHCl) in binding buffer and store at 4 °C.

(3) Once plaques become visible, transfer the plates to a 37 °C room or incubator, place one filter on each plate, label the filters according to the plate number, and unequivocally mark the position of the filter on the plate, e.g. by punching holes through the filter and into the agar using a syringe needle. Avoid cooling the plates below 37 °C (which would reduce protein expression). Incubate for 6 h at 37 °C. This incubation should be ready around 6 a.m.

(4) Remove the first set of filters from the plate, place them with the protein side up on Whatman paper, and place the second filter on the plate. Before labelling the new filters, proceed to Step 5, i.e. with the denaturation/renaturation of the proteins on the first set of filter replicas. For easy subsequent alignment of the two sets of filter replicas, the markings on the second set should be made at the same positions as they were on the first set. Incubate for about 2 h at 37 °C.

(5) All subsequent treatments are carried out at 4 °C and on an orbital shaker. Transfer the first set of filters one by one into 250 ml of 6 M GuHCl/binding buffer and incubate for 10 min. Dilute the remaining 250 ml of 6 M GuHCl/binding buffer with 250 ml of binding buffer (= 3 M GuHCl). Replace the 6 M GuHCl/binding buffer with 250 ml of the 3 M solution and incubate for 10 min. Use the residual 250 ml to prepare the next dilution. Incubate the filters sequentially in 1.5, 0.75, 0.375 and 0.19 M GuHCl/binding buffer and finally twice in binding buffer alone for 10 min at each step. The GuHCl must be completely removed during the last two steps, since otherwise the blocking reagent will precipitate on the filters and this will later cause a strong spotty background. In total, the procedure will take about 2 h.

(6) Remove the second set of filter replicas from the plates and incubate both sets in a fresh container with 250 ml of blocking solution (binding buffer +5% low-fat milk powder). Dip in the filters, one by one, and incubate for 2 h. In third and subsequent screening rounds, this incubation can be performed overnight.

(7) While the filters are soaking in the blocking solution, label the DNA probe (described in Step 4). Also, heat-denature 400 μg of salmon sperm DNA.

(8) Prepare a large Petri dish with about 20 ml of binding buffer containing 0.25% milk powder, and 5 μg/ml (of final volume) denatured salmon sperm competitor DNA, and add the double-stranded labelled DNA probe. Add filters individually, avoiding the transfer of buffer along with the filters. Take into account the fact that approximately 1 ml of buffer is

transferred with each large filter (final volume about 50 ml). Incubate on an orbital shaker for 4 h at 4 °C, allowing proper agitation of the filters. Remember to take appropriate safety precautions, since you are handling quite a lot of radioactivity. In a number of instances we have found that competitor DNA was not required or in fact was detrimental by reducing specific signals.

(9) Remove the filters one by one from the probe solution and transfer them into binding buffer containing 0.25% milk powder. Wash three times (20 min each) in 500–1000 ml of buffer on an orbital shaker. Place filters on a plastic undersurface (e.g. PE-filter paper from S&S), cover with Saran Wrap, and expose for 2–24 h at −80 °C. Keep filters frozen while developing and evaluating the films. It is possible to rewash the filters if the background is too high.

For subsequent rounds, one may reduce the amount of labelled dNTP to 50 μCi each, the volume of the nick-translation reaction to 30 μl, and the binding reaction volume to 40 ml. Also, the second and third rounds can be performed on 90-mm-diameter Petri dishes, and at least for the third round the screening procedure might be modified to restrict working hours to daytime. A typical result of a first-round screening is shown in Fig. 1A.

3.4 Methods confirming the specificity of the cloned gene products

Once phages displaying positive results in the southwestern screening are plaque pure, it must be demonstrated that the candidate(s) binds to the recognition site of interest sequence specifically. The most straightforward approach is the so-called 'pizza experiment' (Fig. 1D): the phage is plated on a 90-mm-diameter plate at a density of 100–200 PFU, and the resulting filter is cut into several pieces. The filter pieces are then probed with wild-type and with mutant versions of the (concatenated) binding site oligonucleotides. Apart from phage clones expressing proteins specific to the binding site (e.g. PRHP and PRHA in Fig. 1B), additional phages expressing proteins which recognise DNA independent of the actual sequence are found at a frequency of up to 50% (Fig. 1B; SW/A). If the binding specificity of the candidate phages is similar to that known from the nuclear factor recognising the sequence element of interest, the cDNA might be directly excised from the phage (lambdaZAP) or conventionally subcloned into a plasmid vector. For more valid data on the binding specificity beyond that obtained by the 'pizza experiment', competition experiments are required. To perform these, the candidate protein needs to be expressed in E. coli either from a lysogen (in the case of λgt11) or from a plasmid carrying an inducible promoter (Izawa et al., 1993). Alternatively, the cDNA can be brought under the control of, for example, the T7 promoter, and the candidate protein expressed in a coupled in vitro transcription/translation system (e.g. da Costa e Silva et al., 1993). We prefer the in vitro transcription/translation approach because of its wider applicability. With the expressed protein in one's hands, EMSAs can be performed, allowing a more

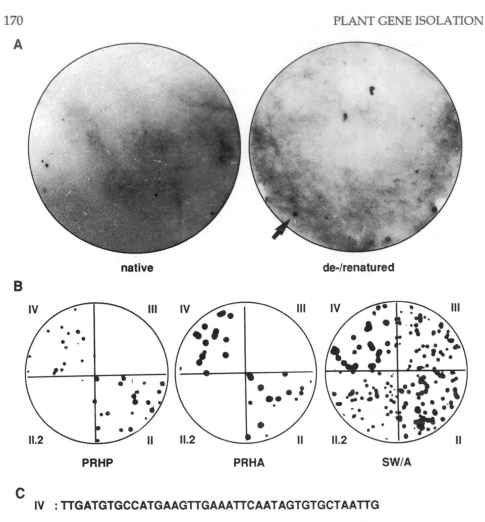

A

native   de-/renatured

B

IV   III   IV   III   IV   III

II.2   II   II.2   II   II.2   II

PRHP   PRHA   SW/A

C

IV : TTGATGTGCCATGAAGTTGAAATTCAATAGTGTGCTAATTG

II :   ATTCAATAGTGTGCTAATTGTTTAAGAGTTG

II.2 :   ATTCAATAGTGTGCTATCCAGTTAAGAGTTG

III : CTCCTGTACAATTCAAACATTGTTCAAACAAGGAACCTAAG

Figure 1. Screening of expression cDNA libraries using recognition site DNA probes. (A) shows one set of filter replicas (native and denatured/renatured) obtained from a first-round screening of an *Arabidopsis* lambdaZAP cDNA expression library using the recognition site sequence probe II depicted in (C). The arrow points to a positive signal of a recombinant phage clone that was subsequently plaque purified and shown to contain a plant homeodomain-protein-encoding cDNA (PRHA) (Korfhage et al., 1994). A typical 'pizza-type experiment' is shown in (B). Three plaque-purified phage clones originally screened with probe II, two isolated from an *Arabidopsis* (PRHA and SW/A) and one from a parsley (PRHP) cDNA library, were checked for sequence-specific binding to the recognition site DNA. Note that PRHP and PRHA, both encoding plant homeodomain proteins (Korfhage et al., 1994), show specific binding to DNA probes containing a TAATTG sequence motif (probes II and IV in (C)). In contrast, clone SW/A binds to all tested DNA probes, including to sequences in which this motif was mutated (II.2) or was lacking (III)

detailed characterisation of the protein–DNA interaction. Additional information, both general and plant specific, on this topic is available (Forster et al., 1992; Lane et al., 1992; Izawa et al., 1993).

4 SUCCESSFUL APPLICATION

The broad and general applicability of screening expression libraries with antibody probes is reflected by the ever-increasing number of genes cloned by this method. Table 1 gives a summary of a number of diverse plant-protein-encoding cDNAs that have been isolated by this immunological screening technique. The table is far from complete and is intended merely to draw attention to the various plant species from which genes have been cloned in this way. It also demonstrates that the majority of such cloned plant genes are derived from plant libraries constructed in bacteriophage lambda-derived prokaryotic expression vectors.

In contrast, Table 2 contains all published plant genes encoding DNA-binding proteins that have been cloned from plant expression libraries using the recognition site binding technique up to July 1994.

5 CONCLUDING REMARKS

Screening of expression libraries constructed with bacteriophage lambda and plasmid vectors using antibody probes has proven to be highly successful in the isolation of eukaryotic and prokaryotic genes. The technique has gained widespread application in nearly every field of molecular biology research. It requires only a specific antibody, thus bypassing the need for protein sequencing. Similarly, screening of expression libraries using DNA probes comprising specific binding sites (recognition site DNA) with which particular protein factors interact removes the need to enrich for, or to purify, transcription factors, which often are present in only very low amounts within the cells. Regardless of whether antibodies or DNA ligands were used as probes, the isolated cDNA clones may not contain the entire transcript. Nevertheless, these cDNA fragments can be used as probes to rescreen the libraries to obtain full-length cDNA clones or to isolate the corresponding genomic sequences. Additionally, the ability to produce large amounts of cloned transcription factors is an important prerequisite for detailed analyses of how such factors fulfil their functions, as activators or repressors, by binding to the regulatory DNA regions of specific genes. However, as mentioned earlier in this chapter both techniques have their limitations. Furthermore, alternative procedures to achieve the same goals exist. Recent advances in microsequence technology, to obtain amino acid sequence information on proteins and peptides, have dramatically reduced the amount of protein required to only 1–5 pmol. Hence, in an increasing number of cases, specific oligonucleotides based on amino acid sequence data are being used as probes to screen DNA libraries in which expression of the product is not required. Development of novel methods such as the one-hybrid system, which allows one to identify genes encoding proteins that recognise a specific DNA sequence in yeast cells (Li & Hersko-witz, 1993; Wang & Reed, 1993), may prove to be a valuable supplement to the recognition site DNA screening of expression libraries in *E. coli*. Since proteins expressed in yeast can undergo post-translational modifications, it is conceivable that genes encoding protein factors which require such modification for in vivo DNA binding can be isolated with this procedure.

ACKNOWLEDGMENT

We thank Dr Robert S. Cormack, Dr Paul Rushton and Michael Feldbrügge for critical reading of this manuscript.

Author for correspondence: I. E. Somssich

ESSENTIAL READING

Benton, W.D. and Davis, R.W. (1977) Screening Lambda gt recombinant clones by hybridization in situ. *Science*, **196**, 180–182.

Huynh, T.V., Young, R.A. and Davis, R.W. (1985) Constructing and screening cDNA libraries in λgt10 and λgt11. In *DNA Cloning: A Practical Approach* (ed. D.M. Glover), pp. 49–78. IRL Press, Oxford, Washington DC.

Singh, H., LeBowitz, J.H., Baldwin, A.S. and Sharp, P.A. (1988) Molecular cloning of an enhancer binding protein: isolation by screening an expression library with a recognition site DNA. *Cell*, **52**, 415–423.

Singh, H., Clerc, R.G. and LeBowitz, J.H. (1989) Molecular cloning of sequence-specific DNA binding proteins using recognition site probes. *BioTechniques*, **7**, 252–261.

Snyder, M., Elledge, S., Sweetser, D. et al. (1987) λgt11: gene isolation with antibody probes and other applications. *Methods Enzymol.*, **154**, 107–128.

REFERENCES

Baum, G., Chen, Y., Arazi, T. et al. (1993) A plant glutamate decarboxylase containing a calmodulin binding domain. Cloning, sequence, and functional analysis. *J. Biol. Chem.*, **268**, 19610–19617.

Bellmann, R. and Werr, W. (1992) Zmhox1a, the product of a novel maize homeobox gene, interacts with the *Shrunken* 26 bp *feedback* control element. *EMBO J.*, **11**, 3367–3374.

Breiteneder, H., Pettenburger, K., Bito, A. et al. (1989) The gene coding for the major birch pollen allergen *Betvl*, is highly homologous to a pea disease resistance response gene. *EMBO J.*, **8**, 1935–1938.

Broglie, K.E., Gaynor, J.J. and Broglie, R.M. (1986) Ethylene-regulated gene expression: molecular cloning of the genes encoding an endochitinase from *Phaseolus vulgaris*. *Proc. Natl. Acad. Sci. USA*, **83**, 6820–6824.

Buffard, D., Breda, C., van Huystee, R.B. et al. (1990) Molecular cloning of complementary DNAs encoding two cationic peroxidases from cultivated peanut cells. *Proc. Natl. Acad. Sci. USA*, **87**, 8874–8878.

Côté, F., Cutt, J.R., Asselin, A. and Klessig, D.F. (1991) Pathogenesis-related acidic β-1,3-glucanase genes of tobacco are regulated by both stress and developmental signals. *Mol. Plant-Microbe Interact.*, **4**, 173–181.

Cowell, I.G. and Hurst, H.C. (1993) Cloning transcription factors from a cDNA expression library. In *Transcription Factors: A Practical Approach* (ed. D.S. Latchman), pp. 105–123. IRL Press, Oxford.

da Costa e Silva, O., Klein, L., Schmelzer, E. et al. (1993) BPF-1, a pathogen-induced DNA-binding protein involved in the plant defense response. *Plant J.*, **4**, 125–135.

de Vetten, N.C., Lu, G. and Ferl, R.J. (1992) A maize protein associated with the G-box binding complex has homology to brain regulatory proteins. *Plant Cell*, **4**, 1295–1307.

Dehesh, K., Bruce, W.B. and Quail, P.H. (1990) A trans-acting factor that binds to a GT-motif in a phytochrome gene promoter. *Science*, **250**, 1397–1399.

DellaPenna, D., Alexander, D.C. and Bennett, A.B. (1986) Molecular cloning of tomato fruit polygalacturonase: analysis of polygalacturonase mRNA levels during ripening. *Proc. Natl. Acad. Sci. USA*, **83**, 6420–6424.

DeLuca, V., Marineau, C. and Brisson, N. (1989) Molecular cloning and analysis of cDNAs encoding a plant tryptophan decarboxylase: comparison with animal dopa decarboxylases. *Proc. Natl. Acad. Sci. USA*, **86**, 2582–2586.

Dunsmuir, P., Bond, D., Lee, K. et al. (1988) Stability of introduced genes and stability in expression. In *Plant Molecular Biology Manual* (eds S.B. Gelvin, R.A. Schilperoort and D.-P.S. Verma), pp. C1/1–17. Kluwer Academic Publishers, Dortrecht.

Efstratiadis, A., Kafatos, F.C., Maxam, A.M. and Maniatis, T. (1976) Enzymatic in vitro synthesis of globin genes. *Cell*, **7**, 279–288.

Ehrlich, K.C., Cary, J.W. and Ehrlich, M. (1992) A broad bean cDNA clone encoding a DNA-binding protein resembling mammalian CREB in its sequence specificity and DNA methylation sensitivity. *Gene*, **117**, 169–178.

Feltkamp, D., Masterson, R., Starke, J. and Rosahl, S. (1994) Analysis of the involvement of *ocs*-like bZIP-binding elements in the differential strength of the bidirectional *mas*1'2' promoter. *Plant Physiol.*, **105**, 259–268.

Foley, R.C., Grossman, C., Ellis, J.G. et al. (1993) Isolation of a maize bZIP protein subfamily: candidates for the ocs-element transcription factor. *Plant J.*, **3**, 669–679.

Forster, R., Gasch, A., Kay, S. and Chua, N.-H. (1992) Analysis of protein/DNA interactions. In *Methods in Arabidopsis Research* (eds C. Koncz, N.-H. Chua and J. Schell), pp. 378–392. World Scientific, Singapore.

Fritzemeier, K.-H., Cretin, C., Kombrink, E. et al. (1987) Transient induction of phenylalanine ammonia-lyase and 4-coumarate:CoA ligase mRNAs in potato leaves infected with virulent and avirulent races of *Phytophthora infestans*. *Plant Physiol.*, **85**, 34–41.

Fromm, H. and Chua, N.-H. (1992) Cloning of plant cDNAs encoding calmodulin-binding proteins using ^{35}S-labeled recombinant calmodulin as a probe. *Plant Mol. Biol. Rep.*, **10**, 199–206.

Garabedian, M.J., LaBaer, J., Liu, W.-H. and Thomas, J.R. (1993) Analysis of protein–DNA interactions. In *Gene Transcription: A Practical Approach* (eds B.D. Hames and S.J. Higgins), pp. 243–293. IRL Press, Oxford.

Gilmartin, P.M., Memelink, J., Hiratsuka, K. et al. (1992) Characterization of a gene encoding a DNA binding protein with specificity for a light-responsive element. *Plant Cell*, **4**, 839–849.

Gowri, G., Paiva, N.L. and Dixon, R.A. (1991) Stress responses in alfalfa (*Medicago sativa* L.) 12. Sequence analysis of phenylalanine ammonia-lyase (PAL) cDNA clones and appearance of PAL transcripts in elicitor-treated cell cultures and developing plants. *Plant Mol. Biol.*, **17**, 415–429.

Grunstein, M. and Hogness, D.S. (1975) Colony hybridization: a method for the isolation of cloned DNAs that contain a specific gene. *Proc. Natl. Acad. Sci. USA*, **72**, 3961–3965.

Gubler, U. and Hoffman, B.J. (1983) A simple and very efficient method for generating cDNA libraries. *Gene*, **25**, 263–269.

Guiltinan, M.J., Marcotte, W.R.J. and Quatrano, R.S. (1990) A plant leucine zipper protein that recognizes an abscisic acid response element. *Science*, **250**, 267–271.

Hai, T., Liu, F., Coukos, W.J. and Green, M.R. (1989) Transcription factor ATF cDNA clones: an extensive family of leucine zipper proteins able to selectively form DNA-binding heterodimers. *Genes Dev.*, **3**, 2083–2090.

Hanahan, D. and Meselson, M. (1980) Plasmid screening at high colony density. *Gene*, **10**, 63–67.

Hauffe, D.K. (1988) Induktion von O-Methyltransferasen der Furanocumarinbiosynthese in *Petroselinum crispum* L. PhD thesis, University of Köln.

Hedrick, S.A., Bell, J.N., Boller, T. and Lamb, C.J. (1988) Chitinase cDNA cloning and mRNA induction by fungal elicitor, wounding, and infection. *Plant Physiol.*, **86**, 182–186.

Higuchi, R., Paddock, G.V., Wall, W. and Salser, W. (1976) A general method for cloning eucaryotic structural gene sequences. *Proc. Natl. Acad. Sci. USA*, **73**, 3146–3150.

Höfte, H., Desprez, T., Amselm, L. et al. (1993) An inventory of 1152 expressed sequence tags obtained by partial sequencing of cDNAs from *Arabidopsis thaliana*. *Plant J.*, **4**, 1051–1061.

Izawa, T., Foster, R. and Chua, N.H. (1993) Plant bZIP protein DNA binding specificity. *J. Mol. Biol.*, **230**, 1131–1144.

Katagiri, F. and Chua, N.-H. (1992) Plant transcription factors—present knowledge and future challenges. *Trends Genet.*, **8**, 22–27.

Katagiri, F., Lam, E. and Chua, N.-H. (1989) Two tobacco DNA-binding proteins with homology to the nuclear factor CREB. *Nature*, **340**, 727–730.

Kawaoka, A., Kawamoto, T., Sekine, M. et al. (1994) A cis-acting element and a trans-acting factor involved in the wound-induced expression of a horseradish peroxidase gene. *Plant J.*, **6**, 87–97.

Korfhage, U., Trezzini, G.F., Meier, I. et al. (1994) Plant homeodomain protein involved in transcriptional regulation of a pathogen defense-related gene. *Plant Cell*, **6**, 695–708.

Lane, D., Prentki, P. and Chandler, M. (1992) Use of gel retardation to analyze protein–nucleic acid interaction. *Microbiol. Rev.*, **56**, 509–528.

Li, J.J. and Herskowitz, I. (1993) Isolation of ORC6, a component of the yeast origin recognition complex by a one-hybrid system. *Science*, **262**, 1870–1874.

Lois, A.F. and West, C.A. (1990) Regulation of expression of the casbene synthetase gene during elicitation of castor bean seedlings with pectic fragments. *Arch. Biochem. Biophys.*, **276**, 270–277.

Lugert, T. and Werr, W. (1994) A novel DNA-binding domain in the *Shrunken* initiator-binding protein (IBP1). *Plant Mol. Biol.*, **25**, 493–506.

Maekawa, T., Sakura, H., Kanei-Ishii, C. et al. (1989) Leucine zipper structure of the protein CRE-BP1 binding to the cyclic AMP response element in brain. *EMBO J.*, **8**, 2023–2028.

Mehdy, M.C. and Lamb, C.J. (1987) Chalcone isomerase cDNA cloning and mRNA induction by fungal elicitor, wounding and infection. *EMBO J.*, **6**, 1527–1533.

Meier, I. and Gruissem, W. (1994) Novel conserved sequence motifs in plant G-box binding proteins and implications for interactive domains. *Nucleic Acids Res.*, **22**, 470–478.

Mikami, K., Sakamoto, A. and Iwabuchi, M. (1994) The HBP-1 family of wheat basic/leucine zipper proteins interacts with overlapping cis-acting hexamer motifs of plant histone genes. *J. Biol. Chem.*, **269**, 9974–9985.

Nakajima, N., Mori, H., Yamazaki, K. and Imaseki, H. (1990) Molecular cloning and sequence of a complementary DNA encoding 1-aminocyclopropane-1-carboxylate synthase induced by tissue wounding. *Plant Cell Physiol.*, **31**, 1021–1029.

Nieto-Sotelo, J., Ichida, A. and Quail, P.H. (1994) PF1: an A-T hook-containing DNA binding protein from rice that interacts with a functionally defined d(AT)-rich element in the oat phytochrome A3 gene promoter. *Plant Cell*, **6**, 287–301.

Oeda, K., Salinas, J. and Chua, N.-H. (1991) A tobacco bZip transcription activator (TAF-1) binds to a G-box-like motif conserved in plant genes. *EMBO J.*, **10**, 1793–1802.

Okayama, H. and Berg, P. (1982) High efficiency cloning of full-length cDNA. *Mol. Cell. Biol.*, **2**, 161–170.

Pabo, C.A. and Sauer, R.T. (1992) Transcription factors: structural families and principles of DNA recognition. *Annu. Rev. Biochem.*, **61**, 1053–1095.

Pasquali, G., Goddijn, O.J.M., de Waal, A. et al. (1992) Coordinated regulation of two indole alkaloid biosynthetic genes from *Catharanthus roseus* by auxin and elicitors. *Plant Mol. Biol.*, **18**, 1121–1131.

Perisic, O. and Lam, E. (1992) A tobacco DNA binding protein that interacts with a light-responsive box II element. *Plant Cell*, **4**, 831–838.

Rabbitts, T.H. (1976) Bacterial cloning of plasmids carrying copies of rabbit globin mRNA. *Nature*, **260**, 221–225.

Rougeon, F., Kourilsky, P. and Mach, B. (1975) Insertion of a rabbit β-globin gene sequence into an *E. coli* plasmid. *Nucleic Acids Res.*, **2**, 2365–2378.

Sambrook, J., Fritsch, E.F. and Maniatis, T. (1989) *Molecular Cloning: A Laboratory Manual*. Cold Spring Harbor Laboratory Press, Cold Spring Harbor, NY.

Sato, T. and Theologis, A. (1989) Cloning the mRNA encoding 1-aminocyclopropane-1-carboxylate synthase, the key enzyme for ethylene biosynthesis in plants. *Proc. Natl. Acad. Sci. USA*, **86**, 6621–6625.

Scharf, K.-D., Rose, S., Zott, W. et al. Three tomato genes code for heat stress transcription factors with a region of remarkable homology to the DNA-binding domain of the yeast HSF. *EMBO J.*, **9**, 4495–4501.

Schindler, U., Beckmann, H. and Cashmore, A.R. (1992) TGA1 and G-box binding factors: two distinct classes of Arabidopsis leucine zipper proteins compete for the G-box-like element TGACGTGG. *Plant Cell*, **4**, 1309–1319.

Schindler, U., Beckmann, H. and Cashmore, A.R. (1993) HAT3.1, a novel Arabidopsis homeodomain protein containing a conserved cysteine-rich region. *Plant J.*, **4**, 137–150.

Scioli, J.R. and Zilinskas, B.A. (1988) Cloning and characterization of a cDNA encoding the chloroplastic copper/zinc-superoxide dismutase from pea. *Proc. Natl. Acad. Sci. USA*, **85**, 7661–7665.

Shao, W. and Pratt, K. (1994) Cloning genes for DNA binding proteins: dot-blotting to optimize cDNA expression library screening. *Anal. Biochem.*, **218**, 465–468,

Sheehy, R.E., Pearson, J., Brady, C.J. and Hiatt, W.R. (1987) Molecular characterization of tomato fruit polygalacturonase. *Mol. Gen. Genet.*, **208**, 30–36.

Short, J.M., Fernandez, J.M., Sorge, J.A. and Huse, W.D. (1988) λ ZAP: a bacteriophage lambda expression vector with *in vivo* excision properties. *Nucleic Acids Res.*, **16**, 7583–7600.

Sikela, J.M. and Hahn, W.E. (1987) Screening an expression library with a ligand probe: isolation and sequence of a cDNA corresponding to a brain calmodulin-binding protein. *Proc. Natl. Acad. Sci. USA*, **84**, 3038–3042.

Singh, K., Dennis, E.S., Ellis, J.G. et al. (1990) OCSBF-1, a maize ocs enhancer binding factor: isolation and expression during development. *Plant Cell*, **2**, 891–903.

Spano, A.J., He, Z., Michel, H. et al. (1992) Molecular cloning, nuclear gene structure, and developmental expression of NADPH: protochlorophyllide oxidoreductase in pea (*Pisum sativum* L.). *Plant Mol. Biol.*, **18**, 967–972.

Stanley, K.K. (1983) Solubilization and immune-detection of b-galactosidase hybrid proteins carrying foreign antigenic determinants. *Nucleic Acids Res.*, **11**, 4077–4092.

Tabata, T., Takase, H., Takayama, S. et al. (1989) A protein that binds to a cis-acting element of wheat histone genes has a leucine zipper motif. *Science*, **245**, 965–967.

Tabata, T., Nakayama, T., Mikami, K. and Iwabuchi, M. (1991) HBP-1a and HBP-1b: leucine zipper-type transcription factors of wheat. *EMBO J.*, **10**, 1459–1467.

Takatsuji, H., Mori, M., Benfey, P.N. et al. (1992) Characterization of a zinc finger DNA-binding protein expressed specifically in *Petunia* petals and seedlings. *EMBO J.*, **11**, 241–249.

Takeuchi, Y., Yoshikawa, M., Takeba, G. et al. (1990) Molecular cloning and ethylene induction of mRNA encoding a phytoalexin elicitor-releasing factor, b-1,3-endogluca-nase, in soybean. *Plant Physiol.*, **93**, 673–682.

Teutsch, H.G., Hasenfratz, M.P., Lesot, A. et al. (1993) Isolation and sequence of a cDNA encoding the Jerusalem artichoke cinnamate 4-hydroxylase, a major plant cytochrome P450 involved in the general phenylpropanoid pathway. *Proc. Natl. Acad. Sci. USA*, **90**, 4102–4106.

Tjaden, G. and Coruzzi, G.M. (1994) A novel AT-rich DNA binding protein that combines an HMG I-like DNA binding domain with a putative transcription domain. *Plant Cell*, **6**, 107–118.

Valenta, R., Duchêne, M., Pettenburger, K. et al. (1991) Identification of profilin as a novel pollen allergen; IgE autoreactivity in sensitized individuals. *Science*, **253**, 557–560.

van Tunen, A.J., Koes, R.E., Spelt, C.E. et al. (1988) Cloning of the two chalcone flavanone isomerase genes from Petunia hybrida; coordinate, light regulated and differential expression of flavonoid genes. *EMBO J.*, **7**, 1257–1263.

Varin, L., DeLuca, V., Ibrahim, R.K. and Brisson, N. (1992) Molecular characterization of two plant flavonol sulfotransferases. *Proc. Natl. Acad. Sci. USA*, **89**, 1286–1290.

Vinson, C.R., LaMarco, K.L., Johnson, P.F. et al. (1988) In situ detection of sequence specific DNA binding activity specified by a recombinant bacteriophage. *Genes Dev.*, **2**, 801–806.

Wang, M.M. and Reed, R.R. (1993) Molecular cloning of the olfactory neuronal tran-scription factor Olf-1 by genetic selection in yeast. *Nature*, **364**, 121–126.

Weißhaar, B., Armstrong, G.A., Block, A. et al. (1991) Light-inducible and constitutively expressed DNA-binding proteins recognizing a plant promoter element with func-tional relevance in light responsiveness. *EMBO J.*, **10**, 1777–1786.

Werr, W., Überlacker, B. and Klinge, B. (1994) Screening of cDNA expression libraries with synthetic oligonucleotides for DNA-binding proteins. In *Plant Molecular Biology Manual* (eds S.B. Gelvin, R.A. Schilperoort and D.-P.S. Verma), pp. F2/1–12. Kluwer Academic Publishers, Dordrecht.

Young, R.A. and Davis, R.W. (1983a) Efficient isolation of genes by using antibody probes. *Proc. Natl. Acad. Sci. USA*, **80**, 1194–1198.

Young, R.A. and Davis, R.W. (1983b) Yeast RNA polymerase II genes: isolation with antibody probes. *Science*, **222**, 778–782.

8 Functional Complementation in Yeast and *E. coli*

JAMES A. H. MURRAY
ALISON G. SMITH
University of Cambridge, Cambridge, UK

Functional complementation involves suppression of a mutant phenotype by introducing a wild-type copy of the corresponding gene. This chapter discusses the principles and practice of isolating plant cDNAs by functional complementation of yeast and *E. coli* mutants.

(1) We present a compilation of the different categories of plant genes isolated by this method, and also describe the use of functional complementation to indicate or verify the function of a cloned plant gene.

(2) Functional complementation using yeast or *E. coli* requires a strain carrying a mutation corresponding to the plant gene of interest, a library of plant cDNAs in an appropriate expression vector, and a suitable selection scheme to identify transformants in which the phenotype has been rescued.

(3) The main advantages of the method are that no DNA or amino acid sequence information about the plant gene is required, nor do these sequences need to be conserved between the plant and the mutant host. The cDNAs isolated uncode functional proteins and are often full-length.

(4) The principal disadvantages are that the strategy cannot be applied to genes for which no analogous mutation exists in yeast or *E. coli*, the most obvious examples of which are those for photosynthetic proteins. Future developments in transformation procedures may open up a wider range of host organisms. Correct targeting of the plant protein within the mutant host cell may also create limitations.

1 INTRODUCTION – A BRIEF HISTORY OF COMPLEMENTATION

Complementation, the restoration of the normal phenotype of a mutant by the introduction of a wild-type allele, is one of the most important techniques in genetic analysis. It is used in conjunction with sexual crossing to prove whether mutations are in the same or different genes, or with genetic transformation techniques to show whether cloned DNA sequences correspond to given mutations. Indeed, it was complementation of a live non-virulent *Pneu-*

Plant Gene Isolation: Principles and Practice. Edited by G. D. Foster and D. Twell.

mococcus strain with a substance from a heat-killed virulent strain that resulted in the original identification of DNA as the genetic material (Avery et al., 1944).

The use of functional complementation to isolate individual genes also has a long history (at least in molecular biology terms), since it was the approach adopted for the isolation of the first *E. coli* genes to be cloned in the late 1970s. The rationale of the approach is to transform a mutant strain that has a readily identifiable phenotype with a library of DNA fragments in a suitable vector, and then select those colonies with a restored wild-type phenotype. With the use of appropriate growth conditions, cells not containing a clone of the wild-type gene cannot grow, whereas cells taking up by chance a vector containing the appropriate gene will be able to grow, divide and form colonies (illustrated for yeast in Fig. 1). In this way, colonies containing putative clones of the gene of interest can be identified, and the transforming DNA can then be prepared for further analysis in vitro.

The approach of functional complementation as a way to identify clones in random libraries corresponding to mutations of interest is not limited to reintroduction of DNA into the original source organism as was the case when *E. coli* genes were cloned using *E. coli* mutants. The conservation of function between organisms means that in many cases the equivalent protein from a different organism is able to function sufficiently well to provide rescue of the mutant phenotype. Probably the first examples of this were the cloning of genomic sequences encoding the yeast *HIS3* and *LEU2* genes (encoding imidazole glycerol phosphate dehydratase and β-isopropylmalate dehydrogenase respectively) by complementation of their equivalent mutations in *E. coli* *hisB* and *leuB* (Ratzkin & Carbon, 1977; Struhl & Davis, 1977). This required that the DNA from yeast was expressed in *E. coli*. Apparently, yeast upstream sequences gave sufficient expression of the genomic sequences in *E. coli* for rescued clones to be obtained with many mutants. Together with the low occurrence of introns in yeast genes, this allowed the yeast *ARG4*, *TRP1*, *TRP5* and *URA3* genes to be cloned by complementation of *E. coli* mutants over the period 1976–1978. In all cases, libraries of yeast DNA fragments cloned in plasmid vectors were used (review: Carbon, 1993).

Despite these early successes in isolating yeast genes by complementing *E. coli* mutants, and the isolation of the *Drosophila* phosphoribosylglycinamide formyl transferase (GART) gene by complementation of the yeast *ade8* mutation with a library of *Drosophila* genomic DNA fragments (Henikoff et al., 1981), the approach of functional complementation was not widely applied until the late 1980s. This was in part due to the lack of appropriate cDNA libraries available in yeast or *E. coli* vectors, and in part to a largely unsubstantiated feeling that the distance between *E. coli* and yeasts on the one hand and multicellular higher eukaryotes on the other was perhaps too great for successful functional complementation, and moreover that only biosynthetic or catabolic enzymes were likely to be identifiable by such a strategy.

A turning point in the general perception of the usefulness and power of functional complementation was the cloning by Lee & Nurse (1987) of a human gene complementing the cell cycle mutation *cdc2* of *Schizosaccharomyces pombe*.

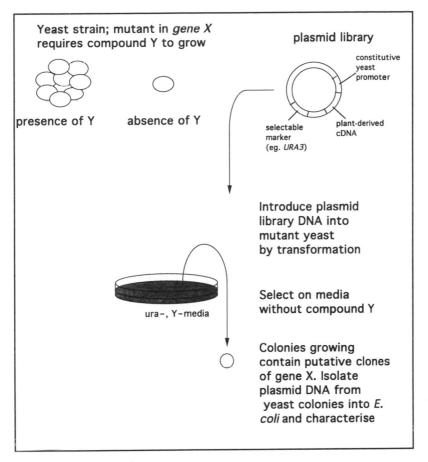

Figure 1. The principle of complementation illustrated for isolation of cDNA clones from a plant corresponding to a yeast mutant in gene X, which results in inability of cells of this yeast strain to grow on media lacking compound Y. A plant cDNA library is constructed in a vector allowing expression of the cDNAs in yeast, and introduced into the yeast strain by transformation. In vector shown here, the yeast marker *URA3* is carried on the plasmid as the selectable marker. Selection can be carried out directly for complementing clones as shown here, by plating on media lacking both uracil and compound Y. Alternatively, the initial plating can select for plasmid uptake on media lacking uracil but containing Y, and colonies are then replica-plated on ura⁻ Y⁻ media. The relative advantages of the two approaches are discussed in the text. In either case, plasmids are rescued back into *E. coli* by preparing yeast DNA and transforming *E. coli*, in order to prepare sufficient plasmid DNA for characterisation of the clones obtained. The principles of using an *E. coli* mutant are the same, except that the final step is clearly not required

In this case, the authors again did not use a library dedicated for expression in fission yeast, but used a 'standard' expression cDNA library designed for use in mammalian cells and carrying the SV40 promoter. They therefore relied on sufficient expression from this promoter in fission yeast, and cotransformation with a separate selectable vector for selection. The isolation of a human homologue of a gene known to have a key role in yeast cell division control not only suggested that cellular processes could be sufficiently well conserved over wide evolutionary spans for functional complementation to be useful for a much greater range of genes than previously considered likely, but also proved that it could be successful in cases where DNA sequence divergence was sufficient to preclude a cloning strategy based on low-stringency hybridisation.

The isolation of the human *cdc2* gene was followed by isolation of cDNAs using yeast complementation for a mammalian cAMP phosphodiesterase (Colicelli et al., 1989), a human GART gene (Schild et al., 1990), a multifunctional purine biosynthetic gene (Minet & Lacroute, 1990), and a human subunit homologous to a yeast CCAAT-binding transcription factor (Becker et al., 1991). In this latter case, as with the *cdc2* cloning, complementation presumably required functional interactions with other yeast proteins involved in a multi-subunit complexes.

In conjunction with this changing perception, and with vectors and libraries suitable for a complementation approach becoming more widely available (Schild et al., 1990; Elledge et al., 1991), the stage was set for a far more widespread use of functional complementation for the isolation of genes involved in a range of cellular and biochemical processes from higher eukaryotes. Interest was further stimulated by the use of *E. coli* and yeast to identify sequence-specific DNA-binding proteins and to identify clones encoding proteins interacting with a target polypeptide using two-hybrid (section 4.3) and related strategies (Elledge et al., 1989; Chien et al., 1991).

Probably the first plant genes cloned by complementation were a phosphoenolpyruvate cDNA from maize (Izui et al., 1986), superoxide dismutase cDNAs from tobacco and *Arabidopsis* (Van Camp et al., 1990), and soybean Δ^1-pyrroline-5-carboxylate reductase (Delauney & Verma, 1990). All were obtained by complementing *E. coli* mutants.

Nowadays, given a suitable mutant, functional complementation is almost always the first choice for the isolation of homologous genes of micro-organisms, including bacteria, fungi and algae such as *Chlamydomonas*, for which high-efficiency transformation systems are available. The approach is also being increasingly used to test the identity of genes isolated by other means, such as DNA homology, chromosome walking or random sequencing of clones (Chapters 5, 10 and 15). A leap of faith is required to use functional complementation to isolate genes from evolutionarily distant organisms, since it requires the conservation of function rather than of nucleotide sequence or epitopes (for DNA probes and antibodies, respectively). The numerous examples of the isolation of heterologous genes by complementation demonstrate its success for a wide range of genes (see Table 1). In this chapter, we outline the use of this method for the isolation of genes from higher plants. We focus on complementation of the Gram-negative bacterium *E. coli* and the budding yeast *Saccharomyces cerevisiae*, for both of which a vast collection of mutant phenotypes is readily available, reflecting not only biosynthetic and catabolic metabolic processes, but also cellular processes such as secretion, cell division, signal transduction and so on. However, the principles are equally applicable to complementation cloning using many host organisms.

2   REQUIREMENTS FOR FUNCTIONAL COMPLEMENTATION

Functional complementation, either for homologous or heterologous gene isolation, has three essential requirements: a mutant organism with an easily detectable phenotype, an efficient transformation procedure and a library of clones which can be expressed in the mutant. There are many examples of plant mutants (review: King, 1986), and plant transformation processes, at least for some species, are relatively straightforward (review: Gheysen et al., 1992). In principle, therefore, it may well be possible to isolate a plant gene by homologous complementation. In practice, however, the rates of transformation of plant cells are too low, and the methods of selection for transformants too laborious, for this to be a sensible strategy to follow at the present time. What is more likely is the use of cyanobacterial and *Chlamydomonas* mutants, both of which can be easily genetically manipulated. This is particularly important for the isolation of genes related to photosynthetic processes, which are very similar in these organisms and higher plants. However, since to date the majority of published examples are for the isolation of plant genes by the functional complementation of *E. coli* and budding yeast mutants (see section 3), we will concentrate on these. Nonetheless, many of the comments we make will be of direct relevance to the use of other organisms. Furthermore, they will be applicable to researchers interested in testing the possible function of cDNAs already isolated and showing homology to yeast or *E. coli* proteins for which corresponding mutant phenotypes are available.

The principal advantages of complementation cloning, when a choice of possible strategies is available, is that the cDNA clones isolated are often full-length or nearly so, and that something is known about the activity of the protein they encode. However, at this point it should be pointed out that functional complementation of a mutant does not necessarily prove that the gene isolated is a direct homologue of the defective gene, but rather may be able to overcome the defect by some other means. A good example of this phenomenon is that of 5-aminolaevulinic acid synthase (ALA synthase), an enzyme responsible for the synthesis of the tetrapyrrole precursor ALA. The enzyme is found in animal and yeast mitochondria, and in some bacteria, and uses succinyl CoA and glycine. Both the human and bacterial genes for ALA synthase can complement an *E. coli hemA* mutant (Schoenhaut & Curtis, 1986; Leong et al., 1982), despite the fact that in *E. coli*, as in most bacteria and all plants and algae, ALA is made by a completely different pathway, starting from glutamate. A further example was the isolation of truncated clones of mitotic cyclins, as well as cyclins involved in G1- to S-phase transition, by complementation of a yeast strain deficient in its own G1 cyclins (Léopold & O'Farrell, 1991; Lew et al., 1991).

Table 1. Plant genes isolated by complementation of yeast (*S. cerevisiae*) and *E. coli* mutants

Gene isolated	Plant	Organism and gene complemented	Authors
Nucleotide biosynthesis enzymes			
Dihydroorotate dehydrogenase	*Arabidopsis*	Yeast *ura1*, also *ura2, 4, 5–10, his3, leu2, ade2* and *trp1*, complemented but not characterised	Minet et al. (1992)
Adenine *P*-ribosyltransferase	*Arabidopsis*	*E. coli purM*	Moffat et al. (1992)
AIR synthetase	*Arabidopsis*	Also *purB*, *purC* and *purH*	Senecoff & Meagher (1993)
5-Aminoimidazole ribonucleotide (AIR) carboxylase and AIR-4-N-succinocarboxamide synthase	*Vigna aconitifolia* (moth bean)	*purE* and *purC*	Chapman et al. (1994)
Glycinamide ribonucleotide (GAR) synthetase, GAR transformylase, aminoimidazole ribonucleotide (AIR) synthetase	*Arabidopsis*	*E. coli purD, purN* and *purM*	Schnorr et al. (1994)
Aspartate transcarbamylase, orotate phosphoribosyltransferase, orotidine-5-phosphate decarboxylase	*Arabidopsis*	yeast *ura3, ura5–ura10* and *ura2*	Nasr et al. (1994)
Amino acid biosynthesis enzymes			
Δ-1-pyrrolidone 5-carboxylase reductase	Soybean	*E. coli proC*	Delauney & Verma (1990)
3-Isopropylmalate dehydrogenase	*Arabidopsis* Brassica napus	*E. coli leuB* Yeast *leu2*	Elledge et al. (1991) Ellerström et al. (1992)
Phosphoribosylanthranilate transferase	*Arabidopsis*	*E. coli trpD*	Elledge et al. (1991), Rose et al. (1992)
Chorismate mutase	*Arabidopsis*	Yeast *aro7*	Eberhard et al. (1993)
Cysteine synthase	Watermelon	*E. coli* Cys⁻ (*cycK cysM*)	Noji et al. (1994)
Other biosynthetic enzymes			
Cycloartenol synthase	*Arabidopsis*	Yeast[a] *erg7*	Corey et al. (1993)
Ferrochelatase	*Arabidopsis*	Yeast *hem15*	Smith et al. (1994)

3-Methyladenine DNA glycosylase	Arabidopsis	E. coli (double mutant deficient in 3-methyladenine glycosylases (tag alkA)	Santerre & Britt (1994)
ATP-sulphurylase	Potato	Yeast met3	Klonus et al. (1994)
	Arabidopsis	Yeast met3	Leustek et al. (1994)
5'-Adenylylphosphosulphate kinase	Arabidopsis	Yeast met14	Jain & Leustek (1994)
Mevalonate kinase	Arabidopsis	Yeast erg12	Riou et al. (1994)
Cholinephosphotransferase	Soybean	Yeast cpt1	Dewey et al. (1994)
Membrane transporters			
K^+ transporter	Arabidopsis	Yeast trk1, trk2	Sentenac et al. (1992), Anderson et al. (1992), Sussman (1992)
Sucrose carrier	Spinach	Yeast	Riesmeier et al. (1992)
	Potato	Yeast	Riesmeier et al. (1993)
General amino acid transporter (AAP2)	Arabidopsis	Yeast mutant in citrulline uptake	Kwart et al. (1993)
Neutral amino acid transporter I (Pro)	Arabidopsis	Yeast mutant in proline uptake	Frommer et al. (1993)
Neutral amino acid transporter II (His and Ala)	Arabidopsis	Yeast hip1 mutant in histidine uptake	Hsu et al. (1993a)
Amino acid/H^+ transporter	Arabidopsis	Yeast	Hsu et al. (1993b)
Other proteins			
Fe superoxide dismutase	Arabidopsis and Nicotiana plumbaginifolia	E. coli sodB	Van Camp et al. (1990)
ER to Golgi traffic	Arabidopsis and S. pombe	Yeast sar1 and sec12	D'Enfert et al. (1992)
Ser/Thr protein kinase (LHCP kinase?)	Arabidopsis	Yeast (GAL4/LCHP fusion protein)	Kohorn et al. (1992)
UV damage-repair enzymes	Arabidopsis	E. coli recA−, Uvr−, Phr−	Pang et al. (1993)
Poly(A)-binding proteins	Arabidopsis	Yeast	Belostotsky & Meagher (1993)
G1 cyclins	Arabidopsis	Yeast	Soni et al. (1995)
Transcription factor	Arabidopsis	Yeast hap2	Edwards & Smith (unpublished)
Peroxisomal biogenesis	Arabidopsis	Yeast	Baker (unpublished)

[a] Assay of transformants by thin-layer chromatography for epoxysqualene mutase activity, not by growth.

3   GENES ISOLATED BY COMPLEMENTATION CLONING

Table 1 is a compilation of examples of plant genes isolated to date by functional complementation of mutants of *E. coli* or yeast. It is not intended to be an exhaustive list, but rather gives a indication of the range of proteins which can complement defects in these micro-organisms. Since heterologous functional complementation has been used to a greater extent in organisms other than plants, Table 2 presents examples of other proteins whose genes have been isolated by this strategy, to suggest further possibilities for cloning of plant homologues. Finally, Table 3 gives examples of yet other proteins, from both plants and other organisms, whose functions have been established by functional complementation of *E. coli* or yeast mutants, again to give an indication of the possibilities of the technique.

4   CONSIDERATIONS

The genes listed in Tables 1–3 represent examples from many different biochemical pathways and cellular processes, including enzymes, transcription factors, components of multi-subunit complexes, and membrane-associated processes, suggesting that complementation has a wide applicability. However, careful thought is required before embarking on a programme to attempt

Table 2. A selection of genes from other organisms isolated by complementation

Gene product isolated	Organism	Mutant complemented	Authors
cdc2 cyclin-dependent protein kinase	Human	*S. pombe cdc2*	Lee & Nurse (1987)
G1/S cyclins	Human	Yeast *cln⁻*	Koff et al. (1991), Lew et al. (1991), Xiong et al. (1991)
Chaperonin-like protein	Human	Yeast *gap1*	Segel et al. (1992)
Adenylosuccinate lyase	Avian	*E. coli purB*	Aimi et al. (1990a, b)
CCAAT-binding protein	Human	Yeast *hap2*	Becker et al. (1991)
CDC25 homologue (ras activator)	Mouse	Yeast *cdc25*	Martegani et al. (1992)
CDC34 homologue	Human	Yeast *mec1*	Plon et al. (1993)
Glycogen branching enzyme	Human	Yeast	Thon et al. (1993)
Dihydroxypolyprenyl-benzoate methyltransferase	Rat	Yeast ubiquinone synthesis mutant (*coq3*)	Marbois et al. (1994)
ALA dehydratase and uroprophyrinogen III synthase	*Anacystis*	*E. coli hemB* and *hemD*	Jones et al. (1994)

Table 3. A selection of genes whose function has been tested or analysed in yeast (*S. cerevisiae*) or *E. coli* mutants

Gene product tested	Organism	Mutant complemented	Authors
Plants			
Glutamine synthetase	Bean	*E. coli glnA*	Bennett & Cullimore (1990)
p34^{cdc2} (cyclin-dependent kinase)	Maize	*S. pombe cdc2*	Colasanti et al. (1991)
Protein kinase	Rye	Yeast *snf1*	Alderson et al. (1991)
Protein phosphatase type I	*Arabidopsis*	*S. pombe dis2–11*	Nitschke et al. (1992)
Phytoene desaturase	Tomato	*E. coli* (not mutant)	Pecker et al. (1992)
Cysteine synthase	Spinach	*E. coli* Cys⁻ (*cycK cysM*)	Saito et al. (1992)
Mn-SOD	Maize	Yeast *sod2*	Zhu & Scandalios (1992)
Protein phosphatase	*Arabidopsis*	*S. pombe cdc25*^{ts}/ *wee 1*⁻	Ferreira et al. (1993)
ER retention receptor	*Arabidopsis*	Yeast (not mutant)	Lee et al. (1993)
GTP-binding protein	*Arabidopsis*	Yeast *ypt6*	Bednarek et al. (1994)
Other organisms			
Phytoene synthase	Synechococcus	*E. coli* (not mutant)	Chamovitz et al. (1992)
Photolyase	*Anacystis*	*E. coli*	Takao et al. (1989)
secE	*Staphylococcus carnosus*	*E. coli secE*	Meens et al. (1994)
Ribonuclease H	Trypanosomatid	*E. coli*	Campbell & Ray (1993)
secA	*Bacillus subtilis*	*E. coli secA*	Klose et al. (1993)
DNA polymerase β	Mammalian	*E. coli*	Sweasy & Loeb (1993)
RCC1 chromosome condensation protein	Human	Yeast *prp20*	Fleischmann et al. (1991)
Multidrug resistance (mdr)	Mammalian	Yeast *ste6*	Raymond et al. (1992)
CDC42	*C. elegans*	Yeast *cdc42*	Chen et al. (1993)
Mitotic activator p80^{cdc25}	Human	*S. pombe cdc25*	Gould et al. (1990)
Metallothionein	Monkey	Yeast *cup1*	Thiele et al. (1986)

the isolation of any specific gene, and in this section we discuss a number of pertinent considerations.

There are two basic reasons for carrying out complementation experiments, leading to slight differences in the approaches and considerations involved, although the experimental techniques are the same. The first is aimed at the isolation of cDNAs complementing a given mutation, and the majority of the chapter will concentrate on this. The second type of experiment aims to obtain functional information concerning genes that have already been cloned or their similarity to genes identified in other species, by expressing them in heterologous organisms and assaying their ability to confer specific phenotypes (Table 3). This frequently requires the complementation of mutations, as in the analysis of the complementation of the yeast *prp20* mutation by the human chromosome condensation gene RCC1 (Fleischmann et al., 1991), but can also involve the specification of additional biochemical processes not normally found in that organism. For example, in plants the first committed step in carotenoid biosynthesis is the conversion of geranylgeranyl pyrophosphate (GGPP) to prephytoene pyrophosphate (PPPP) and thence to phytoene. The expression of a putative phytoene synthase clone in *E. coli*, together with a known GGPP synthase, resulted in *E. coli* cells able to accumulate phytoene, demonstrating the identity of the clone and that a single polypeptide enzyme converts PPPP to phytoene (Chamovitz et al., 1992).

Experiments involving functional testing will normally be more straightforward, in that the principal requirements will normally be to assess whether rescue of a given mutation occurs, and to ascertain the characteristics of the rescued cells.

4.1 Choice of biological process

Despite the many examples presented here, it is probable that some biological processes are likely to be more amenable than others to a complementation cloning strategy. One might suggest that the cloning of an enzyme activity which converts a soluble substrate to a soluble diffusible product should be a relatively straightforward target, but even here one must consider the existence and conservation of the pathway in question in yeast or *E. coli*. Clearly, screening for nitrogen-fixation genes, for example, in either of these organisms is inappropriate, and more subtle differences between biosynthetic pathways may also exist. Moreover, the possibility of alternative or scavenging pathways in these organisms interfering with the proposed screen, and the intracellular location of the pathway in both the source and screening organisms, must be considered. Thus if a desired enzyme activity is located in the plant chloroplast, is it likely to be targeted to the mitochondrion in yeast, and is this the normal site of action of the yeast protein? Furthermore, if post-translational modifications such as processing or glycosylation are required for normal function of the eukaryotic protein, are they likely to occur in the screening organism? Such considerations may exclude *E. coli* for screening in some cases. Types of post-translational modification and expression in *E. coli* and yeast are reviewed in Yarranton (1992) and Barr (1992) respectively.

Similar considerations also apply to screening for genes involved in cellular processes such as division or secretion. Here, more careful consideration of likely conservation of processes is required, and in most cases the use of yeast rather than *E. coli* will be required.

Despite the foregoing words of caution, if there is a lesson to be learned from the applications of functional complementation, it is that major surprises have resulted from bold experiments. There are probably a number of factors that account for this. In yeast, one of these may be the variability of plasmid copy number frequently seen in different yeast transformants using high-copy plasmids (Futcher & Cox, 1984), which is likely to provide a range of expression levels. This variation occurs between separate clones (i.e. colonies) arising from the same transformation with any given high-copy yeast plasmid. Thus, even if a protein is required in stoichiometric rather than catalytic amounts, transformants with the correct expression level may be obtained, providing sufficient are screened. Further variation in expression levels appears to occur as a result of clones inverted with respect to the promoter designed to express them in the screening organism (see section 6).

This variability can also help compensate in cases in which the intracellular destination of the expressed protein is incorrectly specified for correct function in yeast, since high-level expression can overload the normal pathway for that signal, resulting in protein also being present elsewhere in the cell, particularly the cytosol. This has been observed for the TRP1 gene product, normally a cytosolic protein, but which is secreted when artificially fused to a secretion signal. *trp1* mutants expressing this fusion protein at low to moderate levels cannot grow on media lacking tryptophan because the normally cytosolic protein is directed into the endoplasmic reticulum and secreted. However, high-level expression from the strong *GAL1–GAL10* promoter of the secretion signal–TRP1 fusion does allow *trp1* mutants to grow, presumably because the secretion pathway is overloaded, and TRP1 protein accumulates in the cytosol (Toyn et al., 1988).

4.2 Choice of organism and mutant strain

The choice of organism for screening is likely in many cases to be dictated by the availability of the appropriate mutation of interest, by the considerations discussed in the section 4.1, or by the libraries which have been constructed or can be obtained. Local experience available in the laboratory, or from nearby collaborators, may also affect the final decision. In general, the organism of choice is likely to be *E. coli* or the budding yeast *S. cerevisiae* simply because of the wide range of mutants available and the relative ease of transformation. Many mutant strains are obtainable from stock centres, particularly the Yeast Genetic Stock Center at Berkeley, and the Bachmann *E. coli* collection at Yale, or direct from the appropriate researcher. Addresses of these and other culture collections are given in Appendix 1.

However, in certain cases there may be a specific requirement to use other bacterial mutants, yeasts (particularly fission yeast, *Schizosaccharomyces pombe*), filamentous fungi or photosynthetic algae, and it is likely that this range of organisms will be extended in the future. For a limited number of genes a choice of screening host may be available, and in such cases factors such as the higher efficiency of transformation and more rapid growth of *E. coli*, coupled with the ability to prepare large quantities of DNA for putative clones directly, may well bias the researcher towards *E. coli*.

The choice of mutant allele to be used is liable to have a significant effect on the final outcome of the experiment. It is highly desirable to use a non-reverting allele, preferably a deletion of the gene, or at least a double-point mutation, as otherwise a substantial background of revertant colonies can be anticipated. Such mutants can be constructed if the cloned gene is available, particularly in yeast, although here it should be considered whether a gene has been inactivated by targeted insertion of a plasmid that leaves a repeated structure on the chromosome and can therefore be re-excised by general recombination processes in the cell, or whether a 'one-step gene disruption' strategy has been used (Rothstein, 1983), which actually deletes the endogenous gene and is essentially non-reverting.

In any case, before a screening experiment is carried out, it is essential to measure the reversion frequency of the proposed strain (or preferably a selection of alternative strains) under the actual selection conditions to be employed. Ideally this would entail taking them through the transformation procedure using a control vector plasmid, and plating in the conditions to be used for the screen. Reversion should be $< 10^{-6}$. Duplicate platings on media selecting only for the marker conveniently allows the transformation frequencies of the strains in question to be assessed at the same time. If replica plating is to be used as part of the procedure (section 5.2) then this should also be assessed. If practicable, parallel experiments should be performed with the cloned homologous gene, where available, in order to assess the proposed selection conditions.

4.3   Alternative screening strategies

In addition to straightforward screens based on the omission of a required intermediate or of an antimetabolite designed for selection of clones rescuing existing mutants, it is also possible to devise artificial selections based on molecular genetic manipulations. One such strategy involves replacing the promoter of the gene in the screening organism with a tightly regulated promoter, effectively creating a conditional phenotype dependent on the growth media employed. This type of approach is particularly useful for essential genes for which appropriate conditional mutants are not available or if no selection can be devised for them. For example, there are three genes for G1 cyclins in budding yeast (*CLN1*, *CLN2* and *CLN3*) which are functionally redundant so that the activity of all three genes must be removed to stop cell division (Richardson et al., 1989). No such triple loss-of-function mutants had been isolated by conventional mutagenesis, so to devise a selection system to clone heterologous cyclins, strains were constructed in which two of the three *CLN* genes were inactivated by targeted insertional mutagenesis and the third had its normal upstream sequences replaced by a galactose-inducible promoter. Strains of this type show a galactose-dependent growth phenotype, and stop dividing on media containing glucose, since this represses the galactose-inducible promoter responsible for expressing the only G1 cyclin gene they contain. Such strains have been used to isolate G1 cyclins from human (Koff et al., 1991; Lew et al., 1991; Xiong et al., 1991), *Drosophila* (Lahue et al., 1991; Léopold &

O'Farrell, 1991) and *Arabidopsis* (Soni et al., 1995) by transforming with suitable cDNA libraries and selecting on plates containing glucose. In such cases, we have observed that it is also important to maintain selection for the markers used for gene inactivation, as otherwise an unacceptably high level of reversion occurs. It should be borne in mind that it is also possible in principle to isolate cDNAs that activate the repressed promoter in this type of screen (in other words, for the *GAL* promoter, a *GAL4* homologue).

This type of strategy can be used in many situations where the cloned gene is available for the screening organism, but, since it involves molecular genetic manipulation and characterisation of the strain before use, will often be less convenient than conventional mutants if these are available. If it is contemplated, the most important consideration is that a suitably tightly regulated promoter is chosen to control expression of the endogenous gene in the screening organism, so that it can be completely inactivated in the appropriate selection conditions. The *GAL1–GAL10* and similar galactose-inducible promoters satisfy this requirement in *S. cerevisiae* and as a well-tested system probably constitute the first choice. Other possibilities for yeast include systems based on induction by foreign compounds such as copper (Macreadie et al., 1991) or glucocorticoid hormones (Picard et al., 1990). This latter example also operates in *S. pombe*. For *E. coli*, a number of inducible promoter systems are also available, though most are leaky when uninduced (review: Yarranton, 1992). In all cases the suitability of a given promoter will be a function of its leakiness and the level of expression which causes interference in the selection scheme.

A further class of selection is based on activation of a specific promoter construct by a foreign cDNA. In the specific case of cloning transcription factors, it would be possible to establish a selection based, for example, on activation of *lacZ* (β-galactosidase) expression in yeast, using an artificial promoter containing the binding motif for the desired factor. cDNA clones encoding transcription factors binding to these elements and activating transcription would result in a blue colour on appropriate indicator plates. For DNA-binding proteins that do not have transcriptional activation properties, this approach would not be suitable, but an *E. coli* assay for identifying factors binding to a specified DNA sequence has also been described (Elledge et al., 1989).

A more general application of the type of yeast screen described in the previous paragraph is the so-called two-hybrid strategy for studying protein–protein interactions (Chien et al., 1991; Bartel et al., 1993; Guarente, 1993). This powerful approach can be used both to test the existence of a physical association between proteins (Jackson et al., 1993; Li & Fields, 1993; Brown et al., 1994) and to isolate new polypeptides that interact with a defined protein 'bait' (Hannon et al., 1993; Harper et al., 1993; Staudinger et al., 1993). It relies on the ability of transcription factors such as GAL4 to be split into separate polypeptide chains carrying the DNA-binding domain and the transcriptional activation domain. However, these domains do not interact to form a functional transcriptional activator unless the two polypeptides are brought into close physical proximity. Thus if the DNA-binding domain of GAL4 (or another DNA-binding protein such as bacterial lexA) is fused to all or part of one protein (the 'bait'), and a test or 'target' protein is fused to the transcriptional activator domain of GAL4 (a viral *trans*-activator domain such as VP16 can also be used), a functional transcriptional activator is formed if the bait specifically interacts with the target protein. This is detected by the presence of a construct in the cell with multiple GAL4-binding sites upstream of a β-galactosidase gene, resulting in blue colonies when *trans*-activation occurs. The target protein can be either a specific construct, or a fusion library of random cDNAs linked to a GAL4 activation domain.

A number of modifications to the two-hybrid approach can be envisaged, including an interference assay for detecting proteins that block the activation by a defined bait–target pair, for example by specific proteolysis. This was used to isolate an *Arabidopsis* serine-threonine kinase, which interacts with the N-terminus of a light-harvesting chlorophyll *a/b*-binding protein (LHCP). The cDNA was selected in yeast from an *Arabidopsis* expression library for its ability to inhibit a transcriptional activator GAL4–LHCP fusion protein, but not inhibit native GAL4 protein (Kohorn et al., 1992). However, this work highlights the difficulty of interpreting the role of some proteins isolated by this type of approach.

The two-hybrid system has been widely used to analyse mammalian proteins and identify clones encoding interacting proteins. It is likely that it will become a much more important tool for plant molecular biologists than it has been to date, as the appropriate libraries become more widely available.

4.4 Choice of vector and library

The choice of library to use in a screening experiment is another factor liable to have a significant influence on the outcome. The decision has to be made whether an existing library can be used, or whether a new library must be constructed. In Table 4 we present information on a selection of libraries used to obtain the genes listed in Table 1.

4.4.1 Plasmid vectors

Plasmids are required for all yeast and the majority of E. coli complementation experiments. Libraries are therefore either constructed directly in plasmid vectors, or in lambda phage vectors that can be subsequently converted to plasmid form. Plasmid cDNA libraries are generally considered more difficult to construct, largely because of the lower frequency of direct DNA transformation compared to infection by in vitro packaged lambda phage. However, the use of electroporation, offering transformation frequencies of more than 10^{10} transformants per microgram of DNA (Inoue et al., 1990; Zabarovsky & Winberg, 1990), makes the construction of plasmid libraries of large numbers of clones more feasible. The excellent *Arabidopsis* cDNA library in the yeast pFL61 vector (Fig. 2A) is particularly notable (Minet et al., 1992), and has been successfully used for complementation by a number of different groups (see Table 4).

Further plasmid vectors used for construction of plant cDNA expression libraries are listed in Table 4. For *E. coli*, most are simply constructed in the polylinker of vectors of the pUC or pBluescript™ type, making use of the *lac* promoter which normally expresses β-galactosidase and provides the blue–white selection of such vectors. However, it should be noted that the translation initiation codon in these vectors is upstream of the polylinker, so that inserted cDNAs will either be expressed as fusion proteins or as a consequence of reinitiation of translation at initiation codons within the cDNA sequence.

Table 4 also lists a number of yeast plasmid vectors in which plant cDNA libraries have been constructed. Essentially any yeast expression vector with the appropriate selection marker and a promoter of the desired class (see below) can be used to construct a new plasmid library, as will be seen from a survey of yeast references quoted in Table 2. Schild et al. (1990) and Becker et al. (1991) describe useful yeast vectors for human complementing libraries, and yeast expression vectors have been the subject of a detailed review by Romanos et al. (1992). Most cDNA synthesis and cloning kits available commercially are based on phage cloning, because of its generally higher efficiency, but Life Technologies (previously Gibco BRL) market a kit for construction of directional cDNA plasmid libraries between unique *Sal*I and *Xho*I sites. The plasmid vector provided with the kit, pSPORT1, is designed for cloning of cDNAs between T7 and Sp6 promoters. It is therefore not suitable for complementation. We have constructed two vectors suitable for use with this plasmid cDNA synthesis kit which express in budding yeast (*S. cerevisiae*: plasmid pCEX) and fission yeast (*S. pombe*: plasmid pPEX) respectively (J.A.H. Murray & J.P. Carmichael, in preparation). Both vectors contain appropriately placed unique *Sal*I–*Not*I–*Xho*I restriction sites, in pCEX downstream of a galactose-inducible promoter (*GAL10*–*CYC1* fusion; selectable marker *URA3*), and in pPEX downstream of an *S. pombe ADH* promoter (Selectable marker *URA3*; complements *S. pombe ura4*).

4.4.2 Phage vectors

In order to use phage libraries for complementation experiments they must be converted to plasmid form. There are two approaches, both of which can be used to convert quantitatively complete libraries (Short et al., 1988; Delauney & Verma, 1990; Elledge et al., 1991). The first, employed in the λZAP series of vectors (Short et al., 1988) (marketed by Stratagene), has a central 'plasmid' portion of the phage which is equivalent to the widely used plasmid pBluescript (or derivatives thereof), and it is into the polylinker in this region that the cDNAs are cloned. The plasmid sequences are flanked on one side by the initiation sequence for replication by single-stranded phage such as f1 or M13, and on the other by the replication termination sequence. Super-infection with wild-type M13 phage results in replication of the plasmid portion between

Table 4. Selection of plant libraries used to screen *E. coli* and yeast mutants by complementation

Reference	Source of mRNA	Vector type	Vector name	Promoter	Selection[b]	Notes
E. coli						
Van Camp et al. (1990)	*Nicotiana plumbaginifolia* cell suspension	Plasmid	pUC18	*lac*	Ampicillin[R]	
Delauney & Verma (1990)	*Arabidopsis thaliana* Soybean 21-day-old nodules	Plasmid Lambda phage/phagemid	pUC18 λZAPII/ pBluescript II[d]	*lac* *lac*	Ampicillin[R] Ampicillin[R]	Screened as plasmid[a]
Chapman et al. (1994)	Mothbean (*Vigna aconitifolia*) nodules	Plasmid	pcDNAII	*lac*	Ampicillin[R]	
Senecoff & Meagher (1993)	*Arabidopsis thaliana* leaves and stems	Plasmid	pcDNAII	*lac*	Ampicillin[R]	
Elledge et al. (1991)	*Arabidopsis thaliana* whole plants	Lambda phage/plasmid	λYES	*lac*	None as phage Amp.[R] as plasmid	Screened as plasmid
Schnorr et al. (1994)	*Arabidopsis thaliana* whole plants	Lambda phage/plasmid	λYES	*lac*	None	Library of Elledge et al. (1991) Screened directly as phage in λKC lysogen[c]
	Arabidopsis thaliana suspension cells	Lambda phage/phagemid	λZAP/ pBluescript II[d]	*lac*	Ampicillin[R]	Screened as plasmid[a]

Yeast (*S. cerevisiae*)

		Phage/ plasmid (CEN)				
Elledge et al. (1991)	*Arabidopsis thaliana* whole plants	Phage/ plasmid (CEN)	λYES	GAL1[e]	URA3	Screened as plasmid. See also Anderson et al. (1992), Hsu et al. (1993), Leustek et al. (1994) and Jain & Leustek (1994)
Minet et al. (1992)	*Arabidopsis thaliana* seedlings	Plasmid (2μ)	pFL61	PGK[f]	URA3	See also Nasr et al. (1994), Sentenac et al. (1992), Corey et al. (1993), Smith et al. (1994) and Soni et al. (1995)
Riesmeier et al. (1992)	Spinach leaves	Plasmid (2μ)	YEp112A1NE	ADH1[g]	TRP1	Constructed from inserts excised from λZAPII library
Klonus et al. (1994)	Potato leaves	Plasmid (2μ)	YEp112A1NE	ADH1[g]	TRP1	Constructed from inserts excised from λZAPII library
Ellerström et al. (1992)	Rape immature seeds	Plasmid (2μ)	pHR50	GAPDH[h]	URA3	Fusion library to GAL4 activation domain
Smith et al. (1993)	*Arabidopsis* sulphur-starved roots	Plasmid (2μ)	pYES2	GAL1	URA3	

[a]λZAPII library constructed as lambda phage and converted to plasmid form by super-infection with helper phage for screening.

[b]In addition to the screening selection.

[c]The λYES library was constructed as a phage, but can be converted to plasmid form by cre site-specific recombinase, which acts on target *loxP* sites that flank the central portion of the λYES vector. λKC expresses cre, so *E. coli* strains with the required mutations which can be converted to a λKC lysogen can be screened directly by infecting with the λYES library. Infecting phage are converted to plasmid form when they enter the cell, and are maintained as plasmids. Alternatively, for screening other *E. coli* strains, or for transforming yeast, the λYES library can be converted to plasmid form by passage through the λKC lysogen BNN132 (Elledge et al., 1991).

[d]After conversion to plasmid form.

[e]*Gal1* promoter is inducible by galactose; there is little or no uninduced expression. Expression levels can be manipulated by altering glucose/galactose ratios.

[f]*PGK* (phosphoglycerate kinase) promoter is constitutive, but induced to a higher level by glucose.

[g]*ADH1* (alcohol dehydrogenase) promoter is essentially constitutive (Ammerer, 1983).

[h]*GAPDH* (glyceraldehyde-3-phosphate dehydrogenase) promoter is essentially constitutive.

Type of yeast plasmid maintenance sequences is also indicated. CEN, carries centromere for single-copy maintenance; 2μ, carries sequences from 2μ plasmid for high-copy maintenance. See text and Barr (1992) for review.

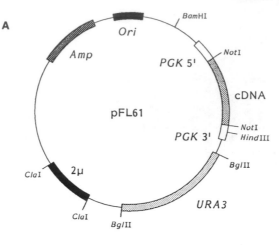

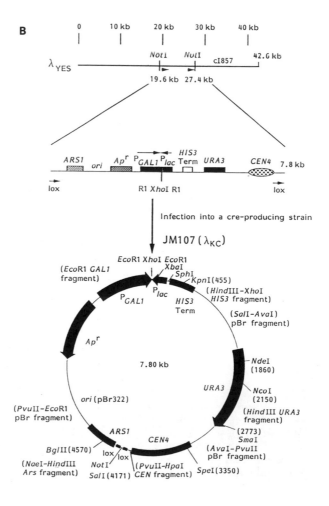

these signals. Packaging signals have been removed, so replicating double-stranded plasmid is produced. Libraries based on λZAP and λZAPII have been converted to plasmid form and used to screen *E. coli* mutants, relying on expression from the *lac* promoter that lies upstream of the polylinker in pBluescript (see references in Table 4). However, to our knowledge a vector of this type designed to express in yeast has not yet been constructed, although Brunelli & Pall (1993a) describe a versatile series of yeast expression vectors suitable for directional recloning of cDNAs from 'ZAP'-type vectors.

The second strategy relies on excision catalysed by cre site-specific recombinase, which acts on *loxP* sites flanking a central plasmid portion. Infection of *E. coli* expressing cre results in excision of the plasmid from the phage in a circular form. This is the basis of the vector λYES (Fig. 2B; Elledge et al., 1991), as well as a family of vectors described by Brunelli & Pall (1993b). All these vectors can express in both yeast and *E. coli*, in the case of λYES from non-directional cDNAs cloned between opposing yeast (*GAL1*) and *E. coli* (*lac*) promoters, or from directionally cloned cDNAs behind yeast promoters also showing activity in *E. coli* (Brunelli & Pall, 1993b). Further vectors similar to λYES, but with the essentially constitutive *ADH1* promoter, are also available (Elledge et al., 1993).

The *Arabidopsis* library in λYES has been widely used after conversion to plasmid form (see references in Table 4) and was made from mRNA extracted from leaves, shoots, stems and flowers of *Arabidopsis* plants at all stages of development.

4.4.3 Features of existing libraries

For experiments designed to clone *Arabidopsis* genes by yeast complementation which do not require library preparation from specific tissues, the choice is essentially between the Elledge λYES (Elledge et al., 1991) and Lacroute pFL61 libraries (Minet et al., 1992). These illustrate different principles that also apply to the choice of other vectors. The λYES-derived library (the plasmid vector after cre-mediated excision from its carrier lambda arms is called pSE936) expresses cDNAs under a galactose-inducible promoter. The vector also carries a yeast centromere sequence, so it is held at single copy in yeast cells. The combination of low copy number and the *GAL1* promoter means that very low levels of cDNA are expressed on glucose plates, with much higher levels when

Figure 2. (A) Map of the plasmid yeast expression vector pFL61 (Minet et al., 1992). A widely used *Arabidopsis* cDNA library is available in this vector, which also illustrates the general principles of yeast plasmid vectors, including a 2μ plasmid-derived sequence for high-copy-number and relatively stable maintenance in yeast, a yeast selectable marker (*URA3*), and cDNA cloning sites lying between a PGK (phosphoglycerate kinase) promoter and 3′ transcription termination signals. (B) Map of the λYES vector designed by Elledge et al. (1991), showing the cre–lox-mediated conversion from a phage to a plasmid vector. cDNAs are cloned between the *GAL1* promoter and the *HIS3* terminator to allow expression in yeast. The initial library can be constructed in the phage form of the vector, and then transformed into yeast after conversion to the plasmid form. Reproduced with permission from Elledge et al. (1991)

the promoter is induced by galactose. In principle, intermediate levels of expression can be achieved by manipulating the glucose/galactose ratio. This means that potential complementing clones can be tested for a galactose-dependent growth phenotype, to check that growth is controlled by the promoter and is not due to reversion. However, the phenomenon of cDNAs expressed from plant 5' untranslated sequences, whatever their orientation, (see above) means that this is not a watertight test. Clearly this vector is not suitable for artificial selection based on placing a chromosomal gene behind a galactose-inducible promoter. It should also be noted that the *GAL* promoter can be leaky when used on a high-copy-number vector (Brunelli & Pall, 1993a).

In contrast, in the pFL61 library cDNAs are expressed behind the phosphoglycerate kinase (PGK) promoter (Fig. 2A), and sequences from the yeast 2μ plasmid provide high copy number, and relatively stable maintenance in yeast strains containing the wild-type 2μ plasmid. Such strains are designated [cir$^+$], and comprise the majority of laboratory yeast strains unless specified plasmid-free [cir^0]. The *PGK* promoter is expressed under most conditions, but is further induced on glucose, the normal carbon source for yeast growth. Thus there are fewer opportunities for manipulating expression with this vector, but the previous comments (regarding the benefits of the clonal variability of plasmid copy numbers discussed in section 4.1) apply.

4.4.4   S. pombe *vectors*

For complementation of *Schizosaccharomyces pombe* (fission yeast), there are a number of specialised vectors available, of which the plasmid pPEX discussed above is one (see Moreno et al. (1991) for a general review of *S. pombe* techniques and expression systems). There are also vectors available that have promoters active in both *S. cerevisiae* and *S. pombe* (Picard et al., 1990), or have potential to express in these two yeasts and mammalian cells (using SV40 promoter) (Camonis et al., 1990), although these plasmids remain untested for complementation library construction.

4.4.5   Two-hybrid *vectors*

A number of vectors suitable for construction of two-hybrid libraries with cDNAs fused to the activation domain of GAL4 or VP16 have been described (Chien et al., 1991; Bartel et al., 1993; Legrain et al., 1994), and a series of vectors are also available from Clontech Laboratories Inc., Palo Alto ('Matchmaker' series). Recently, an *Arabidopsis* fusion library for two-hybrid screening has become available through the *Arabidopsis* Stock Center in Ohio.

5   TRANSFORMATION AND SCREENING PROCEDURES

In this section we outline typical procedures that might be used for library screening by complementation in *E. coli* and yeast. The testing of function of cloned genes will only involve construction of plasmids, transformation and assays of phenotype, so most of the comments below do not apply in these cases.

5.1 Transformation and screening—*E. coli*

Screening of *E. coli* mutants for complementing clones will involve transforming the library into competent cells, and selecting for clones which allow growth on selective media. For example, if clones complementing the *leuB* mutation were desired, minimal media without leucine (but containing other nutritional requirements for the strain) would be used. It is advisable also to include the antibiotic selection for the transforming plasmid (normally ampicillin) to reduce the background of revertant colonies. The concentration of antibiotic required for selection should be tested in advance on minimal plates plus leucine with a test plasmid, since the concentration required may be different than on rich plates. Under certain circumstances, for instance if the rescue is weak, it may be necessary to select primary transformants on rich media containing antibiotic, followed by replica-plating onto media to select for functional complementation. This is more likely to be necessary for yeast mutant hosts (see section 5.2), but we suggest that this strategy is attempted with *E. coli* mutants if the initial double-selection procedure does not work.

Transformation can be carried out by any high-efficiency method, including modified calcium chloride procedures (Hanahan, 1985; Ausubel et al., 1987; Sambrook et al., 1991). However, electroporation is probably the method of choice (Inoue et al., 1990; Zabarovsky & Winberg, 1990), since it appears to give a more linear increase in the number of colonies obtained when increasing amounts of transforming plasmid DNA are present, providing that the conductivity of the solution does not become so high that arcing between the electrodes occurs. For this reason it is important that the plasmid DNA is as free of salts as possible.

Libraries received from other workers may need to be amplified before use. Standard procedures for the amplification of phage libraries are available in cloning manuals (Ausubel et al., 1987; Sambrook et al., 1991). Plasmid libraries should be amplified by plating at least three times the original library size at a density at which individual colonies can just be distinguished. The colonies are washed from the plates for large-scale plasmid DNA preparation. This procedure reduces the risk that plasmids causing slightly lower growth could be lost from the library, as might be the case in liquid culture.

Phage libraries will also need to be converted into plasmid form as discussed above. The only exceptions to this are libraries of the λYES type, which are used in a host that can be made lysogenic for the cre-expressing phage λKC (Elledge et al., 1991). In this case, plasmid excision occurs immediately on entry of the λYES phage into the cell, making it possible to transfect the lambda phage library directly. It should be noted that not all *E. coli* strains can be made into λKC lysogens (Schnorr et al., 1994).

After transforming the library and selecting for functional complementation, large, clearly growing colonies are picked from the plates and plasmid preparations made. The plasmids obtained are retransformed into the test strain to check that the complementing phenotype is plasmid linked. Plasmid clones passing the retransformation test are mapped for restriction endonuclease sites and/or hybridised to one another to allow them to be grouped into one or more classes. At the same time, subclones can be generated to test the location of the complementing sequence in the cDNA clone, and these can also be used for DNA sequencing.

Particularly if polymerase chain reaction (PCR) has been used at any stage in library construction, it is also advisable to carry out Southern blots, hybridising representatives of the isolated clone(s) to genomic DNA extracted from the source organism, to confirm that the clones obtained have the expected origin.

5.2   Transformation and screening—yeast

The typical procedures involved in isolating cDNAs by complementation of yeast are illustrated in Fig. 3, taking the example of the *Arabidopsis* gene for ferrochelatase (Smith et al., 1994). For detailed protocols, see Smith & Murray (1996).

Various methods of yeast transformation are available, including spheroplast/PEG procedures (Burgers & Percival, 1987) and electroporation (Becker & Guarente, 1991; Manivasakam & Schiestl, 1993). However, the most convenient procedure is the treatment of whole cells with lithium salts, and a high-efficiency protocol using single-stranded carrier DNA has been published (Schiestl & Gietz, 1989; Gietz et al., 1992, Elledge et al., 1993). We have found that this procedure gives good results with a variety of genetic backgrounds, usually over 10^4 transformants per microgram of pFL61 plasmid library DNA. Higher frequencies are also reported. It is relatively easy to scale up, either by multiple transformations or by using larger volumes, although in this case the length of heat shock has to be adjusted.

Yeast transformants can be selected only for plasmid uptake (e.g. URA^+ for pFL61), and the colonies obtained then replica-plated onto medium that selects for the functional complementation. In this case, large irregular colonies are picked for further analysis, since these arose from many cells transferred from the original selection plate. Smooth colonies are likely to be contaminants. Alternatively, the transformants can be selected directly on plates demanding both plasmid uptake and functional complementation. (It is important to select for the plasmid as well as complementation, as otherwise a high frequency of revertants may be observed.) This has the advantage that poorly complementing clones are more readily identified, since they will appear over a period of 1–2 weeks, and replica plating is not required. However, for nutritional selections, the demands on the cell of recovery from the transformation, at the same time as growing on minimal medium required for the functional complementation selection, may be too much. These problems can be reduced by allowing transformants to recover in rich medium for 2 h before plating, and by using 'dropout' plates for the selection. These contain all amino acids except those being selected, and give considerably faster growth than minimal plates

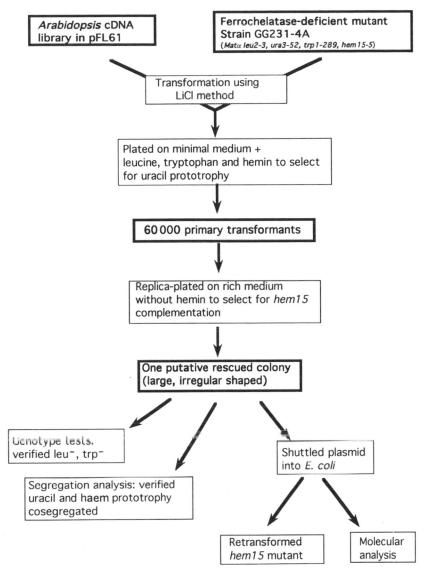

Figure 3. Typical procedure of isolating plant cDNA clones by complementation, illustrated by the example of isolating an *Arabidopsis* cDNA for the haem biosynthesis enzyme ferrochelatase, encoded by *HEM15* (Smith et al., 1994). The yeast strain GG231-4A contains no detectable ferrochelatase activity and so can only grow in the presence of exogenous haemin

containing only amino acids essential for the strain (Ausubel et al., 1987). Our advice would be that for biosynthetic genes replica plating may be the best approach, whilst for cellular processes, where rescue may be weaker, direct selection is better.

Having obtained putative colonies by either approach, it is then necessary to establish that the growth is of the mutant strain rescued by a foreign cDNA.

Contaminant yeast can sometimes be a problem, particularly during autumn, and since these are presumably wild type it is worth testing that putative colonies still have the other auxotrophic requirements that would be expected. In many cases it may then be worth carrying out segregation analysis to show that the phenotype is plasmid linked. Yeast plasmids are lost by 1–10% of cells per generation, so overnight non-selective growth in liquid culture usually results in a mixed population of plasmid-containing and plasmid-free cells. These can be identified by replica plating or picking colonies from rich to minimal media. Alternatively, for certain markers counterselection is available (e.g. 5-fluoroorotic acid for *URA3*) which directly selects for *ura*⁻ (and therefore plasmid-free) cells (Boeke et al., 1987) (see Sikorski & Boeke (1991) for various counterselections available).

When using galactose-inducible libraries in vectors such as λYES, it will be necessary to use galactose in the screening media, and loss of complementing ability due to promoter switch-off after transfer to glucose plates can be used to demonstrate plasmid-associated rescue. However, the comments made above concerning expression from 5′ untranslated sequences in the plant cDNA should be noted.

Further analysis requires transfer of potentially rescuing plasmids into *E. coli* so that DNA can be prepared. A number of protocols for this have been published offering rapid DNA preparations from small culture volumes (Chow, 1989; Soni & Murray, 1992). The best methods are those which use highly competent *E. coli*. If these can be made or purchased, rescue into *E. coli* followed by retransformation of the mutant yeast strain may be more rapid and conclusive than plasmid loss experiments. In any case, rescued plasmids will be tested for retransforming ability (Fig. 3), and it is advisable to do this selecting both directly for complementation and initially for plasmid uptake only. Colonies can then be streaked onto double-selective plates to test for heterologous complementation by the cDNA clone.

In some cases of weak rescue, it can be difficult to assess whether growth is taking place. During the isolation of G1 cyclins from *Arabidopsis* we believed it was likely that a number of potential cyclins might be isolated (Soni et al., 1995). We were therefore potentially interested in clones that gave even very slow growth, and used a micromanipulator to move individual cells to defined positions on selective plates. Microscopic re-examination of these cells after a number of hours of growth allowed the number of cell divisions in a given time period to be assessed, and we were able to identify unequivocably even weakly rescuing clones (Fig. 4).

Further analysis of clones will then involve their grouping by restriction mapping and hybridisation, and DNA sequencing.

6   POSSIBLE PROBLEMS OR WHY DOESN'T IT WORK?

Despite the continuing appearance of a wide variety of genes isolated by functional complementation, it is difficult to assess the frequency with which

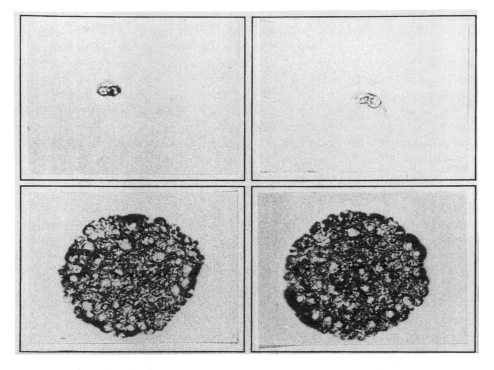

Figure 4. Micromanipulation of individual yeast cells, grown on medium selecting for the presence of the plasmid marker, and then transferred to a plate selecting both for the plasmid marker and for complementation by an *Arabidopsis* cDNA. Cells contain either vector with no insert (top panels), or vector plus complementing cDNA (lower panels). Cells were photographed after overnight growth. Clones complementing the G1 cyclin deficient yeast strain BF305-15d#21 were isolated from the pFL61 library of Minet et al. (1992) as described (Soni et al., 1995). Lower panels show complementation by clones later categorised as cyclin δ3 on left and cyclin δ1 on right

such experiments fail, or what the reasons for such failures may be. One of us (JAHM) has attempted to isolate plant cDNAs complementing the *S. cerevisiae* *dbf2* mutation involved in cell cycle control without success. In this case it is probable that the gene product or cellular process is insufficiently conserved to allow plant clones to rescue. Other possible reasons for failure could include screening insufficient transformants or the quality of the cDNA library. Other issues such as possible mistargeting in the cell were discussed earlier (section 4.1).

It is unknown to what extent expression level is likely to prevent the successful isolation of clones. Presumably a window of acceptable expression levels exist, below which there is insufficient gene product for selected cells to grow, and above which excessive quantities inhibit growth in some way. For certain gene products this window could be relatively narrow, particularly if they are required in precise stoichiometric amounts. Nevertheless, the spectrum of plasmid copy numbers observed with some vectors (section 4.4.3) may well provide sufficient variability to enhance significantly the frequency of isolation of 'difficult' clones. Another factor which presumably affects expression level is the nature and extent of sequence upstream of the ATG translation start on the cDNA clones isolated, and if the level of expression is too high for effective rescue, clones may be selected that have upstream sequence that reduce transcription or translation in the screening organism. Thus we have obtained cyclin clones by complementation in yeast in which considerable upstream non-translated sequence remains intact, including upstream out-of-frame ATGs, which are known to reduce significantly translation efficiency in yeast. We also noted that a number of the clones obtained were inverted with respect to the yeast promoter designed for expression of the cDNA (Soni et al., 1995). We presume that low-level expression occurred from initiation in other plasmid sequences or in the plant non-translated leader sequence. Similar isolation of inverted clones has also been noted by other authors, in these cases working with *E. coli* (Aimi et al., 1990a; Senecoff & Meagher, 1993). Thus, in summary, it would appear that obtaining an appropriate expression level for complementation may be a question of screening a greater number of clones rather than using various different promoters.

The most common problem encountered may be the interpretation of clones obtained, if they do not appear to have the homologies expected, and yet still rescue the mutant. In such cases, it may be helpful to test their ability to rescue other alleles of the same mutation or related mutations that could still be complemented (e.g. Soni et al., 1995), although this is not always the case. For instance, three different plant cDNAs have been obtained by functional complementation of the yeast mutant defective in the CCAAT-binding factor HAP2 (D.B. Edwards and A.G. Smith, unpublished results). Only clone P1 appears to be the plant HAP2 homologue (from sequence similarity), even though another clone (P5) appears to provide similar levels of functional complementation (Fig. 5). The identity of this, and of the third clone P3, remains unknown, but they are unlikely to be general transcription factors since neither of them, nor P1, can complement yeast *hap3* or *hap4* mutants.

A number of other examples of rescue by 'unexpected' clones have been reported, and although these may eventually prove interesting, their analysis is not necessarily straightforward. Such gene products might have rather general roles. Thus Segel et al. (1992) isolated a human chaperonin-like protein while complementing a yeast amino acid transport mutant. Kohorn et al. (1992) isolated a cDNA which was selected in yeast for its ability to inhibit a transcriptional activator GAL4–LHCP fusion protein, but not inhibit native GAL4 protein. The cDNA encoded a 595 amino acid protein with at least two functional domains, one with similarity to the family of protein serine/threonine kinases and another that contains an epidermal growth factor repeat. The function in plants remains obscure, however.

The problem of spurious clones is particularly acute with two-hybrid screens. Since the assay requires only a cDNA–GAL4 activation domain fusion to interact with a target protein, it is possible to imagine identifying pieces of DNA which 'by chance' reading out of frame or from non-coding regions have an activity not expressed in the source organism. Thus 'sticky' regions of

Growth on lactate

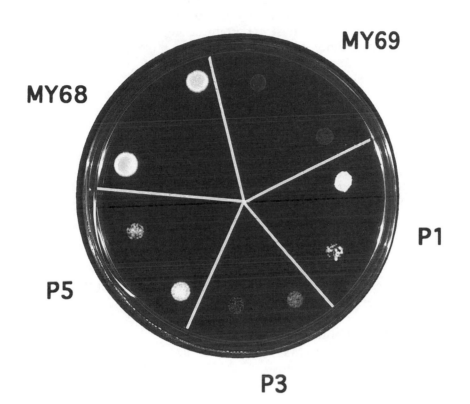

Figure 5. Growth on lactate of two different dilutions of various yeast strains. MY68 is the parent strain with normal mitochondrial function and therefore aerobic respiration. MY69 has been deleted for the gene encoding the CCAAT-binding factor HAP2.and so cannot grow on lactate as the sole carbon source. P1, P3 and P5 are MY69 cells containing three different *Arabidopsis* cDNAs in the yeast expression vector pFL61 (D. B. Edwards and A. G. Smith, unpublished results)

polypeptide sequence can give spurious activation. This type of phenomenon was used by Ptashne to show that random pieces of *E. coli* DNA by chance encoding acidic peptide sequences act as transcriptional activators (Ma & Ptashne, 1987). Bartel et al. (1983) have suggested ways of reducing false positives in two-hybrid screens, by incorporating additional selections.

7 CONCLUDING REMARKS

Functional complementation can be a powerful method both for testing the characteristics of cloned genes and for isolating new genes equivalent to mutations known in genetically tractable organisms such as yeast and *E. coli*. In this chapter we have discussed a number of aspects of this approach which we hope may be of use to other researchers considering this type of strategy. From its early beginnings in the identification of DNA as the genetic material, to the sophisticated rationale of the two-hybrid system, functional complementation has been one of the most important strategies for the investigation of genes and genomes. Since it does not require any molecular information about the gene product, merely conservation of function, it holds considerable potential for the isolation of plant genes.

ACKNOWLEDGMENTS

Work in our laboratories involving cloning of plant genes by functional complementation has been supported by the AFRC/BBSRC. JAHM would particularly like to thank Jeremy Carmichael, who initiated complementation screening in his laboratory. AGS is grateful to Dr Rosine Labbe-Bois, at the Institut Jacques Monod in Paris, who initiated her into the mysteries of yeast molecular biology. We would also like to thank all collaborators who have generously shared mutant strains and libraries, allowing this work to be carried out.

APPENDIX 1   STOCK CENTRES FOR OBTAINING MUTANT STRAINS

Yeast Genetic Stock Center
Department of MCB/Division of Genetics, University of California, Berkeley, California 94720, USA
Phone: (1) 510 642-0815   Fax: (1) 510 642-8589

American Type Culture Collection
12301 Parklawn Drive, Rockville, Maryland 20852-1776, USA
Phone: (1) 301 881 2600   Fax: (1) 301 770-2587   Online: Telnet or Gopher to ATCC.org   World Wide Web: http://www.atcc.org

National Collection of Yeast Cultures
BBSRC Institute of Food Research, Colney Lane, Norwich NR4 7UA, UK
Phone: (44) 1603 56122   Fax: (44) 1603 507723

NCIMB Ltd (National Collection of Industrial and Marine Bacteria)
23 St Machar Drive, Aberdeen, Scotland AB2 1RY, UK
Phone: (44) 1224 273332   Fax: (44) 1224 487658

***E. coli* Genetic Stock Center**
Department of Biology, Yale University, New Haven, Connecticut 06511-7444, USA
Gopher to: cgsc.biology.yale.edu   World Wide Web: http://cgsc.biology.yale.edu/ top.htm1

ESSENTIAL READING

Delauney, A.J. and Verma, D.P.S. (1990) A soybean gene encoding Δ^1-pyrroline-5-carboxylate reductase was isolated by functional complementation in *Escherichia coli* and is found to be osmoregulated. *Mol. Gen. Genet.*, **221**, 299–305.

Elledge, S.J., Mulligan, J.T., Ramer, S.W. et al. (1991) λYES: a multifunctional cDNA expression vector for the isolation of genes by complementation of yeast and *Escherichia coli* mutations. *Proc. Natl. Acad. Sci. USA*, **88**, 1731–1735.

Elledge, S.J., Bai, C. and Edwards, M.C. (1993) Cloning human genes using cDNA expression libraries in *S. cerevisiae*. *Methods: Comparison Methods Enzymol.*, **5**, 96–101.

Minet, M., Dufour, M.-E. and Lacroute, F. (1992) Complementation of *Saccharomyces cerevisiae* auxotrophic mutants by *Arabidopsis thaliana* cDNAs. *Plant J.*, **2**, 417–422.

Sentenac, H., Bonneaud, N., Minet, M. et al. (1992) Cloning and expression in yeast of a plant potassium ion transport system. *Science*, **256**, 663–665.

REFERENCES

Aimi, J., Badylak, J., Williams, J. et al. (1990a) Cloning of a cDNA encoding adenylosuccinate lyase by functional complementation in *Escherichia coli*. *J. Biol. Chem.*, **265**, 9011–9014.

Aimi, J., Qiu, H., William, J. et al. (1990b) *De novo* purine nucleotide biosynthesis: cloning of human and avian cDNAs encoding the trifunctional glycinamide ribonucleotide synthetase-aminoimidazole ribonucleotide synthetase-glycinamide ribonucleotide transformylase by functional complementation in *E. coli*. *Nucleic Acids Res.*, **18**, 6665–6672.

Alderson, A., Sabelli, P.A., Dickinson, J.R. et al. (1991) Complementation of *snf1*, a mutation affecting global regulation of carbon metabolism in yeast, by a plant protein kinase cDNA. *Proc. Natl. Acad. Sci. USA*, **88**, 8602–8605.

Ammerer, G. (1983) Expression of genes in yeast using the *ADC1* promoter. *Methods Enzymol.*, **101**, 192–201.

Anderson, J.A., Huprikar, S.S., Kochian, L.V. et al. (1992) Functional expression of a probable *Arabidopsis thaliana* potassium channel in *Saccharomyces cerevisiae*. *Proc. Natl. Acad. Sci. USA*, **89**, 3736–3740.

Ausubel, F.M., Brent, R., Kingston, R.E. et al. (1987) *Current Protocols in Molecular Biology*. John Wiley & Sons, New York.

Avery, O.T., Macleod, C.M. and McCarthy M. (1944) Studies on the chemical nature of the substance inducing transformation of pneumococcal types. Induction of transformation by a deoxyribonucleic acid fraction isolated from *Pneumococcus* Type III. *J. Exp. Med.*, **79**, 137.

Barr, P.J. (1992) Expression of foreign genes in yeast. In *Transgenesis—Applications of Gene Transfer* (ed. J.A.H. Murray), pp. 55–79. John Wiley and Sons, Chichester.

Bartel, P., Chien, C.-t., Sternglanz, R. and Fields, S. (1993) Elimination of false positives that arise in using the two-hybrid system. *BioTechniques*, **14**, 920–924.

Becker, D.M. and Guarente, L. (1991) High efficiency transformation of yeast by electroporation. *Methods Enzymol.*, **194**, 182–187.

Becker, D.M., Fikes, J.D. and Guarente, L. (1991) A cDNA encoding a human CCAAT-binding protein cloned by functional complementation in yeast. *Proc. Natl. Acad. Sci. USA*, **88**, 1968–1972.

Bednarek, S.Y., Reynolds, T.L., Schroeder, M. et al. (1994) A small GTP-binding protein from *Arabidopsis thaliana* functionally complements the yeast *YPT6* null mutant. *Plant Physiol.*, **104**, 591–596.

Belostotsky, D.A. and Meagher, R.B. (1993) Functional complementation of the poly

(A)⁻ binding protein (pabp) activity in yeast by *Arabidopsis thaliana* floral-specific cDNA—effect on poly(A) tail length and translation initiation. *J. Cell. Biochem.*, **s17b**, 18.

Bennett, M. and Cullimore, J. (1990) Expression of 3 plant glutamine synthetase cDNAs in *Escherichia coil*—formation of catalytically active isoenzymes, and complementation of a *gln*A mutant. *Eur. J. Biochem.*, **193**, 319–324.

Boeke, J.D., Trueheart, J., Natsoulis, G. and Fink, G.R. (1987) 5-Fluoroorotic acid as a selective agent in yeast molecular genetics. *Methods Enzymol.*, **154**, 164–175.

Brown, N.G., Costanzo, M.C. and Fox, T.D. (1994) Interactions among three proteins that specifically activate translation of the mitochondrial *COX3* mRNA in *Saccharomyces cerevisiae*. *Mol. Cell. Biol.*, **14**, 1045–1053.

Brunelli, J.P. and Pall, M.L. (1993a) A series of yeast/*Escherichia coli* λ expression vectors designed for directional cloning of cDNAs and *cre/lox*-mediated plasmid excision. *Yeast*, **9**, 1309–1318.

Brunelli, J.P. and Pall, M.L. (1993b) A series of yeast shuttle vectors for expression of cDNAs and other DNA sequences. *Yeast*, **9**, 1299–1308.

Burgers, P.M. and Percival, K.J. (1987) Transformation of yeast spheroplasts without cell fusion. *Anal. Biochem.*, **163**, 391–397.

Camonis, J.H., Cassan, M. and Rousset, J.-P. (1990) Of mice and yeast: versatile vectors which permit expression in both budding yeast and higher eukaryotic cells. *Gene*, **86**, 263–268.

Campbell, A.G. and Ray, D.S. (1993) Functional complementation of an *Escherichia coli* ribonuclease H mutation by a cloned genomic fragment from the trypanosomatid *Crithidia fasciculata*. *Proc. Natl. Acad. Sci. USA*, **90**, 9350–9354.

Carbon, J. (1993) Genes, replicators, and centromeres: the first artificial chromosomes. In *The Early Days of Yeast Genetics* (eds M.N. Hall & P. Linder), pp. 375–390. Cold Spring Harbor Laboratory Press, Cold Spring Harbor.

Chamovitz, D., Misawa, N., Sandmann, G. and Hirschberg, J. (1992) Molecular cloning and expression in *Escherichia coli* of a cyanobacterial gene coding for phytoene synthase, a carotenoid biosynthesis enzyme. *FEBS Lett.*, **296**, 305–310.

Chapman, K.A., Delauney, A.J., Kim, J.H. and Verma, D.P.S. (1994) Structural organization of *de novo* purine biosynthesis enzymes in plants: 5-aminoimidazole ribonucleotide carboxylase and 5-aminoimidazole-4-N-succinocarboxamide ribonucleotide synthetase cDNAs from *Vigna aconitifolia*. *Plant Mol. Biol.*, **24**, 389–395.

Chen, W., Lim, H.H. and Lim, L. (1993) The *CDC42* homologue from *Caenorhabditis elegans*. Complementation of yeast mutation. *J. Biol. Chem.*, **268**, 13280–13285.

Chien, C., Bartel, P.L., Sternglanz, R. and Fields, S. (1991) The two-hybrid system: a method to identify and clone genes for proteins that interact with a protein of interest. *Proc. Natl. Acad. Sci. USA*, **88**, 9578–9582.

Chow, T.Y.-K. (1989) Purification of yeast – *E. coli* shuttle plasmid suitable for high transformation frequency in *E. coli*. *Nucleic Acids Res.*, **17**, 8391.

Colasanti, J., Tyers, M. and Sundaresan, V. (1991) Isolation and characterization of cDNA clones encoding a functional p34^{cdc2} homologue from *Zea mays*. *Proc. Natl. Acad. Sci. USA*, **88**, 3377–3381.

Colicelli, J., Birchmeier, C., Michaeli, T. et al. (1989) Isolation and characterization of a mammalian gene encoding a high-affinity cAMP phosphodiesterase. *Proc. Natl. Acad. Sci. USA*, **86**, 3599–3603.

Corey, E.J., Matsuda, S.P.T. and Bartel, B. (1993) Isolation of an *Arabidopsis thaliana* gene encoding cycloartenol synthase by functional expression in a yeast mutant lacking lanosterol synthase by the use of a chromatographic screen. *Proc. Natl. Acad. Sci. USA*, **90**, 11628–11632.

D'Enfert, C., Gensse, M. and Gaillardin, C. (1992) Fission yeast and a plant have functional homologues of the Sar1 and Sec12 proteins involved in ER to Golgi traffic in budding yeast. *EMBO J.*, **11**, 4205–4211.

Dewey, R.E., Wilson, R.F., Novitzky, W.P. and Goode, J.H. (1994) The *AAPT1* gene of soybean complements a chlolinephosphotransferase-deficient mutant of yeast. *Plant Cell*, **6**, 1495–1507.

Diffley, J.F.X., Cocker, J.H., Dowell, S.J. and Rowley, A. (1994) Two steps in the assembly of complexes at yeast replication origins in vivo. *Cell*, **78**, 303–316.

Eberhard, J., Raesecke, H.-R., Schmid, J. and Amrhein, N. (1993) Cloning and expression in yeast of a higher plant chorismate mutase: molecular cloning, sequencing of the cDNA and characterization of the *Arabidopsis thaliana* enzyme expressed in yeast. *FEBS Lett.*, **334**, 233–236.

Elledge, S.J., Sugiono, P., Guarente, L. and Davis, R.W. (1989) Genetic selection for genes encoding sequence-specific DNA-binding proteins. *Proc. Natl. Acad. Sci. USA*, **86**, 3689–3693.

Ellerström, M., Josefsson, I.-G., Rask, L. and Ronne, H. (1992) Cloning of a cDNA for rape chloroplast3-isopropylmalate dehydrogenase by genetic complementation in yeast. *Plant Mol. Biol.*, **18**, 557–566.

Ferreira, P.C.G., Hemerly, A.S., Van Montagu, M. and Inzé, D. (1993) A protein phosphatase 1 from *Arabidopsis thaliana* restores temperature sensitivity of a *Schizosaccharomyces pombe cdc25ts/wee1−* double mutant. *Plant J.*, **4**, 81–87.

Fleischmann, M., Clark, M.W., Forrester, W. et al. (1991) Analysis of yeast *prp20* mutations and functional complementation by the human homologue *RCC1*, a protein involved in the control of chromosome condensation. *Mol. Gen. Genet.*, **227**, 417–423.

Frommer, W.B., Hummel, S. and Riesmeier, J.W. (1993) Expression cloning in yeast of a cDNA encoding a broad specificity amino acid permease from *Arabidopsis thaliana*. *Proc. Natl. Acad. Sci. USA*, **90**, 5944–5948.

Futcher, A.B. and Cox, B.S. (1984) Copy number and the stability of 2-μm circle-based artificial plasmids of *Saccharomyces cerevisiae*. *J. Bacteriol.*, **157**, 283–290.

Gheysen, G., Angenon, G. and Van Montagu, M. (1992) Transgenic plants: *Agrobacterium tumefaciens*-mediated transformation and its use for crop improvement. In *Transgenesis—Applications of Gene Transfer* (ed. J.A.H. Murray), pp. 187–232. John Wiley and Sons, Chichester.

Gietz, D., St Jean, A., Woods, R.A. and Schiestl, R.H. (1992) Improved method for high efficiency transformation of intact yeast cells. *Nucleic Acids Res.*, **20**, 1425.

Gould, K.L., Moreno, S., Tonks, N.K. and Nurse, P. (1990) Complementation of the mitotic activator, p80cdc25, by a human protein-tyrosine phosphatase. *Science*, **250**, 1573–1576.

Guarente, L. (1993) Strategies for the identification of interacting proteins. *Proc. Natl. Acad. Sci. USA*, **90**, 1639–1641.

Hanahan, D. (1985) Techniques for transformation of *E. coli*. In *DNA Cloning*, Vol. 1 (ed. D.M. Glover), pp. 109–135. Oxford University Press, Oxford.

Hannon, G.J., Demetrick, D. and Beach, D. (1993) Isolation of the Rb-related p130 through its interaction with CDK2 and cyclins. *Genes Dev.*, **7**, 2378–2391.

Harper, J.W., Adami, G.R., Wei, N. et al. (1993) The p21 Cdk-interacting protein Cip1 is a potent inhibitor of G1 cyclin-dependent kinases. *Cell*, **75**, 805–816.

Henikoff, S., Tatchell, K., Hall, B.D. and Nasmyth, K.A. (1981) Isolation of a gene from *Drosophila* by complementation in yeast. *Nature*, **289**, 33–37.

Hsu, L.-C., Chiou, T.-J., Chen, L. and Bush, D.R. (1993a) Cloning a plant amino acid transporter by functional complementation of a yeast amino acid transport mutant. *Proc. Natl. Acad. Sci. USA*, **90**, 7441–7445.

Hsu, L.C., Chiou, T.J., Chen, L.S. and Bush, D.R. (1993b) Cloning and characterization of a plant proton-amino-acid symport by functional complementation of a yeast amino-acid transport mutant. *Plant Physiol.*, **102**, 387–399.

Inoue, H., Nojima, H. and Okayama, H. (1990) High efficiency transformation of *Escherichia coli* with plasmids. *Gene*, **96**, 23–28.

Izui, K., Ishijima, S., Yamaguchi, Y. et al. (1986) Cloning and sequence analysis of cDNA encoding active phosphoenolpyruvate carboxylase of the C4-pathway from maize. *Nucleic Acids Res.*, **14**, 1615–1628.

Jackson, A.L., Pahl, P.M.B., Harrison, K. et al. (1993) Cell cycle regulation of the yeast Cdc7 protein kinase by association with the Dbf4 protein. *Mol. Cell. Biol.*, **13**, 2899–2908.

Jain, A. and Leustek, T. (1994) A cDNA clone for 5'-adenylylphosphosulfate kinase from *Arabidopsis thaliana*. *Plant Physiol.*, **105**, 771–772.

Jones, M.C., Jenkins, J.M., Smith, A.G. and Howe, C.J. (1994) Cloning and characterisation of genes for tetrapyrrole biosynthesis from the cyanobacterium *Anacystis nidulans* R2. *Plant Mol. Biol.*, **24**, 435–448.

Kieber, J.J., Tissier, A.F. and Signer, E.R. (1992) Cloning and characterization of an *Arabidopsis thaliana* topoisomerase I gene. *Plant Physiol.*, **99**, 1493–1501.

King, J. (1986) Plant cells and isolation of lethal variants. *Enzyme Microb. Technol.*, **9**, 514–522.

Klonus, D., Höfgen, R., Willmitzer, L. and Riesmeier, J.W. (1994) Isolation and characterization of two cDNA clones encoding ATP-sulfurylases from potato by complementation of a yeast mutant. *Plant J.*, **6**, 105–112.

Klose, M., Schimz, K.L., Vandervolk, J. et al. (1993) Lysine-106 of the putative catalytic-ATP binding site of the *Bacillus subtilis sec*A protein is required for functional complementation of *Escherichia coli sec*A mutants *in vivo*. *J. Biol. Chem.*, **268**, 4504–4510.

Koff, A., Cross, F., Fisher, A. et al. (1991) Human cyclin E, a new cyclin that interacts with two members of the *CDC2* gene family. *Cell*, **66**, 1217–1228.

Kohorn, B.D., Lane, S. and Smith, T.A. (1992) An *Arabidopsis* serine/threonine kinase homologue with an epidermal growth factor repeat selected in yeast for its specificity for a thylakoid membrane protein. *Proc. Natl. Acad. Sci. USA*, **89**, 10989–10992.

Kwart, M., Hirner, B., Hummel, S. and Frommer, W.B. (1993) Differential expression of two related amino acid transporters with differing substrate specificity in *Arabidopsis thaliana*. *Plant J.*, **4**, 993–1002.

Lahue, E.E., Smith, A.V. and Orr-Weaver, T.L. (1991) A novel cyclin gene from *Drosophila* complements *CLN* function in yeast. *Genes Dev.*, **5**, 2166–2175.

Lee, H.I., Gal, S., Newman, T.C. and Raikel, N.V. (1993) The *Arabidopsis* endoplasmic reticulum retention receptor functions in yeast. *Proc. Natl. Acad. Sci. USA*, **90**, 11433–11437.

Lee, M.G. and Nurse, P. (1987) Complementation used to clone a human homologue of the fission yeast cell cycle control gene *cdc2*. *Nature*, **327**, 31–35.

Legrain, P., Dokhelar, M.-C. and Transy, C. (1994) Detection of protein–protein interactions using different vectors in the two-hybrid system. *Nucleic Acids Res.*, **22**, 3241–3242.

Leong, S.A., Ditta, G.S. and Helinski, D.R. (1982) Heme biosynthesis in *Rhizobium*. Identification of a cloned gene coding for 5-aminolevulinic acid synthetase from *Rhizobium meliloti*. *J. Biol. Chem.*, **257**, 8724–8730.

Léopold, P. and O'Farrell, P.H. (1991) An evolutionarily conserved cyclin homolog from Drosophila rescues yeast deficient in G1 cyclins. *Cell*, **66**, 1207–1216.

Leustek, T., Murillo, M. and Cervantes, M. (1994) Cloning of a cDNA encoding ATP sulfurylase from *Arabidopsis thaliana* by functional expression in *Saccharomyces cerevisiae*. *Plant Physiol.*, **105**, 897–902.

Lew, D.J., Dulic, V. and Reed, S.I. (1991) Isolation of three novel human cyclins by rescue of G1 cyclin (Cln) function in yeast. *Cell*, **66**, 1197–1206.

Li, B. and Fields, S. (1993) Identification of mutations in p53 that affect its binding to SV40 large T antigen by using the yeast two-hybrid system. *FASEB J.*, **7**, 957–963.

Ma, J. and Ptashne, M. (1987) A new class of yeast transcriptional activators. *Cell*, **51**, 113–119.

Macreadie, I.G., Horaitis, O., Verkuylen, A.J. and Savin, K.W. (1991) Improved shuttle vectors for cloning and high-level Cu^{2+}-mediated expression of foreign genes in yeast. *Gene*, **104**, 107–111.

Manivasakam, P. and Schiestl, R.H. (1993) High efficiency transformation of *Saccharo-*

myces cerevisiae by electroporation. *Nucleic Acids Res.*, **21**, 4414–4415.

Marbois, B.N., Hsu, A., Pillai, R. et al. (1994) Cloning of a rat cDNA encoding dihydroxypolyprenylbenzoate methyltransferase by functional complementation of a *Saccharomyces cerevisiae* mutant deficient in uniquinone biosynthesis. *Gene*, **138**, 213–217.

Martegani, E., Vanoni, M., Zippel, R. et al. (1992) Cloning by functional complementation of a mouse cDNA encoding a homologue of *CDC25*, a *Saccharomyces cerevisiae* RAS activator. *EMBO J.*, **11**, 2151–2157.

Meens, J., Klose, M. and Freudl, R. (1994) The *Staphylococcus carnosus* secE gene-cloning, nucleotide sequence and functional complementation in *Escherichia coli* secE mutant strains. *FEMS Microbiol. Lett.*, **117**, 113–119.

Minet, M. and Lacroute, F. (1990) Cloning and sequencing of a human cDNA coding for a multifunctional polypeptide of the purine pathway by complementation of the *ade2-101* mutant in *Saccharomyces cerevisiae*. *Curr. Genet.*, **18**, 287–291.

Moffatt, B.A., McWhinnie, E.A., Burkhart, W.E. et al. (1992) A complete cDNA for adenine phosphoribosyltransferase from *Arabidopsis thaliana*. *Plant Mol. Biol.*, **18**, 653–662.

Moreno, S., Klar, A. and Nurse, P. (1991) Molecular genetic analysis of fission yeast *Schizosaccharomyces pombe*. *Methods Enzymol.*, **194**, 795–823.

Nasr, F., Bertauche, N., Dufour, M.-E. et al. (1994) Heterospecific cloning of *Arabidopsis thaliana* cDNAs by direct complementation of pyrimidine auxotrophic mutants of *Saccharomyces cerevisiae*. I. Cloning and sequence analysis of two cDNAs catalysing the second, fifth and sixth steps of the de novo pyrimidine biosynthesis pathway. *Mol. Gen. Genet.*, **244**, 23–32.

Nitschke, K., Fleig, U., Schell, J. and Palme, K. (1992) Complementation of the cs *dis2-11* cell cycle mutant of *Schizosaccharomyces pombe* by a protein phosphatase from *Arabidopsis thaliana*. *EMBO J.*, **11**, 1327–1333.

Noji, M., Murakoshi, I. and Saito, K. (1994) Molecular cloning of a cysteine synthase cDNA from *Citrullus vulgaris* (watermelon) by genetic complementation in an *Escherichia coli* Cys⁻ auxotroph. *Mol. Gen. Genet.*, **244**, 57–66.

Pang, Q., Hays, J.B., Rajagopal, I. and Schaefer, T.S. (1993) Selection of *Arabidopsis* cDNAs that partially correct phenotypes of *Escherichia coli* DNA-damage-sensitive mutants and analysis of two plant cDNAs that appear to express UV-specific dark repair activities. *Plant Mol. Biol.*, **22**, 411–426.

Pecker, I., Chamovitz, D., Linden, H. et al. (1992) A single polypeptide catalyzing the conversion of phytoene to ζ-carotene is transcriptionally regulated during tomato fruit ripening. *Proc. Natl. Acad. Sci. USA*, **89**, 4962–4966.

Picard, D., Schena, M. and Yamamoto, K.R. (1990) An inducible expression vector for both fission and budding yeast. *Gene*, **86**, 257–261.

Plon, S.E., Leppig, K.A., Do, H.-N. and Groudine, M. (1993) Cloning of the human homolog of the *CDC34* cell cycle gene by complementation in yeast. *Proc. Natl. Acad. Sci. USA*, **90**, 10484–10488.

Ratzkin, B. and Carbon, J. (1977) Functional expression of cloned yeast DNA in *E. coli*. *Proc. Natl. Acad. Sci. USA*, **74**, 487–491.

Raymond, M., Gros, P., Whiteway, M. and Thomas, D.Y. (1992) Functional complementation of yeast ste6 by a mammalian multidrug resistance mdr gene. *Science*, **256**, 232–234.

Richardson, H.E., Wittenberg, C., Cross, F. and Reed, S.I. (1989) An essential G1 function for cyclin-like proteins in yeast. *Cell*, **59**, 1127–1133.

Riesmeier, J.W., Willmitzer, L. and Frommer, W.B. (1992) Isolation and characterization of a sucrose carrier cDNA from spinach by functional expression in yeast. *EMBO J.*, **11**, 4705–4713.

Riesmeier, J.W., Hirner, B. and Frommer, W.B. (1993) Potato sucrose transporter expression in minor veins indicates a role in phloem loading. *Plant Cell*, **5**, 1591–1598.

Riou, C., Tourte, Y., Lacroute, F. and Karst, F. (1994) Isolation and characterization of a cDNA encoding *Arabidopsis thaliana* mevalonate kinase by genetic complementation in yeast. *Gene*, **148**, 293–297.

Romanos, M.A., Scorer, C.A. and Clare, J.J. (1992) Foreign gene expression in yeast: a review. *Yeast*, **8**, 423–488.

Rose, A.B., Casselman, A.L. and Last, R.L. (1992) A phosphoribosylanthranilate transferase gene is defective in blue fluorescent *Arabidopsis thaliana* tryptophan mutants. *Plant Physiol.*, **100**, 582–592.

Rothstein, R. (1983) One-step gene disruption in yeast. *Methods Enzymol.*, **101**, 202–211.

Saito, K., Miura, N., Yamazaki, M. et al. (1992) Molecular cloning and bacterial expression of cDNA encoding a plant cysteine synthase. *Proc. Natl. Acad. Sci. USA*, **89**, 8078–8082.

Sambrook, J., Fritsch, E.F. and Maniatis, T. (1989) *Molecular Cloning—A laboratory Manual*. Cold Spring Harbor Laboratory Press, Cold Spring Harbor, New York.

Santerre, A. and Britt, A.B. (1994) Cloning of a 3-methyladenine-DNA glycosylase from *Arabidopsis thaliana*. *Proc. Natl. Acad. Sci. USA*, **91**, 2240–2244.

Schiestl, R.H. and Gietz, R.D. (1989) High efficiency transformation of intact yeast cells using single stranded nucleic acids as a carrier. *Curr. Genet.*, **16**, 339–346.

Schild, D., Brake, A.J., Kiefer, M.C. et al. (1990) Cloning of three human multifunctional *de novo* purine biosynthesis genes by functional complementation of yeast mutants. *Proc. Natl. Acad. Sci. USA*, **87**, 2916–2920.

Schnorr, K.M., Nygaard, P. and Lalone, M. (1994) Molecular characterization of *Arabidopsis thaliana* cDNAs encoding three purine biosynthetic enzymes. *Plant J.*, **6**, 113–121.

Schoenhaut, D.S. and Curtis, P.J. (1986) Nucleotide sequence of mouse 5-aminolevulinic acid synthase cDNA and expression of its gene in hepatic and erythroid tissue. *Gene*, **48**, 55–63.

Segel, G.B., Boal, T.R., Cardillo, T.S. et al. (1992) Isolation of a gene encoding a chaperonin-like protein by complementation of yeast amino acid transport mutants with human cDNA. *Proc. Natl. Acad. Sci. USA*, **89**, 6060–6064.

Senecoff, J.F. and Meagher, R.B. (1993) Isolating the *Arabidopsis thaliana* genes for de novo purine synthesis by suppression of *Escherichia coli* mutants. I. 5'-Phosphoribosyl-5-aminoimidazole synthetase. *Plant Physiol.*, **102**, 387–399.

Short, J.M., Fernandez, J.M., Sorge, J.A. and Huse, W.D. (1988) λZAP—a bacteriophage λ expression vector with *in vivo* excision properties. *Nucleic Acids Res.*, **16**, 7583–7600.

Sikorski, R.S. and Boeke, J.D. (1991) *In vitro* mutagenesis and plasmid shuffling: from cloned gene to mutant yeast. *Methods Enzymol.*, **194**, 302–318.

Smith, F.W., Hawkesford, M.J., Prosser, I.M. and Clarkson, D.T. (1993) Approaches to cloning genes encoding for nutrient transporters in plants. *Plant Soil*, **155/156**, 139–142.

Smith, A.G. and Murray, J.A.H. (1996) *Methods Molecular Cell. Biol.*, in press.

Smith, A.G., Santana, M.A., Wallace-Cook, A.D.M. et al. (1994) Isolation of a cDNA encoding chloroplast ferrochelatase from *Arabidopsis thaliana* by functional complementation of a yeast mutant. *J. Biol. Chem.*, **269**, 13405–13413.

Soni, R. and Murray, J.A.H. (1992) A rapid and inexpensive method for isolation of shuttle vector DNA from yeast for the transformation of *E. coli*. *Nucleic Acids Res.*, **20**, 5852.

Soni, R., Carmichael, J.P., Shah, Z.H. and Murray, J.A.H. (1995) A family of cyclin D homologs from plants differentially controlled by growth regulators and containing the conserved retinoblastoma protein interaction motif. *Plant Cell*, **7**, 85–103.

Staudinger, J., Perry, M., Elledge, S.J. and Olson, E.N. (1993) Interactions among vertebrate helix-loop-helix proteins in yeast using the two-hybrid system. *J. Biol. Chem.*, **268**, 4608–4611.

Struhl, K. and Davis, R.W. (1977) Production of a functional eukaryotic enzyme in *Escherichia coli*: cloning and expression of the yeast structural gene for imidazole glycerol phosphate dehydratase (*his3*). *Proc. Natl. Acad. Sci. USA*, **74**, 5255–5259.

Sussman, M.R. (1992) Shaking *Arabidopsis thaliana*. *Science*, **256**, 619.

Sweasy, J.B. and Loeb, L.A. (1993) Detection and characterization of mammalian DNA polymerase β mutants by functional complementation in *Escherichia coli*. *Proc. Natl. Acad. Sci. USA*, **90**, 4626–4630.

Takao, M., Oikawa, A., Eker, A.P.M. and Yasui, A. (1989) Expression of an *Anacystis nidulans* photolyase gene in *Escherichia coli*—functional complementation and modified action spectrum of photoreactivation. *Photochem. Photobiol.*, **50**, 633–637.

Thiele, D.J., Walling, M.J. and Hamer, D.H. (1986) Mammalian metallothionein is functional in yeast. *Science*, **231**, 854–856.

Thon, V.J., Khalil, M. and Cannon, J.F. (1993) Isolation of human glycogen branching enzyme cDNAs by screening complementation in yeast. *J. Biol. Chem.*, **268**, 7509–7513.

Toyn, J., Hibbs, A.R., Sanz, P. et al. (1988) *In vivo* and *in vitro* analysis of *ptl1*, a yeast *ts* mutant with a membrane-associated defect in protein translocation. *EMBO J.*, **7**, 4347–4353.

Van Camp, W., Bowler, C., Villarroel, R. et al. (1990) Characterization of iron super-oxide dismutase cDNAs from plants obtained by genetic complementation in *Escherichia coli*. *Proc. Natl. Acad. Sci. USA*, **87**, 9903–9907.

Xiong, Y., Connolly, T., Futcher, B. and Beach, D. (1991) Human D-type cyclin. *Cell*, **65**, 691–699.

Yarranton, G.T. (1992) High-level gene expression in *Escherichia coli*. In *Transgenesis—Applications of Gene Transfer* (ed. J.A.H. Murray), pp. 1–29. John Wiley and Sons, Chichester.

Zabarovsky, E.R. and Winberg, G. (1990) High efficiency electroporation of ligated DNA into bacteria. *Nucleic Acids Res.*, **18**, 5912.

Zhu, D. and Scandalios, J.G. (1992) Expression of the maize Mn*Sod* (*Sod3*) gene in MnSOD-deficient yeast rescues the mutant yeast under oxidative stress. *Genetics*, **131**, 803–809.

III Map-based Cloning

9 Classical Mutagenesis and Genetic Analysis

IGOR VIZIR
GLENN THORLBY
BERNARD MULLIGAN
University of Nottingham, Nottingham, UK

(1) This chapter includes a description of classical mutagenesis methods in plants together with a summary of the types of molecular lesions produced by chemical and irradiation mutagenesis. The particular advantages of classical and insertional mutagenesis approaches and the parameters affecting the efficiency and optimisation of mutagenesis experiments are considered.

(2) Procedures for linkage detection, and preliminary and fine genetic mapping of new mutations, are outlined together with factors affecting the results of mapping experiments. In particular, methods employing different mapping populations, trisomics, translocations, classical and molecular markers are discussed.

(3) New developments in genetic mapping are considered, including the use in *Arabidopsis* of T-DNA reporter genes as fine mapping markers and the incorporation of the quartet mutation for tetrad analysis. Finally, new methods for the generation and application of deletions and deficiencies for gene cloning and genome analysis are discussed.

1 INTRODUCTION

Mutagenic agents have been used in attempts to induce useful phenotypic variation in plants for more than 70 years. The classical mutagenic techniques developed during this period remain of fundamental importance in the genetic dissection of plant biology. Large collections of mutant lines have been isolated and extensively used in many different fields of plant research and crop breeding. Fundamental and applied aspects of classical mutagenesis research have been extensively reviewed (Evans, 1962; Auerbach, 1976; Gottschalk, 1983). A useful collection of methods in plant mutagenesis has been published by the International Atomic Energy Agency (International Atomic Energy Agency, 1977).

The phenotypic consequences of spontaneous and induced gene mutations in plants provide a rich resource for studies on the function of wild-type genes. *Arabidopsis* is arguably the most versatile model species for research into plant gene function and genome organisation and will be the subject of much of the

Plant Gene Isolation: Principles and Practice. Edited by G. D. Foster and D. Twell.

discussion below. In *Arabidopsis* and other well-established genetic model species, e.g. rice and tomato, for which detailed genetic maps are available (e.g. O'Brien, 1993), the chromosomal location of new gene mutations may be readily mapped by classical procedures. The continuing development in these species of resources such as high-density, integrated genetic maps based on different types of DNA markers and marker mutations and the availability of genomic libraries in yeast artificial chromosome (YAC), cosmid and phage vectors (Chapter 4) have opened the way, as described in Chapter 10, for the cloning by map-based methods of many of the genes identified by mutation. In *Arabidopsis*, the considerable progress currently being made in constructing contigs of overlapping YAC clones covering much of the genome and the gathering pace of genome sequencing efforts will provide improved opportunities for cloning classically mutagenized and mapped genes (National Science Foundation, 1995).

Insertional mutagenesis or gene-tagging, based on the activity of endogenous transposons, has long been one of the most important techniques in plant developmental biology. This approach and its extension by the application of maize transposons in, for example, *Arabidopsis* and tomato is described in detail in Chapter 12. The use of *Agrobacterium* T-DNA as an alternative insertional mutagen has proved a spectacular success in *Arabidopsis* and is described in Chapter 11. There is no doubt that most genes should be isolatable through such gene-tagging approaches. However, populations representing thousands of independent mutations will need to be screened many times over to reach this goal (Feldmann & Marks, 1987; Feldmann, 1991; Koncz et al., 1992; Bechtold et al., 1993). In an extension of this methodology, novel insertion alleles of genes of known sequence have been isolated from pools of hundreds of T-DNA-transformed plants by using polymerase chain reaction (PCR) with oligonucleotide primers for the T-DNA and the target gene (Feldmann et al., 1994; McKinney et al., 1995). This method is also appropriate for screening pools of transposon-tagged lines (Ballinger & Benzer, 1989; Rushforth et al., 1993). Transposon systems have considerable potential for the development of further sophisticated methods for gene cloning (Jones et al., 1989; Dean et al., 1992; Bancroft et al., 1992; Coupland, 1992; Fedoroff & Smith, 1993; Aarts et al., 1993; Klimyuk et al., 1995).

Despite the demonstrated success of insertional mutagenesis in *Arabidopsis*, however, the question still remains as to whether it is possible to clone every gene through such approaches. For example, insertion mutants of essential genes may not always be recoverable, due to complete elimination. In contrast, point mutations causing resistant or leaky alleles of essential genes might be very useful. Mutant alleles of the *csr* gene carrying point mutations, for example, can be easily detected, but insertion mutations cannot be obtained due to the lethal nature of the null phenotype (Haughn & Somerville, 1987; Mourad et al., 1994). In cases where insertion mutants cannot be found, cloning can be based on a classical mutant through map-based approaches. Taken together, gene tagging and classical mutagenesis methods provide a multifaceted and powerful tool for plant gene cloning.

With these considerations in mind, therefore, it seems probable that in order to reach genes lying in very different chromosomal environments, a wide range of mutagenic methods will be required to achieve anything like saturation mutagenesis of even a small plant genome like that of *Arabidopsis*. Recently, about 30% saturation of the *Drosophila* genome with mutations has been achieved, i.e. 4000 genes have been identified and mapped out of a total assumed number of 15 000 genes (Ashburner, 1993). The levels of mutation saturation of the *Arabidopsis* genome and genomes of other plant species are relatively much lower, though significant progress towards increasing mutation density is being made (Cherry, 1994; National Science Foundation, 1995; Polacco, 1995).

Chromosomal mutations such as translocations and trisomic and telotrisomic lines are of considerable value for linkage analysis of new mutations. However, the availability and stage of development of such tools in different species are very variable. In maize, for example, such lines have been widely exploited (Beckett, 1978; Carlson, 1988; Coe et al., 1988a). In contrast, in *Arabidopsis*, such materials have been little used despite several demonstrations of their usefulness (Koornneef et al., 1982b; Richards & Ausubel, 1988; Petrov & Vizir, 1987; Koncz et al., 1990). The availability of small deletions can be helpful for gene cloning by genomic subtraction methods (Straus & Ausubel, 1990; Sun et al., 1992; Lisitsyn et al., 1993; Kang et al., 1994). Deficiencies (large deletions covering two or more genes) can be used to analyse the functional importance of different chromosomal regions, fine-scale deletional mapping and integration of the physical and classical genetic maps (Khush & Rick, 1968; Wessler & Varagona, 1985; Rinchik, 1994; Ogihara et al., 1994; Van Wordragen et al., 1994; Kulkarni et al., 1994).

Novel methods to generate chromosomal mutations with predetermined breakpoint positions using heterologous site-specific exonucleases and recombination sites are also under development (Odell et al., 1990; Onouchi et al., 1991; Bayley et al., 1992; Kilby et al., 1993). It should be possible to disperse such mutation sites over different chromosomal regions via transposition events (Osborne et al., 1993; Van Haaren & Ow, 1993; Onouchi et al., 1995).

2 CLASSICAL MUTAGENESIS METHODS AND TYPES OF MUTATIONS

Seeds, pollen and a variety of plant tissue and cell systems have been employed as subjects for mutagenesis in different species (Rédei, 1970; Lindgren, 1975; Feldmann et al., 1994). Mutation with a view to gene cloning is most often focused on generating point mutations and small deletions, and seed mutagenesis is usually employed to this end. Highly efficient methods of chemical and irradiation mutagenesis of seeds have been developed (Redei, 1970; International Atomic Energy Agency, 1977; Koornneef et al., 1982a; Haughn & Somerville, 1987).

Tissue-culture-based mutagenesis has also been employed for many years to generate novel plant mutants and cell lines of agricultural and industrial interest (Collin & Dix, 1990). Somaclonal variation, which can induce a range of gross chromosomal alternations, as well as more limited gene mutations, has been studied as a possible route for generation of novel genetic variation (Evans & Sharp, 1986).

In plants, as in animals and bacteria, classical mutagenesis techniques have included the use of physical mutagens, i.e. the ionising radiations (X-rays, gamma rays and neutrons) and UV light, and a huge variety of chemical agents. One of the most advantageous features of these classical mutagens is their high efficiency at generating mutations. Furthermore, the most widely employed mutagenic methods are at least partly predictable in their outcomes. The bias of a mutagen towards producing a particular class of molecular lesion is a useful feature in planning mutagenesis experiments. Recently, as discussed below, the physical mapping and DNA sequencing of loci mutagenised by irradiation and chemicals in plants has provided more precise information about these different types of DNA alterations.

A particular advantage of chemical mutagenesis lies in the range and subtlety of genetic effects that may be caused by point mutations at different locations in a particular gene, from complete loss of gene function, to altered levels or timing of mRNA expression, or variations in encoded protein activity. Insertional mutagenesis or radiation-induced deletions are usually more robust in effect and null mutations will often be the result.

In the best studied plant species, irradiation and chemical mutagenesis have yielded large collections of mutant lines, and many of these are available through genetic stock centres. There is easy access to information about these plant resources and related genetic and molecular data through the Internet (Cherry et al., 1992; Hoover & Kristofferson, 1992; Cherry, 1994; Anderson, 1995; Polacco, 1995).

Pollen-irradiation mutagenesis can be used for generating deficiencies in a variety of plant species (McClintock, 1944; Stadler & Roman, 1948; Khush & Rick, 1968; Mottinger, 1970; Rhoades & Dempsey, 1973; Kindiger et al., 1991; Timpte et al., 1994; Vizir et al., 1994; Van Wordragen et al., 1994). In order to simplify the isolation of specific deficiencies, pollen irradiation of T-DNA lines carrying negatively selectable genetic markers located in different chromosomal regions can also be employed (Vizir et al., 1995). Biological approaches to mutagenesis are also possible in certain species. For example, the gametocidal genes of *Aegilops speltoides* can be used for generation of deletions in common wheat (Ogihara et al., 1994).

3 IRRADIATION MUTAGENESIS

A large body of cytological evidence from plant and animal studies suggests that ionising radiation generates chromosomal breaks which, following DNA repair, result in a variety of chromosomal aberrations. Among the latter, 'gene mutations' are less frequent than 'chromosomal mutations', which include translocations, inversions, deletions and deficiencies. As yet, relatively few sequences of irradiation-mutagenised genes in plants have been published. However, recent work with irradiation mutants of *Arabidopsis* supports the general assumption that irradiation causes single- and double-strand breaks which, following DNA repair, result in chromosomal rearrangements. These studies have also indicated that deletions may be accompanied by other local chromosomal alterations, such as insertions and inversions, not detectable at the cytological level.

Fast-neutron-induced alleles of the *GA1* locus of *Arabidopsis* include a 5-kb deletion (*ga1–3*) and an insertion or inversion (*ga1–2*) (Koornneef et al., 1983a; Sun et al., 1992). With the X-ray-induced transparent testa mutant, *tt3* (Koornneef, 1990), of *Arabidopsis*, Shirley et al. (1992) found a deletion of 7.4 kb of the *DFR* gene (dihydroflavonol 4-reductase). This deletion at one chromosomal breakpoint was accompanied by a 52-bp deletion at another breakpoint and a 2.8-cM inversion between the two. Similarly, Shirley et al. (1992) found a complex rearrangement involving an inversion and an accompanying insertion induced by fast-neutron radiation at the *TT5* locus.

Following gamma-irradiation (90 krad from ^{137}Cs) mutagenesis of seeds of the *Arabidopsis* ecotype Landsberg *erecta*, Whitelam et al. (1993) found that either an insertion or an inversion could explain the changed restriction map of a mutant phytochrome A gene (*fhy2-1*). Similarly, Wilkinson & Crawford (1991) induced a deletion at the *CHL3* gene of the ecotype Columbia by irradiating seeds with 30 krad of gamma rays. Peng & Harberd (1993) induced a new allele at the *GAI* locus (*gai-d3*) with gamma rays and found that transmission of this mutation through the pollen was impaired, a common observation, particularly for larger deletions (Rhoades & Dempsey, 1973, Kindiger et al., 1991).

In certain genomic regions, genes may be more densely organised than in others, and therefore a relatively small deletion that may act genetically as a point mutation has the potential for encompassing several genes (Shirley & Goodman, 1993). In *Arabidopsis*, for example, a 6.4-kb genomic DNA fragment was shown to contain reading frames for four possible genes (Le Guen et al., 1994).

UV radiation (wavelength 254 nm) has little power to penetrate plant tissues and therefore has most often been employed in mutagenesis of mature pollen or cells in tissue culture. Cytological and genetic studies with pollen or microspores of maize and *Tradescantia*, respectively, showed that UV irradiation principally induces terminal deletions (Auerbach, 1976). In micro-organisms, UV produces base pair substitutions (usually GC to AT or C to T transitions) and frameshift mutations, while more rarely transversions are also induced (Negrutiu, 1990).

The work described above provides a demonstration of the complex nature of irradiation mutagenesis. Mutation events can be mediated by irradiation-stimulated homologous recombination between direct or inverted repeats, resulting in formation of deletions or inversions, respectively (Peterhans et al., 1990; Tovar & Lichtenstein, 1992; Swoboda et al., 1993; Kusano et al., 1994;

Chedin et al., 1994). However, very little is known about the detailed mechanisms of formation of deficiencies, translocations and large inversions (Pologe & Ravetch, 1988; Wesley & Eanes, 1994; Aoki et al., 1995). Recently, a number of radiation-sensitive mutants of *Arabidopsis* have been isolated (Britt et al., 1993; Davies et al., 1994; Harlow et al., 1994; Jenkins et al., 1995) and a number of plant genes involved in DNA repair and recombination have been cloned (Kieber et al., 1992; Batschauer, 1993; Butt et al., 1993; Pang & Hays, 1993; Smith et al., 1995). Such developments should lead to an improved understanding of the relationship between DNA repair and recombination and the generation of chromosomal mutations in plants.

4   CHEMICAL MUTAGENESIS

In *Arabidopsis*, the most commonly employed chemical for seed mutagenesis is ethyl methane sulphonate (EMS). EMS alkylates DNA, causing base pair transitions, most often GC to AT, due to the mispairing of O^6-alkyl-G with T. Alternatively, the formation of O^4-ethyl-T may result in TA to CG transitions (Griffiths et al., 1993). It is generally assumed that this mechanism operates in *Arabidopsis* and other plants and that EMS mutagenesis results primarily in point mutations. Recent sequencing data support this assumption (see below). The extensive earlier literature on chemical mutagenesis in *Arabidopsis* is summarised in Rédei (1970). More recently, Koornneef et al. (1982a) and Haughn & Somerville (1987) have described straightforward methods for EMS mutagenesis of seeds. Other mutagenic chemicals and methods for their use in plants have been discussed in detail (International Atomic Energy Agency, 1977).

There are numerous variations possible in the EMS mutagenesis procedure (Rédei, 1970; Rédei & Koncz, 1992). Fortunately, EMS is such an effective mutagen that many of these variations are more of academic than practical importance. Koornneef et al. (1982a), summarising previous work in *Arabidopsis* by Muller and Van der Veen, point out that the mutagenic effect of EMS is related to the concentration of mutagen (over a dose range of 3–17.5 mM) and the time of exposure. Variations in temperature also influence the efficiency of mutagenesis.

Evidence from sequencing of mutagenised genes in *Arabidopsis* shows that EMS does generate point mutations. Dolferus et al. (1990) sequenced two mutant *ADH* genes; one contained a C to T transition, and the other a change from G to A. Orozco et al. (1993) found a G to A transition in a mutant rubisco activase gene. Similarly, Yanovsky (Arabidopsis Electronic Bulletin Board, 1993) sequenced 12 EMS-mutated alleles of the *AP1* locus; all were single-nucleotide substitutions. Niyogi et al. (1993) sequenced two different alleles of *trp4* (*trp4-1* and *trp4-2*; β subunit of anthranilate synthase), both of which were G to A changes. In addition, Giraudat et al. (1992) found a C to T transition in the *abi3-4* (abscisic acid insensitive) mutation. EMS-induced point mutations have also been found in the *csr* (chlorsulphuron-resistant acetolactate synthase) gene (Haughn & Somerville, 1987). Single C to T base pair changes were also found in EMS-mutagenised *Agrobacterium rhizogenes rolA* genes in transgenic *Arabidopsis* (Dehio & Schell, 1993).

Despite the weight of sequencing evidence, caution should perhaps be exercised in assuming that EMS only induces point mutations in plants. Parallel studies in *Drosophila* have shown that, not infrequently, EMS may also produce *N*-alkylation of G and A, leading to depurination of DNA, and this in turn results in chromosomal aberrations (Ashburner, 1989).

Other chemical mutagens may be of value in specific cases. For example, azide mutagenesis is particularly effective in barley, because of the conversion of this promutagen into azidoalanine (Plewa & Wagner, 1993). In some plant species, e.g. *Arabidopsis*, maize and chick peas, azide treatment is not mutagenic. Either the enzymology or other biochemical factors for promutagen conversion may be missing or expressed at different stages of the growth cycle, or perhaps any DNA damage is repaired with high efficiency (Negrutiu, 1990). Nitrosomethylurea (NMU) is an interesting chemical which efficiently produces plastid, as well as nuclear, mutations (Hagemann, 1982). Certain chemicals, e.g. diepoxybutane, may be useful alternatives to ionising radiation as a way of generating deletion mutations and are currently under investigation in plant systems (Kreizinger, 1960; Salmeron et al., 1994a,b).

5   MUTATION FREQUENCY

A number of factors have a bearing on mutation frequency, such as the mechanism of mutagen action (Sparrow, 1961; Ashburner, 1989; Griffiths et al., 1993), target gene size and nucleotide composition (Haughn & Somerville, 1987; Bichara et al., 1995), genomic location (Swoboda et al., 1993; Brown & Sundaresan, 1991), chromatin structure (Jackson, 1991; Shaffer et al., 1993), replication timing (Salganik, 1983), efficiency of DNA repair (Britt et al., 1993; Pang & Hays, 1991; Veleminsky & Gichner, 1978) and transcriptional activation (Schlissel & Baltimore, 1989; Zehfus et al., 1990; Lindahl, 1991). The frequency of a particular mutation can be underestimated if a degree of elimination occurs (Butler, 1977; Dellaert, 1980; Vizir et al., 1994).

The recovery of mutants induced by high levels of mutagens is limited by somatic effects, such as reduced viability, growth abnormalities and reduced fertility. Therefore, every mutagen has a most effective dose, which produces the maximum level of mutagenesis with minimal somatic effects. For example, nitrosomethyl biuret (NMB) and NMU are both highly mutagenic, but the efficiency of NMB mutagenesis is greater, due to a lower level of somatic effects (Usmanov & Sokhibnazarov, 1975).

The ability to predict mutation frequencies when using a particular mutagen is an important factor in planning a mutagenesis experiment. Mutation frequency can be estimated by a number of different methods, which differ in their statistical and genetic assumptions (Li & Rédei, 1969; Yonezawa & Yamagata, 1975; Haughn & Somerville, 1987; Koornneef et al., 1982b; Dellaert,

1982; Mednik, 1988). On the basis of data for 16 EMS-mutagenized loci, Koornneef et al. (1982b) have estimated that the average mutation frequency is 0.2×10^{-3} per locus in *Arabidopsis*. However, in these experiments, average mutation frequencies for smaller subsets of loci showed considerable deviations from the overall average value and ranged from 0.5×10^{-4} to 0.4×10^{-3}. With EMS, the *GA1*, *GA4*, *HY3*, *CER1* and *CER3* loci showed particularly high mutation frequencies. Average mutation frequencies for the same loci were different when using X-rays or fast neutrons. Thus, individual loci may vary considerably in their susceptibility to attack by a particular mutagen and to different types of mutagenic treatments. In another study, Haughn & Somerville (1987) estimated an average mutation frequency of 0.5×10^{-3} for EMS mutagensis in *Arabidopsis*. Similar estimations have been made for a number of other mutagens (Rédei, 1970).

Recently, Okubara et al. (1994) found that the downy mildew resistance locus, *Dm3*, of lettuce was particularly prone to mutagenesis by fast neutrons. It was suggested that the *Dm3* locus was either inherently susceptible to DNA damage, consisted of multiple tightly linked genes or was a large gene spread over an extended region of the chromosome. A similar conclusion concerning the more highly mutable loci of *Arabidopsis* was reached by Koornneef et al. (1982a).

Estimations of mutation frequencies have also been obtained for pollen-irradiation mutagenesis in *Arabidopsis*. Recessive multimarker lines were pollinated with gamma-irradiated wild-type pollen, and mutant phenotypes were observed in the F1. An average mutation frequency of 0.9×10^{-2} was found for the 12 loci analysed (Vizir et al., 1994). A value of 1.1×10^{-2} was obtained for the *AUX2* locus by Timpte et al. (1994). These high values (compare the mutation frequencies for EMS mutagenesis of seeds) reflect the fact that gamma irradiation of pollen causes large deletions and deficiencies.

6 FREQUENCY OF APPEARANCE OF MUTANT PLANTS WITHIN INDIVIDUAL M2 FAMILIES

The probability of isolating a particular mutation through seed mutagenesis is a product not only of factors affecting the frequency of mutation but also the frequency of mutant detection.

Many factors may affect the frequency of appearance of mutant plants within individual M2 families. One of the most important factors is the fate of a mutagenised shoot apical cell (Furner & Pumfrey, 1992; Irish & Sussex, 1992) and the contribution of its cell lineage to the inflorescence of the mature plant (Furner & Pumfrey, 1993). The number of shoot apical initial cells (genetically effective cell number; GECN) in *Arabidopsis* seeds has been estimated as 1–4 cells (Li & Rédei, 1969; Grinikh et al., 1974; Koornneef et al., 1982a; Feldmann et al., 1994). Therefore, the frequency of appearance of mutant plants within individual M2 families can range from 25% in the former case to 6.3% in the latter extreme. Any diplontic selection against the cells carrying a mutation would decrease these proportions (Langridge, 1958). Another factor is the possibility of reduced transmission of mutant gametes (Dellaert, 1980; Meinke, 1982). Also, variations in the expression of mutant phenotypes and in the viability of mutant plants might influence the efficiency of mutant detection (Griffiths et al., 1993).

7 OPTIMISATION OF MUTAGENESIS EXPERIMENTS

Despite the high efficiency of classical mutagensis, the screening of mutagenised plant populations can be a labour-intensive exercise and, furthermore, can occupy glasshouses, growth rooms and other resources for considerable periods. Careful planning of optimum population sizes for mutant screening programmes can, however, help to increase the efficiency of isolation of desired mutants.

The probability (P) of isolating a particular mutation is composed of the probability of inducing the mutation (P_m) and the probability of finding the mutation in an individual family (P_d), and can be estimated from the following formula, related to that used in Rédei and Koncz (1992):

$$P = P_m P_d = [1 - (1 - m)^{N_1}][1 - (1 - d)^{N/N_1}]$$

where m is the frequency of M1 plants carrying the particular mutation, d (assumed GECN = 2, therefore d is 0.125 for the graphs presented in Figures 1–3) is the frequency of appearance of mutant plants in a mutant M2 family, N is the M2 population size and N_1 is the number of M1 plants.

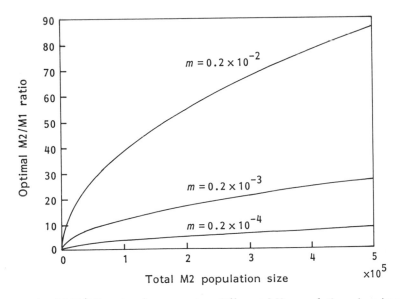

Figure 1. Optimal M2/M1 ratios for screening different M2 population sizes in order to achieve the maximum probability of mutant isolation. (m = mutation frequency)

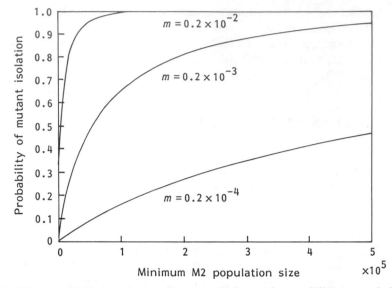

Figure 2. Minimum M2 population sizes needed to achieve different probabilities of mutant isolation. (m = mutation frequency)

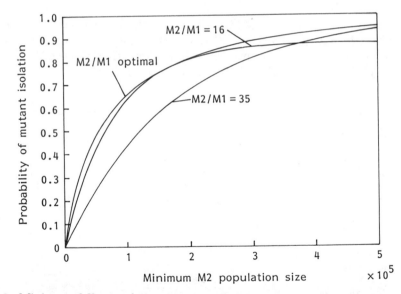

Figure 3. Minimum M2 population sizes needed to achieve different probabilities of mutant isolation when using alternative screening strategies

7.1 Maximising the efficiency of isolation of mutants

7.1.1 Optimal M2/M1 ratio

The optimal M2/M1 ratio (number of M2 seeds harvested per individual M1 plant) has been considered previously by Rédei & Koncz (1992) and Lehle

Seeds (1993). Taking as examples a mutation frequency per locus of 0.2×10^{-3} (Koornneef et al., 1982b) and the theoretical values of 0.2×10^{-4} and 0.2×10^{-2}, which cover a range of possible mutation frequencies, the relationship between the M2/M1 ratio and M2 population size is presented in Fig. 1. The optimal M2/M1 ratio changes somewhat depending on mutation frequency and on the total size of the M2 population analysed. For example, if an M2 population of 500 000 plants is to be screened, the optimum M2/M1 ratios to obtain the maximum number of mutants at the given (0.2×10^{-4}, 0.2×10^{-3}, 0.2×10^{-2}) mutation frequencies are 8, 27 and 86, respectively. For a population of 100 000, the ratios are 3, 10 and 35, respectively, while for populations smaller than 100 000 plants the optimal M2/M1 ratio falls rapidly (Fig. 1).

7.1.2   Minimum M2 population size required for the isolation of a new mutant

The minimum M2 population size has been considered by many authors. Different methods have been used for these estimations (e.g. Haughn & Somerville, 1987; Rédei & Koncz, 1992).

Calculations based on mutation frequencies   Taking the same mutation frequencies used above as examples (0.2×10^{-4}, 0.2×10^{-3} and 0.2×10^{-2}), the relationship between the probability of isolation of a particular mutation and M2 population size is presented in Fig. 2. For example, using the optimal M2/M1 ratio and a mutation frequency of 0.2×10^{-3}, for an M2 population size of 500 000 there is a 95% probability of isolating a particular mutant. At the higher mutation frequency (0.2×10^{-2}), approximately 50 000 plants must be screened to reach this probability (Fig. 2). However, at the lower mutation frequency (0.2×10^{-2}), a population of 500 000 M2 plants will give only a 47% probability of finding the desired mutant (Fig. 2). In contrast to the above examples, which are based on EMS seed mutagenesis data, an M1 population of only 500 up to a few thousand plants is required to isolate one or more deficiency mutations at a specific locus using gamma irradiation of pollen (Vizir et al., 1994; Timpte et al., 1994).

Calculations based on gene/genome structure data   The calculation presented by Haughn & Somerville (1987) is based on empirical observations for EMS mutagenesis of *Arabidopsis* seeds. Under their standard conditions, the average rate of recovery of any given loss-of-function (recessive) mutation was 1 in 2000 M1 plants. Allowing for the percentage of GC to AT transitions (assumed to be most frequently caused by EMS) which cause amino acid substitutions (60.4%) and chain termination (5.5%), and assuming no bias in codon usage, in principle 65.6% of EMS-induced lesions could cause loss-of-function mutations. It was estimated, however, from the sequences of EMS-mutagenised genes in bacteriophage lambda that only about 31% of EMS-induced changes would result in loss of protein function. For a 1500-bp gene in the *Arabidopsis* genome, which has a GC content of 41.4%, therefore, 621 EMS-induced changes are possible. For a loss-of-function mutation in a single gene, $621 \times 0.414 = 193$ mutations are possible. The frequency of inducing a mutation at a specific nucleotide was, therefore, $1/(2000 \times 193)$ or 1 in 386 000 M2 plants. It was estimated that, under standard conditions of mutagenesis, 125 000 M1 plants were needed for saturation of the *Arabidopsis* genome with recessive mutations.

For pollen-irradiation mutagenesis, assuming that the *Arabidopsis* genome is about 100×10^3 kb in size (Meyerowitz, 1992) and the average size of a deficiency is less than 90 kb (Vizir et al., 1994), $100 \times 10^3/90 = 1.1 \times 10^3$ deficiency lines are required to cover most regions of the *Arabidopsis* genome.

7.1.3 Comparison of different approaches towards maximising mutant isolation efficiency

The probability of obtaining a mutant by following the optimal M2/M1 approach will be reduced somewhat when the practicalities of harvesting and sampling M1 populations are considered. For the average mutation frequency, Fig. 3 shows the probability of isolating a particular mutant when employing the optimal M2/M1 ratio approach. The pooling of M2 progenies from individual M1 plants does not have a serious impact on the probability of mutant isolation, if the contribution of each M1 plant in the pools is almost the same.

Pooling can considerably simplify seed sampling and this can be one of the most important practical factors in a large-scale mutagenesis experiment. However, the optimal M2/M1 ratio can be disturbed to some extent as a result of the pooling of M2 progenies, depending on the viability and seed production capacity of individual M1 plants. For example, weak M1 plants may not contribute proportionally to the M2. In this case, the best strategy is to harvest M1 plants individually and to screen M2 families separately. For a 99% probability of finding one or more mutant plants in a '7 : 1' segregating M2 family, 35 M2 seeds per M1 plant must be sown.

Compared to use of the optimal M2/M1 ratio scheme, the probability of recovering a particular mutant by screening individual M2 families (at an M2/M1 = 35), is lower, as shown in Fig. 3. The latter method, however, offers some advantages. For example, mutations are easily detected if they segregate repeatedly in a small group. Furthermore, sterile mutants or plants with low viability can be saved and maintained as heterozygotes. In this regard, by using an M2/M1 ratio of 16, we can be 99% sure that one or more heterozygote plants, carrying a detected mutation, can be found in the population. This ratio still provides a high probability of isolating a mutation and is very close to the maximum probability obtainable using the optimal M2/M1 ratios (Fig. 3).

Pooling progenies from individual M1 plants into large groups can significantly decrease the probability of finding heterozygous siblings of a detected mutant. Small pooling groups might be a better strategy (Feldmann et al., 1994).

Care must be taken to select the most suitable plant varieties for experimental mutagenesis work in order to maximise the use of growth space and to ease the harvest of seeds. With *Arabidopsis*, for example, the Landsberg *erecta* ecotype, which has a compact and erect morphology, is most widely used in mutagenesis experiments. This has the advantage that plants can be grown at higher density and still be identified individually for scoring and harvesting purposes.

7.2 The problem of multiple induced mutations and obtaining plants carrying a single-gene mutation

Mutagenised M1 plants may be chimeric for a large number of heterozygous mutations, many of which will be transmitted to the M2. Multiple mutations can influence the phenotypic expression of a wanted mutation and complicate further analysis. Newly isolated mutants must, therefore, be repeatedly back-crossed (see below) to the original wild-type line. This ensures that the new mutant phenotype is 'clean' and due to a single-gene mutation. According to Haughn & Somerville (1987), a newly isolated mutant of *Arabidopsis* may carry more than 70 homozygous mutations and a larger number of heterozygous ones. However, in practice one does not observe so many obvious mutations segregating out when a new recessive is backcrossed to wild type and the F2 examined. Perhaps most induced mutations are silent or have only a small impact on plant phenotype.

If the new mutation and the contaminating mutation are unlinked (or linked but lie more than 50 cM apart) then six backcrosses will give a 98% probability of separating the new mutation from the contaminating mutation. This is a time-consuming procedure, even with *Arabidopsis*, but much more so for other plants, which have a much longer life-cycle. It is likely that even after intensive backcrossing some contaminating mutations closely linked to a new mutation will still remain. If these contaminating mutations show a phenotypic effect, it can lead to an incorrect interpretation of the pleiotropy of a new mutation, e.g. pseudopleiotropy (Gottschalk, 1983). Only recombination between the new gene and the contaminating mutation (Koornneef, 1981) or the generation of a number of mutants allelic to the new mutation can resolve this problem (Clark et al., 1993).

8 GENETIC ANALYSIS OF A NEW MUTATION

According to the classical scheme (Griffiths et al., 1993; Koornneef, 1994), genetic analysis begins with an analysis of the mutant phenotype and its inheritance through subsequent generations, tests for allelism with known mutations which show a similar phenotype, reciprocal crosses to check the possibility of non-nuclear location of the mutation, segregation analysis and cosegregation analysis with various tester lines. This sequence of steps helps to establish the nature of the genetic factor causing particular alterations in phenotype.

8.1 Linkage detection and preliminary mapping

The localisation of a new mutation to a particular chromosome can be efficiently done using trisomics (Rédei, 1970; Koornneef & Stam, 1992; Rédei & Koncz, 1992; Koornneef, 1994). The advantage of this method is that chromosome assignments can be made with near certainty and small population sizes. For *Arabidopsis* (haploid chromosome number of 5), crosses to a minimum of four trisomic lines must be made to detect linkage, for which an F2 population of 16 plants gives a probability of linkage detection of 99%. If no linkage is detected, then the mutation is linked to the 5th chromosome not included in the tester set of four trisomics. However, the use of trisomics requires some experience to ensure that the segregation of the trisomic phenotypes can be reliably detected (Koornneef, 1994).

Alternatively, reciprocal translocations, can be employed. For a self-fertilising plant such as *Arabidopsis*, a line heterozygous for a translocation appears as a semisterile plant, and a homozygote as a normal plant (Koornneef et al., 1982b). In rare cases, a line heterozygous for a translocation might be segregating lethal embryos and have a distinctive phenotype as a homozygote (Petrov & Vizir, 1987). For *Arabidopsis*, a minimum set of three translocation tester lines must be used.

As an alternative to mapping new mutations with trisomics and translocations, mutations may be mapped in relation to existing classical (visible viable or lethal) mutations (Koornneef et al., 1983b; Franzmann et al., 1993) and molecular markers: restriction fragment length polymorphisms (RFLPs) (Chang et al., 1988; Nam et al., 1989; Chang & Meyerowitz, 1991), random amplified polymorphic DNA (RAPDs) (Reiter et al., 1992), co-dominant cleaved amplified polymorphic sequences (CAPS) (Konieczny & Ausubel, 1993), and simple sequence length polymorphisms (SSLPs) (Bell & Ecker, 1994). Several genetic mapping schemes can be considered in this regard (Fig. 4).

Linkage analysis and mapping procedures take into account the ability to distinguish between all or some genotypic classes and the selection or elimination of some classes from the analysed segregating populations (Allard, 1956; Mather, 1957; Bailey, 1961; Coe, 1982; Koornneef & Stam, 1987; Korol et al., 1994). The following types of markers can be employed: (1) recessive markers, i.e. a homozygous visible mutation or the absence of a dominant marker; (2) dominant markers, where the heterozygous class cannot be distinguished from the homozygous, e.g. RAPDs, or a positively or negatively selectable reporter gene, e.g. in a T-DNA line; (3) co-dominant molecular markers, e.g. RFLPs and all CAPS and SSLPs; (4) classical co-dominant markers such as embryo lethality or lethal colour markers (*emb*/EMB plants produce normal and abnormal, inviable seeds in their siliques/pods etc. and so such plants can be easily distinguished from EMB/EMB plants) (Servitova & Cetl, 1985; Patton et al., 1991).

To assign a new recessive mutation to a chromosome location, F2(R) cosegregation analysis using classical markers may be employed. The cross between a mutant and a marker line will establish a heterozygote in which the new mutation and the mapping markers are in repulsion (R) phase. F2(R) analysis

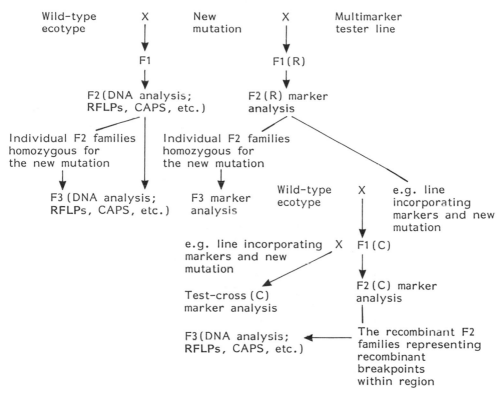

Figure 4. Optional schemes for genetic mapping of a new mutation. ↓ = successive generations; | = selected subpopulation. R represents a cross in which a new mutation and a mapping marker are in repulsion. C represents a cross in which a new mutation and a mapping marker are in coupling

becomes less efficient as recombination frequency increases. For example, for a recombination frequency greater than 40%, an F2(R) population of over 300 plants will be required to detect linkage (Koornneef & Stam, 1987).

To improve the accuracy of linkage detection with classical markers, alternative methods to F2(R) can be used. The F2(R) obtained from the first linkage test can be used to isolate multiple mutant plants which carry the chromosomes with the new mutation in combination with one or more mapping markers. These plants can be crossed with wild type to obtain coupling-phase heterozygous plants, which in turn can be crossed to the multiple mutant in order to obtain a test-cross (TC) progeny. With a test-cross, very few plants are required to detect linkage. Alternatively, coupling (C)-phase heterozygotes can be selfed to obtain an F2(C). However, the use of alternative methods to F2(R) analysis would be redundant if the available mapping markers provide good coverage of the genome for linkage detection and preliminary mapping in any chromosomal region.

In *Arabidopsis*, an extensive set of classical multimarker lines for mapping has been constructed in such a way that, in principle, linkage analysis over most regions of the genome is an easy task (e.g. Koornneef et al., 1983b; Patton et al., 1991). The same is true for several other monocot and dicot species (see maps and references in O'Brien (1993).

There are cases where co-dominant classical markers, such as embryo lethals or lethal colour mutations, can be employed (Mednik & Usmanov, 1982; Servitova & Cetl, 1985; Patton et al., 1991). These markers give a better mapping accuracy than recessive markers. The resolution power of linkage detection methods decreases with the type of marker or mapping population in the following order: TC > F2 (co-dominant mapping marker) > F2(C) > F2(R) (Bailey, 1961; Koornneef et al., 1983).

The advantage of molecular markers is that they do not interfere with the expression of a new mutation and further analysis of DNA from the same F2 populations is possible whenever new molecular markers become available. Large numbers of mapped molecular markers are available in many species, including *Arabidopsis* (Chang et al., 1988; Nam et al., 1989; DeVisente & Tanksley, 1991; Reiter et al., 1992; Lister & Dean, 1993; Jacobs et al., 1995; Polacco, 1995). The *Arabidopsis* RFLP mapping set (ARMS), incorporating co-dominant markers, allows the detection of linkage with a total F2 population as small as 16 plants (Fabri & Schaffner, 1994). ARMS can be used for efficient genetic mapping of any new mutation which has been generated or introduced in the genetic background of Landsberg *erecta*, Columbia or Enkheim ecotype. Similar sets have been developed for the RAPDs, CAPS and SSLPs (Reiter et al., 1992; Konieczny & Ausubel, 1993; Bell & Ecker, 1994). For preliminary mapping of a new mutation employing PCR-based markers, pools of mutant plants derived from F2 populations can be used (Williams et al., 1993).

8.2   Fine mapping

Once linkage has been established, fine mapping of a mutation will be required as a prerequisite for map-based gene cloning. There are two main options to be considered for fine recombinational mapping of a new mutation (Fig. 4): (1) employing only molecular marker analysis; (2) employing easy-to-score mapping markers for selection of recombinants in the chromosomal interval flanking the new mutant locus and further analysis of these recombinants with any molecular markers previously mapped in this region.

F2(R) analysis cannot give an accurate map position within a workable population size (Fig. 5). The resolving power of genetic mapping methods decreases with the type of marker or mapping population in the following order (see Fig. 5): (a) F2 (mapping marker and a new mutation are co-dominant); (b) TC approximately equal to (c) F2 (mapping marker is co-dominant); (d) F2(C); (e) F2(R) (employing the F2/F3 mapping scheme) (Koornneef & Stam, 1988); (f) F2(R).

In order to achieve a higher accuracy when fine mapping a new mutation, larger sizes of segregating populations must be analysed (Fig. 5). In these cases, the possible disadvantage of molecular markers for some laboratories is the scale of DNA isolations and manipulations required for analysis of the

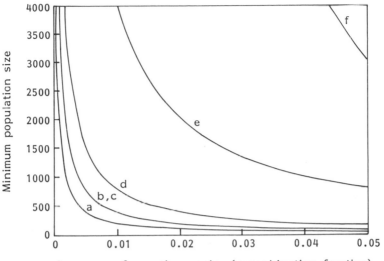

Figure 5. Minimum population sizes to be used in order to achieve different levels of mapping accuracy with alternative mapping methods (see text for details; $P < 0.05$; estimations made by Student's t-test)

mapping populations. With PCR-based markers (RAPDs, CAPS and SSLPs), however, this obstacle can be overcome (Klimyuk et al., 1993). Moreover, the scale of analysis of DNA samples representing individual plants of F2 populations can be greatly reduced by selection of F2 plants showing the phenotype of the new mutant (Koornneef and Stam, 1988) and by using a pooling scheme (Churchill et al., 1993).

Preliminary mapping work provides an opportunity to identify a set of markers flanking a gene of interest for further fine mapping within a defined map region. The F2/F3 populations can provide (ideally in a uniform genetic background) a multimarker line carrying the new mutation and two flanking, easily scored, mapping markers. In some cases, extra crosses may be required for the construction of such a line. This multimarker line can then be crossed to another ecotype for further RFLP, RAPD, CAPS and SSLP mapping.

From the F2 of the latter cross, recombinants with breakpoints on both sides of the gene of interest are readily selected. The ratio of the two classes of recombinants is a measure of the map position of the new mutation within the interval between flanking markers. The highest accuracy in mapping a mutation can be achieved through this approach. For example, Fig. 6 shows that less than 40 recombinants are required to achieve an accuracy of 1/10 part of the interval length between the mapping markers (i.e. the mutations flanking the gene of interest). A set of recombinants between the genetic markers either side of a new mutation can be used for a fine mapping with molecular markers which have been previously mapped in this chromosomal region. When some recombination breakpoints are found which lie between the closest molecular

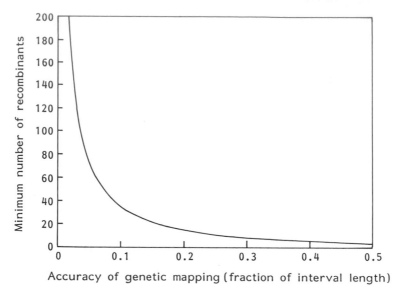

Figure 6. The minimum number of recombinants required in order to achieve different levels of accuracy when mapping within a chromosomal interval (see text for details; $P < 0.05$; estimations made by Student's t-test)

marker and a new mutation, this will indicate that the density of recombination breakpoint has overtaken the level of saturation of the region with molecular markers. Several examples illustrating the application of this method for cloning of a new gene defined by mutation are presented in Chapter 10.

There are various computer programs available which can be used for: analysing segregation and cosegregation data, analysing linkage, estimating recombination values and their errors and the integration, assembly and drawing of genetic maps, e.g. MAPMAKER (Lander et al., 1987); LINKAGE (Suiter et al., 1983); CPROP (Letovsky, 1992); Recombinant Inbred Plant Manager (Manly, 1992); RECF2 (Koornneef & Stam, 1992); JoinMap (Stam, 1993); DRAWMAP (Van Ooijen, 1994); LINKEM (Vowden et al., 1995); and aspects of human genome mapping (Bishop, 1994).

8.3 Factors affecting genetic mapping

A number of factors affecting meiotic recombination can potentially influence the results of genetic mapping experiments. For example, the frequency of meiotic recombination may vary naturally depending on the sex (Vizir & Korol, 1990; Burt et al., 1991; De Visente & Tanksley, 1991; Groover et al., 1995; Lagercrantz & Lydiate, 1995), age (Griffing & Langridge, 1963; Couzin & Fox, 1974) and line of plants (Stadler, 1926; Rasmusson, 1927; Koornneef & Hanhart, 1988; Paredes & Gepts, 1995) and gene and chromosomal environment (Rhoades, 1968; Cornu & Maizonnier, 1979). A number of chemical and physical agents can also modify meiotic recombination (review: Dishler, 1983 and in Zhuchenko and Korol, 1985) and may perhaps provide an opportunity for isolating extra recombination breakpoints in chromosomal regions, where the frequency of recombination events is normally low.

The gametes representing certain genotypes might be inviable during gametogenesis or be less able to participate in fertilisation. Such effects can obviously disturb segregation ratios and ultimately complicate genetic mapping (Koornneef and Stam, 1992; Korol et al., 1994). For example, certation (reduced transmission of mutations through pollen) has been demonstrated in *Arabidopsis* by Dellaert (1980); gametes carrying the *dis1* allele were transmitted less efficiently than those carrying the wild-type gene. Many other examples of selective elimination have been discovered (Rick, 1966; Meinke, 1982; Timpte et al., 1994; Sano et al., 1994; Paredes & Gepts, 1995). Distortion of segregation might also be due to meiotic drive causing unequal representation of gametes carrying a particular mutant allele (Lyttle, 1993). Algorithms have been developed which can accommodate such distortions (Koornneef & Stam, 1992; Korol et al., 1994).

Linkage results can also be inconclusive due to quasi-linkage, in which mutations located on different chromosomes demonstrate dependent cosegregation as a result of non-random meiotic chromosome behaviour (Mike, 1977). In such cases, genetic analysis can become so complicated that either other genetic materials should be used or alternative strategies considered.

The best results from mapping experiments are obtained when a new mutation shows clear phenotypic expression and complete penetrance. In some cases, the phenotypic expression of a new mutation can be uncertain and difficult to score; this obstacle can usually be overcome by the optimisation of growth conditions. There is also an algorithm which can take into account such disturbances (Bailey, 1961; Korol et al., 1994).

8.4 Developments in genetic mapping

Considerable progress in genetic mapping in many species continues to be made on the basis of classical markers (Koornneef et al., 1983b; Franzmann et al., 1993; O'Brien, 1993). But the most spectacular progress has come about through the use of molecular markers (Chang et al., 1988; Nam et al., 1989; Chang & Meyerowitz, 1991; Reiter et al., 1992; Konieczny & Ausubel, 1993; Bell & Ecker, 1994). Collectively, classical and molecular mapping approaches are helping to increase the saturation of genetic maps (Hauge et al., 1993; Lister & Dean, 1993). Such advances in turn provide further opportunities for the genetic analysis of new mutations with eventual gene cloning in mind.

In many cases, when initiating a fine-scale genetic mapping programme, the use of simple, easy-to-score visible genetic markers is preferable. Unfortunately, in some species, such markers are not always available for every region of the genome. This problem might be overcome through the introduction into chromosomal locations of interest of T-DNA insertions carrying scoreable markers. Reporter genes have been expressed under different promoters in different types of cells, tissues, and organs, and provide reliable and convenient visible genetic markers. Suitable genes isolated from *Arabidopsis* include (among many others!): *gl1* (Marks & Feldmann, 1989), *csr* (Haughn et al., 1988), *adh* (Chang & Meyerowitz, 1986) and *ch42* (Koncz et al., 1990). Other eukaryotic and prokaryotic genes include: *codA* (Perera et al., 1993), *Basta*[R] (Bechtold et al., 1993), *gus* (Jefferson et al., 1987), *lacZ* (Teeri et al., 1989), *P450* (O'Keefe et al., 1994) and *GFP* (Niedz et al., 1995). Genetic mapping of the transgenes has been achieved by cosegregation analysis of T-DNA and by recovering plant genomic DNA flanking T-DNA by plasmid rescue and mapping the cloned DNA as an RFLP in recombinant inbred lines (Whittier et al., 1993; Bancroft & Dean, 1993; Lister & Dean, 1993; Jacobs et al., 1995).

In *Arabidopsis*, the recent isolation of the *quartet* mutant, in which functional pollen grains remain attached in a tetrad (Fig. 7A,B), provides an opportunity to perform tetrad analysis (Preuss et al., 1994). The availability of pollen-specific genetic markers and reporters (Twell et al., 1990) has provided another tool for genetic mapping of new mutations, including those disrupting pollen development (Fig. 7C). These tools should also be useful for mapping deficiencies and other chromosomal rearrangements, which frequently affect pollen development (Rhoades & Dempsey, 1973; Kindiger et al., 1991; Sano et al., 1994).

Small and large deletions (deficiencies) have proved to be useful for fine-scale deletion mapping and analysis of function, such as centromere mapping, along particular chromosomal regions (Van Wordragen et al., 1994). Furthermore, deficiencies can be employed for fine mapping of newly identified genes (Rinchik, 1994; Kulkarni et al., 1994). New methods permitting the generation of deletions in specific chromosomal locations might be very useful for this purpose (Kilby et al., 1993; Van Haaren & Ow, 1993). Homozygous deletions have been used directly for gene cloning by genomic subtraction (see Chapter 14) in plants and microbes (Straus & Ausubel, 1990; Sun et al., 1992; Kang et al., 1994). Also, in chromosome walking, deletion alleles with or without other genome rearrangements may help to 'flag' a locus.

9   CONCLUDING REMARKS

The most direct application of genetic mapping lies in gene cloning by map-based methods as described in Chapter 10. More generally, genetic methods can be applied for mapping any localised genomic features, including visible mutations, T-DNA inserts, deletions and deficiencies. Similarly, given the knowledge that mobile elements transpose to linked sites, an accurate map position of the target gene and nearby transposons should help to refine the tagging methods (Chapter 12). Classical and molecular genetic approaches are helping to increase the saturation and accuracy of genetic and physical maps. Large deletions and deficiencies will be of value for deletional mapping of chromosomal regions and the integration of classical and molecular genetic maps and will provide information about the functional importance of different chromosomal regions. Small deletions can be used for cloning of the corresponding wild-type genes by genomic subtraction. In genetic model species

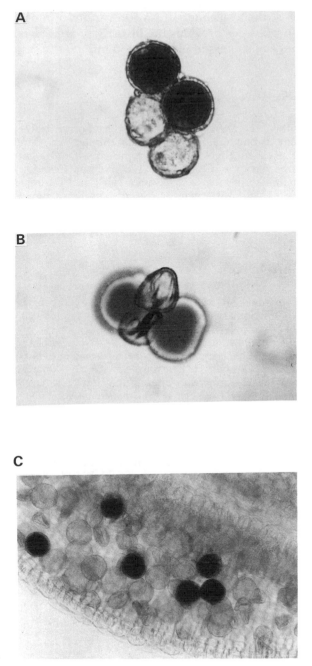

Figure 7. (A) Pollen tetrad of the *quartet* mutant of *Arabidopsis*, showing segregation of alcohol dehydrogenase staining. (B) *quartet* pollen segregating a deficiency (collapsed grains). (C) Mapping a deficiency by cosegregation with a T-DNA carrying a pollen-expressed *GUS* reporter gene. Dark (blue) grains show recombinant pollen. The deficiency results in collapsed pollen and almost complete suppression of GUS expression

such as *Arabidopsis*, the availability of a wide variety of mutation methods and genetic analysis procedures provides the breadth of approach needed to ensure a reasonable chance of success in plant gene cloning.

ACKNOWLEDGMENTS

We would like to thank Dr David Twell and Dr Gary Foster for their constant support and encouragement during the writing of this chapter, Dr Gareth Jones, Dr John Brookfield and Dr Abraham Korol for useful comments, Dr Ottoline Leyser for providing us with an early version of Chapter 10 and Mr Brian Case for assistance with the preparation of illustrative materials. Drs David Twell and Daphne Preuss kindly provided the *Arabidopsis GUS* transformant and *quartet* mutant (shown in Fig. 7), respectively.

Research in the authors' laboratory is funded by the BBSRC and EC.

ESSENTIAL READING

Feldmann, K.A., Malmberg, R.L. and Dean, C. (1994) Mutagenesis in Arabidopsis. In *Arabidopsis* (eds C.R. Somerville and E.M. Meyerowitz), pp. 137–172. Cold Spring Harbor Laboratory Press, New York.

Haughn, G. and Somerville, C.R. (1987) Selection for herbicide resistance at the whole-plant level. In *Biotechnology in Agricultural Chemistry* (eds H.M. Le Baron, R.O. Mumma, R.C. Honeycutt and J.H. Duesing, pp. 98–107. American Chemical Society Symposium Series No. 334, American Chemical Society, Washington DC.

International Atomic Energy Agency (1977) *Manual on Mutation Breeding*, 2nd edn. Technical Reports Series No. 119. IAEA, Vienna.

Koornneef, M., Dellaert, L.W.M. and Van der Veen, J.H. (1982a) EMS- and radiation-induced mutation frequencies at individual loci in *Arabidopsis thaliana* (L.) Heynh. *Mutat. Res.*, **3**, 109–123.

Koornneef, M. and Stam, P. (1992) Genetic analysis. In *Methods in Arabidopsis Research* (eds C. Koncz, N.H. Chua and J. Schell), pp. 83–99. World Scientific, Singapore.

REFERENCES

Aarts, M.G.M., Dirkse, W.G., Stiekema, W.J. and Pereira, A. (1993) Transposon tagging of a male sterility gene in Arabidopsis. *Nature*, **363**, 715–717.

Allard, R.W. (1956) Formulas and tables to facilitate the calculation of recombination values in heredity. *Hilgardia*, **24**, 235–278.

Anderson, M.L. (1995) Weeds World. Arabidopsis newsletter on WWW. *Probe*, **5**, 2–4.

Aoki K., Suzuki, K., Sugano, T. et al. (1995) A novel gene, *Translin*, encodes a recombination hotspot binding protein associated with chromosomal translocations. *Nature Genet.*, **10**, 167–173.

Ashburner, M. (1989) *Drosophila—a Laboratory Handbook*. Cold Spring Harbor Laboratory Press, New York.

Ashburner, M. (1993) Flybase. *Genome News*, **13**, 19–20.

Auerbach, Ch. (1976) *Mutation Research*. Chapman and Hall, London.

Bailey, N.T.J. (1961) *An Introduction to the Mathematical Theory of Genetic Linkage*. Clarendon Press, Oxford.

Ballinger, D.G. and Benzer, S. (1989) Targeted gene mutations in Drosophila. *Proc. Natl. Acad. Sci. USA*, **86**, 9402–9406.

Bancroft, I. and Dean, C. (1993) Transposition pattern of the maize element Ds in *Arabidopsis thaliana*. *Genetics*, **134**, 1221–1229.

Bancroft, I., Bhatt, A.M., Sjodin, C. et al. (1992) Development of an efficient two-element transposon tagging system in *Arabidopsis thaliana*. *Mol. Gen. Genet.*, **233**, 449–461.

Batschauer, A. (1993) A plant gene for photolyase, an enzyme catalyzing the repair of UV-light-induced DNA damage. *Plant J.*, **4**, 705–709.

Bayley, C.C., Morgan, M., Dale, E.C. and Ow, D.W. (1992) Exchange of gene activity in transgenic plants catalyzed by the *Cre-lox* site-specific recombination system. *Plant Mol. Biol.*, **18**, 353–361.

Bechtold, N., Ellis, J. and Pelletier, G. (1993) In planta Agrobacterium mediated gene transfer by infiltration of adult *Arabidopsis thaliana* plants. *C. R. Acad. Sci.*, **316**, 1188–1193.

Beckett, J.B. (1978) B-A translocations in maize. I. Use in locating genes by chromosome arms. *J. Hered.*, **69**, 27–36.

Bell, C.J. and Ecker, J.R. (1994) Assignment of 30 microsatellite loci to the linkage map of Arabidopsis. *Genomics*, **19**, 137–144.

Bichara, M., Schumacher, S. and Fuchs, R.P.P. (1995) Genetic instability within monotonous runs of CpG sequences in Escherichia coli. *Genetics*, **140**, 897–907.

Birchler, J.A. (1982) The mapping of genes by the use of simple and compound translocations. In *Maize for Biological Research* (ed. W.F. Sheridan), pp. 75–83. University Press of North Dakota, Grand Forks, North Dakota.

Bishop, M.J. (1994) *Guide to Human Genome Computing*. Academic Press, London.

Britt, A.B., Chen, J.J., Wykoff, D. and Mitchell, D. (1993) A UV-sensitive mutant of Arabidopsis defective in the repair of pyrimidine–pyrimidinone (6-4) dimers. *Science*, **261**, 1571–1574.

Brown, J. and Sundaresan, V. (1991) A recombination hotspot in the maize A1 intragenic region. *Theor. Appl. Genet.*, **81**, 185–188.

Burt, A., Bell, G. and Harvey, P.H. (1991) Sex differences in recombination. *J. Evol. Biol.*, **4**, 259–277.

Butler, L. (1977) Viability estimates for sixty tomato mutants. *Can. J. Genet. Cytol.*, **19**, 31–38.

Butt, N. J., Watts, F. Z. and Moore, A.L. (1993) Isolation of genes involved in nucleotide excision repair in Arabidopsis thaliana. In *Abstracts of the Fifth International Conference on Arabidopsis Research*, p. 265. Ohio State University, Columbus, Ohio.

Carlson, W.R. (1988) The cytogenetics of corn. In *Corn and Corn Improvement* (eds G.F. Sprague and J.W. Dudley), pp. 259–343. American Society of Agronomy, Madison, Wisconsin.

Chang, C. and Meyerowitz, E.M. (1986) Molecular cloning and DNA sequence of the *Arabidopsis thaliana* alcohol dehydrogenase gene. *Proc. Natl. Acad. Sci. USA*, **83**, 1408–1412.

Chang, C. and Meyerowitz, E.M. (1991) Plant genome studies, restriction fragment length polymorphism and chromosome mapping information. *Curr. Opin. Genet. Dev.*, **1**, 112–118.

Chang, C., Bowman, J.L., De John, A.W. et al. (1988) Restriction fragment length polymorphism linkage map for *Arabidopsis thaliana*. *Proc. Natl. Acad. Sci. USA*, **85**, 6856–6860.

Chedin, F., Dervyn, E., Dervyn, R. et al. (1994) Frequency of deletion formation decreases exponentially with distance between short direct repeats. *Mol. Microbiol.*, **12**, 561–569.

Cherry, J.M. (1994) The Internet and Electronic Arabidopsis Information Service. In *Arabidopsis* (eds C.R. Somerville and E.M. Meyerowitz), pp. 137–172. Cold Spring Harbor Laboratory Press, New York.

Cherry, J.M., Cartinhour, S.W. and Goodman, H.M. (1992) AAtDB, an *Arabidopsis thaliana* database. *Plant Mol. Biol. Rep.*, **10**, 308–309, 409–411.

Churchill, G.A., Giovannoni, J.J., and Tanksley, S.D. (1993). Pooled-sampling makes high-resolution mapping practical with DNA markers, *Proc. Natl. Acad. Sci. USA*, **90**, 16–20.

Clark, S.E., Running, M.P. and Meyerowitz, E.M. (1993) *CLAVATA1*, a regulator of meristem and flower development in Arabidopsis. *Development*, **119**, 397–418.

Coe, E.H. (1982) Planning progeny sizes and estimating recombination percentages. In *Maize for Biological Research* (ed. W.F. Sheridan), pp. 89–89. University Press of North Dakota, Grand Forks, North Dakota.

Coe, E., Hoisington, D. and Chao, S. (1988a) The genetics of corn. In *Corn and Corn Improvement* (eds G.F. Sprague and J.W. Dudley), pp. 81–258. American Society of Agronomy, Madison, Wisconsin.

Coe, E.H., Jr, Neuffer, M.G. and Hoisington, D.A. (1988b) The genetics of corn. In *Corn and Corn Improvement* (eds G.F. Sprague and J.W. Dudley), pp. 81–258. American Society of Agronomy, Madison.

Collin, A. and Dix, P.J. (1990) Culture systems and selection procedures. In *Plant Cell Line Selection. Procedures and Applications*, pp. 3–18. Weinheim and VCH Publishers, New York.

Cornu, A. and Maizonnier, D. (1979) Enhanced non-disjunction and recombination as consequences of gamma-induced deficiencies in *Petunia hybrida*. *Mutat. Res.*, **61**, 57–63.

Coupland, G. (1992) Transposon tagging in Arabidopsis. In *Methods in Arabidopsis Research* (eds C. Koncz, N.-H. Chua and J. Schell), pp. 291–309. World Scientific, Singapore.

Couzin, D.A. and Fox, D.P. (1974) Variation in chiasma frequency during tulip anther development. *Chromosoma*, **46**, 173–179.

Davies, C., Howard, D., Tam, G. and Wong, N. (1994) Isolation of *Arabidopsis thaliana* mutants hypersensitive to gamma radiation. *Mol. Gen. Genet.*, **243**, 660–665.

De Visente, M.C. and Tanksley, S.D. (1991) Genome-wide reduction in recombination of backcross progeny derived from male versus female gametes in an interspecific cross of tomato. *Theor. Appl. Genet.*, **83**, 173–178.

Dean, C., Sjodin, C., Page, T. et al. (1992) Behaviour of the maize transposable element Ac in *Arabidopsis thaliana*. *Plant J.*, **2**, 69–81.

Dehio, C. and Schell, J. (1993) Stable expression of a single copy *rol A* gene in transgenic *Arabidopsis thaliana* plants allows an exhaustive mutagenesis analysis of the transgene-associated phenotype. *Mol. Gen. Genet.*, **241**, 359–366.

Dellaert, L.M.W. (1980) Segregation frequencies of radiation-induced viable mutants in *Arabidopsis thaliana* (L.) Heynh. *Theor. Appl. Genet.*, **57**, 137–143.

Dellaert, L. (1982) The use of induced mutations in improvement of seed propagated crops: selection methods. In *Induced Variability in Plant Breeding* (ed. C. Broertjes), pp. 14–18. Centre for Agricultural Publishing and Documentation, Wageningen.

Dishler, V.Y. (1983) *Induced Recombinogenesis in Higher Plants*. Zeenatie, Riga (in Russian).

Dolferus, R., Van den Bossche, D. and Jacobs, M. (1990) Sequence analysis of two null-mutant alleles of the single Arabidopsis *adh* locus. *Mol. Gen. Genet.*, **224**, 297–302.

Evans, D.A. and Sharp, W.R. (1986) Applications of somaclonal variation. *Bio/Technology*, **4**, 528–532.

Evans, H.J. (1962) Chromosome aberrations induced by ionising radiations. *Int. Rev. Cytol.*, **13**, 221–308.

Fabri, C.O. and Schaffner, A.R. (1994) An *Arabidopsis thaliana* RFLP mapping set to localize mutations to chromosomal regions. *Plant J.*, **5**, 149–156.

Fedoroff, N.V. and Smith, D.L. (1993) A versatile system for detecting transposition in Arabidopsis. *Plant J.*, **3**, 273–290.

Feldmann, K.A. (1991) T-DNA insertion mutagenesis in Arabidopsis: mutational spectrum. *Plant J.*, **3**, 273–290.

Feldmann, K.A. and Marks, M.D. (1987) Agrobacterium-mediated transformation of germinating seeds of *Arabidopsis thaliana*: a non-tissue culture approach. *Mol. Gen. Genet.*, **208**, 1–9.

Franzmann, I.., Yoon, E. and Meinke, D. (1993) Saturation of the linkage map of *Arabidopsis thaliana* with embryonic mutations. In *Abstracts of the Fifth International Conference on Arabidopsis Research*, p. 182. Ohio State University, Columbus.

Furner, I.J. and Pumfrey, J.E. (1992) Cell fate in the shoot apical meristem of *Arabidopsis thaliana*. *Development*, **115**, 755–764.

Furner, I.J. and Pumfrey, J.E. (1993) Cell fate in the inflorescence meristem and floral buttress of *Arabidopsis thaliana*. *Plant J.*, **4**, 917–931.

Giovannoni, J.J., Wing, R.A., Ganal, M.W. and Tanksley, S.D. (1991) Isolation of molecular markers from specific chromosomal intervals using DNA pools from existing mapping populations. *Nucleic Acids Res.*, **19**, 6553–6558.

Giraudat, J., Hauge, B.M., Valon, C. et al. (1992) Isolation of the *Arabidopsis thaliana* ABI3 (abscisic acid insensitive) gene by positional cloning. *Plant Cell*, **4**, 1251–1261.

Gottschalk, W. (1983) *Induced Mutations in Plant Breeding*. Springer, Berlin.

Griffing, B. and Langridge, J. (1963) Factors affecting crossing-over in the tomato. *Aust. J. Biol. Sci.*, **16**, 826–837.

Griffiths, A.J.F., Miller, J.H., Suzuki, D.T. et al. (1993) *An Introduction to Genetic Analysis*. W.H. Freeman and Company, New York.

Grinikh, L.I., Shevchenko, V.V., Grigoreva, G.A. and Draginskaya, L.Y. (1974) Study of chimerism in the reproductive tissue of *Arabidopsis thaliana* plants following irradiation of seeds. *Genetika*, **10**, 18–28.

Groover, A.T., Williams, C.G., Devey, M.E. et al. (1995) Sex-related differences in meiotic recombination frequency in *Pinus taeda*. *J. Hered.*, **86**, 157–158.

Hagemann, R. (1982) Induction of plastome mutations by nitroso-urea-compounds. In *Methods in Chloroplast Molecular Biology*, pp. 119–127. Elsevier Biomedical Press, Amsterdam.

Halfter, U., Morris, P.C. and Willmitzer, L. (1992) Gene targeting in *Arabidopsis thaliana*. *Mol. Gen. Genet.*, **231**, 186–193.

Harlow, G., Jenkins, M.E., Pittalawala, T.S. and Mount, D.W. (1994) Isolation of *uvh1*, an Arabidopsis mutant hypersensitive to ultraviolet light and ionizing radiation. *Plant Cell*, **6**, 227–235.

Hauge, B.M., Hanley, S.M., Cartinhour, S. et al. (1993) An integrated genetic/RFLP map of the *Arabidopsis thaliana* genome. *Plant J.*, **3**, 745–754.

Haughn, G.W., Smith, J., Mazur, B. and Somerville, C. (1988) Transformation with a mutant Arabidopsis acetolactate synthase gene renders tobacco resistant to sulfonylurea herbicides. *Mol. Gen. Genet.*, **211**, 266–271.

Hoover, K. and Krisrofferson, D. (1992) Electronic communications for plant biology. *Plant Mol. Biol. Rep.*, **10**, 228–231.

Irish, V.F. and Sussex, I.M. (1992) A fate map of the Arabidopsis embryonic shoot apical meristem. *Development*, **115**, 745–753.

Jackson, D.A. (1991) Structure–function relationships in eukaryotic nuclei. *BioEssays*, **13**, 1–10.

Jacobs, J.M.E., Van Eck, H.J., Arens, P. et al. (1995) A genetic map of potato (*Solanum tuberosum*) integrating molecular markers, including transposons, and classical markers. *Theor. Appl. Genet.*, **91**, 289–300.

Jefferson, R.A., Kavanagh, T.A. and Bevan, M.W. (1987) GUS fusions: beta-glucuronidase as a sensitive and versatile gene fusion marker in higher plants. *EMBO J.*, **6**, 3901–3907.

Jenkins, M.E., Harlow, G.R., Lui, Z. et al. (1995) Radiation-sensitive mutants of *Arabidopsis thaliana*. *Genetics*, **140**, 725–732.

Jones, J.D.G., Carland, F.M., Maliga, P. and Dooner, H.K. (1989) Visual detection of transposition of the maize element Activator (Ac) in tobacco seedlings. *Science*, **244**, 204–207.

Kang, S., Chumley, F.G. and Valent, B. (1994) Isolation of the mating-type genes of the phytopathogenic fungus *Magnaporthe grisea* using genomic subtraction. *Genetics*, **138**, 289–296.

Khush, G.S. and Rick, C.M. (1968) Cytogenetic analysis of the tomato genome by means of induced deficiencies. *Chromosoma*, **23**, 452–484.

Kieber, J.J., Tissier, A.F. and Signer, E.R. (1992) Cloning and characterization of an *Arabidopsis thaliana* topoisomerase I gene. *Plant Physiol.*, **99**, 1493–1501.

Kilby, N.J., Snaith, M.R. and Murray, J.A.H. (1993) Site-specific recombinases: tools for genome engineering. *TIG*, **9**, 413–421.

Kindiger, B., Beckett, J.B. and Coe, E.H. (1991) Differential effects of specific chromosomal deficiencies on the development of the maize pollen grain. *Genome*, **34**, 579–594.

Klimyuk, V.I., Carroll, B.J., Thomas, C.M. and Jones, J.D.G. (1993) Alkali treatment for rapid preparation of plant material for reliable PCR analysis. *Plant J.*, **3**, 493–494.

Klimyuk, V.I., Harrison, K., Tissier, A. and Jones, J.D.G. (1995) The isolation and characterisation of the meiosis-specific Arabidopsis thaliana *DMC1* gene (*AtDMC1*). In *Abstracts of the Sixth International Conference on Arabidopsis Research*, p. 71. Madison, Wisconsin.

Koncz, C., Mayerhofer, R., Koncz-Kalman, Z. et al. (1990) Isolation of a gene encoding a novel chloroplast protein by T-DNA tagging in *Arabidopsis thaliana*. *EMBO J.*, **9**, 1337–1346.

Koncz, C., Nemeth, K., Redei, G.P. and Schell, J. (1992) T-DNA insertional mutagenesis in Arabidopsis. *Plant Mol. Biol.*, **20**, 963–976.

Konieczny, A. and Ausubel, F.M. (1993) A procedure for mapping Arabidopsis mutations using codominant ecotype-specific PCR-based markers. *Plant J.*, **4**, 403–410.

Koornneef, M. (1981) The complex syndrome of *ttg* mutants. *Arab. Inform. Serv.*, **18**, 45–51.

Koornneef, M. (1990) Mutations affecting the testa colour in Arabidopsis. *Arab. Inf. Serv.*, **27**, 1–4.

Koornneef, M. (1994) Arabidopsis genetics. In *Arabidopsis* (eds E.M. Meyerowitz and C.R. Somerville), pp. 89–120. Cold Spring Harbor Laboratory Press, New York.

Koornneef, M. and Hanhart, C.J. (1988) The effect of genetic background on recombination in Arabidopsis. *Arab. Inf. Serv.*, **26**, 73–78.

Koornneef, M. and Stam, P. (1987) Procedures for mapping by using F2 and F3 populations. *Arab. Inf. Serv.*, **25**, 35–40.

Koornneef, M., Dresselhuys, H.C. and Ramulu, K.S. (1982b) The genetic identification of translocations in Arabidopsis. *Arab. Inf. Serv.*, **19**, 93–99.

Koornneef, M., Van Eden, J., Hanhart, C.J. and De Jongh, A.M.M. (1983a) Genetic fine-structure of the *GA-1* locus in the higher plant *Arabidopsis thaliana* (L.) Heynh. *Genet. Res.*, **41**, 57–68.

Koornneef, M., Van Eden, J., Hanhart, C.J. et al. (1983b) Linkage map of *Arabidopsis thaliana*. *J. Hered.*, **74**, 265–272.

Koornneef, M., Hanhart, C.J., Van Loenen Martinet, E.P. and Van der Veen, J.H. (1987) A marker line, that allows the detection of linkage on all Arabidopsis chromosomes. *Arab. Inf. Serv.*, **25**, 46–50.

Korol, A.B., Preigel, I.A. and Preigel, S.I. (1994) *Recombination, Variability and Evolution: Algorithms of Estimation and Population-Genetic Models*. Chapman & Hall, London.

Kreizinger, J.D. (1960) Diepoxybutane as a chemical mutagen in *Zea mays*. *Genetics*, **45**, 143–154.

Kulkarni, J.Sh., Newby, L.M. and Jackson, F.R. (1994) Drosophila GABAergic systems. II. Mutational analysis of chromosomal segment 64AB, a region containing the glutamic acid decarboxylase gene. *Mol. Gen. Genet.*, **243**, 555–564.

Kusano, K., Sunohara, Y., Takahashi, N. et al. (1994) DNA double-strand break repair: genetic determinants of flanking crossing-over. *Proc. Natl. Acad. Sci. USA*, **91**, 1173–1177.

Lagercrantz, U. and Lydiate, D.J. (1995) RFLP mapping in *Brassica nigra* indicates differing recombination rates in male and female meioses. *Genome*, **38**, 255–264.

Lander, E.S., Green, P., Abrahamson, J. et al. (1987) MAPMAKER: an interactive computer package for constructing primary genetic linkage maps of experimental and natural populations. *Genomics*, **1**, 174–181.

Langridge, J. (1958) A hypothesis of developmental selection exemplified by lethal and semi-lethal mutants of Arabidopsis. *Aust. J. Biol. Sci.*, **11**, 58–68.

Le Guen, L., Thomas, M. and Kreis, M. (1994) Gene density and organization in a small region of the Arabidopsis genome. *Mol. Gen. Genet.*, **245**, 390–396.

Lehle Seeds (1993) *Arabidopsis Catalog*, Arizona, Tucson.

Letovsky, S. (1992) CPROP: a rule-based program for constructing genetic maps. *Genomics*, **12**, 435–446.

Li, S.L. and Rédei, G.P. (1969) Estimation of mutation rate in autogamous diploids. *Radiation Botany*, **9**, 125–131.

Lindahl, K.F. (1991) HIS and HERS recombinational hotspots. *TIG*, **7**, 272–276.

Lindgren, D. (1975) Sensitivity of premeiotic and meiotic stages to spontaneous and induced mutation in barley and maize. *Hereditas*, **79**, 227–238.

Lisitsyn, N., Lisitsyn, N. and Wigler, M. (1993) Cloning the difference between two complex genomes. *Science*, **259**, 946–951.

Lister, C. and Dean, C. (1993) Recombinant inbred lines for mapping RFLP and phenotypic markers in *Arabidopsis thaliana*. *Plant J.*, **4**, 745–750.

Lyttle, T.W. (1993) Cheaters sometimes prosper: distortion of mendelian segregation by meiotic drive. *TIG*, **9**, 205–210.

Manly, K. (1992) RI plant manager: a microcomputer program for genetic mapping with recombinant inbred strains. *Maize Genet. Coop. Newslett.*, **66**, 29.

Marks, M.D. and Feldmann, K.A. (1989) Trichome development in *Arabidopsis thaliana*. I. T-DNA tagging of the *GLABROUS 1* gene. *Plant Cell*, **1**, 1043–1050.

Mather, K. (1957) *The Measurement of Linkage in Heredity*. Methuen and Co., Ltd, London.

McClintock, B. (1944) The relation of homologous deficiencies to mutations and allelic series in maize. *Genetics*, **29**, 478–502.

McKinney, E.C., Ali, N., Traut, A. et al. (1995) Identification of T-DNA mutation in Arabidopsis based on target gene sequence: mutations in actin genes *ACT2* and *ACT4*. In *Abstracts of the Sixth International Conference on Arabidopsis Research*, p. 394. Madison, Wisconsin.

Mednik, I.G. (1988) A method for evaluating the frequencies of induced mutations in Arabidopsis based on embryo-test data. *Arab. Inf. Serv.*, **26**, 67–72.

Mednik, I.G. and Usmanov, P.D. (1982) On an interesting case of incomplete dominance in *Arabidopsis thaliana*. *Arab. Inf. Serv.*, **19**, 87–92.

Meinke, D.W. (1982) Embryo-lethal mutants of *Arabidopsis thaliana*: evidence for gametophytic expression of the mutant genes. *Theor. Appl. Genet.*, **63**, 381–386.

Meyerowitz, E.M. (1992) Introduction to the Arabidopsis genome. In *Methods in Arabidopsis Research* (eds C. Koncz, N.H. Chua and J. Schell), pp. 100–118. World Scientific, Singapore.

Mike, V. (1977) Theories of quasi-linkage and 'affinity': some implications for population structure. *Proc. Natl. Acad. Sci. USA*, **74**, 3513–3517.

Mottinger, J.P. (1970) The effects of X rays on the bronze and shrunken loci in maize. *Genetics*, **64**, 259–271.

Mourad, G., Haughn, G. and King, J. (1994) Intragenic recombination in the *CSR1* locus of Arabidopsis. *Mol. Gen. Genet.*, **243**, 178–184.

Müller, A.J. (1965) Comparative studies on the induction of recessive lethals by various mutagens. In *Arabidopsis Research* (ed. G. Robbelen), pp. 192–199. Gottingen University, Gottingen.

Nam, H.G., Giraudat, J., den Boer, B. et al. (1989) Restriction fragment length polymorphism linkage map of *Arabidopsis thaliana*. *Plant Cell*, **1**, 699–705.

National Science Foundation (1995) *Multinational coordinated Arabidopsis thaliana Genome*

Research Project. Progress Report: year four (ed. A. Moffat). NSF 95-43. National Science Foundation, Arlington.

Negrutiu, I. (1990) In vitro mutagenesis. In *Plant Cell Line Selection. Procedures and Applications* (ed. P. Dix), pp. 19–38. VCH Publishers, Weinheim.

Niedz, R.P., Sussman, M.R. and Satterlee, J.S. (1995) Green fluorescent protein—an in vivo reporter of plant gene expression. *Plant Cell Rep.*, **14**, 403–406.

Niyogi, K.K., Last, R.L., Fink, G.R. and Keith, B. (1993) Suppressors of *trp1* fluorescence identify a new Arabidopsis gene, *TRP4*, encoding the anthranilate synthase beta subunit. *Plant Cell*, **5**, 1011–1027.

O'Brien, S.J. (1993) *Genetic Maps, Book 6 Plants*, 6th edn., Cold Spring Harbor Laboratory Press, New York.

Odell, J., Caimi, P., Sauer, B. and Russell, S. (1990) Site-directed recombination in the genome of transgenic tobacco. *Mol. Gen. Genet.*, **223**, 369–378.

Ogihara, Y., Hasegawa, K. and Tsujimoto, H. (1994) High-resolution cytological mapping of the long arm of chromosome 5A in common wheat using a series of deletion lines induced by gametocidal (Gc) genes of *Aegilops speltoides*. *Mol. Gen. Genet.*, **244**, 253–259.

O'Keefe, D.P., Tepperman, J.M., Dean, C. et al. (1994) Plant expression of a bacterial cytochrome P450 that catalyzes activation of a sulfonylurea pro-herbicide. *Plant Physiol.*, **105**, 473–482.

Okubara, P.A., Anderson, P.A., Ochoa, O.E. and Michelmore, R.W. (1994) Mutants of downy mildew resistance in *Lactuca sativa*. *Genetics*, **137**, 867–874.

Onouchi, H., Yokoi, K., Machida, C. et al. (1991) Operation of an efficient site-specific recombination system of *Zygosaccharomyces rouxii* in tobacco cells. *Nucleic Acids Res.*, **19**, 6373–6378.

Onouchi, H., Nishihama, R., Kudo, M. et al. (1995) Vizualization of site-specific recombination catalyzed by a recombinase from *Zygosaccharomyces rouxii* in *Arabidopsis thaliana*. *Mol. Gen. Genet.*, **247**, 653–660.

Orozco, B.M., McClung, C.R., Werneke, J.M. and Ogren, W.L. (1993) Molecular basis of the ribulose-1,5-bisphosphate carboxylase/oxygenase activase mutation in *Arabidopsis thaliana* is a guanine-to-adenine transition at the 5'-splice junction of intron 3'. *Plant Physiol.*, **102**, 227–232.

Osborne, B.I., Wirtz, U. and Baker, B. (1993) A method for mutation, genomic rearrangement and molecular cloning based on the Ds transposon and the *Cre-Lox* recombination system. In *Abstracts of the Fifth International Conference on Arabidopsis Research*, p. 188. Ohio State University, Columbus.

Pang, Q. and Hays, J.B. (1991) UV-B-inducible and temperature-sensitive photoreactivation of cyclobutane pyrimidine dimers in *Arabidopsis thaliana*. *Plant Physiol.*, **95**, 536–543.

Pang, Q. and Hays, J. (1993) UV-B-inducible and constitutive genes that mediate repair and toleration of UV-damaged DNA in Arabidopsis. In *Abstracts of the Fifth International Conference on Arabidopsis Research*, p. 59. Ohio State University, Columbus.

Paredes, O.M. and Gepts, P. (1995) Segregation and recombination in intergene pool crosses of *Phaseolus vulgaris* L. *J. Hered.*, **86**, 98–106.

Patton, D.A., Franzmann, L.H. and Meinke, D.W. (1991) Mapping genes essential for embryo development in Arabidopsis thaliana. *Mol. Gen. Genet.*, **227**, 337–347.

Peng, J. and Harberd, N.P. (1993) Derivative alleles of the Arabidopsis gibberellin-insensitive (*gai*) mutation confer a wild type phenotype. *Plant Cell*, **5**, 351–360.

Perera, R.J., Liard, C.G. and Signer, E.R. (1993) Cytosine deaminase as a negative selective marker for Arabidopsis. *Plant Mol. Biol.*, **23**, 793–799.

Peterhans, A., Schlupmann, H., Basse, C. and Paszkowski, J. (1990) Intrachromosomal recombination in plants. *EMBO J.*, **9**, 3437–3445.

Petrov, A.P. and Vizir, I.Y. (1987) The application of genetic translocation for establishment of linkage groups. *Tsitol. Genet.*, **21**, 22–25.

Plewa, M.J. and Wagner, E.D. (1993) Activation of promutagens by green plants. *Annu. Rev. Genet.*, **27**, 93–113.

Polacco, M. (1995) Major new data in MaizeDB and how to access. *Probe*, **5**, 4–6.

Pologe, L.G. and Ravetch, J.V. (1988) Large deletions result from breakage and healing of *P. falciparum* chromosomes. *Cell*, **55**, 869–874.

Preuss, D., Rhee, S.Y. and Davis, R.W. (1994) Tetrad analysis possible in Arabidopsis with mutation of the *QUARTET* (*QRT*) genes. *Sciences*, **264**, 1458–1460.

Rasmusson, J. (1927) Genetically changed linkage values in Pisum. *Hereditas*, **10**, 1–152.

Rédei, G.P. (1970) *Arabidopsis thaliana* (L.) Heynh. A review of genetics and biology. *Bibliogr. Genet.*, **20**, 1–150.

Rédei, G.P. and Koncz, C. (1992) Classical mutagenesis. In *Methods in Arabidopsis Research* (eds C. Koncz, N-H. Chua and J. Schell), pp. 16–82. World Scientific, Singapore.

Reiter, R.S., Williams, J.G.K., Feldmann, K.A. et al. (1992) Global and local genome mapping in *Arabidopsis thaliana* by using recombinant inbreds and random amplified polymorphic DNAs. *Proc. Natl. Acad. Sci. USA*, **89**, 1477–1481.

Rhoades, M.M. (1968) Genetic effects of heterochromatin in maize. In *Maize Breeding and Genetics* (ed. D.B. Walden), pp. 641–671. Wiley, New York.

Rhoades, M.M. and Dempsey, E. (1973) Cytogenetic studies on a transmissible deficiency in chromosome 3 of maize. *J. Hered.*, **64**, 125–128.

Richards, E.J. and Ausubel, F.M. (1988) Isolation of a higher eukaryotic telomere from *Arabidopsis thaliana*. *Cell*, **53**, 127–136.

Rick, C.M. (1966) Abortion of male and female gametes in the tomato determined by allelic interaction. *Genetics*, **53**, 85–96.

Rinchik, E.M. (1994) Molecular genetics of the Brown (b)-locus region of mouse chromosome 4. II. Complementation analyses of lethal brown deletion. *Genetics*, **137**, 855–865.

Rushforth, A.M., Saari, B. and Anderson, P. (1993) Site-selected insertion of the transposon Tc1 into a *Caenorhabditis elegans* myosin light chain gene. *Mol. Cell. Biol.*, **13**, 902–910.

Salganik, R.I. (1983) Directed chemical treatments on the particular genes. *Vestnick Acad. Nauk USSR*, **11**, 58–67 (in Russian).

Salmeron, J.M., Barker, S.J., Carland, F.M. et al. (1994a) Tomato mutants altered in bacterial disease resistance provide evidence for a new locus controlling pathogen recognition. *Plant Cell*, **6**, 511–520.

Salmeron, J., Rommens, C., Barker, S. et al. (1994b) Genetic dissection of bacterial speck disease resistance in tomato. *Euphytica*, **79**, 195–200.

Sano, Y., Sano, R., Eiguchi, M. and Hirano, H.-Y. (1994) Gamete eliminator adjacent to the *wx* locus as revealed by pollen analysis in rice. *J. Hered.*, **85**, 310–312.

Schlissel, M.S. and Baltimore, D. (1989) Activation of immunoglobulin kappa gene rearrangement correlates with induction of germline kappa gene transcription. *Cell*, **58**, 1001–1007.

Servitova, J. and Cetl, I. (1985) The use of recessive lethal chlorophyll mutants for linkage mapping of *Arabidopsis thaliana* (L.) Heynh. *Arab. Inf. Serv.*, **22**, 59–63.

Shaffer, S.D., Wallrath, L.L. and Elgin, S.C.R. (1993) Regulating genes by packaging domains: bits of heterochromatin in euchromatin? *TIG*, **9**, 35–37.

Shirley, B.W. and Goodman, H.M. (1993) An Arabidopsis gene homologous to mammalian and insect genes encoding the largest proteasome subunit. *Mol. Gen. Genet.*, **241**, 586–594.

Shirley, B.W., Hanley, S. and Goodman, H.M. (1992) Effects of ionizing radiation on a plant genome, analysis of two Arabidopsis transparent testa mutations. *Plant Cell*, **4**, 333–347.

Smith, K.N., Davies, C.S. and Signer, E.R. (1995) Arabidopsis *RAD51* is a member of a conserved family of DNA repair and recombination genes. In *Abstracts of the Sixth International Conference on Arabidopsis Research*, p. 329. Madison, Wisconsin.

Sparrow, A.H. (1961) Types of ionizing radiation and their cytogenetic effects. In *Symposium on Mutation and Plant Breeding*, pp. 55–119. National Academy of Sciences, National Research Council, Washington DC.

Stadler, L.J. (1926) The variability of crossing-over in maize. *Genetics*, **11**, 1–37.

Stadler, L.J. and Roman, H. (1948) The effects of X-rays upon mutation of the gene A in maize. *Genetics*, **33**, 273–303.

Stam, P. (1993) Construction of integrated genetic linkage maps by means of a new computer package: JoinMap. *Plant J.*, **3**, 739–744.

Straus, D. and Ausubel, F.M. (1990) Genomic subtraction for cloning DNA corresponding to deletion mutations. *Proc. Natl. Acad. Sci. USA*, **87**, 1889–1893.

Suiter, K.A., Wendel, J.F. and Case, J.S. (1983) LINKAGE-1: a pascal computer program for the detection and analysis of genetic linkage. *J. Hered.*, **74**, 203–204.

Sun, T., Goodmann, H.M. and Ausubel, F.M. (1992) Cloning the Arabidopsis *GA1* locus by genomic subtraction. *Plant Cell*, **4**, 119–128.

Swoboda, P., Hohn, B. and Gal, S. (1993) Somatic homologous recombination in plants: the recombination frequency is dependent on the allelic state of recombining sequences and may be influenced by genomic position effects. *Mol. Gen. Genet.*, **237**, 33–40.

Teeri, T.H., Lehvaslaiho, H., Franck, M. et al. (1989) Gene fusions to lacZ reveal new expression patterns of chimeric genes in transgenic plants. *EMBO J.*, **8**, 343–350.

Timpte, C., Wilson, A.K. and Estelle, M. (1994) The axr2-1 mutation of *Arabidopsis thaliana* is a gain-of-function mutation that disrupts an early step in auxin response. *Genetics*, **138**, 1239–1249.

Tovar, J. and Lichtenstein, C. (1992) Somatic and meiotic chromosomal recombination between inverted duplications in transgenic tobacco plants. *Plant Cell*, **4**, 319–332.

Twell, D., Yamaguchi, J. and McCormick, S. (1990) Pollen-specific gene expression in transgenic plants: coordinate regulation of two different tomato gene promoters during microsporogenesis. *Development*, **109**, 705–713.

Usmanov, P.D. and Sokhibnazarov, Sh. (1975) Genetic and somatic effects of N-nitrosomethylbiuret on *Arabidopsis thaliana* (L.) Heynh. *Genetika*, **11**, 51–58.

Van Haaren, M.J.J. and Ow, D.W. (1993) Prospects of applying a combination of DNA transposition and site-specific recombination in plants: a strategy for gene identification and cloning. *Plant Mol. Biol.*, **23**, 525–533.

Van Ooijen, J.W. (1994) DRAWMAP—a computer program for drawing genetic linkage maps. *J. Hered.*, **85**, 66.

Van Wordragen, M.F., Weide, R., Liharska, T. et al. (1994) Genetic and molecular organization of the short arm and pericentromeric region of tomato chromosome 6. *Euphytica*, **79**, 169–174.

Veleminsky, J. and Gichner, T. (1978) DNA repair in mutagen-induced higher plants. *Mutat. Res.*, **55**, 71–84.

Vizir, I.Y. and Korol, A.B. (1990) Sex difference in recombination frequency in Arabidopsis. *Heredity*, **65**, 379–383.

Vizir, I.Y., Anderson, M.L., Wilson, Z.A. and Mulligan, B.J. (1994) Isolation of deficiencies in the Arabidopsis genome by gamma-irradiation of pollen. *Genetics*, **137**, 1111–1119.

Vizir, I., Chiurazzi, M., Mulligan, B. and Signer, E. (1995) Isolation of specific deficiencies in the Arabidopsis genome via deletion mutagenesis of T-DNA insertions. In *Abstracts of the Sixth International Conference on Arabidopsis Research*, p. 24. Madison, Wisconsin.

Vowden, C.J., Ridout, M.S. and Tobutt, K.R. (1995) LINKEM: a program for genetic linkage analysis. *J. Hered.*, **86**, 249–250.

Wesley, C.S. and Eanes, W.F. (1994) Isolation and analysis of the breakpoint sequences of chromosome inversion *In(3L)Payne* in Drosophila melanogaster. *Proc. Natl. Acad. Sci. USA*, **90**, 6621–6625.

Wessler, S.R. and Varagona, M.J. (1985) Molecular basis of mutations at the *waxy* locus of maize, correlation with the fine structure genetic map. *Proc. Natl. Acad. Sci. USA*, **82**, 4177–4181.

White, M., Dominska, M. and Petes, T.D. (1993) Transcription factors are required for the meiotic recombination hotspot at the *HIS4* locus in *Saccharomyces cerevisiae*. *Proc. Natl. Acad. Sci. USA*, **91**, 3132–3136.

Whitelam, G.C., Johnson, E., Peng, J. et al. (1993) Phytochrome A null mutants of Arabidopsis display a wild type phenotype in white light. *Plant Cell*, **5**, 757–768.

Whittier, R.F., Liu, Y.-G., Oosumi, T. and Mitsukawa, N. (1993) Mapping of T-DNA insertion sites on Arabidopsis chromosomes using flanking sequence probes obtained by thermal asymmetric interlaced (TAIL) PCR. In *Abstracts of the Fifth International Conference on Arabidopsis Research*, p. 197. Ohio State University, Columbus.

Wilkinson, J.Q. and Crawford, N.M. (1991) Identification of the Arabidopsis *CHL3* gene as the nitrate reductase structural gene *NIA2*. *Plant Cell*, **3**, 461–471.

Williams, J.G.K., Reiter, R.S., Young, R.M. and Scolnik, P.A. (1993) Genetic-mapping of mutations using phenotypic pools and mapped RAPD markes, *Nucleic Acids Res.*, **21**, 2697–2702.

Yonezawa, K. and Yamagata, H. (1975) Comparison of the scoring methods for mutation frequency in self-pollinating disomic plants. *Radiat. Biol.*, **15**, 241–256.

Zehfus, B.R., McWilliams, A.D., Lin, Y.-H. et al. (1990) Genetic control of RNA polymerase I-stimulated recombination in yeast. *Genetics*, **126**, 41–52.

Zhuchenko, A.A., and Korol, A.B. (1985). *Recombination in evolution and selection.* Nauka, Moscow (in Russian).

10 Chromosome Walking

OTTOLINE LEYSER
University of York, York, UK

CAREN CHANG
University of Maryland, MD 20742, USA

Gene isolation by chromosome walking involves the construction of a series of overlapping clones starting from a DNA segment known to map close to the desired gene, and proceeding in steps until the gene is reached. The main advantage of this approach is that it can be used to isolate any gene for which there are two alleles conferring different phenotypes. No prior knowledge about the gene or its product is required. Because of this, chromosome walking is particularly applicable to genetic model systems such as *Arabidopsis*, where it can be used to isolate mutationally defined genes. Unlike most of the gene isolation approaches described in this book, chromosome walking does not depend on any single technique but rather it relies on a wide range of genetic and molecular biological methods. Many of these methods are detailed in other chapters. This chapter describes the strategies which should be adopted to clone a gene by chromosome walking.

1 INTRODUCTION

The previous chapter described how plant genes can be identified by mutation. The characterisation of phenotypes caused by such mutations gives important information on the wild-type function of the genes concerned. However, a more complete understanding requires gene isolation. Chromosome walking is a very powerful technique for the isolation of genes defined only by mutation or naturally occurring polymorphism. It involves the construction of a series of overlapping clones starting from a DNA segment known to map close to the desired gene, and proceeding in steps until the gene is reached. Any gene that can be mapped can be cloned by chromosome walking. No prior knowledge of the gene product is required, and the nature of the mutation or polymorphism defining the gene is irrelevant. Because of this lack of restrictions, chromosome walking has been used in animal systems for many years, and to great effect. For example, a 200-kb chromosome walk resulted in the isolation of the developmental regulatory genes in the *Ubx* gene cluster of *Drosophila melanogaster* (Bender et al., 1983). Similarly, chromosome walking has been used to

Plant Gene Isolation: Principles and Practice. Edited by G. D. Foster and D. Twell.
© 1996 John Wiley & Sons Ltd.

clone many genes from the nematode *Caenorhabditis elegans*, another important genetic model system in animal biology.

Despite this impressive track record, it is only recently that chromosome walking has become a viable option for plant gene isolation. This is because, for reasons that will become apparent, chromosome walking is much easier in organisms with a short generation time, a high-density genetic map, a small simple genome, a high map unit/kb ratio, and good transformation technology. Historically, plant geneticists and plant breeders have worked on species which have none or few of these properties. The cloning of mutationally defined genes in plants was therefore largely restricted to species with a transposon tagging system such as maize. Fortunately, the situation has been rapidly improving. A major reason for this has been the widespread adoption of *Arabidopsis thaliana* as a genetic study system for plant biology (Meyerowitz, 1987). This species fulfils all the criteria for efficient chromosome walking. The international focus on *Arabidopsis* is making walking continually easier, as more markers are added to the map and the *Arabidopsis* genome project progresses. Furthermore, walking in several traditional study species has become increasingly feasible due to the growing density of molecular marker maps, helped particularly by the advent of polymerase chain reaction (PCR)-based markers (Chapter 13). There is hope for even the most intractable systems, because the genomes of several related species have been shown to be highly syntenic, i.e. the order of genes along their chromosomes is conserved (Ahn & Tanksley, 1993; Kurata et al., 1994). This opens the possibility of using a species with a small, simple genome to walk to a mutationally defined gene in a related species with a large, complex genome (Moore et al., 1993).

These developments have led to the first successful chromosome walks in plants. At the time of writing, there are nine published examples (Table 1) (Giraudat et al., 1992; Arondel et al., 1992; Leyser et al., 1993; Chang et al., 1993; Martin et al., 1993; Leung et al., 1994; Meyer et al., 1994; Pepper et al., 1994; Mindrinos et al., 1994) with many more walks at an advanced stage (e.g. Putterill et al., 1993). Eight of these nine walks were carried out in *Arabidopsis*, which remains the most tractable system. However, walking in tomato is becoming increasingly feasible, as demonstrated by the ninth example, the tomato *Pto* gene (Martin et al., 1993). Walk targets to date include genes involved in plant processes as diverse as photomorphogenesis (Pepper et al., 1994) and fatty acid biosynthesis (Arondel et al., 1992). Already the technology has resulted in the isolation of the first disease-resistance genes (Martin et al., 1993; Mindrinos et al., 1994) and several of the first genes known to be involved in plant hormone signalling (Giraudat et al., 1992; Leyser et al., 1993; Chang et al., 1993; Leung et al., 1994; Meyer et al., 1994). These are central areas of plant biology which have proved difficult to investigate by conventional biochemical approaches. Many of the genes involved are defined only by dominant mutations which are unlikely to arise by insertional mutagenesis, so gene-tagging approaches would be more difficult. These early successes show the potential power of chromosome walking to elucidate fundamental areas of plant biology. As walking becomes progressively easier, it is expected to make an ever-increasing impact.

Table 1. A summary of the published accounts of plant genes cloned by chromosome walking at the time of writing

Gene	Plant species	Phenotype of mutant or varient allele	Length of contig assembled	Gene homologies	Reference
ABI3	Arabidopsis	Abscisic acid insensitivity	Mapping showed gene on known contig	Transcription factor involved in action in maize	Giraudat et al. (1992)
FAD3	Arabidopsis	Reduced linoleic acid desaturation	340 kb	Bacterial fatty acid desaturases	Arondel et al. (1992)
AXR1	Arabidopsis	Auxin resistance	250 kb	N-terminal half of Ubiquitin-activating enzyme, E1	Leyser et al. (1993)
ETR1	Arabidopsis	Ethylene resistance	230 kb	Two-component regulators	Chang et al. (1993)
Pto	Tomato	Disease resistance	100 kb	Serine-theorine protein kinase	Martin et al. (1993)
ABI1	Arabidopsis	Abscisic acid insensitivity	500 kb	Calcium-modulated protein phosphatase	Leung et al. (1994)
ABI1	Arabidopsis	Abscisic acid insensitivity	500 kb	Calcium-modulated protein phosphatase	Meyer et al. (1994)
DET1	Arabidopsis	Impaired etiolation response	370 kb	Novel nuclear protein	Pepper et al. (1994)
RPS2	Arabidopsis	Disease resistance	Mapping showed gene on known contig	Novel nucleotide-binding protein with leucine-rich repeats	Mindrinos et al. (1994)

2   THE BASIC PROCEDURE

Standard chromosome walking proceeds in four stages. First, DNA markers closely linked to, and preferably flanking, the gene of interest are identified. These flanking markers delimit a DNA segment that contains the gene. Second, recombinational breakpoints flanking the gene are obtained to allow the gene to be delimited further during the course of the walk. Third, a series of overlapping clones covering the delimited region is isolated using the flanking markers as start points and using the recombinational breakpoints to both orient and monitor the progress of the walk. Fourth, the gene is identified from within the series of clones. Each of these stages is described in detail below.

2.1   Identifying flanking DNA markers

2.1.1   Preliminary mapping and the choice of markers

Initially, an approximate chromosomal location for the gene of interest must be established. Visible markers or molecular markers can be used for mapping. Mapping with visible markers is discussed in Chapter 9. An advantage of visible markers is that no DNA extractions are required. Even if the strategy for preliminary mapping is molecular, it is advisable to make crosses with a range of visible marker lines so that the F2 material is available for later use (see below).

   Two advantages of using molecular markers for preliminary mapping are that they are usually co-dominant, and many markers can be scored in the progeny of a single cross. Molecular markers are mapped based on their ability to detect DNA polymorphisms between different inbred plant lines. There is an ever-increasing number of methods for the detection of such polymorphisms. The earliest DNA marker maps are based on randomly isolated DNA clones which are then used as probes to detect restriction fragment length polymorphisms (RFLPs) between inbred lines on Southern blots (Botstein et al., 1980). In most systems, RFLP maps are still the densest molecular marker maps, but the increased use of PCR-based markers is likely to lead to the replacement of this technology (Chapter 13). RFLP markers have the advantage that they are a permanent collection of DNA clones, so once linkage to a particular marker is established, the clone is already available as a walk start point. The major drawback with standard RFLPs is that they can only be detected on Southern blots, so preparation of sufficient quantities of restrictable DNA is required for linkage analysis, and the detection procedure is relatively slow and laborious. When working with *Arabidopsis*, the requirement for large amounts of DNA can slow things down further, since another generation of plants will be required for adequate quantities of tissue to be collected. With PCR-based markers, relatively crude, low-yield DNA preparations are adequate. However, particularly when the strategy is based on primers which amplify DNA segments from several areas of the genome, the linked DNA must be isolated and/or cloned before it can be used as a walk start point. Cleaved amplified polymorphic sequence (CAPS) markers (Konieczny & Ausubel, 1993) provide a useful bridge between conventional RFLPs and PCR-based markers. Any RFLP can be converted into a CAPS marker by designing PCR primers which flank the polymorphic restriction site. The primers can then be used to amplify the intervening DNA segment. The restriction site polymorphism can be scored by cutting the amplified products with the relevant enzyme, subjecting the products to agarose gel electrophoresis, staining with ethidium bromide, and visualising the restriction fragments directly by UV fluorescence. Most PCR-based marker are, however, distinctly different from conventional RFLPs in that the polymorphism is detected directly in the PCR reaction and does not rely on restriction site differences. For example, random amplified polymorphic DNA (RAPD) markers use random primers to amplify DNA sequences. Polymorphisms are detected because different plant lines may or may not contain the primer hybridization site and so a DNA segment may or may not be amplified.

2.1.2   Generating mapping populations

Regardless of the marker system selected, the basic mapping procedure involves crossing a line mutant in the gene of interest with a wild-type line of a different genetic background. In the F2 of this cross, DNA polymorphisms between the two parental genetic backgrounds will be segregating, as will the phenotype conferred by the mutation. Polymorphisms which are genetically linked to either the mutant or the wild-type phenotype can then be identified. The most straightforward way to do this is to take a random sample of F2 individuals, score them with respect to their visible phenotype, and score them with respect to their DNA polymorphisms. The more individuals included in the sample, the more accurate, and the higher the resolution of, the mapping data obtained. Recombination frequencies can then be calculated to give an estimate of the map distances involved (Chapter 9). For most DNA polymorphisms, all three genotypes can be scored (both classes of homozygote and the heterozygote). If possible, the genotype at the locus of interest should also be completely ascertained by scoring phenotypes in the next generation. This means that both chromosomes in each F2 plant are completely scored, increasing the amount of information obtained. If scoring the next generation is not practical, then the entire F2 sample can be taken from the double recessive class, thus achieving the same end. The disadvantage of this selective approach is that the accuracy of mapping data obtained could be impaired in the unlikely event that the molecular markers scored do not show normal Mendelian segregation.

2.1.3   More detailed mapping

Once a general chromosomal location has been identified, more detailed mapping is carried out to identify the most closely linked DNA markers available and their relative order with respect to the gene of interest. Plants from the original F2 sample that carry chromosomes with recombinational breakpoints between the gene of interest and linked molecular or visible markers are scored with respect to other DNA markers known to map to the region. The number of such available markers depends on the species concerned and on luck. As maps are becoming denser, the probability of finding closely linked markers is rapidly increasing. For some species there are several different maps, mainly consisting of one particular marker type each. These maps have been loosely integrated by mapping markers from one map relative to markers from other maps. For example, in *Arabidopsis*, existing maps include the visible marker map (O'Brien, 1990), two independent RFLP maps (Chang et al., 1988; Nam et al., 1989), a RAPD map (Reiter et al., 1992), a CAPS map (Konieczny & Ausubel, 1993), and a microsatellite map (Bell & Ecker, 1994). Markers in common between various of these maps allow approximate correlations of map positions. It must be emphasised, however, that the relative positions of closely linked markers from different maps cannot be reliably predicted. Attempts to do so, such as the integrated and unified maps for *Arabidopsis* (Hauge et al., 1993), are only best approximations. The order of markers shown on such maps should definitely not be relied on for fine-structure mapping around the gene of interest.

2.2   Obtaining linked recombinational breakpoints

As soon as an approximate map position for the gene has been determined, fine-structure mapping should be initiated. The purpose of fine-structure mapping is to establish a collection of meiotic recombinant lines known to have recombinational breakpoints on either side of the gene. The importance of this collection cannot be overemphasised. First, the direction and extent of the chromosome walk can be established in stage 3 of the basic procedure described above by mapping the DNA clones in the walk relative to the recombinational breakpoints. Second, the desired gene can be tightly delimited by mapping the positions of the two recombinational breakpoints that most closely flank the gene. Assuming that such breakpoints are randomly distributed, the probability of a breakpoint lying close to the desired gene increases as the size of the recombinant collection increases. It is therefore useful to generate a substantial number of recombinants at the outset (section 2.2.3), particularly since at least three plant generations are required for their isolation.

2.2.1   Using flanking visible markers to reduce the workload

There are a number of useful strategies for identifying individuals carrying recombinational breakpoints near the gene of interest. One of the most efficient methods is to preselect plants known to have breakpoints flanking the gene of interest using flanking visible markers. Briefly, recombinant individuals are identified, based on phenotype, from the F2 of crosses between a line mutant in the gene of interest, and lines carrying proximal and/or distal visible mutations. Not only are these plants known to have recombinational break-points in the region, but also the orientation of these breakpoints with respect to the gene of interest is known. This allows molecular markers obtained in the walk to be mapped to positions proximal or distal to the gene. Using this method, the *ABI3*, *ETR1* and *RPS2* genes were delimited to relatively short DNA segments of 35 kb and 47 kb respectively (Giraudat et al., 1992; Chang et al., 1993; Mindrinos et al., 1994). Similarly, the *AXR1* (Leyser et al., 1993), *Pto* (Martin et al., 1993) and *ABI1* (Leung et al., 1994) genes were each assigned to a specific yeast artificial chromosome (YAC) (Burke et al., 1987) clone within their respective walks.

There are some general points concerning the visible markers selected. An obvious prerequisite is that two genetic backgrounds must be included in the cross to allow DNA polymorphisms to be scored. If the most appropriate visible marker is in the same genetic background as the mutant gene of interest, then a double mutant must first be constructed. This line can then be outcrossed to wild-type plants of a different genetic background. The precise way in which these crosses are set up will differ on a case-by-case basis. The optimal set-up is to start with both recessive alleles (mutant or wild type) on the same chromosome, i.e. in *cis* or coupling, as shown in Fig. 1. In this way recombinant individuals in the F2 will be those showing the recessive pheno-type at one locus and the dominant phenotype at the other. If the recessive alleles are originally on opposite chromosomes, i.e. in *trans* or in repulsion, then the recombinant class will be the double recessive class. This class is extremely rare because it requires a plant to inherit two independent recombi-nant chromosomes.

Although the *cis* configuration gives a far higher proportion of recombinants than the *trans* configuration, a disadvantage is that it can be time-consuming to construct the lines needed. In most cases, the mutant alleles of the gene of interest and the flanking marker gene are both recessive. In order to place them in *cis*, three generations are required. In all, then, six generations will be needed to isolate the recombinants. If, however, either the gene of interest or the flanking marker gene have dominant mutant alleles, then the recessive alleles are already in *cis*, and recombinant isolation will take only three generations in total.

It is very helpful if the phenotype of the flanking mutations chosen can be scored early, so that nonrecombinants can be weeded out. If the *cis* configura-tion can be used, early lethal phenotypes are of particular value. This automa-tically eliminates all individuals inheriting two parental chromosomes from the recessive parent. It also eliminates one of the classes of recombinant, but this is a small price to pay, since it dramatically reduces the complexity of the F2 population. Only two phenotypes will be segregating, the double dominant phenotype and the rare, single recessive recombinant phenotype.

2.2.2 *Alternative methods for recombinant isolation*

In some situations, visible markers flanking the gene may not be available to help target the isolation of recombinants. An alternative is to use DNA markers

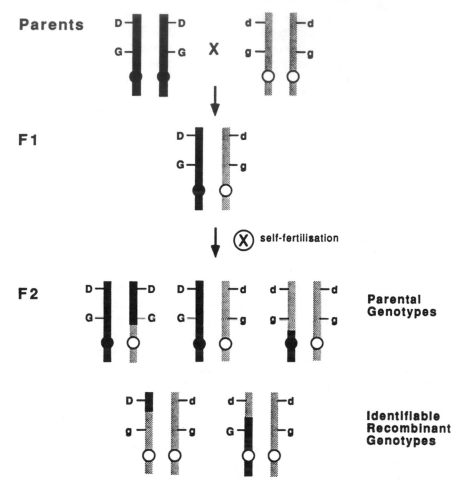

D, d = Distal Visible Marker (dominant, recessive allele)

G, g = Gene of Interest (dominant, recessive allele)

Figure 1. Genetic crosses used to isolate recombinational breakpoints between the gene of interest and a flanking visible marker. Parent lines in differing genetic backgrounds carry the dominant and recessive alleles in *cis*, respectively. F2 individuals containing recombinant genotypes are identified based on the visible phenotypes at both loci

to score F2 individuals for recombinational breakpoints between the gene of interest and the particular DNA marker. In order to obtain an adequate number of recombinants, hundreds of F2 plants may have to be analysed for DNA polymorphisms. PCR-based markers are particularly suited for this task because they allow the use of crude DNA samples which are rapidly prepared from small amounts of tissue. The amount of work can be reduced by pooling DNA from F2 individuals of the recessive phenotype. When a recombination event is detected in a pool, sub-pools can be screened until the recombinant F2 individual/s are identified. Many F2s can be pooled and a grid system in microtitre dishes can be used to identify the recombinant individual when a recombinant is detected in any particular pool. In some cases, this approach for obtaining recombinants can actually be faster than the traditional visible marker approach, because with co-dominant DNA markers it is not necessary to construct lines which place recessive alleles of the marker genes and the gene of interest in *cis*. This step alone takes several generations. The pooling technique is simple enough that it can be used to screen random DNA markers for linkage to the gene of interest if, for example, there are very few known markers in the region where the gene maps.

A variation on this theme is the case where the mutation in the gene of interest has been introgressed into a different genetic background from that in which it was isolated. The gene will be on an island of DNA derived from the line in which the mutation was generated, in a sea of DNA derived from the strain into which it was introgressed. This allows the identification of linked DNA markers by looking for polymorphisms associated with the progenitor strain in the introgressed mutant line. This approach, sometimes called chromosome landing, was key to the isolation of the tomato disease-resistance gene, *Pto* (Martin et al., 1993), and was also used to fine map the *Arabidopsis DET1* gene (Pepper et al., 1994). Many such introgressed lines pre-exist in crop species as a result of centuries of plant breeding. These introgressed genes are, of course, of considerable agronomic importance, so this approach offers tremendous potential for the future, especially since the technique could be applied to species where there are few or no pre-existing map data.

Another option for delimiting the desired gene may be to isolate an inter-allelic recombinant. The breakpoint in such a recombinant will be within the gene of interest, so the position of the gene can be pinpointed by determining as closely as possible the position of the breakpoint by mapping with molecular markers. This method only works if mutant alleles have been isolated in different genetic backgrounds. The frequency of interallelic recombination will be extremely low, and so this strategy may be impractical unless the screen is simple and can be performed at, say, the seedling stage.

2.2.3 Self-incompatible species

The methods described so far assume that the gene of interest was identified in a species in which self-fertilisation is much easier to achieve than outcrossing. If outcrossing is easier, then the basic strategy would remain the same, but the way in which the crosses are set up would clearly be modified. For example, recombination events would be detected by backcrossing F1 plants to a double recessive stock.

2.2.4   How many recombinants?

When isolating recombinants, a desirable number to aim for might be one that provides approximately one DNA breakpoint every 10–20 kb. The average breakpoint distribution is estimated by simply dividing the number of base pairs separating the gene of interest and the flanking visible marker used by the number of recombinants isolated. If the distance between the two recombinant loci is only known in centiMorgans, which is usually the case, then an estimate of the physical distance in base pairs can be obtained using the ratio of the genome size in base pairs and the genome size in centiMorgans. However, because recombination frequencies vary greatly throughout genomes (e.g. Dooner, 1986; Ganal et al., 1989; Brown & Sundaresan, 1991; Segal et al., 1992; Schmidt & Dean, 1993), it would be better to base the estimate of breakpoint distribution on a correlation of recombination frequencies and physical distances in the local region of the walk.

2.3   Isolating a series of contiguous DNA clones between the flanking markers

Once the most closely linked flanking markers have been identified, walking can begin. This involves isolating the DNA between the flanking markers in a contiguous series of overlapping clones (a contig). In *Arabidopsis*, extensive regions of the genome have already been placed in contigs, anchored to the marker maps (Hauge et al., 1991; Dennis et al., 1993). If the gene of interest lies in such a region, it may be that no walking is necessary, or that at least part of the desired contig is already available. As genome projects progress, this happy state of affairs will become more frequent. Currently, however, it is likely that some walking will be necessary.

2.3.1   The first walk step

The first step is to screen genomic libraries with the flanking marker probes. At this stage, the library chosen should be the one with the largest inserts, usually a YAC library (Burke et al., 1987; see Chapter 4). The method of probe preparation will depend on the nature of the flanking markers used. If the flanking markers are conventional RFLPs, then the probes used to detect the RFLPs can be used directly to screen the library (Fig. 2). If PCR-based markers are used, the precise method for probe preparation will depend on the type of marker used and the type of library to be screened. For markers which involve the amplification of many unlinked DNA fragments, only the relevant, linked fragment should be used as a probe. For methods that result in the amplification of a single, unique DNA fragment, such as CAPS, the primers can be used directly in PCR-based gridded YAC library screens (Krol et al., 1987).

Figure 2. The results of a typical *Arabidopsis* YAC library screen using a unique-sequence, radiolabelled probe. The entire library can be replicated onto a single filter for probing. In this example a single YAC clone hybridised to the probe. This figure was kindly provided by Caroline Dean and Renate Schmidt

2.3.2   *Assessing the progress of the walk*

The first screens will usually yield one cluster of YACs for each flanking probe. If the flanking probes are close and/or the YAC library has a large insert size, it may be that a single YAC will hybridise to both probes, indicating that it includes the gene of interest. If the two YAC clusters do not connect, further walk steps may be needed. However, it is perfectly possible that the YACs include the gene even though the walk does not cover the whole region. It is better to gain as much information as possible about the YACs already isolated before proceeding to the next step. It should be possible to orient the YACs, assess the progress of the walk, eliminate chimeric YACs (section 2.3.3) and choose the most effective probes if another step is needed. This crucial information can only be gained if there is an adequate collection of lines with recombinational breakpoints of known orientation with respect to the gene.

Each YAC cluster should consist of a set of overlapping genomic DNA segments. A general picture of a YAC cluster can be derived from Southern analysis of total yeast DNA extracted from each YAC strain and probed with the DNA marker used to isolate the cluster. This will confirm that the YACs are true positives, and ascertain whether the whole DNA segment used to isolate the YACs is included in each YAC. These blots can also be probed with other mapped markers in the region, which, if they show hybridisation, might allow the YAC cluster to be oriented with respect to the gene of interest.

A more detailed picture can be built up using probes isolated from the ends of the YAC insert DNA. End clones can be isolated by several methods, depending on the vector used in the construction of the YAC library. Left ends can usually be cloned by plasmid rescue, whilst right ends as well as left ends can be cloned by vectorette (Riley et al., 1990) or inverse PCR (Silverman et al., 1989) (Chapter 13). The end clones can be used as probes on the YAC strain Southern blots. This will demonstrate where the ends of the YACs map with respect to the other YACs in the cluster. All the ends except two should hybridise to at least one other YAC. Those two ends represent the extreme proximal and distal ends of that YAC cluster. If more than two ends do not hybridise to other YACs, then this suggests that there are chimeric YACs containing DNA segments from unrelated parts of the genome. Alternatively, it is possible that two YACs may not share common sequences because they include sequences from opposite ends of the probe used to isolate the YACs, and extend in opposite directions. If this is the case, then one of the ends from each YAC will map to the DNA probe used to isolate the YACs.

Once candidate extreme end clones have been identified, they can be mapped with respect to the gene of interest using the recombinant line collection. For this to be achieved, they must be able to detect a DNA polymorphism that is segregating in the recombinant line collection. A variety of approaches can be used to identify such a polymorphism. For example, the end clones can be used to detect conventional RFLPs on genomic Southern blots. Since the clones are usually relatively short, the probability of finding such a RFLP using standard six-cutter restriction enzymes is quite low. To improve the odds, a longer segment of DNA can be isolated from a lambda or cosmid library (Chapters 3 and 4), using the end clone as a probe. Alternatively, four-cutter enzymes can be used, either with Southern blots or after primer design, to identify a CAPS marker from the end clone. Once a suitable polymorphism has been identified, the collection of lines known to carry recombination events near the gene of interest can be scored for the polymorphism.

The collection of recombinant lines will have breakpoints known to be either between the polymorphism detected by the original flanking marker and the gene of interest (class 1), or on the other side of the gene (class 2) (Fig. 3). The relative map positions of the recombinational breakpoints and the end clone polymorphism must then be determined. This is achieved by scoring the recombinant line collection for the polymorphism detected by the end clone. If there are any class 1 breakpoints which map between the end clone polymorphism and the original flanking marker, then the end clone must be closer to the gene of interest than the flanking marker (Fig. 3). In this way, the orientation of the YAC cluster on the chromosome can be established. If there is still at least one class 1 breakpoint mapping between the end clone polymorphism and the gene of interest, then the YAC cluster does not extend as far as the gene (Fig. 3). If there are no breakpoints of either class which map between the gene and the end clone polymorphism, then the clone must be very close to the gene but its exact position with respect to the gene cannot be

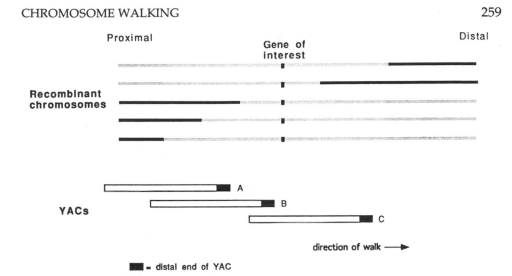

Figure 3. Monitoring a chromosome walk by mapping YAC ends relative to recombinational breakpoints both proximal and distal to the gene of interest. The distal end of YAC A maps proximal of a proximal breakpoint, which indicates that YAC A has not reached the gene of interest. The distal end of YAC B cosegregates with the locus, so it cannot be determined whether YAC B contains the gene. The distal end of YAC C maps distal to a breakpoint that is distal to the gene, and, therefore, the contig must contain the gene. Proximal YAC ends should be mapped similarly

determined (Fig. 3). If any class 2 breakpoints map between the end clone polymorphism and the gene, then the YAC cluster must include the gene (Fig. 3).

If the segregation pattern of the end probe polymorphism appears random and does not fit with the segregation of the flanking marker, then it is probable that the YAC from which it is derived is chimeric, and this part of the YAC is from an unlinked site in the genome.

If the extreme ends of both the distal and proximal YAC clusters can be mapped, and it is found that the clusters do not include the gene of interest, then the end clones closest to the gene can be used to probe the YAC library to isolate two new YAC clusters. These next walk steps can be analysed in exactly the same way as the first YAC steps. The walk then proceeds in this manner, building up a YAC contig towards the gene, until the gene has been crossed from one direction or the other.

2.3.3   Potential problems

YAC problems   Chimeric, deleted or rearranged YACs can cause considerable problems (see also Chapter 4). A large proportion of YAC clones within a library can be aberrant, displaying physical linkages that misrepresent the actual genome (Schmidt & Dean, 1993; Putterill et al., 1993). Because of this it is important to obtain multiple independent clones so that all portions of the contig are represented more than once. Unless each step is confirmed with independent clones, aberrant clones will seriously confound the map data. In addition, any DNA clone that is used for taking a step should be mapped back to the locus to prevent a blind jump to an unrelated locus.

The proportion of chimeric YACs will differ for each YAC library but it can be as high as 40% (Putterill et al., 1993). Chimeric YACs can be identified by several methods. First, as described above, their ends will map to different parts of the genome. Second, if a YAC is suspected of being chimeric, it is worth probing it with chloroplast DNA and any common non-dispersed repetitive DNAs available for the species concerned. For *Arabidopsis*, whole YAC libraries have been screened with these DNAs, so this information is already available through the *Arabidopsis* database (Chapter 9).

The most straightforward method for the detection of deletions or rearrangements in the YAC insert is to compare the restriction patterns obtained using YAC strain DNA and plant genomic DNA on Southern blots probed with whole YAC DNA. YACs can be isolated for probe preparation using pulsed field gels (Schwartz & Cantor, 1984). If the YAC is large, restriction analysis should be carried out with rare-cutter enzymes, and the fragments separated by pulsed field gel electrophoresis (Chapter 4). If the YAC is small, say less than 100 kb, then the conventional gels can be used.

An alternative method is to construct a restriction map of the YAC using partial digests. Samples of YAC strain DNA are subjected to restriction digestion with increasing amounts of enzyme. These partially digested fragments are then separated by gel electrophoresis, blotted and probed with the YAC end probes, or end-specific probes from the YAC vector. As before, rare cutters and pulsed field gels are needed for large YACs, whilst six-cutters and conventional gels are also useful when working with smaller YACs. Only restriction fragments attached to the YAC end will be detected. In a complete digest, only the end fragment itself will be detected. The next biggest fragment, common in a near-complete digest, will represent the end fragment plus the neighbouring fragment. The sizes of the successive fragments can be determined using standards, and in this way a restriction map of the YAC can be built up. This can be compared with other YACs in the walk and with plant genomic DNA probed with whole YAC DNA. If inconsistencies are detected, then it is likely that the YAC is rearranged, deleted or chimeric.

YAC gaps   Another potential problem is that not all genomic DNA regions will be represented in the YAC libraries. To get around this it may be necessary to walk a few steps in a cosmid or lambda library. Alternatively, special jumping and linking libraries can be constructed (Poustka et al., 1986).

Multiple YACs in one yeast strain   Occasionally a single YAC strain will contain more than one YAC, which can cause confusion. The hybridising YAC can be identified by Southern analysis of pulsed field gels prepared from the YAC strain DNA. End-cloning strategies will yield all possible ends, so the appropriate ends must also be identified in this way.

Repetitive DNA   A serious problem for walkers is repetitive DNA, particularly dispersed repeats (Chapter 1). During the walk, this is a problem if the probe used to take a step includes repeated sequences, and hence hybridises to YACs derived from other areas of the genome. If many YACs are detected, then it is immediately obvious that something is amiss. If only a few YACs are detected, more work is needed before it becomes clear that some are from other areas of the genome, and this can take up a lot of time. These problems can usually be reduced if small end probes of only a few kilobases are used, since these are less likely to contain repeats. If an end probe does prove to contain repetitive DNA, then an alternative probe from within the walk can be used. If there are long stretches of interspersed repetitive DNA, large YACs are needed to step over these regions. Repetitive DNA can cause further problems when whole YACs are used as probes on Southern blots (see above), or in library screens (see below). If one repeat type is known to be the main difficulty, excess, unlabelled DNA of this type can be used to block hybridisation to these sequences. In general, problems with repetitive DNA are rather intractable and this is one of the main reasons why walking in species with high repetitive DNA content is very difficult.

Highly monomorphic regions   Another difficult problem arises when polymorphisms cannot be found with YAC end probes, and so cannot be mapped. When all search methods fail (see above), then it may be necessary to proceed to the next step and hope for better luck with mapping the next end. If this is the case, it may be better to try to map the next most extreme end probe and use that to isolate the next step. Using unmapped extreme end probes is dangerous, since there is no direct evidence, other than faith in the integrity of the YAC, that the probe DNA is from the desired region of the genome.

2.4   Identifying the desired gene within the contig

There are two approaches for identifying the desired gene within the contig. One is to identify a genomic DNA fragment within the contig that confers the function of the desired gene using transformation. For example, a wild-type genomic fragment which rescues the phenotype of a loss-of-function mutation in the gene of interest must include the gene. The other approach is to obtain gene candidates within the contig, and then determine which of the candidates is the desired gene based on transformation and/or comparisons of the DNA sequences from wild-type and mutant alleles. Usually, a combination of these approaches is taken, with an initial emphasis on one or the other.

2.4.1   *Plant transformation*

The desired gene, which has been delimited by fine-structure mapping, can be delimited further based on gene function using transformation. In this approach, overlapping DNA fragments derived from the contig are tested in vivo for the ability to complement a mutation or impose a phenotype such as pathogen resistance. This was the principal approach used in isolating the genes *AXR1* and *ABI1*. Complementation of an *axr1-3* mutation by transformation narrowed down the position of the *AXR1* gene from a YAC to a 16.5-kb DNA fragment (Leyser et al., 1993) (Fig. 4). Similarly, imposition of the

Figure 4. An example of a functional complementation experiment. A DNA segment from a walk was shown to contain the gene of interest by its ability to restore the wild-type phenotype to mutant plants. In this example, a 16.5-kb segment of DNA was shown to restore auxin sensitivity to the *axr1* auxin-resistant mutant of *Arabidopsis* (Leyser et al., 1993). The photograph shows seedlings which have been germinated on concentrations of auxin that are inhibitory to root elongation. The wild-type plant on the left shows severely inhibited root growth, whilst the *axr1* mutant plant in the centre is able to elongate a root. The right-hand plant is mutant at *axr1* but has been transformed with a 16.5-kb segment of DNA from a chromosome walk covering the *AXR1* locus. This DNA segment restores wild-type auxin sensitivity to the mutant plant, demonstrating that it contains the *AXR1* gene

dominant *abi1-1* mutant phenotype on wild-type plants narrowed down the position of the *ABI1* gene from a YAC to a 0.6-kb fragment (Meyer et al., 1994). *ABI3* was also further delimited from a 35-kb cosmid clone to an 11-kb DNA fragment (Giraudat et al., 1992).

Transformation is now routine for many plants. The most common technique relies on the DNA transfer system of the pathogenic soil bacterium *Agrobacterium tumefaciens*, whose wide host range allows transformation of many dicot and several monocot species (Zambryski, 1992; Bytebier et al., 1987; Schafer et al., 1987). This method makes use of the natural ability of *Agrobacterium* to infect plant cells and transfer a specific DNA fragment called T-DNA to the plant nucleus, where it is stably integrated at one or more random sites within the plant genome (Zambryski, 1992). The procedure typically involves infection of plant tissue explants by agrobacteria harbouring a specially designed transformation vector that carries in place of the T-DNA a drug-resistance marker and the DNA to be transferred. Following infection, plants are regenerated from the callus of transformed plant cells, which were selected based on drug resistance. For *Arabidopsis*, methods have been developed for directly transforming the plant, eliminating the time-consuming regeneration step (Bechtold et al., 1993; Feldmann & Marks, 1987). Other transformation methods utilise physical means to introduce DNA into plant cells. These procedures include particle-gun transformation (Klein et al., 1987), electroporation (Fromm et al., 1982), ultrasonication (li-Jian et al., 1991), and microinjection (Reich et al., 1986).

If *Agrobacterium* is used to transform plants with DNA from the walk, it would be convenient if the vector of the original walking library is a transformation vector, because then the DNA clones can be used directly in plant transformation as they are isolated (after transfer to *Agrobacterium*). Most likely, however, the contig will consist of YACs that must be subcloned into an appropriate vector. Cosmid transformation vectors are commonly used because they carry large inserts of up to about 45 kb (Chapter 4). DNA clones for transformation should be large and overlapping to ensure that the entire gene is contained on at least one clone. In addition, the overlaps are useful for determining the location of the gene, as illustrated nicely in Meyer et al. (1994). Subcloning a YAC requires pulsed field gel electrophoresis (Schwartz & Cantor, 1984) to separate the YAC from the other yeast chromosomes. The gel-purified YAC is then subcloned using routine methods. Another option is to subclone the entire genome of the YAC-containing yeast strain, avoiding large-scale preparation of purified YAC DNA. As an alternative to subcloning the contig, the contig DNA can be used to screen existing transformation-vector-based libraries such as the one described by Olszewski et al. (1988) for *Arabidopsis*. One cosmid transformation library for *Arabidopsis* has been propagated in an *Agrobacterium* host strain for immediate use in transformation (Lazo et al., 1991). For the case in which a dominant phenotype is to be conferred to wild-type plants, a library containing DNA from the dominant strain must be constructed in a transformation vector. The contig, which typically derives from wild-type YACs, can be re-isolated from this library. Otherwise, the walk can be initially carried out using such a library, but it may not be worth the effort because the steps would be too small for most walks.

Functional tests will generally involve rescue of a mutant phenotype. A potential obstacle is that the mutation may not be in an easily transformable line. To get around this, either the mutant is backcrossed to a transformable line and then subsequently transformed, or the construct of interest is crossed from a transformable line into the mutant. This is not a problem when the test involves imposition of a phenotype using a dominant gene. Examples of phenotype imposition are pathogen resistance, which was conferred to a susceptible tomato line by the tomato *Pto* gene (Martin et al., 1993), and insensitivity to the hormones ethylene and abscisic acid conferred to wild-type *Arabidopsis* by the *ETR1* (Chang et al., 1993) and *ABI1* (Leung et al., 1994; Meyer et al., 1994) genes, respectively. A caution here is that the introduced gene may not confer the expected dominant phenotype in the presence of two wild-type copies of the gene. It may be informative to check the dominant phenotype in a triploid that contains one copy of the dominant gene and two copies of the wild-type gene (Leung et al., 1994). A triploid can be obtained as the progeny of a cross between a diploid mutant and a tetraploid wild type.

The main problem presented by the usual transformation procedure is that it typically requires several months to regenerate transformed plants and obtain seed. Because of this length of time, transformation has not been used as the exclusive approach for identifying the relevant gene. Candidate genes within the contig should be simultaneously isolated and characterised (as described in section 2.4.2). In fact, transformation is not necessarily employed at all at this stage if the functional assay is not a simple one, e.g. gas chromatographic analysis of fatty acid composition for the *FAD3* gene (Arondel et al., 1992). Furthermore, if recombination breakpoints have already delimited the gene to a sufficiently small region, then the gene will probably be located faster using other methods. To be able to screen clones more rapidly, it would be valuable to have a transient expression assay and/or a reporter gene system for the gene of interest. For example, if the gene of interest is required to activate transcription of a cloned gene, then the promoter of this second gene, fused to a reporter gene such as β-glucuronidase (*GUS*), could be used to assay for the presence of the gene of interest.

2.4.2   *Isolating candidate genes*

There are several strategies for identifying candidate genes within a delimited DNA segment—the segment being either the contig or the DNA fragment that has been defined by transformation. A useful first step is to analyse RNA blots probed with DNA fragments of the delimited region in order to identify the transcribed regions. Next, a cDNA library is typically screened in order to obtain cDNAs that potentially represent the desired gene. The *Arabidopsis* *FAD3* gene was isolated in this manner by using entire YAC clones (comprising the contig) to probe a cDNA library made from developing seeds (Arondel et al., 1992). The *FAD3* mRNA was expected to be moderately abundant within developing seeds, and, in fact, more than half of the clones hybridising to one YAC represented a single gene that was later established to be *FAD3*, based on genetic complementation of the mutation. (An interesting aside in this example is that the cDNA clones were obtained from a *Brassica napus* cDNA library, because the appropriate library in *Arabidopsis* was not available.) The same cDNA isolation strategy was used to clone the tomato *Pto* gene (Martin et al., 1993). For this approach to succeed, it is critical that the desired gene is represented in the cDNA library. Whenever possible, the choice of library should be based on knowledge of, or predictions about, the gene's expression pattern. In addition, cDNAs isolated in this way should be genetically mapped to verify that they cosegregate with the relevant locus.

For a gene that is expressed at a low level, there are alternative strategies for identifying candidate genes within the contig. A genomic DNA fragment carrying the gene might detect an RFLP if an allele of the relevant gene contains a DNA rearrangement, an insertion/deletion, or a base change within a restriction site. There are also a number of methods for detecting single-base mutations (Prosser, 1993). If the desired gene is highly conserved among species, then hybridisation of contig DNA fragments to genome blots of other species can be used to identify conserved DNA segments. Genomic DNA sequencing might reveal a long open reading frame or sequence similarity to entries in the expressed sequence tag (EST) database (Chapter 15).

Once candidate genes are obtained, certain clues may distinguish a particular gene and eliminate others. The DNA sequence may display revealing characteristics. For example, the *FAD3* gene, which encodes omega-3 desaturase, was found to be very similar in sequence to a bacterial desaturase (Arondel et al., 1992). Other clues might be obtained by hybridising the genes to RNA blots in order to analyse patterns of expression. There may be a response to an inducing factor, or the transcript may differ between mutant and wild-type alleles. For example, *AXR1* transcripts differed in both size and abundance in mutant alleles compared to the wild type (Leyser et al., 1993).

In order to establish that a gene is indeed the one of interest, the gene is tested for the ability to complement a mutation or impose a phenotype using transformation. In addition, the DNA sequences of mutant alleles can be determined and compared to the wild type. The presence of mutations is strong evidence that the gene of interest is in hand.

3 GENERAL POINTS

Some general points concerning the chromosome walking procedure are emphasised here. As genome projects advance, it will be possible to obtain DNA contigs for any area of the genome of some plant species. Therefore, careful mapping of the gene of interest will be paramount for rapid isolation of the gene. It is essential to map the gene of interest accurately relative to the other markers in the region. For this to be accomplished, the gene of interest must be included in the same cross as the other markers. The relative order of markers, and the nearest marker to the gene of interest, cannot be inferred by aligning maps of different crosses.

The value of recombinational breakpoints flanking the desired gene cannot be overemphasised. Although mapping of recombinational breakpoints is not essential for cloning a gene by walking (as exemplified by the cloning of the *Arabidopsis FAD3* gene described in Arondel et al. (1992)), it elegantly delimits a gene with certainty to a relatively small DNA fragment. In turn, the time and effort required to isolate the gene are greatly reduced. Recombinants are generally easy to obtain, especially when visible markers are employed. It is therefore worthwhile to collect as many recombinants as possible. The larger the recombinant collection, the easier it is to monitor the progress of the walk, and the more narrowly the gene is delimited. Optimally, with enough recombinants, the breakpoints will delimit a single gene — the gene of interest.

It is useful to carry out the various tasks of chromosome walking simultaneously, because the pace of a walk cannot be predicted in advance. For instance, a walk might be quite advanced before any recombinants are isolated. If the walk is sluggish, it might be possible to identify the desired gene by DNA sequencing or by transformation before the contig is formally completed. To avoid spending several years on a walk, crosses to obtain recombinants should be started immediately. Gene identification strategies can begin even during the walk, and cDNAs should be isolated and characterised while waiting for transformants.

Every walk will be different. The most effective strategies for a particular locus may be obvious only in hindsight. The course chosen for each walk will depend upon available resources, personal preferences, and largely upon luck.

4 PROSPECTS

Chromosome walking is rapidly becoming an accessible and indispensable approach for plant gene isolation. Increasingly, technological advances are making the entire procedure faster and easier. At the forefront of this progress is the model plant *Arabidopsis thaliana*. Various tools to facilitate walking have been developed for this plant. High-density linkage maps composed of RFLP markers (Chang et al., 1988; Nam et al., 1989; Hauge et al., 1993) and RAPD markers (Reiter et al., 1992) are integrated with the classical visible marker map. Recombinant inbred lines are available, so that new DNA markers can be easily integrated into existing maps (Reiter et al., 1992; Lister & Dean, 1993). At least 380 RFLPs have been mapped, which translates into an average walking distance of about 65 kb to any locus in *Arabidopsis*, starting from the nearest RFLP marker (Hwang et al., 1991). Additionally, there is a growing collection of mapped PCR-based markers, which consists of RAPDs (Williams et al., 1990), CAPs (Konieczny & Ausubel, 1993), and simple sequence length polymorphisms (SSLPs) (Bell & Ecker, 1994). Several different YAC libraries are available (Grill & Somerville, 1991; Ward & Jen, 1990; Ecker, 1990). The most recent YAC library contains 2300 clones with an average insert size of 250 kb, which is equivalent to almost six haploid genomes (Schmidt & Dean, 1993).

A significant advance for map-based gene cloning will be the completion of physical maps for entire plant genomes. Construction of one such map is a major effort of the international *Arabidopsis* genome project (Dennis et al., 1993; Schmidt & Dean, 1993; Hwang et al., 1991). The map will be a clonal reconstruction of the genome, so that DNA clones corresponding to any map position will be immediately available. A similar library is close to completion for the nematode *Caenorhabditis elegans*, whose genome size is comparable to that of *Arabidopsis* (Coulson et al., 1991). So far in *Arabidopsis*, 750 cosmid contigs representing 90–95% of the genome have been aligned using restriction digest fingerprinting (Hauge et al., 1991), and more than 100 mapped YAC contigs covering approximately 60% of the genome have been assembled (Dennis et al., 1993). Construction of the physical map is aided by several factors, such as the growing density of mapped DNA markers, the data derived from numerous walks that are in progress, and the availability of several YAC libraries.

Chromosome walking in plants other than *Arabidopsis* is also within reach. Genome projects for tomato, maize and rice are quite advanced; there are YAC libraries and well-developed classical (O'Brien, 1990) and molecular (Tanksley et al., 1992; Burr & Burr, 1991; Coe et al, 1990; Nagamura, 1993) genetic maps for each of these plants. The tomato and maize YAC libraries each contain about three haploid genomes; the tomato library consists of 22 000 YAC clones with an average insert size of 140 kb (Martin et al., 1992), while the maize library contains 79 000 YAC clones with an average insert size of 145 kb (Edwards et al., 1992). The rice library represents an estimated 5.5 genomes and contains 12 000 YAC clones with an average insert size of 350 kb (Umehara et al., 1992). At the time of writing, there is one example of a successful chromosome walk in tomato (Martin et al., 1993), but walks to several other tomato genes are not far behind (e.g. Giovannoni et al., 1991; Wing et al., 1994). Molecular linkage maps for a number of other plants, such as lettuce and soybean, are being developed (Chang & Meyerowitz, 1991). These maps represent a first step towards achieving map-based cloning in these other plants. In the absence of a high-density linkage map, positional cloning is still feasible because loci can be introgressed into a different genetic background and random markers can be screened for linkage to the loci of interest. The ability to walk to genes in a variety of plants is particularly important for obtaining genes related to crop improvement; although *Arabidopsis* is useful for isolating numerous plant genes, it does not serve as a model for some agronomically important processes.

Technological advances in several areas are continually improving the chromosome walking procedure. One area of notable progress is the recent development of new types of DNA markers such as various PCR-based markers. Compared to traditional RFLPs, these simpler methods for detecting polymorphisms significantly speed up the mapping of mutations and recombinational breakpoints, as well as the screening of random markers for linkage to the gene of interest. Inclusion of these recent markers (along with newly cloned genes) in linkage maps can rapidly increase the density of DNA markers, resulting in shorter, more defined walks. The isolation of overlapping clones is likely to be made easier with improvements in YAC vectors, increases in YAC insert size, and possibly the use of different types of vector. Longer YACs will be especially useful for large genomes containing a high proportion of repetitive DNA. YAC inserts as large as 700 kb have been cloned in *Arabidopsis* (Ecker, 1990); however, improved plant DNA isolation methods should lead to even longer YACs. The task of gene identification, once the contig is isolated, is likely to be facilitated by new transformation protocols that are simpler and faster (e.g. Bechtold et al., 1993).

As the available molecular tools improve, increasingly the rate-limiting stage will be fine-structure mapping around the gene of interest. In the fruit fly, *Drosophila*, genetic mapping is greatly facilitated by the existence of defined chromosomal deletions and rearrangements, which can be cytologically as well as genetically defined with the use of polytene chromosomes. Similar chromosomal abnormalities have even made possible chromosome walking to human genes which, considering the complexity of the human genome, is a remarkable achievement. It would clearly be a great advantage to have such tools available to plant walkers. Although large deletions are frequently lethal in the haploid phase of the plant life-cycle, a wide variety of chromosomal rearrangements have been obtained in maize using irradiation mutagenesis of pollen (Coe et al., 1988), and, recently, chromosomal deficiencies in *Arabidopsis* have been obtained by the same method (Vizir et al., 1994). The continued generation of rearrangements and small deletions in plant study species could significantly improve the efficiency of chromosome walking.

5　CONCLUDING REMARKS

The power of chromosome walking lies in the fact that it can be used to isolate any gene for which there are two alleles conferring different phenotypes. The approach is thus widely applicable. It requires no knowledge of the gene product, gene expression, or the nature of mutations in the gene. In many instances, it may be the only way to obtain a particular gene. Fortunately, chromosome walking is fast on its way to becoming a routine procedure in plants due to directed genome studies and technological improvements. Clearly, there will soon be a proliferation of plant genes isolated by walking. The importance of this for plant biology is profound, because analysis of the isolated genes will elucidate many fundamental and previously undescribed plant processes.

ESSENTIAL READING

Chang, C., Kwok, S.F., Bleecker, A.B. and Meyerowitz, E.M. (1993) *Arabidopsis* ethylene-response gene *ETR1*: similarity of product to two-component regulators. *Science*, **262**, 539–544.
Leyser, H.M.O., Lincoln, C.A., Timpte, C. et al. (1993) *Arabidopsis* auxin-resistance gene

AXR1 encodes a protein related to ubiquitin-activating enzyme E1. *Nature*, **364**, 161–164.

Martin, G.B., Brommonschenkel, S.H., Chunwongse, J. et al. (1993) Map-based cloning of a protein kinase gene conferring disease resistance in tomato. *Science*, **262**, 1432–1435.

Pepper, A., Delaney, T., Washburn, T. et al. (1994) DET1, a negative regulator of light-mediated development and gene expression in *Arabidopsis*, encodes a novel nuclear-localized protein. *Cell*, **78**, 109–116.

Putterill, J., Robson, F., Lee., K. and Coupland, G. (1993) Chromosome walking with YAC clones in *Arabidopsis*: isolation of 1700 kb of contiguous DNA on chromosome 5, including a 300 kb region containing the flowering-time gene. *CO. Mol. Gen. Genet.*, **239**, 145–157.

REFERENCES

Ahn, S. and Tanksley, S.D. (1993) Comparative linkage maps of the rice and maize genomes. *Proc. Natl. Acad. Sci. USA*, **90**, 7980–7984.

Arondel, V., Lemieux, B., Hwang, I. et al. (1992) Map-based cloning of a gene controlling omega-3 fatty acid desaturation in *Arabidopsis*. *Science*, **258**, 1353–1355.

Bechtold, N., Ellis, J. and Pelletier, G. (1993) In planta *Agrobacterium* mediated gene transfer by infiltration of adult *Arabidopsis thaliana* plants. *C.R. Acad. Sci. Paris*, **316**, 1194–1199.

Bell, C.J. and Ecker, J.R. (1994) Assignment of 30 microsatellite loci to the linkage map of *Arabidopsis*. *Genomics*, **19**, 137–144.

Bender, W., Akam, M., Karch, F. et al. (1983) Molecular genetics of the bithorax complex of *Drosophila melanogaster*. *Science*, **221**, 23–29.

Botstein, D., White, R.L., Skolnick, M. and Davis, R.W. (1980) Construction of a genetic linkage map in man using restriction fragment length polymorphisms. *Am. J. Human. Genet.*, **32**, 314–331.

Brown, J. and Sundaresan, V. (1991) A recombination hotspot in the maize A1 intragenic region. *Theor. Appl. Genet.*, **81**, 185–188.

Burke, D.T., Carle, G.F. and Olson, M.V. (1987) Cloning of large segments of exogenous DNA into yeast by means of artificial chromosome vectors. *Science*, **236**, 806–812.

Burr, B. and Burr, F.A. (1991) Recombinant inbreds for molecular mapping in maize: theoretical and practical considerations. *Trends Genet.*, **7**, 55–60.

Bytebier, B., Deboeck, F., DeGreve, H. et al. (1987) T-DNA organisation in tumour cultures and transgenic plants of the monocotyledon *Asparagus officinalis*. *Proc. Natl. Acad. Sci. USA*, **84**, 5345–5349.

Chang, C. and Meyerowitz, E.M. (1991) Plant genome studies: restriction fragment length polymorphism and chromosome mapping information. *Curr. Opin. Genet. Dev.*, **1**, 112–118.

Chang, C., Bowman, J., Dejohn, A. et al. (1988) Restriction fragment length polymorphism linkage map of *Arabidopsis thaliana*. *Proc. Natl. Acad. Sci. USA*, **85**, 57–87.

Coe Jr E.H., Neuffer, M.G. and Hoisington, D.A. (1988) The genetics of corn. In *Corn and Corn Improvement* (eds G.F. Sprague and J.W. Dudley), pp. 81–258. American Society of Agronomy, Madison, Wisconsin.

Coe, E., Hoisington, D. and Chao, S. (1990) Gene list and working maps. *Maize Genet. Coop. Newslett.*, **64**, 154–163.

Coulson, A., Kozono, Y., Lutterbach, B. et al. (1991) YACs and the *C. elegans* genome. *Bioessays*, **13**, 413–417.

Dennis, L., Dean, C., Flavell, R. et al. (1993) The multinational coordinated *Arabidopsis thaliana* genome research project progress report: year three. Publication NSF 93–173. National Science Foundation, Washington, DC.

Dooner, H.K. (1986) Genetic fine structure of the Bronze locus in maize. *Genetics*, **113**, 1021–1036.

Ecker, J.R. (1990) PFGE and YAC analysis of the *Arabidopsis* genome. *Methods* **1**, 186–194.

Edwards, K.J., Thompson, H., Edwards, D. et al. (1992) Construction and characterisation of a yeast artificial chromosome library containing three haploid maize genome equivalents. *Plant Mol. Biol.*, **19**, 299–308.

Feldmann, K.A. and Marks, M.D. (1987) *Agrobacterium*-mediated transformation of germinating seeds of *Arabidopsis thaliana*: a non-tissue culture approach. *Mol. Gen. Genet.*, **208**, 1–9.

Fromm, M., Taylor, L.P. and Walbot, V. (1982) Expression of genes transferred into monocot and dicot plant cells by electroporation. *Proc. Natl. Acad. Sci. USA*, **82**, 5824–5828.

Ganal, M.W., Young, N.D. and Tanksley, S.D. (1989) Pulsed field gel electrophoresis and physical mapping of large DNA fragments in the Tm-2a region of chromosome 9 in tomato. *Mol. Gen. Genet.*, **215**, 395–400.

Giovannoni, J.J., Wing, R.A., Ganal, M.W. and Tanksley, S.D. (1991) Isolation of molecular markers from specific chromosomal intervals using DNA pools from existing mapping populations. *Nucleic Acids Res.*, **19**, 6553–6558.

Giraudat, J., Hauge, B.M., Valon, C. et al. (1992) Isolation of the *Arabidopsis ABI3* gene by positional cloning. *Plant Cell*, **4**, 1251–1261.

Grill, E. and Somerville, C. (1991) Construction and characterization of yeast artificial chromosome library of *Arabidopsis* which is suitable for chromosome walking. *Mol. Gen. Genet.*, **226** 484–490.

Hauge, B.M., Hanley, S., Giraudat, J. and Goodman, H.M. (1991) Mapping the *Arabidopsis* genome. In *Molecular Biology of Plant Development* (eds G.I. Jenkins and W. Schuch), pp. 45–56. The Company of Biologists, Cambridge.

Hauge, B.M., Hanley, S.M., Cartinhour, S. et al. (1993) An integrated genetic/RFLP map of the *Arabidopsis thaliana* genome. *Plant J.*, **3**, 745–754.

Hwang, I., Kohchi, T., Hauge, B.M. et al. (1991) Identification and map position of YAC clones comprising one-third of the *Arabidopsis* genome. *Plant J.*, **1**, 367–374.

Klein, T.M., Wolf, E.D., Wu, R. and Sanford, J.C. (1987) High velocity microprojectiles for delivering nucleic acids into living cells. *Nature*, **327**, 70–73.

Konieczny, A. and Ausubel, F.M. (1993) A procedure for mapping *Arabidopsis* mutations using codominant ecotype-specific PCR-based markers. *Plant J.*, **4**, 403–410.

Krol, A., Carbon, P., Ebel, J-P. and Appel, B. (1987) *Xenopus tropicalis* U6 snRNA genes transcribed by PolIII contain the upstream promoter elements used by PolII dependent U snRNA genes. *Nucleic Acids Res.*, **15**, 2463–2478.

Kurata, N., Moore, G., Yoshiaki, N. et al. (1994) Conservation of genome structure between rice and wheat. *Biotechnology*, **12**, 276–278.

Lazo, G.R., Stein, P.A. and Ludwig, R.A. (1991) A DNA transformation-competent *Arabidopsis* genomic library in *Agrobacterium*. *Biotechnology*, **9**, 963–967.

Leung, K., Bouvier-Durand, M., Morris, P.-C. et al. (1994) *Arabidopsis* ABA response gene ABI1: features of a calcium-modulated protein phosphatase. *Science*, **264**, 1448–1452.

Li-Jian, Z., Le-Mei, C., Ning, X. et al. (1991) Efficient transformation of tobacco by ultrasonication. *Biotechnology*, **9**, 996–997.

Lister, C. and Dean, C. (1993) Recombinant inbred lines for mapping RFLP and phenotypic markers in *Arabidopsis thaliana*. *Plant J.*, **4**, 745–750.

Martin, G.B., Ganal, M.W. and Tanksley, S.D. (1992) Construction of a yeast artificial chromosome library of tomato and identification of cloned segments linked to two disease resistance loci. *Mol. Gen. Genet.*, **233**, 25–32.

Meyer, K., Leube, M.P. and Grill, E. (1994) A protein phosphatase 2C involved in ABA signal transduction in *Arabidopsis thaliana*. *Science*, **264**, 1452–1455.

Meyerowitz, E. (1987) *Arabidopsis thaliana*. *Annu. Rev. Genet.*, **21**, 93–111.

Mindrinos, M., Katagiri, F., Yu, G-L. and Ausubel, F.M. (1994) The *A. thaliana* disease resistance gene *RPS2* encodes a protein containing a nucleotide-binding site and leucine-rich repeats. *Cell*, **78**, 1089–1099.

Moore, G., Gale, M.D., Kurata, N. and Flavell, R.B. (1993) Molecular analysis of small grain cereal genomes: current status and prospects. *Biotechnology*, **11**, 584–589.

Nagamura, Y. (1993) A high density STS and EST linkage map of rice. *Rice Genome*, **2**, 11–13.

Nam, H.-G., Giraudat, J., Boer, B.D. et al. (1989) Restriction fragment length polymorphism linkage map of *Arabidopsis thaliana*. *Plant Cell*, **1**, 699–705.

O'Brien, S. (1990) *Genetic Maps*, 5th edn. Cold Spring Harbor Laboratory Press, New York.

Olszewski, N.E., Martin, F.B. and Ausubel, F.M. (1988) Specialized binary vector for plant transformation: expression of the *Arabidopsis thaliana* AHAS gene in *Nicotiana tabacum*. *Nucleic Acids Res.*, **16**, 10765–10782.

Poustka, A., Pohl, T., Barlow, D.P. et al. (1986) Molecular approaches to mammalian genetics. *Cold Spring Harbor Symp. Quant. Biol.*, **51**, 131–139.

Prosser, J. (1993) Detecting single-base mutations. *Trends Biotechnol.*, **11**, 238–246.

Reich, T.J., Iyer, V.N. and Miki, B.L. (1986) Efficient transformation of alfalfa protoplasts by the intranuclear microinjection of Ti plasmids. *Biotechnology*, **4**, 1001–1003.

Reiter, R.S., Williams, J.G.K., Feldmann, K.A. et al. (1992) Global and local genome mapping of *Arabidopsis thaliana* by using recombinant inbred lines and random amplified polymorphic DNAs. *Proc. Natl. Acad. Sci. USA*, **89**, 1477–1481.

Riley, J., Butler, R., Ogilive, D. et al. (1990) A novel, rapid method for the isolation of terminal sequences from yeast artificial chromosome (YAC) clones. *Nucleic Acids Res.*, **18**, 2887–2890.

Schafer, W., Gîrz, A. and Kahl, G. (1987) T-DNA integration and expression in a monocot crop plant after induction of *Agrobacterium*. *Nature*, **327**, 529–532.

Schmidt, R. and Dean, C. (1993) Towards construction of an overlapping YAC library of the *Arabidopsis thaliana* genome. *Bioessays*, **15**, 63–69.

Schwartz, D.C. and Cantor, C.R. (1984) Separation of yeast chromosome-sized DNAs by pulsed field gel electrophoresis. *Cell*, **37**, 67–75.

Segal, G., Sarfatti, M., Schaffer, M.A. et al. (1992) Correlation of genetic and physical structure in the region surrounding the I2 *Fusarium oxysporum* resistance locus in tomato. *Mol. Gen. Genet.*, **231**, 179–185.

Silverman, G.A., Ye, R.D., Pollock, K.M. et al. (1989) Use of yeast artificial chromosome clones for mapping and walking within human-chromosome segment 18Q21.3. *Proc. Natl. Acad. Sci. USA*, **86**, 7485–7489.

Tanksley, S.D., Ganal, M.W., Prince, J. et al. (1992) High density molecular linkage maps of the tomato and potato genomes: biological inferences and practical applications. *Genetics*, **132**, 1141–1160.

Umehara, Y., Miyazaki, A., Tanoue, H. et al. (1992) Construction of rice genome libraries in yeast artificial chromosomes for physical mapping. Meeting Abstract. *Plant Genome I*. San Diego, California.

Vizir, I.Y., Anderson, M.L., Wilson, Z.A. and Mulligan, B.J. (1994) Isolation of deficiencies in the *Arabidopsis* genome by γ-irradiation of pollen. *Genetics*, **137**, 1111–1119.

Ward, E.R. and Jen, G.C. (1990) Isolation of single copy-sequence clones from a yeast artificial chromosome library of randomly-sheared *Arapidopsis thaliana* DNA. *Plant Mol. Biol.*, **14**, 561–568.

Williams, J.G.K., Kubelik, A.R., Livak, K.J. et al. (1990) DNA polymorphisms amplified by arbitrary primers are useful as genetic markers. *Nucleic Acids Res.*, **18**, 6531–6535.

Wing, R.A., Zhang, H.-B. and Tanksley, S.D. (1994) Map-based cloning in crop plants. Tomato as a model system: I. Genetic and physical mapping of jointless. *Mol. Gen. Genet.*, **242**, 681–688.

Zambryski, P. (1992) Chronicles from the *Agrobacterium*–plant cell DNA transfer story. *Annu. Rev. Plant Physiol. Plant Mol. Biol.*, **43**, 465–490.

IV   Insertional Mutagenesis

11 T-DNA-mediated Insertional Mutagenesis

KEITH LINDSEY and JENNIFER F. TOPPING
University of Durham, UK

This chapter describes the use of T-DNA-mediated insertional mutagenesis to identify plant genes. This approach has a number of useful features for gene identification and isolation:

(1) Insertion of T-DNA may disrupt a gene in such a way as to prevent its expression, and this may result in the generation of an aberrant phenotype.

(2) Specialised T-DNAs carrying reporter genes can generate functional tags of genes. Here, reporter genes fused to weak promoters or lacking promoters altogether ('interposons') have been used, respectively, to detect genes by their activation following their integration either adjacent to native enhancer elements ('enhancer trapping'), downstream of native gene promoters ('promoter trapping') or at exon–intron boundaries ('gene trapping'). A fourth type of vector, carrying strong enhancer elements, can be used to induce the overexpression of native genes, to generate a dominant mutant phenotype.

(3) Insertion of T-DNA sequences into or adjacent to genes therefore facilitates gene identification on the basis of mutant phenotype or reporter transgene activity, and the cloning of the gene may be achieved by plasmid rescue, inverse PCR or following the construction and screening of a genomic library made from transformants of interest.

Using this cloning approach, a number of genes have been isolated, most commonly on the basis of a mutant phenotype.

1 INTRODUCTION

It has been estimated that, during the life-cycle of higher plants, between approximately 30 000 and 100 000 genes are expressed, according to the species (Goldberg, 1988; Gibson & Somerville, 1993). It is the selective expression of these genes, and the enormously complex interactions of the gene products, that orchestrate cell metabolism, growth, differentiation and development. Genes are being identified and isolated, for example, that are required for the determination of cell fate and organ identity (e.g. Bradley et al., 1993; Meyerowitz, 1994), and many other genes, which may not play a determinative role in

Plant Gene Isolation: Principles and Practice. Edited by G. D. Foster and D. Twell.
© 1996 John Wiley & Sons Ltd.

development, are known to be expressed in spatially or temporally restricted patterns, to maintain tissue or organ function (Edwards & Coruzzi, 1990). One of the principal aims of developmental biology is to understand the relationship between gene expression and the co-ordination of growth and development in multicellular organisms, but to date we have isolated and characterised only relatively few genes.

A cursory glance through the 'Contents' pages of this book will be enough to convince the reader that a large number of techniques are now available to provide access to a wide range of genes of interest. One of the greatest problems at present, however, is in identifying genes that are expressed in very restricted cell types, such that the transcript or protein represents a low proportion of the total RNA/protein molecules in a tissue or organ, and perhaps is present during a very short time window of development. Differential cDNA library screening, for example, has allowed the identification of tissue-specific or tissue-abundant transcripts (Chapter 6), but the approach favours the isolation of genes encoding relatively abundant mRNAs. Many regulatory genes that play key roles in developmental or metabolic pathways are expected to encode low-abundance transcripts. cDNA and genomic subtraction techniques provide one solution (Chapter 14), but an alternative approach, which is now well proven within a relatively short period in plants, is that of insertional mutagenesis using T-DNAs, and this is the subject of this chapter.

Insertional mutagenesis occurs naturally in a number of plant species, through the excision and re-integration of endogenous transposable elements, which may correspondingly insert into a gene, disrupting it both structurally and, potentially, functionally (Chapter 12). If the disrupted gene is not functionally redundant, that is, if other genes present are unable to complement partially or completely the function of that gene, progeny homozygous for the mutation may exhibit a phenotypic abnormality which may be detectable either morphologically, biochemically or physiologically (or, of course, by a combination of these). Mutagenesis is therefore a powerful approach to the identification of genes through their loss of, or change in, function, an approach already well established to analyse the growth and development of a range of other organisms, from bacteria to invertebrates and mammals (e.g. Priess, 1994; Rijkers et al., 1994; Rubin, 1988; Schüpbach & Roth, 1994; see also Chapters 9 and 12). Mutants can be generated by chemical and physical means, but insertional mutagenesis, using known DNA sequences as insertion elements, has the added advantage of greatly facilitated gene cloning, since the insertion element, be it a transposon or a T-DNA, acts as a tag of the mutant gene (Walbot, 1992).

The engineering of T-DNAs to incorporate screenable and/or selectable markers has greatly improved their value as insertional mutagens. The presence of a constitutively expressed selectable or screenable marker gene, such as the *nptII* gene that confers resistance to aminoglycoside antibiotics (Reynaerts et al., 1988) or the screenable *uidA* (*gusA*) gene encoding β-glucuronidase (Jefferson et al., 1987), has the advantage that the probability of having tagged a gene can be determined with relative ease, by investigating the cosegregation of the mutant phenotype and the transgene in crossing experiments.

A limitation of the mutagenesis approach to gene identification and isolation is the possible functional redundancy of genes. If a gene function can be replaced, fully or even partially, by a different gene (as would be the case, for example, if the mutant gene represented one member of a functionally related gene family, or as would occur in polyploid species), then there may be no, or a much reduced, phenotypic aberration evident, and the mutation may be difficult or impossible to detect. One way around this difficulty, and a strategy which we have been developing and characterising in our laboratory, is to use as insertion elements foreign reporter genes that will be activated only upon insertion into or adjacent to functional genes. Since the reporter gene expression is a genetically dominant trait, and of non-plant origin, detection of functionally redundant genes is possible. Such functional gene tags have been described by us as 'interposons' (Topping et al., 1991), and one can consider broadly three types: enhancer traps, promoter traps and gene traps (Fig. 1).

Enhancer trap vectors (Fig. 1A) comprise a dominant reporter gene that is fused to a 'minimal' or 'naive' promoter, containing, for example, a TATA box, but no major upstream elements that direct tissue specificity or level of expression. The rationale is that the vector acts as a gene tag following its activation by the enhancer elements associated with native genes close to the site of insertion. Such enhancers may direct tissue specificity of expression of the reporter gene, and within a transgenic population, a diversity of expression patterns is detectable (Goldsbrough & Bevan, 1991; Topping et al., 1991). Activation of the enhancer trap is not expected to be dependent on insertion directly into a gene sequence, since enhancers can function over relatively large distances, irrespective of orientation relative to the direction of transcription (Ptashne, 1986).

Promoter trap vectors (Fig. 1B) are distinct from enhancer traps in that the reporter gene lacks a minimal promoter. Activation in this case is expected to require insertion within a gene, downstream of the promoter, to generate a functional gene fusion which, depending on the design of the vector, may be either a transcriptional or translational fusion (section 4.1).

Gene trap vectors (Fig. 1C), like promoter traps, lack a functional promoter, but are equipped with an intron splice acceptor site. The rationale is that a translational fusion with exon sequence is generated preferentially, putatively avoiding the potential problem of excision of the tag, should it insert into an intron.

A further type of gene-tagging vector relies on a different approach to that described above. Here, the T-DNA itself carries strong enhancer sequences, the objective being that overexpression of a native gene, adjacent to the site of insertion, is induced, resulting in the generation of a dominant mutant phenotype (Fig. 1D; Hayashi et al., 1992).

Before describing experimental procedures involved in gene-tagging experiments, we will discuss some background to, and practical successes of, the technique of T-DNA insertional mutagenesis.

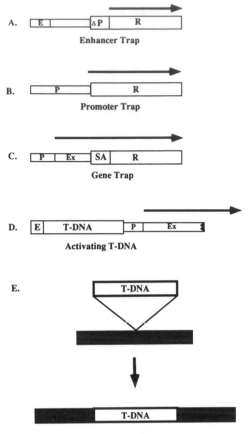

Figure 1. Schematic representation of gene-tagging vectors. (A) Enhancer trap vector, comprising a minimal promoter (ΔP) linked to a reporter gene. Insertion adjacent to a native enhancer element (E) results in the activation of transcription of the reporter (indicated by arrow). (B) Promoter trap vector, comprising a promoterless reporter gene (R). Insertion downstream of a native gene promoter (P) results in the activation of transcription of the reporter (indicated by arrow). (C) Gene trap vector, comprising a promoterless reporter gene (R) fused to a splice acceptor site (SA). Insertion within a native gene at the exon/intron splice site results in translational fusion between native gene exon (Ex) and the reporter gene (indicated by arrow). (D) Activating T-DNA, comprising a T-DNA carrying a strong enhancer element (E). The enhancer can activate a promoter (P) associated with a native gene (Ex) to induce overexpression (indicated by arrow). (E) Inactivation of a native gene (shaded box) by a T-DNA is possible following insertion, allowing gene detection on the basis of a mutant phenotype

2   THEORY AND APPLICATION OF T-DNA-MEDIATED INSERTIONAL MUTAGENESIS TECHNIQUES

2.1   Where and how does T-DNA integrate?

The soil-borne plant pathogen, *Agrobacterium tumefaciens*, has been exploited in plant genetic engineering because it is itself a natural genetic engineer of plants. Under the inductive effects of phenolics released by wounded dicotyle-donous plant tissues, to which the bacteria become attached, the virulence region of the Ti plasmid is activated, and endonucleases encoded by the *virD* operon cleave the borders of the transferred DNA or 'T-DNA' region (review: Zambryski et al., 1989). The VirD2 and VirE2 proteins mediate T-DNA translo-cation and integration into the plant genome, and the transfer process exhibits some similarities to bacterial conjugation (Stachel & Zambryski, 1986; Lessl & Lanka, 1994).

Wild-type strains of *A. tumefaciens* cause the disease crown gall, character-ised by the formation of tumours at the site of infection. The use of this bacterium to genetically modify plants in an experimentally controlled fashion has required, first, the modification of the Ti plasmid itself, principally by recognising, through mutagenesis studies, that the virulence functions that mediate T-DNA transfer can be physically separated from the sequences responsible for disease symptoms; indeed, the latter can be removed entirely from the Ti plasmid, and can be replaced by non-oncogenic genes of interest. A large number of so-called 'disarmed' vectors are available for plant transforma-tion by *Agrobacterium*, and these have been well reviewed elsewhere (Gelvin, 1993).

The integration process itself has received considerable attention, both from the point of view of understanding its nature and mechanism, and also with the aim of manipulating the process, to introduce a greater level of predictabil-ity and indeed to permit gene targeting and replacement. Unlike the situation in lower eukaryotes such as yeast, transgene integration in plants is by illegitimate, rather than homologous, recombination (Mayerhofer et al., 1991; Gheysen et al., 1991), though homologous recombination does occur at low frequency (Lee et al., 1990). T-DNA is transferred as a single-stranded molecule, associated with the VirD2 and VirE2 proteins comprising the so-called T-complex (Gelvin, 1993). Breaks in the chromosomal DNA are required for insertion of the T-DNA strand, initially at the right border region, and analysis of insertion sites indicates sequence similarities between T-DNA and plant DNA of only a few nucleotides (Mayerhofer et al., 1991; Gheysen et al., 1991). There may be some rearrangements or deletions induced in the host DNA during this process (Gheysen et al., 1987), and precise insertion at the T-DNA borders appears to be the exception rather than the rule; recent studies suggest that up to approximately 30% of insertion events may involve transfer of Ti plasmid sequences beyond T-DNA borders (Martineau et al., 1994). It is also the case that T-DNA molecules may commonly insert in tandem arrays at a single or multiple loci, as inverted or direct repeats (Spielmann & Simpson, 1986; Jones et al., 1987; Jorgensen et al., 1987).

The illegitimacy of recombination events has implications for gene-tagging strategies, using either T-DNAs or heterologous transposon systems. Random, or virtually random, integration suggests that all genes potentially are suscept-ible to the mutagenic effects of foreign DNA, and saturation mutagenesis is, in theory at least, possible. The question of whether T-DNA integration is strictly random has been illuminated to some extent by the data from promoter trapping experiments, to be described in some detail below. First, however, we will review progress to date in identifying and cloning genes by what might be considered conventional methods of T-DNA insertional mutagenesis: that is, methods that utilise transformation vectors unmodified by the presence of gene entrapment reporters.

3   INSERTIONAL MUTAGENESIS BY T-DNAs

It is possible to exploit a number of plant species for the identification and cloning of genes by T-DNA insertional mutagenesis, the principal requirements being ease of transformation and a diploid chromosome complement. *Nicotiana plumbaginifolia*, for example, has been used for a number of years in chemical mutagenesis studies (e.g. Pythoud & King, 1990), and more recently for insertional mutagenesis. Further progress in the use of T-DNA-mediated gene-tagging to identify developmentally interesting genes has taken place over the last five years, largely due to advances that have been made in genetically transforming *Arabidopsis thaliana* by *A. tumefaciens*. There are a number of reasons for choosing *Arabidopsis* for such purposes, most of which have been reviewed numerous times previously (e.g. Meyerowitz, 1987), and which need not be exhaustively discussed here. Suffice it to say that this species, a weedy crucifer, has been used in mutagenesis studies for many decades (mutations having been generated in earlier years by chemicals such as ethyl methane sulphonate, and radiation), resulting in the production of a detailed integrated genetic/RFLP linkage map (Hauge et al., 1993) (currently with over 200 phenotypic markers mapped); the genome is small (the haploid size is approximately 70 000–100 000 kb), facilitating the screening of genomic libraries with cloned T-DNA flanking sequences as probes; it is self-fertile, so that mutant lines can be maintained easily; and genetic transformation is now routine in many laboratories. It is now possible to generate several thousand individual transformants within a period of a few months (Feldmann, 1991; Clarke et al., 1992; Bechtold et al., 1993). The establishment of a physical map (Ward & Jen, 1990; Grill & Somerville, 1991), and programmes aimed at sequencing the genome and cDNAs (Chapter 15), further provide a strong rationale for adopting *Arabidopsis* as the model of choice for higher plant biology.

3.1   Tagging strategy

The generalised approach to identifying genes by T-DNA tagging is presented in Fig. 2. A large population of transgenic plants is generated that contains T-DNA harbouring selectable and, in some cases, screenable reporter genes, e.g. *nptII* (Koncz et al., 1989) and *gusA* (Kertbundit et al., 1991; Topping et al., 1991). This population is then screened for a phenotype of choice, including, for example, activation of a promoter trap on the T-DNA, and its stability over a number of generations ascertained. The segregation of the mutant pheno-type, and its cosegregation with the T-DNA, is characterised, with a view to

281

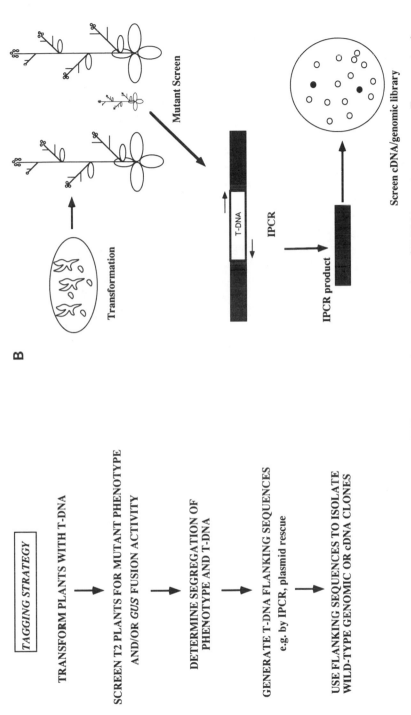

Figure 2. (A) Flow diagram of T-DNA tagging strategy. (B) Diagrammatic representation of T-DNA tagging strategy. Transgenic plants are screened for mutant phenotypes or the activation of interposon tagging vectors; T-DNA flanking sequences can be cloned, e.g. by IPCR, and used to screen cDNA and/or genomic libraries made from wild types plants

determining whether the phenotype is due to a single mutation, and whether that mutation is the consequence of a T-DNA insertion event. If the evidence is convincing that phenotype and T-DNA cosegregate 100% in a large sample population (typically at least 100–200 progeny following both selfing and backcrossing), then cloning can be carried out by one of a number of techniques. Strategies include inverse polymerase chain reaction (PCR) (section 6.4.1; Chapter 13), plasmid rescue or the screening of genomic libraries made from individual transgenic lines of interest. Proof that the gene cloned is allelic to the mutant gene of interest can only be obtained by genetic complementation, in which the wild-type allele is used to transform the mutant, and rescue of the phenotype is demonstrated. Here, the introduction of a wild-type allele of a mutant gene is carried out by *Agrobacterium*-mediated transformation of a line homozygous for the mutation, with the expectation being that regenerants will be heterozygous and phenotypically wild type. Self-fertilisation of the complemented line will result in segregation of the mutant and wild-type alleles in a Mendelian fashion.

Transformation of *Arabidopsis* using *A. tumefaciens* is now possible with a number of explant systems: (1) the inoculation of explanted cotyledons and leaf tissue (Schmidt & Willmitzer, 1988); (2) the inoculation of explanted roots (Valvekens et al., 1988; Clarke et al., 1992; Márton & Browse, 1991); (3) the inoculation of intact seeds (Feldmann & Marks, 1987); (4) the inoculation of wounded inflorescence meristems (Chang et al., 1994); and (5) the vacuum infiltration of whole plants with a liquid suspension of agrobacteria (Bechtold et al., 1993). Arguably the most reproducible of these approaches, and the method most commonly used, is root tissue inoculation, and this is the technique we will describe in some detail later in this chapter. While regeneration of shoots from leaf, cotyledon and root explants appears to be equally efficient, a potential advantage of using roots is that cells of this organ in *Arabidopsis* are predominantly diploid, while aerial tissues are more commonly polyploid (Galbraith et al., 1991). This is an important consideration if the aim of the work is to generate insertional mutants, unmasked by redundant gene copies. In this laboratory, we have found that all regenerants from root explants that we have analysed have proven to be diploid (Clarke et al., 1992).

The seed infection method devised by Feldmann & Marks (1987) has the advantage that it avoids any tissue culture steps, so bypassing potential problems associated with somaclonal variation in the transgenic population, a source of potentially non-tagged mutations. In this method, seeds are imbibed and then incubated for 24 h with agrobacteria on an orbital shaker, dried and sown out under greenhouse conditions. Seed is collected from the mature plants, and tested for the presence of T-DNAs, by plating on media containing kanamycin. While the mechanism by which the technique operates is unclear, it is suggested that the agrobacteria infect sporophytic cells that eventually contribute to the gametes, or possibly the zygote itself. This method has proved to be very successful for the generation of insertional mutants, but has proved to be rather difficult to reproduce beyond the Feldmann laboratory, and, to our knowledge, no other laboratory has managed to get the method working efficiently. The reasons for this problem are currently unknown. Interestingly, despite the avoidance of a tissue culture phase during the transformation procedure, a significant proportion of observed mutant phenotypes have been found not to cosegregate with a detectable T-DNA (e.g. Errampalli et al., 1991). A study of the Feldmann transgenics found that approximately 10% of a total of 8000 individual transformants exhibited a mutant phenotype which was demonstrated to be linked to the T-DNA (Feldmann, 1991). This phenomenon also remains unresolved. The Feldmann collection of transgenic *Arabidopsis* is being made available to the wider scientific community through seed stock centres based in Nottingham, UK and in Ohio, USA. Contact names and addresses are provided at the end of this chapter.

Meristem inoculation (Chang et al., 1994) involves first the excision of the inflorescence from the rosette with a scalpel incision, followed by inoculation of the wound site with *Agrobacterium*. Secondary inflorescences develop, but these too are excised and the wound reinoculated. These plants are then allowed to mature and flower, and seeds are plated out on selective media to identify transformants.

The vacuum infiltration procedure of Bechtold et al., (1993) appears to be even simpler. Flowering plants are immersed in a high-density suspension of *Agrobacterium*, and subjected to vacuum infiltration. The plants are then returned to the greenhouse and allowed to set seed, and the seeds are then screened for transformants.

Both the meristem injection and vacuum infiltration methods are relatively new and are yet to be thoroughly assessed in a large number of laboratories, and to date no isolated mutants have been described in publications. However, their reported success and apparent simplicity are expected to promote their utilisation, particularly in laboratories not expert in tissue culture techniques.

The key step in using a mutational approach to gene identification is undoubtedly in designing an appropriate screen for the expected phenotype of interest. While a number of morphological and pigmentation mutants are relatively easy to detect, and have been catalogued in previous screens of T-DNA populations (Feldmann, 1991; Forsthoefel et al., 1992), certain types of mutant may not be visually obvious and so may require specialist techniques to allow their detection. A good illustration of this point is provided by *Arabidopsis* mutants deficient in phytochrome A. Such plants look perfectly normal in screens of seedlings grown in white light, since phytochrome A is naturally labile in white light; an aberrant phenotype is only detectable when the mutants are grown in far-red light (review: Deng, 1994). Identification of some types of metabolic mutant may require the biochemical analysis of populations of plants, such has been performed for the detection of lipid mutants (Okuley et al., 1994). The use of activator T-DNAs, carrying enhancer elements to induce overexpression of native genes, has allowed the identification of genes involved in hormone metabolism, but this was again dependent on the development of a specialised screen for hormone autotrophy in vitro (Hayashi et al., 1992).

3.2   Spectrum of mutants

A large range of phenotypes have now been detected in screens of T-DNA-transformed populations of *Arabidopsis*, and these are listed in Table 1. As indicated above, the majority of classes identified have been morphologically obvious, and it is certain that as screens become more sophisticated, new classes of mutant will be found. Despite the significant frequency at which

Table 1. Summary of tagged mutants identified in T-DNA-transformed populations of *Arabidopsis thaliana*

	Genes cloned?	Reference
Phenotype		
Dwarf	Yes	Feldmann et al. (1989)
Abnormal flower		
agamous	Yes	Yanofsky et al. (1990)
apetala 2	Yes	Okamuro et al. (1993)
Abnormal flowering time		
luminidependens	Yes	Lee et al. (1994)
Abnormal embryo		
FUS6	Yes	Castle & Meinke (1994)
EMB30	Yes	Shevell et al. (1994)
lethals	No	Errampalli et al. (1991)
Glabrous		
glabrous 1,2	Yes	Marks & Feldmann (1989)
		Larkin et al. (1993)
Abnormal vegetative growth		
pointed first leaf	Yes	van Lijsebettens et al. (1991)
		van Lijsebettens et al. (1994)
tousled	Yes	Roe et al. (1993)
spindly	Yes	Jacobsen & Olszewski (1993)
Pale		
ch42	Yes	Koncz et al. (1990)
Abnormal photomorphogenesis		
cop9	Yes	Wei & Deng (1993)
hy4	Yes	Ahmad & Cashmore (1993)
Auxin autotrophy	Yes	Hayashi et al. (1992)
Promoter trap lines:		
GUS fusion activity		
Embryo	Yes	Topping et al. (1994)
Vascular tissues		
phloem	No	Kertbundit et al. (1991)
leaf, root, anther	Yes	Wei & Lindsey (unpublished)
leaf, root	Yes	Muskett & Lindsey (unpublished)
Flowers		
pollen, stigma	No	Lindsey et al. (1993)
Roots		
nematode-induced	Yes	Goddijn et al. (1993)
pericycle, cortex, epidermis	No	Topping, Wei and Lindsey (unpublished)
Leaves		
hydathodes	No	Lindsey et al. (1993)

non-tagged mutants appear to arise in transgenic populations, irrespective of the method of transformation, a number of genes have now been cloned, many of which are involved in the regulation of morphogenesis and cell–cell signalling, and, more recently, in metabolism. We will now summarise the progress that has been made in this regard.

A number of genes that regulate flowering have now been cloned, following tagging by T-DNAs. These include: *agamous* (*ag2*), a homoeotic gene, mutation of which results in sterility (Yanofsky et al., 1990); *flower* (*Fl1*), which is a homoeotic mutation allelic to the EMS-induced *apetala* 2 mutation, in which the sepals are transformed into carpels, petals are absent and the number of stamens is reduced (Okamuro et al., 1993); and *LUMINIDEPENDENS* (*LD*), involved in the control of flowering time (Lee et al., 1994). Other genes have been cloned that control aspects of vegetative plant growth. A gene that rescues a dwarf mutant (*dwf*) has been isolated (Feldmann et al., 1989); the *glabrous* (*gl1* and *gl2*) genes, required for the normal development of trichomes (Mark & Feldmann, 1989; Herman & Marks, 1989; Larkin et al., 1993), have been cloned and found to encode *myb*-like transcription factors; and *pointed first leaf* (*pfl*) (Van Lijsebettens et al., 1991), which encodes an S18 ribosomal protein (Van Lijsebettens et al., 1994), is necessary for the control of leaf shape in younger leaves.

A large number of embryonic mutants have been identified previously in screens of plants treated with chemical mutagens, and T-DNA tagging is now beginning to provide access to genes important in cell signalling and pattern formation during embryogenesis. Several T-DNA-induced embryonic mutants have been generated (Errampalli et al., 1991), although to date only a small number have been cloned. Recently, a gene which, when mutated, results in grossly aberrant patterning of the embryo, named *GNOM* or *EMB30* (Mayer et al., 1993), has been cloned following its identification in the Feldmann collection of transgenics (Shevell et al., 1994). A further example of a gene involved in embryonic and postembryonic development is the *FUS6-1* gene (Castle & Meinke, 1994). *FUSCA* mutants, which are readily detectable by the accumulation of high levels of anthocyanins in the embryo, also exhibit a similar phenotype to mutants which are defective in light transduction, e.g. *COP1*, *COP9* and *DET1*. The precise function of this class of gene remains to be determined, but it is suggested that it is involved in the light signal transduction pathway (Castle & Meinke, 1994; Misera et al., 1994).

Screens for the *COP* mutants have been carried out in T-DNA populations, and tagged mutants have been identified (Hou et al., 1993). One such gene, *COP9*, has been characterised and it is suggested that it may form part of the signal transduction pathway from light detection via phytochromes to a photomorphogenic response (Wei & Deng, 1993). *HY4*, a gene which is similarly involved in the photomorphogenic response to blue light, has also been T-DNA tagged and cloned, and the protein is a DNA photolyase (Ahmad & Cashmore, 1993). Another gene isolated by T-DNA tagging which appears to be involved in a signalling pathway is the *tousled* (*TSL*) gene. Plants homozygous for the mutant gene are affected in both vegetative growth, with aberrations in leaf number and shape, and also in flower development. Isolation of the tagged gene revealed it to encode a novel protein kinase which is most abundantly expressed in developing floral meristems (Roe et al., 1993). A T-DNA-tagged gene involved in phytohormone signalling has also been characterised; this is *spindly* (*SPY*), in which the mutant has an altered GA signal transduction pathway (Jacobsen & Olszewski, 1993).

Progress in using T-DNA-tagging techniques to identify genes involved in metabolic regulation has been less rapid than for genes of morphogenesis, because of the problem of screening. Recently, however, *FAD2*, involved in fatty acid desaturation, has been tagged and cloned (Okuley et al., 1994).

4 MODIFIED INSERTIONAL MUTAGENESIS: GENE ENTRAPMENT TECHNIQUES

While the success of T-DNA insertional mutagenesis has been dramatic during the relatively short period of time it has been available, there are clearly some limitations inherent in the approach: gene redundancy can mask phenotypes; multiple T-DNA inserts can complicate gene isolation; and it is more difficult to demonstrate linkage between the T-DNA insertion and the observed phenotype than is the case for transposons, since reversion studies are not possible (cf. Chapter 15). Linkage studies alone are insufficient evidence of tagging per se, and the tagging of a gene can only be demonstrated unequivocally by the genetic complementation of the mutant phenotype following transformation with the isolated wild-type homologue of the tagged sequence.

Entrapment techniques, notably promoter trapping and gene trapping, can avoid some of these difficulties, since functional gene fusions are expected to be generated only following insertion directly into genes. Since the reporters are dominant, gene detection is not dependent upon the generation of a mutant phenotype, allowing the tagging of redundant genes, and in cases where a mutant phenotype is generated, cosegregation analysis with the T-DNA is facilitated if the reporter is detected readily, such as is the case for the *gusA* gene (Jefferson et al., 1987). The generation of active gene fusions in cell types of interest also facilitates screening for subtle mutant phenotypes, by focusing the screening process on a subpopulation of transgenics. Furthermore, the expression of genes which, when mutated, result in a lethal phenotype in the homozygous state can be studied in the hemizygous state, so allowing deductions to be made as to their possible role within the plant.

4.1 Promoter trapping

Promoter trap vectors, comprising promoterless selectable or screenable marker genes, have been used widely in animal research (Skarnes, 1990) and have led to the identification of novel patterns of cell-specific and developmentally regulated gene expression (Kothary et al., 1988; Gossler et al., 1989; Rossant &

Joyner, 1989). In plants, either transposon (Fedoroff & Smith, 1993) or T-DNA vectors (Koncz et al., 1989; Topping et al., 1991) can be modified to contain a promoter trap, but T-DNA vectors have been used most widely.

The first experiments in promoter trapping in plants utilised constructs carrying the promoterless selectable marker gene aph(3')II, encoding amino-glycoside phosphotransferase II, which confers resistance to kanamycin. This gene was cloned into T-DNA adjacent to the right border (André et al., 1986; Teeri et al., 1986; Koncz et al., 1989) and introduced, via Agrobacterium-medi-ated transformation, into Nicotiana sp. protoplasts or leaf explants; plants were regenerated in the presence of kanamycin. The rationale was that, under this selection pressure, the only calluses or plants which regenerated were derived from a single cell in which a gene fusion event had taken place. Using this approach, Teeri et al. (1986) identified transgenic lines showing spatially restricted patterns of NPTII activity, but the tagged gene promoters were not isolated.

In general there are certain limitations in the use of selectable reporter genes to monitor in vivo gene fusions. In the first case, it was observed that a high incidence of multiple (5–20) inserts of the T-DNA occurred, perhaps due to the selection pressure applied (Koncz et al., 1989). Second, gene fusions with native promoters that are active in very restricted cell types, or that are of low activity, would be difficult to detect. There were also technical problems relating to the isolation of putatively tagged genes from N. plumbaginifolia, due to the high level of methylation and high abundance of GC-rich repeats in the nuclear DNA. Therefore, most efforts towards gene tagging using promoter traps have exploited Arabidopsis as a model species, and constructs have been designed in which the expression of the selectable marker gene is driven by a linked and constitutive promoter which is independent of gene fusion for activation. Promoterless screenable marker genes replaced selectable genes as promoter traps.

The value of promoter trapping techniques for gene tagging and isolation is, of course, directly related to the frequency at which it is possible to detect reporter gene activation. Data obtained in a number of experimental systems have revealed that the frequency of activation of a promoterless gusA gene is high (Kertbundit et al., 1991; Topping et al., 1991), perhaps even surprisingly so. Kertbundit et al. (1991) built two promoter trap vectors, designed to generate either transcriptional or translational fusions in vivo. In both cases, a promoterless gusA gene was placed adjacent to the right border of a T-DNA. For the 'transcriptional fusion vector', the GUS ATG was located 4 bp from the T-DNA right border, while the 'translational fusion vector' differed only in a 4-bp deletion which removed the ATG initiation codon of the gusA gene. Following transformation of Arabidopsis with each of these vectors, it was found that the frequency at which functional transcriptional gene fusions was generated was 54% but for translational fusions the frequency was much reduced, at 1.6%. This difference presumably reflects the reduced probability of generating in-frame translational fusions, and also perhaps of the reduced activity of GUS fusion proteins.

In this laboratory we have designed a vector in which the *gusA* gene is adjacent to the T-DNA left border and is expected to generate primarily transcriptional gene fusions (Fig. 3; Topping et al., 1991). We have characterised the activation frequencies in various organs of tobacco, *Arabidopsis* and potato (Topping et al., 1991; Lindsey et al., 1993). For example, frequencies in leaf were 22% for tobacco, 28% for *Arabidopsis* and 23% for potato; in roots; 75% for tobacco, 30% for *Arabidopsis* and 9% for potato. Fobert et al. (1991) have used a vector similar in design to that described by Kertbundit et al. (1991), but found a very low level of activation: here, only 5% of transgenic tobacco leaves exhibited expression. The reasons for this discrepancy are unclear at present.

T-DNA-mediated promoter trapping is still a relatively new approach to the isolation of novel plant genes, and to date there are few results available. Kertbundit et al. (1991) have isolated a phloem-specific promoter from *Arabidopsis*: reintroduction of the homologous wild-type genomic sequence demonstrated the specificity of the putative promoter. However, it remains to be shown formally that the isolated sequence regulates a native gene expressed in

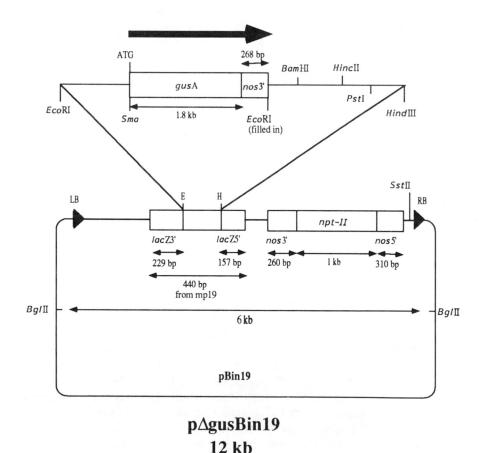

pΔgusBin19
12 kb

Figure 3. Promoter trap vector pΔgusBin19

the phloem. A novel chloroplast protein gene (*cs*), allelic with an X-ray-induced mutation, *ch42*, has been isolated from an *Arabidopsis* line transformed with a promoterless *nptII* gene (Koncz et al., 1990). However, in this example there was no clear expression of a *nptII* gene fusion product in the mutant, but the transformant exhibited a pale phenotype (*chlorata*), and the gene was isolated on the basis of its aberrant phenotype.

In this laboratory we have isolated genes from diverse tagged *Arabidopsis* lines, variously exhibiting *gusA* fusion expression and mutant phenotypes in zygotic embryos (Topping et al., 1994), vascular tissues and anthers (Muskett et al., 1994), and a tuber-expressed gene in potato (L. Rooke, J. Topping & K. Lindsey, unpublished). Although most of these genes are novel and show no significant homology to known genes, one expressed primarily in the tapetum and vascular tissues of *Arabidopsis* is a RNA or DNA helicase encoded by a very low abundance transcript (W. Wei, D. Twell & K. Lindsey, unpublished), and a tagged seed-expressed gene encodes an enzyme of lipid metabolism (P. Rocha, J. Topping & K. Lindsey, unpublished). Promoter trapping has also proved to be very useful for the identification of genes inducible by specific physiological or environmental factors. For example, Goddijn et al. (1993) have demonstrated that, by screening a population of promoterless *gusA*-tagged *Arabidopsis* lines after infection with nematodes, it has been possible to identify lines in which GUS activity is localised to the infection site. Thus it should be possible to isolate genes which are specifically induced by nematodes, and the approach is expected to lead towards both a better understanding of the infection process, and also eventually engineered resistance to nematodes.

A number of the *Arabidopsis* promoter trap lines generated in this laboratory are available to the wider scientific community, and have been passed on to the Nottingham Seed Centre for further distribution. Other lines will be passed on in due course. The promoter trap plasmid pΔgusBin19 (Topping et al., 1991) is also freely available, from this laboratory.

4.2   Enhancer trapping

Within the eukaryotic genome are scattered thousands of tissue-specific enhancer elements associated with, and regulating the transcription of, individual genes or clusters of genes. Enhancers are known to exert their effects over several kilobases (Ptashne, 1986; Stockhaus et al., 1989; Ganss et al., 1994), and can have a significant effect on the quantitative and qualitative expression of introduced transgenes, contributing to so-called position effects. This effect provides a means of tagging enhancers, through the activation of enhancer trap vectors (Fig. 1A; O'Kane & Gehring, 1987; Bellen et al., 1989; Wilson et al., 1989; Topping et al., 1991). The identification and characterisation of these enhancers by T-DNA tagging may provide access to the associated genes, although the potential long-distance action of some enhancers (and so potential distance from the tagging vector) may prove a hindrance in cloning. In *Drosophila*, this problem has proved to be relatively insignificant, since P-element vectors introduce enhancer traps predominantly into genes, rather

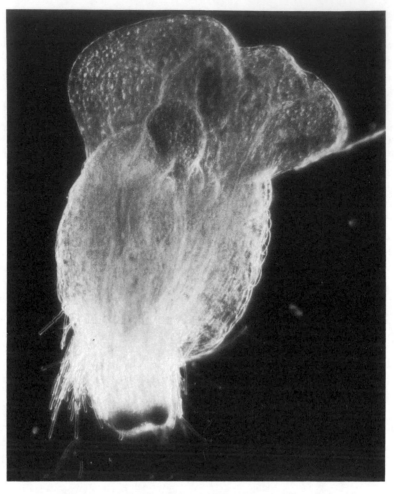

Figure 4. The mutant *hydra* (Topping & Lindsey, unpublished). This mutant was identified in a screen for insertional mutants in embryonic pattern formation, and exhibits defective root and shoot development. It has been crossed with a promoter trap line that exhibits GUS fusion activity in the root meristem. This cell-specific marker reveals meristem activity in the abnormal *hydra* 'root'

than at a distance from them (Engels, 1988). It remains to be determined to what extent this will be the case in plants.

The best studied enhancer trap vector utilises a deleted version of the cauliflower mosaic virus (CaMV) 35S RNA gene promoter, comprising the −90-bp region fused to the *gusA* coding region, in the binary vector Bin19 (Topping et al., 1991; Goldsbrough & Bevan, 1991). We have introduced this vector into populations of tobacco and *Arabidopsis*, to investigate the tagging of tissue-specific enhancer sequences. GUS activities were scattered over a 300-fold range within the transgenic population, and the patterns and levels of GUS expression were found to be unique to individual transformants, even in lines demonstrated to contain single-locus insertions (Topping et al., 1991). Similar results have been reported in transgenic potato (Goldsbrough & Bevan, 1991). Our enhancer trap vector, p − 90gusBin19 (Topping et al., 1991), is freely available, from this laboratory.

An alternative strategy to identify enhancers is to clone random genomic sequences into an enhancer trap vector, comprising a reporter gene fused to a naive promoter. The rationale is that a cloned fragment of DNA, containing enhancer sequences, would activate transcription of the reporter following plant transformation (Ott & Chua, 1990).

5   WHAT DO GENE-TAGGING STUDIES TELL US ABOUT T-DNA INTEGRATION?

The high frequency of activation of promoter and enhancer traps has led to the suggestion that the integration of T-DNA in the plant genome does not occur randomly, as had been thought previously, but that there is preferential integration into regions of the genome which are actively transcribed, including, perhaps, genes being actively transcribed at the time of transformation (Koncz et al., 1989; Topping et al., 1991; Walden et al., 1991). This hypothesis is consistent with observations of promoterless *gusA* activities in plant species with widely varying genome sizes and levels of repetitive DNA. Thus, in tobacco, potato and *Arabidopsis* there is a similarly high level of activation (Lindsey et al., 1993). However, it is perhaps significant that, although in each of the three species investigated GUS activity was detected in every organ and tissue type screened, a species-specific bias in gene fusion expression was observed. This may suggest that the simplest interpretation, namely that genes that are *actively* transcribed at the time of transformation are targeted by T-DNA molecules, is inadequate, since genes can be tagged that are expressed, not only in dividing cells, but specifically in terminally differentiated cell types such as pollen (Lindsey et al., 1993). We propose that *potentially* transcribed regions of the genome are targeted by T-DNA, as well as actively transcribed regions.

6   PRACTICAL ASPECTS OF T-DNA-MEDIATED GENE-TAGGING

6.1   Vectors

A number of vectors have been constructed that are designed to tag plant genes by generating gene fusions in vivo. They are described in the following publications: Koncz et al. (1989); Kertbundit et al. (1991); Topping et al. (1991); and Fobert et al. (1991). We will describe our own, which has been demonstrated to function efficiently in a number of species (Lindsey et al., 1993), for illustrative purposes.

6.1.1   Promoter trap vector pΔgusBin19

The full sequence of the T-DNA of this vector (Topping et al., 1991) is available in GenBank (accession number U12638). It comprises a selectable *nptII* gene near the T-DNA right border, under the transcriptional control of the *nos* promoter and polyadenylation sequences. The promoter trapping sequences are located at the left border, and comprise a promoterless *uid*A (*gusA*) coding region (Jefferson et al., 1987) (encoding β-glucuronidase) plus a *nos* polyadenylation sequence (Fig. 3). The 5′ end of the *gusA* gene is adjacent to the left border and has its own translation initiation codon. There are translational stop codons in the left border region in all reading frames, and so gene fusions generated upon integration are expected to be primarily transcriptional rather than translational, and this is observed (Topping et al., 1994).

6.2   Transformation of *Arabidopsis thaliana*

As described above, there exist a number of different protocols for the stable transformation of *Arabidopsis* by *Agrobacterium tumefaciens*. These have been published elsewhere in detail, and need not be repeated here. Our favoured method, which allows the production of hundreds or thousands of independent transformants, is based upon the root explant transformation systems originally described by Valvekens et al. (1988), but is much more efficient (Clarke et al., 1992). For more information on this technique, see also Lindsey & Wei (1996).

6.3   Screening for GUS fusion expression and mutants

Following plant transformation with the promoter trap vector pΔgusBin19, we screen for lines exhibiting both GUS fusion activity and mutant phenotypes that cosegregate with the T-DNA insert. Depending on the developmental process being studied, we screen for GUS activity either by histochemistry, or, in some cases, by fluorimetry in the first instance (which is sensitive and allows the rapid screening of relatively large organs such as siliques) (Topping et al., 1994) followed by the more detailed histochemical analysis of 'positives' identified fluorimetrically. Methods for GUS enzyme assay and histochemistry are discussed in detail elsewhere (Lindsey & Wei, 1996).

   The method of screening for mutant phenotypes depends on the type of mutant (developmental, biochemical, physiological) of interest, as described above. In view of the low numbers and poor germinability of T1 seeds, genetic analysis of the transformants (e.g. antibiotic resistance segregation) is carried out on T3 and subsequent generations. Kanamycin resistance/mutant cosegregation analysis is carried out as follows.

PROTOCOL 1: MUTANT ANALYSIS

---

(1)   Plate out approximately 200 sterilised T2 seeds on both kanamycin-selection plates and on kanamycin-free plates or sterile perlite or compost. A

T-DNA present as a single active locus will segregate 3 kan^r : 1 kan^s in the progeny. The mutant phenotype, if determined by a single recessive mutation, will appear in 25% of the seedlings on the kanamycin-free plates/perlite/compost which lack kanamycin. If the mutant is embryonic lethal, all the viable progeny will be phenotypically wild type (heterozygous for the mutation or wild type). Kanamycin-sensitive seedlings bleach, and fail to produce either true leaves or roots, on kanamycin-selection plates.

(2) Collect T3 seed from the mutants grown up on non-selective media (assuming they are not embryonic or seedling lethal or sterile) and plate out approximately 200 on kanamycin-selection plates. If the mutant phenotype and T-DNA are linked, 100% of the seedlings will be kanamycin resistant (homozygous for the *nptII* gene) and 100% will be mutant phenotype (homozygous for the mutant gene). Kanamycin-resistant but phenotypically wild-type T3 seedlings are expected to be hemizygous for the T-DNA, and if the T-DNA is causing the mutation one expects a segregation, on selfing, of 25% kanamycin-resistant mutants (homozygous for T-DNA and mutant gene), 50% kanamycin-resistant wild-type phenotypes (heterozygous for T-DNA and mutant gene), and 25% kanamycin-sensitive wild-type phenotypes (lacking both T-DNA and mutant gene). Southern blots should be carried out to confirm the presence/copy numbers of T-DNAs.

(3) To determine whether an embryonic-lethal mutation is caused by a T-DNA insertion event, collect T3 seed from a kanamycin-resistant plant (which is expected to be heterozygous for both the mutant gene and hemizygous for the T-DNA) and plate out on kanamycin-selection plates. Twenty-five per cent of the seeds are expected to be homozygous for the embryonic-lethal mutation and should fail to germinate, while of the germinating seeds, 66% are expected to be kanamycin-resistant wild-type phenotypes (heterozygous for T-DNA and mutant gene) and 33% are expected to be kanamycin-sensitive wild-type phenotypes (lacking both T-DNA and mutant gene).

(4) Further evidence of cosegregation can be obtained by outcrossing lines putatively homozygous for both T-DNA and mutant gene to wild types. Collect F1 seed (which is expected to be heterozygous for both the mutant gene and hemizygous for the T-DNA) and plate out on kanamycin-selection plates. Twenty-five per cent of the seedlings are expected to be kanamycin-resistant mutants (homozygous for T-DNA and mutant gene), 50% kanamycin-resistant wild-type phenotypes (heterozygous for T-DNA and mutant gene), and 25% kanamycin-sensitive wild-type phenotypes (lacking both T-DNA and mutant gene). Again, Southern blots should be carried out to confirm the presence and copy numbers of T-DNAs.

6.4 Cloning strategies

As indicated earlier, three techniques have been used to isolate genes tagged by T-DNA insertional mutagenesis: plasmid rescue (Behringer & Medford,

1992), genomic library construction from individual transformed lines (e.g. Kertbundit et al., 1991; Shevell et al., 1994), and inverse PCR (IPCR; see Chapter 13). Plasmid rescue to isolate T-DNA flanking sequences requires plant transformation with T-DNAs containing a bacterial origin of replication and selectable marker (e.g. ampicillin resistance), such that genomic fragments containing T-DNA can be excised by restriction digestion, religated to form circles and cloned directly in *E. coli*. We, however, routinely use IPCR (Lindsey et al., 1993; Topping et al., 1994), which we find an efficient and reliable method to generate T-DNA flanking sequences, which can then be used as molecular probes in screening wild types cDNA and genomic libraries. We will outline the IPCR methodology here.

6.4.1   IPCR amplification of T-DNA flanking sequences

The first step in amplifying T-DNA flanking sequences by IPCR requires the identification of restriction fragments containing T-DNA sequences and plant flanking DNA, such that the expected PCR product is no greater than approximately 2.5–3 kb. Larger products are difficult to amplify. Purified genomic DNA from transgenic plants is digested with a range of restriction enzymes designed to generate appropriate T-DNA/plant DNA border fragments, and analysed by Southern hybridisation. These fragments are then religated under conditions designed to generate monomeric circles (Collins & Weissman, 1984). Use a test plasmid at single-copy levels mixed with untransformed genomic DNA as a positive control to optimise the reaction conditions.

PROTOCOL 2: INVERSE PCR

---

(1)   Ligate digested DNA fragments in ligation buffer plus T4 ligase (1 unit/μl; BRL Gibco, Middlesex, UK), made up to a final volume of 1 ml with sterile distilled water.

(2)   Extract ligated DNA with phenol/chloroform, precipitate and resuspend to a concentration of 20 ng/μl.

(3)   Use 50 ng of DNA as the substrate for IPCR amplification, in a reaction mixture of PCR buffer (contains 1.5 mM Mg^{2+}; Promega, Southampton, UK), primers (100 ng/ml), 25 mM MgCl$_2$, and *Taq* polymerase (0.5 units/ 0.1 μl; Promega, Southampton, UK). The concentration of DNA and Mg^{2+} used in the amplification reaction is critical, and it is advisable to test empirically a range of concentrations of each in preliminary experiments.

(4)   Place the reaction mixture in a thermal cycler, holding the temperature at 80 °C, and start the reaction by adding dNTP.

(5)   Carry out PCR amplification over 30 cycles. We use the following conditions: denaturation at 95 °C for 4 min during the first cycle and for 1 min in subsequent cycles; primer annealing at 55 °C for 2 min; extension by *Taq* polymerase at 72 °C for 10 min during the first cycle and for 3 min in subsequent cycles; denaturation at 95 °C for 1 min; and final extension at 72 °C for 10 min. The extension time is estimated to be 1 min per 1 kb of expected amplification product.

(6)   Separate the reaction products in a 1% (w/v) agarose thin gel and stain with ethidium bromide to visualise the DNA.

(7)   The PCR products can be cloned directly into T/A cloning vectors such as are available from Invitrogen and Promega (see Chapter 13 for further details). Flanking sequences are identified following sequencing of the PCR products, and can be used in library screens.

7   CONCLUDING REMARKS

Although T-DNA-mediated insertional mutagenesis has been exploited in gene identification programmes for a relatively short period of time, it has been notably successful, not merely in the number of genes cloned, but also in the information that has been generated concerning basic aspects of plant biology, from the genetic control of morphogenesis to the mechanism of foreign gene integration. The even more recent demonstration of successful promoter and enhancer trapping in plants opens the possibility of isolating functionally redundant genes, and genes expressed in very restricted cell types which may be inaccessible by other cloning methods. We fully expect that this approach to understanding the genetic control of plant growth and development will increase dramatically in popularity in the coming years.

ACKNOWLEDGMENTS

We gratefully acknowledge present and past financial support for our work on gene-tagging from AFRC, SERC, BBSRC, CEC (BRIDGE), Dalgety plc and Shell Research Ltd.

ESSENTIAL READING

Feldmann, K.A. (1991) T-DNA insertion mutagenesis in Arabidopsis: mutational spectrum. *Plant J.*, **1**, 71–82.
Koncz, C., Martini, N., Mayerhofer, R. et al. (1989) High frequency T-DNA-mediated gene tagging in plants. *Proc. Natl. Acad. Sci. USA*, **86**, 8467–8471.
Topping, J.F., Wei, W. and Lindsey, K. (1991) Functional tagging of regulatory elements in the plant genome. *Development*, **112**, 1009–1019.
Topping, J.F., Agyeman, F., Henricot, B. and Lindsey, K. (1994) Identification of molecular markers of embryogenesis in *Arabidopsis thaliana* by promoter trapping. *Plant J.*, **5**, 895–903.
Walden, R., Hayashi, H. and Schell, J. (1991) T-DNA as a gene tag. *Plant J.*, **1**, 281–288.

REFERENCES

Ahmad, M. and Cashmore, A.R. (1993) The *HY4* gene involved in blue light sensing in *Arabidopsis thaliana* encodes a protein with the characteristics of a blue light photoreceptor. *Nature*, **366**, 162–166.

André, D., Colau, D., Schell, J. et al. (1986) Gene tagging in plants by a T-DNA insertion mutagen that generates APH (3')II-plant gene fusions. *Mol. Gen. Genet.*, **204**, 512–518.

Bechtold, N., Ellis, J. and Pelletier, G. (1993) In planta *Agrobacterium* mediated gene transfer by infiltration of adult *Arabidopsis thaliana* plants. *C.R. Acad. Sci. Paris, Sciences de la Vie*, **316**, 1194–1199.

Behringer, F.J. and Medford, J.I. (1992) A plasmid rescue technique for the recovery of plant DNA disrupted by T-DNA insertion. *Plant Mol. Biol. Rep.*, **10**, 190–198.

Bellen, H.I., O'Kane, C.J., Wilson, C. et al. (1989) P-element-mediated enhancer detection: a versatile method to study development in *Drosophila*. *Genes Dev.*, **3**, 1288–1300.

Bradley, D., Carpenter, R., Elliott, R. et al. (1993) Gene regulation of flowering. *Phil. Trans. R. Soc. Lond. B*, **339**, 193–197.

Castle, L.A. and Meinke, D.W. (1994) A *FUSCA* gene of *Arabidopsis* encodes a novel protein essential for plant development. *Plant Cell*, **6**, 25–41.

Chang, S.S., Park, S.K., Kim, B.C. et al. (1994) Stable genetic transformation of *Arabidopsis thaliana* by *Agrobacterium* inoculation *in planta*. *Plant J.*, **5**, 551–558.

Clarke, M.C., Wei, W. and Lindsey, K. (1992) High frequency transformation of *Arabidopsis thaliana* by *Agrobacterium tumefaciens*. *Plant Mol. Biol. Rep.*, **10**, 178–189.

Collins, F.S. and Weissman, S.M. (1984) Directional cloning of DNA fragments at a large distance from an initial probe: a circularized method. *Proc. Natl. Acad. Sci. USA*, **81**, 6812–6816.

Deng, X.-W. (1994) Fresh view of light signal transduction in plants. *Cell*, **76**, 423–426.

Edwards, J.W. and Coruzzi, G.M. (1990) Cell-specific gene expression in plants. *Annu. Rev. Genet.*, **24**, 275–303.

Engels, W.R. (1988) P elements in *Drosophila*. In *Mobile DNA* (eds D.E. Berg and M.M. Howe), pp. 437–484. American Society for Microbiology, Washington, DC.

Errampalli, D., Patton, D., Castle, L. et al. (1991) Embryonic lethals and T-DNA insertional mutagenesis in *Arabidopsis*. *Plant Cell*, **3**, 149–157.

Fedoroff, N.V. and Smith, D.L. (1993) A versatile system for detecting transposition in *Arabidopsis*. *Plant J.*, **3**, 273–289.

Feldmann, K.A. and Marks, M.D. (1987) *Agrobacterium*-mediated transformation of germinating seeds of *Arabidopsis thaliana*: a non-tissue culture approach. *Mol. Gen. Genet.*, **208**, 1–9.

Feldmann, K.A., Marks, M.D., Christianson, M.L. and Quatrano, R.S. (1989) A dwarf mutant of *Arabidopsis* generated by T-DNA insertion mutagenesis. *Science*, **243**, 1351–1354.

Fobert, P.R., Miki, B.L. and Iyer, V.N. (1991) Detection of gene regulatory signals in plants revealed by T-DNA-mediated fusions. *Plant Mol. Biol.*, **17**, 837–851.

Forsthoefel, N.R., Wu, Y., Schulz, B. et al. (1992) T-DNA insertion mutagenesis in *Arabidopsis*: prospects and perspectives. *Aust. J. Plant Physiol.*, **19**, 353–366.

Galbraith, D.W., Harkins, K.R. and Knapp, S. (1991) Systemic endopolyploidy in *Arabidopsis thaliana*. *Plant Physiol.*, **96**, 985–989.

Gamborg, O.L., Miller, R.A. Ojima, K. (1968) Nutrient requirements of suspension cultures of soybean root cells. *Exp. Cell Res.*, **50**, 151–158.

Ganss, R., Montoliu, L., Monaghan, A.P. and Schütz, G. (1994) A cell-specific enhancer far upstream of the mouse tyrosinase gene confers high level and copy number-related expression in transgenic mice. *EMBO J.*, **13**, 3083–3093.

Gelvin, S.B. (1993) Molecular genetics of T-DNA transfer from *Agrobacterium* to plants. In *Transgenic Plants*, Vol. 1: *Engineering and Utilization* (eds S. Kung and R. Wu), pp. 49–87. Academic Press, San Diego.

Gheysen, G., Van Montagu, M. and Zambryski, P. (1987) Integration of *Agrobacterium tumefaciens* transfer DNA (T-DNA) involves rearrangements of target plant DNA sequences. *Proc. Natl. Acad. Sci. USA*, **84**, 6169–6173.

Gheysen, G., Villaroel, R. and Van Montagu, M. (1991) Illegitimate recombination in plants: a role for T-DNA integration. *Genes Dev.*, **5**, 287–297.

Gibson, S. and Somerville, C. (1993) Isolating plant genes. *Trends Biotechnol.*, **11**, 306–313.

Goddijn, O.J.M., Lindsey, K., van der Lee, F.M. et al. (1993) Differential expression in nematode-induced feeding structures of transgenic plants harbouring promoter-*gus*A fusion constructs. *Plant J.*, **4**, 863–873.

Goldberg, R.B. (1988) Plants: novel developmental processes. *Science*, **240**, 1460–1467.

Goldsbrough, A. and Bevan, M. (1991) New patterns of gene activity in plants detected using an *Agrobacterium* vector. *Plant Mol. Biol.*, **16**, 263–269.

Gossler, A., Joyner, A.L., Rossant, J. and Skarnes, W.C. (1989) Mouse embryonic stem cells and reporter constructs to detect developmentally regulated genes. *Science*, **244**, 463–465.

Grill, E. and Somerville, C. (1991) Construction and characterization of a yeast artificial chromosome library of *Arabidopsis* which is suitable for chromosome walking. *Mol. Gen. Genet.*, **226**, 484–490.

Hauge, B.M., Hanley, S.M., Cartinhour, S. et al. (1993) An integrated genetic/RFLP map of the *Arabidopsis thaliana* genome. *Plant J.*, **3**, 745–754.

Hayashi, H., Czaja, I., Lubenow, H. et al. (1992) Activation of a plant gene by T-DNA tagging: auxin-independent growth *in vitro*. *Science*, **258**, 1350–1353.

Herman, P.L. and Marks, M.D. (1989) Trichome development in *Arabidopsis thaliana*. II. Isolation and complementation of the GLABROUS 1 gene. *Plant Cell*, **1**, 1051–1055.

Hou, Y., von Arnim, A.G. and Deng, X.-W. (1993) A new class of Arabidopsis constitutive photomorphogenic genes involved in regulating cotyledon development. *Plant Cell*, **5**, 329–339.

Jacobsen, S.E. and Olszewski, N.E. (1993) Mutations at the *SPINDLY* locus of *Arabidopsis* alter gibberellin signal transduction. *Plant Cell*, **5**, 887–896.

Jefferson, R.A., Kavanagh, T.A. and Bevan, M.W. (1987) GUS fusions: β-glucuronidase as a sensitive and versatile gene fusion marker in higher plants. *EMBO J.*, **6**, 3901–3907.

Jones, J.D.G., Gilbert, D.E., Grady, K.L. and Jorgensen, R.A. (1987) T-DNA structure and gene expression in petunia plants transformed by *Agrobacterium tumefaciens* C58 derivatives. *Mol. Gen. Genet.*, **207**, 478–485.

Jorgensen, R., Snyder, C. and Jones, J.D.G. (1987) T-DNA is organized predominantly in inverted repeat structures in plants transformed with *Agrobacterium tumefaciens* C58 derivatives. *Mol. Gen. Genet.*, **207**, 471–477.

Kertbundit, S., De Greve, H., DeBoeck, F. et al. (1991) In vivo random β-glucuronidase gene fusions in *Arabidopsis thaliana*. *Proc. Natl. Acad. Sci. USA*, **88**, 5212–5216.

Koncz, C., Mayerhofer, R., Koncz-Kalman, Z. et al. (1990) Isolation of a gene encoding a novel chloroplast protein by T-DNA tagging in *Arabidopsis thaliana*. *EMBO J.*, **9**, 1337–1346.

Kothary, R.S., Clapoff, S., Brown, A. et al. (1988) A transgene containing *lacZ* inserted into the dystonia locus is expressed in the neural tube. *Nature*, **335**, 435–437.

Larkin, J.C., Oppenheimer, D.G., Pollock, S. and Marks, M.D. (1993) Arabidopsis *GLABROUS 1* gene requires downstream sequences for function. *Plant Cell*, **5**, 1739–1748.

Lee, I., Aukerman, M.J., Gore, S.L. et al. (1994) Isolation of *LUMINIDEPENDENS*: a gene involved in the control of flowering time in Arabidopsis. *Plant Cell*, **6**, 75–83.

Lee, K.Y., Lund, P., Lowe, K. and Dunsmuir, P. (1990) Homologous recombination in plant cells after *Agrobacterium*-mediated transformation. *Plant Cell*, **2**, 415–425.

Lessl, M. and Lanka, E. (1994) Common mechanisms in bacterial conjugation and Ti-mediated T-DNA transfer to plant cells. *Cell*, **77**, 321–324.

Lindsey, K. and Wei, W. (1996) Tissue culture, transformation and transient gene expression in *Arabidopsis*. In *Arabidopsis: A Practical Approach* (ed. B. Mulligan), Oxford University Press, Oxford, in press.

Lindsey, K., Wei, W., Clarke, M.C. et al. (1993) Tagging genomic sequences that direct transgene expression by activation of a promoter trap in plants. *Transgenic Res.*, **2**, 33–47.

Marks, M.D. and Feldmann, K.A. (1989) Trichome development in *Arabidopsis thaliana*. I. T-DNA tagging of the GLABROUS 1 gene. *Plant Cell*, **1**, 1043–1050.

Martineau, B., Voelker, T.A. and Sanders, R.A. (1994) On defining T-DNA. *Plant Cell*, **6**, 1032–1033.

Márton, L. and Browse, J. (1991) Facile transformation of *Arabidopsis*. *Plant Cell Rep.*, **10**, 235–239.

Mayer, U., Büttner, G. and Jürgens, G. (1993) Apical-basal pattern formation in the *Arabidopsis* embryo: studies on the role of the *gnom* gene. *Development*, **117**, 149–162.

Mayerhofer, R., Koncz-Kalman, Z., Nawrath, C. et al. (1991) T-DNA integration: a mode of illegitimate recombination in plants. *EMBO J.*, **10**, 697–704.

Meyerowitz, E.M. (1987) *Arabidopsis thaliana*. *Annu. Rev. Genet.*, **21**, 93–111.

Meyerowitz, E.M. (1994) Pattern formation in plant development: four vignettes. *Curr. Opin. Genet. Dev.*, **4**, 602–608.

Misera, S., Muller, A.J., Weilandheidecker, U. and Jürgens, G. (1994) The fusca genes of *Arabidopsis*: negative regulators of light responses. *Mol. Gen. Genet.*, **244**, 242–252.

Muskett, P., Wei, W. and Lindsey, K. (1994) Promoter trapping to identify and isolate genes expressed in vascular tissues of *Arabidopsis thaliana*. In *Abstracts of the 4th International Congress on Plant Molecular Biology*, p. 2079. International Society for Plant Molecular Biology, Amsterdam.

Okamuro, J.K., den Boer, B.G.W. and Jofuku, K.D. (1993) Regulation of Arabidopsis flower development. *Plant Cell*, **5**, 1183–1193.

O'Kane, C.J. and Gehring, W.J. (1987) Detection in situ of genomic regulatory elements in *Drosophila*. *Proc. Natl. Acad. Sci. USA*, **84**, 9123–9127.

Okuley, J., Lightner, J., Feldmann, K. et al. (1994) Arabidopsis *FAD2* gene encodes the enzyme that is essential for polyunsaturated lipid synthesis. *Plant Cell*, **6**, 147–158.

Ott, R.W. and Chua, N.-H. (1990) Enhancer sequences from *Arabidopsis thaliana* obtained by transformation of *Nicotiana tabacum*. *Mol. Gen. Genet.*, **223**, 169–179.

Priess, J.R. (1994) Establishment of initial asymmetry in early *Caenorhabditis elegans* embryos. *Curr. Opin. Genet. Dev.*, **4**, 563–568.

Ptashne, M. (1986) Gene regulation by proteins acting nearby and at a distance. *Nature*, **322**, 697–701.

Pythoud, F. and King, P.J. (1990) Auxotrophic, temperature-sensitive and hormone mutants isolated *in vitro*. In *Plant Cell Line Selection: Procedures and Applications* (ed. P.J. Dix), pp. 233–255. VCH, Weinheim.

Reynaerts, A., De Block, M., Hernalsteens, J.-P. and van Montagu, M. (1988) Selectable and screenable markers. In *Plant Molecular Biology Manual* (eds S.B. Gelvin and R.A. Schilperoort), pp. A9: 1–16. Kluwer Academic Publishers, Dordrecht.

Rijkers, T., Peetz, A. and Rüther, U. (1994) Insertional mutagenesis in transgenic mice. *Transgenic Res.*, **3**, 203–215.

Roe, J.L., Rivin, C.J., Sessions, R.A. et al. (1993) The *tousled* gene in A. thaliana encodes a protein kinase homolog that is required for leaf and flower development. *Cell*, **75**, 939–950.

Rossant, J. and Joyner, A.L. (1989) Towards a molecular-genetic analysis of mammalian development. *Trends Genet.*, **5**, 277–283.

Rubin, G.M. (1988) *Drosophila melanogaster* as an experimental organism. *Science*, **240**, 1453–1459.

Schmidt, R. and Willmitzer, L. (1988) High efficiency *Agrobacterium tumefaciens*-mediated transformation of *Arabidopsis thaliana* leaf and cotyledon explants. *Plant Cell Rep.*, **7**, 583–586.

Schüpbach, T. and Roth, S. (1994) Dorsoventral patterning in *Drosophila* oogenesis. *Curr. Opin. Genet. Dev.*, **4**, 502–507.

Shevell, D.E., Leu, W.-M., Gillmour, C.S. et al. (1994) *EMB30* is essential for normal cell division, cell expansion, and cell adhesion in Arabidopsis and encodes a protein that has similarity to Sec7. *Cell*, **77**, 1051–1062.

Skarnes, W.C. (1990) Entrapment vectors: a new tool in mammalian genetics. *Bio/Technology*, **8**, 827–831.

Spielmann, A. and Simpson, R.B. (1986) T-DNA structure in transgenic tobacco plants with multiple independent integration events. *Mol. Gen. Genet.*, **205**, 34–41.

Stachel, S.E. and Zambryski, P. (1986) *Agrobacterium tumefaciens* and the susceptible plant cell: a novel adaptation of extracellular recognition and DNA conjugation. *Cell*, **47**, 155–157.

Stockhaus, J., Schell, J. and Willmitzer, L. (1989) Identification of enhancer elements in the upstream region of the nuclear photosynthetic gene ST-LS1. *Plant Cell*, **1**, 805–813.

Teeri, T.H., Herrera-Estrella, L., Depicker, A. et al. (1986) Identification of plant promoters *in situ* by T-DNA-mediated transcriptional fusions to the *npt*II gene. *EMBO J.*, **5**, 1755–1760.

Valvekens, D., van Montagu, M. and van Lijsebettens, M. (1988) *Agrobacterium tumefaciens*-mediated transformation of *Arabidopsis thaliana* root explants by using kanamycin selection. *Proc. Natl. Acad. Sci. USA*, **85**, 5536–5540.

Van Lijsebettens, M., Vanderhaeghen, R. and Van Montagu, M. (1991) Insertional mutagenesis in *Arabidopsis thaliana*: isolation of a T-DNA-linked mutation that alters leaf morphology. *Theor. Appl. Genet.*, **81**, 277–284.

Van Lijsebettens, M., Vanderhaeghen, R., De Block, M. et al. (1994) An S18 ribosomal protein gene copy at the *Arabidopsis PFL* locus affects plant development by its specific expression in meristems. *EMBO J.*, **13**, 3378–3388.

Walbot, V. (1992) Strategies for mutagenesis and gene cloning using transposon tagging and T-DNA insertional mutagenesis. *Annu. Rev. Plant Physiol. Plant Mol. Biol.*, **43**, 49–82.

Ward, E.R. and Jen, G.C. (1990) Isolation of single-copy-sequence clones from a yeast artificial chromosome library of randomly-sheared *Arabidopsis thaliana* DNA. *Plant Mol. Biol.*, **14**, 561–568.

Wei, N. and Deng, X.-W. (1993) Characterisation and molecular cloning of COP9, a genetic locus involved in light-regulated development and gene expression in *Arabidopsis*. In *Proceedings of the 5th International Conference on Arabidopsis Research* (eds K. Hangarter, R. Scholl, K. Davis and K. Feldmann), p. 14. Ohio State University, Columbus, Ohio.

Wilson, C., Pearson, R.K., Bellen, H.J. et al. (1989) P-element-mediated enhancer detection: an efficient method for isolation and characterizing developmentally regulated genes in *Drosophila*. *Genes Dev.*, **3**, 1201–1313.

Yanofsky, M.F., Ma, H., Bowman, J.L. et al. (1990) The protein encoded by the Arabidopsis homeotic gene agamous resembles transcription factors. *Nature*, **346**, 35–39.

Zambryski, P., Tempe, J. and Schell, J. (1989) Transfer and function of T-DNA genes from *Agrobacterium* Ti and Ri plasmids in plants. *Cell*, **56**, 193–201.

ARABIDOPSIS SEED RESOURCE CENTRES

Nottingham *Arabidopsis* Seed Centre (NASC)
Department of Life Science, University of Nottingham, University Park, Nottingham NG7 2RD, UK
Contact: Dr Mary Anderson
E-mail: PLZMLH@VAX.CCC

Ohio State *Arabidopsis* **Biological Resource Centre (ABRS)**
Ohio State University, 1735 Neil Avenue, Columbus, Ohio 43210, USA
Contact: Dr Randy Scholl

12 Transposon-mediated Insertional Mutagenesis

MICHAEL ROBERTS

University of Leicester, Leicester, UK

The insertion of a transposable element into a gene and an accompanying alteration in phenotype provides an association between a locus (defined by the presence of the transposable element) and the change in phenotype (mutation). The exploitation of this association in order to clone genes from such loci is known as transposon tagging. This chapter discusses theoretical and practical aspects of transposon tagging in a variety of plant hosts, using both native and introduced transposons. Considerations for the generation of mutants, the identification of transposon-containing alleles and the isolation of mutant and wild-type genes are given. Particular attention is paid to the wide range of systems available for the control, detection and monitoring of transposable element activity during mutagenesis, features which are especially relevant in transgenic plants. Finally, new applications for transposable elements in molecular genetics will be highlighted.

1 INTRODUCTION

1.1 Transposable elements

The identification of mutants has been important for many years, since it was recognised that their study could provide a great deal of information regarding diverse aspects of biology. The subject of this chapter, transposable elements, represent valuable tools in the generation and study of mutants. Indeed, Gregor Mendel first established the basic principles of inheritance by his studies of the segregation of a mutant phenotype, the wrinkled pea character, which has since been found to be the result of transposon insertion into a pea gene (Bhattacharyya et al., 1990). Genes are generally considered to be stably maintained from one generation to the next, but, since the late nineteenth century, genes with high mutation rates have been studied extensively. Most famous are those of *Zea mays* (maize) and *Antirrhinum majus* (snapdragon), in which many mutable alleles of pigment genes are known which result not in uniform but in sectored phenotypes within individuals (Fig. 1). In 1946 Barbara McClintock published evidence that distinct genetic elements, which she termed controlling elements, regulated some of these mutable loci via chromosome breakage (McClintock, 1946). A locus termed *Dissociation* (*Ds*) marked the

Plant Gene Isolation: Principles and Practice. Edited by G. D. Foster and D. Twell.
© 1996 John Wiley & Sons Ltd.

A

B

Figure 1. Photographs showing phenotypes of mutable alleles in maize and *Antirrhinum*. (A) Maize kernels showing sectors of pigment resulting from reversion of a transposon-induced mutation in a gene for anthocyanin biosynthesis. (B) Spike of *Antirrhinum* flowers showing sectors of anthocyanin pigmentation following excision of *Tam3* from a mutable *pallida* locus. (Photograph kindly provided by Dr Rosemary Carpenter, John Innes Centre, Norwich, UK)

breakpoint, whilst an independent locus, *Activator* (*Ac*), controlled the timing of these events. Soon after, McClintock reported that many *Ds* loci were not fixed in the maize genome but that *Ds* was transposable (McClintock, 1948), as she also later found *Ac* to be. Despite the scepticism of many of her peers, McClintock continued her characterisation of maize controlling elements and was eventually awarded a Nobel prize for her contributions to plant genetics.

To date, transposable genetic elements have been identified in all classes of organism, ranging from the simple insertion sequences and the drug-resistance-carrying transposons of *Escherichia coli* to the numerous elements identified in plants and genetically well studied animals such as *Drosophila* and humans. It is believed that a significant proportion of the repetitive DNA found in most higher eukaryotes consists of, or is related to, transposable element sequences (Smyth, 1991). Molecular analysis of transposons from various organisms has provided evidence that transposable elements are of ancient origin, since transposons fall into families which span the plant and animal kingdoms. The *Ac/Dc* elements, for example, are homologous not only to other plant transposons but also to the *Hobo* element of *Drosophila* (Atkinson et al., 1993).

Two broad categories of transposable element exist, comprising the DNA-based transposons, which transpose via DNA intermediates, and the retro-transposons, or retroelements, which transpose via RNA intermediates and characteristically encode a reverse transcriptase enzyme. Retroelement families are also conserved throughout eukaryotes, members of the *Ty-copia* group, for example, being found in yeasts, animals and plants (Stucka et al., 1992; Flavell et al., 1992). All types of transposon possess certain features which enable their molecular identification as potential mobile elements. The most notable of these are the presence of short terminal repeat sequences which mark the borders of the element and the creation of short direct duplications of the target sequence following insertion. One important feature pertinent particularly to plant transposable elements is that they occur in families of autonomous and non-autonomous elements. This is demonstrated by the *Ac/Ds* family, in which *Ac* is the autonomous element, being capable of the production of active transposase enzyme and thus transposition. The term *Ds* is applied to any element which contains sufficient homology at the element termini that it is mobile when *Ac* transposase is supplied in *trans*, but which cannot produce transposase itself due to deletion, rearrangement or mutation of the coding region.

1.2 Transposon tagging

The activity of transposable elements has provided us with the tool of 'transposon tagging' as a method for gene isolation (Walbot, 1992a) This technique relies upon a recognisable phenotype being caused by the insertion into a gene of a transposon of known sequence. In this situation, the transposable element is referred to as a 'tag', and the mutated gene is said to be 'tagged' by its presence. The mutated gene can be isolated using a transposon sequence as a molecular probe. Cloning via insertional mutagenesis is a particularly useful approach for cloning genes for which no information on the gene product exists. Transposon tagging was first successful in maize (Fedoroff et al., 1984) and *Antirrhinum* (Martin et al., 1985), two species in which a number of mutable alleles had already been characterised. Early experiments related to genes involved in anthocyanin pigment biosynthesis, since changes in coloration of maize kernels and *Antirrhinum* flowers were readily identified (Fig. 1). The creation of mutable alleles as opposed to stable mutations by transposons is by virtue of the ability of DNA-based elements to excise from the site of insertion, causing a reversion to the wild-type phenotype. This is an important diagnostic feature, and transposon tagging in plants has therefore concentrated on these elements. Retroelements, which are recognised as major mutagens in yeast, *Drosophila* and humans, have received less attention in plants to date, since the mutations they cause are not observed as mutable alleles.

Recently, transposons from maize have been used to create mutable alleles and to clone tagged genes from unrelated species following genetic transformation. The application of this technology has greatly expanded the potential for tagging important plant genes and will be discussed in section 5.

2 REQUIREMENTS FOR AND LIMITATIONS OF TRANSPOSON TAGGING

Whilst transposon tagging has undoubtedly facilitated the cloning of important genes from maize and *Antirrhinum*, for a large number of characters transposon tagging would not be the most appropriate method for gene cloning. Transposon tagging combines both molecular and genetic approaches, and before deciding to use this method for gene isolation a number of factors require careful consideration.

2.1 Targets for mutagenesis

First, one must ensure that an effective screening procedure is available to identify mutants of interest. Visible mutations are easiest to identify, especially if a phenotype is evident in seedlings. More growth space is required to screen for characters in mature plants. Screens which require selection in any form of tissue culture, or which require biochemical assays, etc. for identification of mutants, will be highly labour-intensive. Disruptions of genes essential for basic cellular function will give lethal mutations, and thus will be difficult to distinguish from one another. It is possible to work with lethal mutations, as long as heterozygous siblings which do not display the phenotype can be identified. A second problem arises when genes are present in families, such that the activity of other family members can compensate for the loss of function of a particular member. It would be possible to tag such genes, however, if members of the family exhibited different developmental expression patterns for example, such that screening could be concentrated on a particular organ or stage. Such redundancy means that it is usually not possible to tag genes in organisms with polyploid genomes.

Recently, experiments using a novel type of mutant screen have been developed which utilise transgenic plants. For example, in order to identify mutants in signal transduction pathways or particular transcription factors, plants carrying reporter genes (i.e. a transgene activated in response to the targeted factor or transduction pathway) are mutagenised with the aim of characterising lines in which the reporter gene is no longer expressed, or is expressed abnormally (e.g. Li et al., 1994).

In situations where mutants with the desired phenotype have not been described, it may be prudent to initially carry out chemical or radiation mutagenesis experiments in order to identify and characterise mutants affected in the gene or process of interest (Chapter 9). Not only will this approach permit fine tuning of screening procedures, but any mutants obtained can be extremely useful in the production and analysis of tagged allelic mutants.

2.2 Choice of tagging system

Having established that the production of a mutant phenotype by transposon tagging of the gene or genes of interest should be possible, an appropriate transposable element system must be chosen. If the gene of interest can be cloned from plant species in which well-characterised transposons exist—realistically only maize and *Antirrhinum*—then one should consider the use of one of these species. A major drawback with maize, and to a lesser extent with *Antirrhinum*, however, is the practical difficulty in growing large populations of plants. It should now be possible to overcome such difficulties by using other species. First, the literature could be examined with regard to the discovery of active transposable elements in the species of interest. Examples of transposons found in species apart from maize and *Antirrhinum*, and methods for identifying new elements, are discussed in section 4.3. Alternatively, the introduction of transposons from heterologous species by genetic transformation facilitates a tagging programme with certain advantages over using endogenous elements. Engineered elements which allow selection and monitoring of excision and reinsertion events in transgenic plants are now available, and are of particular use in *Arabidopsis*, tomato and *Nicotiana plumbaginifolia*. These systems will be considered in section 5.

2.3 Mutation frequency

An important consideration for mutagenesis is the frequency of occurrence of mutants of interest. For experiments aimed at tagging a defined genetic locus, i.e. a single gene, the mutation frequency will be low. This approach is often termed targeted mutagenesis. Non-targeted mutagenesis would aim to create mutants in a particular process, or stage of development, rather than in a particular gene, so that any gene involved in the process of interest might be cloned. The greater the number of genes involved in that process, the greater the chance of one of these being tagged. Chemical- or radiation-based mutagenesis enable estimations of the relative frequency of mutation at loci controlling different aspects of plant growth and development. It is often assumed that mutagenesis in organisms with large genomes must be less efficient than in organisms with small genomes, and this is one reason for the extensive investigation of the behaviour of heterologous transposons in *Arabidopsis*. This is not necessarily the case for transposon mutagenesis, because recent research has shown that transposons are non-randomly distributed in plant genomes, and tend to be associated with the same fraction of the genome as the genes themselves (Capel et al., 1993).

Equally important in governing mutation frequency is the choice of transposable element. In maize, where a number of elements have been used for tagging, lines exist where there is only one known active element family, such as *Ac/Ds*, *En/I* (*Spm/dSpm*) or *Mutator*. Each of these families exhibits different properties influencing mutation rate which can be exploited. The Robertson's *Mutator* family (Robertson, 1978) causes particularly high mutation frequencies, mainly as a result of the high copy number of *Mu* elements found in most active lines (Bennetzen, 1984). However, the high mutation frequency caused by *Mu* elements is compromised by the difficulty in establishing genetic linkage between an individual element and a mutant allele. To overcome this problem, stocks containing low-copy-number *Mu* elements are available which still give rise to relatively high rates of mutation (Hardeman & Chandler, 1989). In targeted mutagenesis experiments, typical frequencies of new mutants induced by *Mu* in a population are in the range 4×10^{-4} to 1×10^{-5} (Table 1; Robertson, 1985). Mutation frequencies with the *En/Spm* and *Ac* families tend to be in the order of 10^{-5} to 10^{-6}, although this varies depending upon the genetic background and the target locus. It should be noted that many of the genes cloned by association with these elements were isolated from lines historically known to be tagged, rather than being isolated following directed mutagenesis, so that data concerning the frequencies of mutation at single loci using these elements are relatively scarce.

In the case of *Ac* and *Ds*, transposition is non-random, with around 50–60% of new insertions occurring at sites linked to the original locus (e.g. Greenblatt, 1984; Dooner & Belachew, 1989). This can be used to advantage when the chromosomal location of the target locus has been mapped. By choosing stock lines in which *Ac* or *Ds* elements have been mapped proximal to the target, one can expect to significantly increase the chances of tagging the desired gene. This strategy is applicable to any plant containing *Ac* or *Ds* elements,

Table 1. Examples of transposon mutagenesis in maize and *Antirrhinum*

Tagging rationale	Target locus	Frequency of mutant recovery[a]	Reference
Screens in Robertson's *Mutator* background	*Bronze-1*	3.9×10^{-4}	Taylor et al. (1986)
		4.1×10^{-5}	Brown et al. (1989)
		4.3×10^{-5}	Hardeman & Chandler (1989)
		6.6×10^{-5}	Hardeman & Chandler
	Bronze-2	1.7×10^{-5}	McLaughlin & Walbot (1987)
Locus targeted in active *Spm* maize stock	*Opaque-2*	3.8×10^{-6}	Schmidt et al. (1987)
Ac/Ds tagging	*Opaque-2*	2×10^{-5}	Motto et al. (1986)
Linked *Ds* transposition	*Waxy*	6.8×10^{-4}	Weil et al. (1992)
Translocation-mediated linkage to *Ac*	*R-nj*	5×10^{-5}	Dellaporta et al. (1988)
Tam3 tagging	Rust-resistance gene	$\geqslant 5.4 \times 10^{-4}$	Aitken et al. (1992)
Tam3 tagging	Floral morphology	$\sim 2 \times 10^{-4}$	Carpenter & Coen (1990)
Tam3 tagging	Any visible mutation	1.2×10^{-2}	Cited in Walbot (1992a)

[a]Mutant populations were generated by a variety of means; the figures shown here represent the number of mutants found per seed planted for screening.

and is especially useful in maize, where nearly 900 chromosome translocation stocks exist, allowing selection of lines in which the target locus has been brought closer to an active element as a result of chromosome rearrangement (Dellaporta ct al., 1988; Walbot, 1992a). Transposable elements introduced into transgenic plants (section 5) are also being mapped to provide a resource of stocks of tomato (e.g. Thomas et al., 1994; Knapp et al., 1994) and *Arabidopsis* (Nottingham Arabidopsis Stock Centre, UK) from which lines can be selected containing a linked transposon and target allele. The rationale is that suitable elements can be activated to enable efficient targeted tagging. There is also evidence that transposition tends to be linked for the *Spm* family (Peterson, 1970; Cardon et al., 1993b), so the same strategy can be applied. Other elements may have certain sequence requirements or preferences ('hot spots') for insertion, as many prokaryotic transposons do.

These strategies may significantly increase the likelihood of tagging the gene of interest. At a frequency of around 10^{-5} to 10^{-6}, it would be necessary to screen several hundred thousand plants, but at frequencies in the order of 10^{-4}, obtained by using linked transposition of high-activity *Mutator* stocks, the required population size is reduced to around ten thousand. In non-targeted tagging experiments, the frequency of obtaining a mutant of interest is dependent upon the number of mutable alleles identifiable under the screen employed. In one example from *Antirrhinum* (Carpenter & Coen, 1990), from an initial screen of 13 000 M1 and 40 000 M2 plants, 8–11 independent loci were tagged affecting floral development: a frequency of $1.5-2.0 \times 10^{-4}$ for the entire population (one dominant mutant was identified in the M1 generation). In a study where all visible mutants were counted, a frequency for mutagenesis of around 1.2×10^{-2} was observed (Table 1). Tagging frequencies at some loci can be unusually high, as indicated by an experiment in which Aitken et al. (1992) tagged a race-specific rust-resistance gene at least 5 and as many as 15 times in a screen of 11 153 M1 *Antirrhinum* plants produced by crossing to a tester stock.

With regard to the frequencies mentioned above, it is important to appreciate that different loci exhibit different susceptibilities to the insertion of individual transposons. This is highlighted by Walbot (1992a) for the maize bronze allele *Bz1*, where frequencies ranged from $< 2 \times 10^{-5}$ (no mutants in 49 500 progeny) to 3.9×10^{-4}, a difference of at least 20-fold. Environmental conditions also influence transposon-induced mutation rates. Most striking is the effect of temperature on the *Antirrhinum Tam3* element. Excision of *Tam3* from the *nivea* and *pallida* loci is increased 1000-fold at 15 °C compared to 25 °C (Carpenter et al., 1987). Other factors believed to induce transposition are events causing 'genome stress', such as UV and ionising radiation, viral infection and tissue culture (Johns et al., 1985; Peschke et al., 1987; Walbot, 1988, 1992b; Hirochika, 1993). Whether any of these factors could be exploited in a tagging programme is as yet untested. The activation of tobacco retroelements in tissue culture seems particularly pronounced, and may be a contributory factor in the phenomenon of somaclonal variation (Hirochika, 1993). The potential use of retroelements presents an interesting future direction for tagging in plants.

2.4   Germinal versus somatic activity

Transposition events can be classified as either somatic or germinal, depending upon the tissue in which the event occurred. Somatic events are recognised as sectors of tissue containing excised or transposed elements, e.g. pigmented spots on colourless kernels in maize anthocyanin mutants. Germinal events are recognised as progeny genotypically uniform for an excised or transposed element, which would be represented by a fully pigmented kernel in the above example. Seeds or seedlings carrying germinally transposed elements may be formed following either transmission of a somatic sector through the germ line or transpositions in the gametophyte proper. Clearly, germinal transposition is necessary for the production of non-sectored mutants in tagging experiments, which are more readily identified and analysed. In maize, germinal excisions of the *Ac* element are represented in around 1–5% of seedlings (e.g. Dooner & Belachew, 1989). This frequency is similar to that observed for the germinal excision of *Ac* in transgenic tobacco (Jones et al., 1990) and *Arabidopsis* (e.g. Dean et al., 1992). Experimental systems which permit selection for germinal transpositions are thus of particular value.

3   METHODOLOGY OF TRANSPOSON TAGGING

The following sections describe what is required at a practical level for completion of a successful transposon tagging experiment.

3.1   Generation of mutants

3.1.1   *Targeted and non-targeted mutagenesis*

Whether using endogenous or introduced transposons, the first stage of the programme is the generation of a mutated seed population. The starting material should be an isogenic inbred line containing an active transposable element, these plants being termed the M0 generation. M0 plants are grown and selfed to produce the M1 generation, which should contain a range of new transposon insertions. The size of the M1 population should be around twice the minimum size required based on the statistical likelihood of tagging the desired target if this is known (e.g. by reference to previous experiments). Since the majority of mutations caused by transposon insertion will be recessive, loss-of-function mutations, no phenotype will be evident in the M1 plants which will be heterozygous for the mutant allele. Hence an M2 population is generated by selfing the M1 plants such that in one quarter of the seed from each plant, new transposon insertions will be made homozygous, allowing expression of the mutant phenotype (Fig. 2A). Seeds from each selfed M1 plant are collected and a few planted for screening. It is easiest to pool all the seeds, but doing this means that it is not possible to clone genes with lethal mutations. If seeds from each M1 plant are kept separate or in small pools from a few plants, heterozygotes can be recovered from the M1 seed when lethal mutations arise.

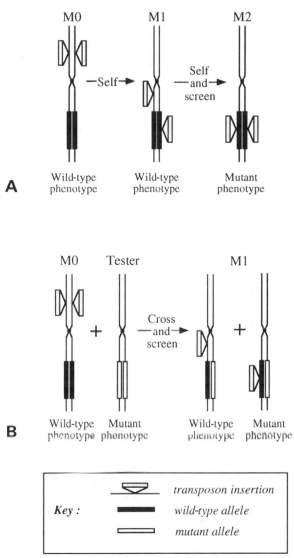

Figure 2. Genetics of strategies for transposon-mediated mutagenesis. (A) Mutagenesis of novel alleles. Elements activated during the development of M0 plants are transmitted as new transposon insertions into the M1 generation. Mutant phenotypes are screened in the M2 following selfing to make alleles homozygous. (B) Tagging genes with known mutant alleles. Transposon and tester stocks containing the untagged mutant allele are crossed to produce an M1 population which can be screened directly for the mutant phenotype

3.1.2 Tagging genes with known mutant phenotypes

In many cases, the target locus will be a gene for which characterised mutants already exist. When this is true, it is not necessary to produce a selfed M2 population. Rather, the M1 is produced by crossing a tester stock homozygous for a mutant allele of the target with a line homozygous for the wild-type allele which also contains an active transposon family. The M1 seed produced will thus be heterozygous for the target locus, such that a mutant phenotype can be observed if the wild-type allele is disrupted (Fig. 2B).

3.2 Cosegregation analysis

Having identified a mutant or mutants of interest, the next step is to demonstrate cosegregation of the phenotype with a particular transposon insertion. This is generally achieved after generating populations segregating for the mutant phenotype by crossing the mutant with several different wild-type land races to produce heterozygotes. A 3:1 segregation in the selfed progeny of these plants confirms that the mutation is at a single locus, with mutant (m/m) and wild-type (M/M) siblings being identified and selfed to bulk up material. In the special cases of lethals and sterile mutants, heterozygotes are bulked by crossing siblings. Whilst undertaking this genetic analysis, crosses to any existing similar mutants should be carried out to test for allelism (Chapter 9).

To perform the cosegregation analysis, Southern blots are carried out using probes to the transposable element suspected of having caused the mutation. DNA isolated from mutant m/m and wild-type M/M lines is digested using several restriction enzymes which do not cut within the transposon sequence and subjected to Southern blotting. Hybridising bands are sought which are always present in m/m but never in M/M DNA samples (Fig. 3). This type of analysis aims to establish genetic linkage between the locus represented by the hybridising band and the locus represented by the phenotype, so the more mutant and wild-type samples from different sources which can be tested, the better. In an ideal case, there will be only one transposon locus in each mutant, which segregates and is not present in homozygous wild-type siblings. However, this case is not the norm, and when using maize there are likely to be several or many segragating bands, especially if tagging with the *Mutator* system. Similar problems arise when using *Tam3* in *Antirrhinum*. These problems can be tackled by using probes specific to certain sub families of *Mu*, *Ac/Ds* or *Spm/dSpm* (*En/I*) elements, but cosegregation in maize can often be very laborious. In the case of transgenic plants containing introduced elements, constructs which include a selectable marker gene within the transposon sequence simplify genetic analysis by establishing linkage of the mutant phenotype and the marker gene.

Any plants reverting to a wild-type phenotype among the progeny of homozygous mutants are particularly valuable in demonstrating cosegregation. The occurrence of revertant wild-type sectors of tissue in mutant plants is good evidence that the mutation is caused by a transposable element, reversion being due to excision (e.g. Aarts et al., 1993). Germinal revertants are especially useful, since they will contain only one or two alleles (plants will be $M\text{-}r/M\text{-}r$ or $M\text{-}r/m$ where $M\text{-}r$ represents the revertant allele), as opposed to somatic revertants, which will contain sectors of both mutant and revertant tissue. Germinal revertants can be used in Southern analysis to demonstrate

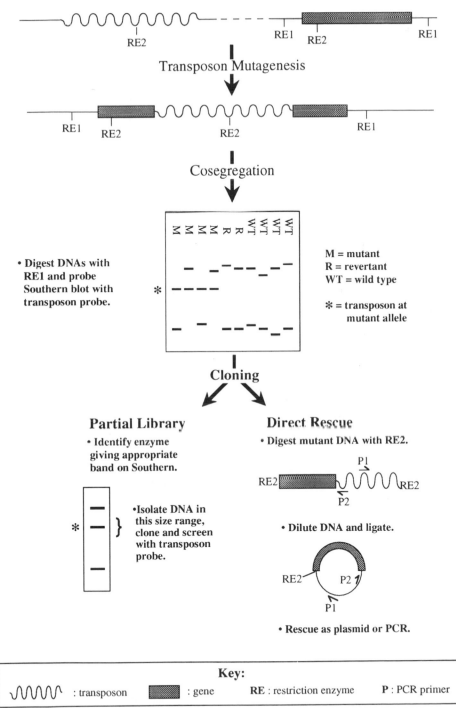

Figure 3. Molecular characterisation of transposon-tagged alleles. Schematic diagram outlining the process of gene isolation by transposon tagging as described in section 3

the loss of any band (or marker gene) identified as cosegregating with the mutant phenotype (e.g. Chuck et al., 1993).

3.3   Isolation of flanking DNA

Once a cosegregating transposon has been identified, its flanking DNA may be cloned. This can be achieved by one of a number of different strategies, as for the T-DNA-tagged mutants discussed in Chapter 11. Complete genomic libraries can be made from mutant lines followed by the isolation of clones hybridising to transposon probes, but it is probably better to construct partial libraries. A partial library is constructed following the identification of a restriction digest using one or two enzymes, which results in the band of interest being clearly separated from other transposon-containing bands. DNA in the appropriate size range is then purified from digested genomic DNA of the mutant and cloned into the vector of choice (Chapter 3). Such size-selected libraries are important in lines containing many copies of elements, and in all cases reduces the total number of clones needed to be screened. Once clones hybridising to the transposon probe have been isolated, inserts are subcloned and sequenced to identify the flanking DNA. Possible alternatives to making genomic libraries from the mutant are to obtain flanking sequences directly by plasmid rescue or by inverse polymerase chain reaction (IPCR) (Chapter 13). In cases where plasmid replication origins and bacterial selection markers are included in the transposon, the DNA can be rescued directly by transformation into bacteria after circularising digested mutant genomic DNA (e.g. Rommens et al., 1992a). Alternatively, oligonucleotide primers are made facing opposite directions out of the terminus of the transposable element sequence (see Fig. 3) to facilitate IPCR amplification of flanking DNA (e.g. Earp et al., 1990; Osborne et al., 1991; Thomas et al., 1994).

Once flanking sequences have been isolated, experiments should be carried out to test whether these sequences correspond to the mutant allele. Initially, flanking sequences are used as probes in Southern blots in order to determine which hybridise to the cosegregating band identified with the transposon probe. A new fragment of a different size should be detected in the wild-type DNA if the flanking probe does represent a mutant allele containing a transposon insertion. Flanking probes can also be used in northern hybridisation experiments to compare mutant and wild-type gene expression. A range of effects on transcription of tagged genes have been observed (review: Weil & Wessler, 1990), depending upon the gene and type of insertion. For example, insertions into non-transcribed regions of genes generally cause alterations in the levels or tissue specificity of expression (e.g. Chen et al., 1987; Coen et al., 1986). Insertions into transcribed sequences tend to result either in decreases in steady-state message levels due to altered stability following aberrant splicing, or premature transcription termination. Results showing reduction or alteration of message expression in tissues affected by the mutation provide good evidence that the flanking sequence obtained is indeed cognate to the mutated allele. However, northern blots may show no difference between mutant and wild-type lines, since transposons in mutant alleles are often spliced to leave an apparently normal-sized transcript (e.g. Dennis et al., 1988; Weil & Wessler, 1990).

3.4 Cloning the wild-type gene

Having obtained the flanking sequence associated with the mutated locus, the final objective is to analyse the wild-type allele corresponding to this sequence. This is achieved by screening either genomic or cDNA libraries, or more simply by PCR amplification of the gene from wild-type DNA. However, a genomic clone will almost always yield the greatest amount of information, since it is more likely to contain the complete coding sequence and regulatory elements associated with the gene of interest. Whilst sequence information derived from such clones might aid in an identification of gene function by comparison with sequence databases, the primary aim, having obtained a clone, is to confirm that it is cognate to the mutant allele. This is established by complementation, whereby the cloned gene is transformed into the mutant with the expectation that a wild-type phenotype will result via genetic complementation. For genes expected to confer phenotypes easily detectable in a small number of cells (e.g. pigment synthesis), complementation may be achieved by transient expression following introduction of the gene by microprojectile bombardment or other DNA-mediated uptake techniques (e.g. Nash et al., 1990).

4 TAGGING USING ENDOGENOUS TRANSPOSABLE ELEMENTS IN MAIZE AND *ANTIRRHINUM*

4.1 Choice of M0 stocks

When attempting an endogenous transposon tagging experiment, the most useful species are maize and *Antirrhinum*. When selecting an M0 population from these species, there are several considerations. First, a compromise must be made between using high-copy-number transposable element stocks, which will maximise the likelihood of obtaining a tagged mutant, and low-copy-number stocks, which will simplify genetic analysis once mutants are identified. To date, the majority of succesful tagging experiments have used *Mu* elements in maize and *Tam3* in *Antirrhinum*. A second important consideration concerns the numbers of plants which must be screened. Wherever possible, stocks should be chosen in which a readily scorable marker for transposon activity exists. For example, stocks containing an element inserted into an anthocyanin biosynthesis gene, such that activity can be monitored by the occurrence of sectors of pigment in *Antirrhinum* flowers or maize kernels, are often employed. In maize it is possible to select M1 plants on the basis of germinal reversion scored in kernels, so that all M2 plants generated are likely to contain a new transposon insertion. In *Antirrhinum*, lines with unstable alleles of the chalcone synthase gene such as lines T53 (*niv-53::Tam1*) (Bonas et al., 1984) and *nivea-recurrens* (*niv-98::Tam3*) (Sommer et al., 1985) have been used tagging experiments to clone floral homeotic genes. The final choice of material will in most cases by determined by the frequency of mutation, since generation and screening of mutants is the most laborious task.

4.2 Cloning tagged loci

4.2.1 *Examples from maize*

Following the cloning of transposable elements from maize, several genes in the anthocyanin biosynthesis pathway were initial targets for transposon tagging, since characterised mutable alleles already existed. These early examples demonstrate a range of strategies employed to facilitate gene isolation, and indicate the variety of approaches which can be undertaken depending upon the characteristics of the mutant involved. In the first application of transposon tagging in plants, Fedoroff et al. (1984) cloned the *Bronze-2* gene via its association with an autonomous *Ac* element in a *bz* mutant line. Restriction analysis of *Ac*-hybridising genomic clones from the mutant was used to eliminate clones containing (*Ds*) elements with restriction patterns different to the cloned *Ac9* element. Flanking DNA from the remaining *Ac*-containing clones was then used as a probe in Southern analysis of mutant and revertant *Bronze* alleles. A similar strategy was applied by Paz-Ares et al. (1986), who clone the maize *c* locus from a line containing and *En* element-induced mutable allele. Genomic clones hybridising to each of three different probes representing the two terminal domains and a central portion of the isolated autonomous *En* element were identified and restriction mapped, clones containing various *dSpm/I* elements being eliminated by this procedure. O'Reilly et al. (1985) used two different mutable alleles of the *a1* locus in order to clone this gene. The mutations were induced by *Mu1* and *En1*, and genomic clones containing these elements from the respective mutants were cross-hybridised to find a common flanking sequence which was shown to contain the *a1* gene. A different approach was taken by McClaughlin & Walbot (1987), who used the predicted expression pattern of the maize *Bronze-2* gene to enable its isolation from a *Mutator*-induced mutable allele. Mutants with the fewest copies of *Mu* elements were identified by Southern analysis, and genomic clones containing *Mu* sequences isolated. Inserts from these clones were labelled and used to probe northern blots containing RNA from wild-type and *bz-2* mutant plants. One clone hybridised to a message present in wild-type but not mutant kernel RNA, and was shown to include the *Bronze-2* gene.

4.2.2 *Examples From* Antirrhinum

In recent years, transposon tagging in *Antirrhinum* has contributed significantly to our understanding of floral development through the cloning of genes controlling organ and meristem identity. The first of these to be cloned was *deficiens*, mutations in which cause the homeotic conversion of petals to sepals and stamens to carpels (Sommer et al., 1990). An unstable *defA-1* mutation was identified with the characteristics of a *Tam* element insertion, but rather than use conventional tagging strategies, a novel approach was devised. A flower-specific cDNA library from wild-type plants was constructed and screened for differentially expressed clones. These were then tested against *defA-1* and *defA-1*-revertant flower RNA in order to identify the defA cDNA clone. A second floral homeotic gene, *floricaula*, which acts earlier in development and is involved in the transition from inflorescence to floral meristems, was also cloned by transposon tagging (Coen et al., 1990). In this case, a typical transposon tagging protocol was followed, with cosegregation of *Tam3* elements in wild-type and mutant lines used to identify and clone a putative *flo* sequence, which was confirmed by the analysis of revertants. *Tam3*-induced mutations have the added advantage that they can easily be attributed to this particular element on the basis of differential temperature-dependent reversion rates.

The *deficiens* and *floricaula* genes, along with many of the maize genes cloned by transposon tagging, such as *opaque-2*, *c1* and *R*, are regulatory genes which encode transcription factors. Transcription factors are often difficult to clone by conventional molecular techniques, since mRNA and protein may be present at low levels. These examples therefore highlight the power of insertional muta-genesis by transposon tagging as a method for cloning important genes without prior biochemical characterisation.

4.3   Native transposons in other species

Whilst the previous sections have highlighted the use of transposable elements in maize and *Antirrhinum*, a large number of transposon-like sequences have been identified in many other plant species. Most of these are apparently immobile, but some retain the ability to transpose. Some of these elements have now been used as tags for gene cloning. The earliest examples of tagging outside maize or *Antirrhinum* were carried out using the *dTph1* element family of petunia, a non-autonomous mobile element present at high copy number in some petunia varieties (Gerats et al., 1990).

The *dTph1* element was cloned following the characterisation of an unstable *An1* locus, which carries the petunia *DFR-C* gene, but most new elements are discovered fortuitously as sequences in genomic clones. However, active elements in other species can be identified and cloned using 'transposon traps'. Such traps have enabled the isolation of active elements from tobacco and *Arabidopsis* (Table 2). Transposon traps rely on insertional mutation of endo-genous or introduced gene constructs such that mutant alleles can readily be recovered and analysed. This method was first successfully used in the isolation of the *Tnt1* retroelement from tobacco (Grandbastien et al., 1989). In these experiments, nitrate reductase mutants selected in tissue culture were characterised, and insertion of the *Tnt1* element found to be responsible in around one-third of all cases.

Table 2. Active transposable elements in important plant species

Species	Element	Isolation	Type	Reference
Arabidopsis	*Tat1*	Inserts in cloned genes	DNA	Peleman et al. (1991)
	Tag1	Targeted tagging/trap- ping	DNA	Tsay et al. (1993)
Petunia	*dTph1*	From mutable *An1* allele	DNA	Gerats et al. (1990)
Tobacco	*Tnt1*	Trapping	Retro	Grandbastien et al. (1989)
	Tto1	PCR	Retro	Hirochika (1993)
	Tto2	PCR	Retro	Hirochika (1993)
	Ac-like	Trapping	DNA	Grappin & Grandbas- tien (1994)
N. plumbaginifolia	*dTnp1*	Trapping	DNA	Meyer et al. (1994)
Rice	*Tos17*	PCR	Retro	Hirochika et al. (1994)

5 TRANSPOSABLE ELEMENTS IN TRANSGENIC PLANTS

5.1 Transposable elements in heterologous hosts

Since the *Ac* and *En* elements are genetically autonomous and capable of *trans*-activation of non-autonomous elements, it was suspected that they must encode their own transposase enzyme. To test the idea that *Ac* could mediate its own transposition, Baker et al., (1986) transferred the element into tobacco, where it was shown to excise from the transforming DNA and re-integrate at different sites in the tobacco genome. Subsequently, constructs were designed in which the *Ac* element, and a derivative containing a 1.6-kb internal deletion, were located in the untranslated leader sequence of a neomycin phosphotransferase II (*nptII*) gene. Reconstitution of the *nptII* gene following transposon excision was expected to confer kanamycin resistance on transgenic tissues. It was found that the full *Ac* element, but not the deleted derivative (artificial *Ds*), was capable of excision in transgenic tobacco, with the constructs successfully enabling antibiotic selection for excision (Baker et al., 1987). It has now been shown that *Ac* is active in a number of mono- and dicotyledonous plant species (Table 3). Subsequently, it was demonstrated that *Ds* elements could be *trans*-activated by *Ac* in transgenic plants (Hehl & Baker, 1989). The *En/Spm*, *Tam3* and *Mu1* elements have also been transferred into non-host plant species, with *En/Spm* and *Tam3*, but not *Mu1*, showing activity (Table 3).

5.2 Advantages of heterologous transposon tagging

The potential for transposon tagging in species other than maize and *Antirrhinum* was an exciting prospect, and considerable effort has been put into establishing tagging systems in model species such as tomato and *Arabidopsis*.

Table 3. Activity of transposable elements in transgenic plants

Element	Species	Transposition	Trans-activation	Reference
Ac	Tobacco	Yes		Baker et al. (1986)
	Arabidopsis	Yes		van Sluys et al. (1987)
	Carrot	Yes		van Sluys et al. (1987)
	Potato	Yes		Knapp et al. (1988)
	Tomato	Yes		Yoder et al. (1988)
	Soybean	Yes		Zhou & Atherly (1990)
	Flax	Yes		Roberts et al. (1990)
	Rice	Yes		Murai et al. (1991)
	N. plumbaginifolia	Yes		Marion-Poll et al. (1993)
	Lettuce	Yes		Yang et al. (1993a)
	Datura innoxia	Yes		Schmidtrogge et al. (1994)
Ds	Tobacco	No	Yes	Hehl & Baker (1989)
	Potato	No	Yes	Lassner et al. (1989)
	Arabidopsis	No	Yes	Masterson et al. (1989)
	Lettuce	No	Yes	Yang et al. (1993b)
En/Spm	Tobacco	Yes		Masson & Fedoroff (1989)
	Potato	Yes		Frey et al. (1989)
	Arabidopsis	Yes		Cardon et al. (1993a)
dSpm	Tobacco	No	Yes	Frey et al. (1990)
	Arabidopsis	No	Yes	Cardon et al. (1993a)
Tam3	Tobacco	Yes		Martin et al. (1989)
	Petunia	Yes		Haring et al. (1989)
dTam3	Tobacco	No	Yes	Haring et al. (1991)
Mu1	Arabidopsis	No	No	Zhang & Summerville (1987)
	Tobacco	No	No	Delgiudice et al. (1991)

The advantages of using transgenic plants containing heterologous transposons for tagging are the simplicity of the genetics involved and the ability to manipulate transposon activity. If known transposable element constructs are introduced, it is a simple task to follow the behaviour of the elements by Southern blot analysis. Furthermore, the inclusion of selectable markers in T-DNAs and/or transposable elements provides a direct method for segregation analysis. However, it can be argued that the low copy number of introduced transposons may negate any advantage by reducing the frequency of mutagenesis.

In order to increase transposition frequencies of introduced elements, the relationship between autonomous and non-autonomous elements has been exploited. A number of laboratories have developed controllable two-element systems for applications in tagging experiments. These systems employ a non-autonomous element which acts as the mutagenic agent and a transposase gene construct to *trans*-activate it. These transposase 'helpers' can consist simply of terminally deleted, and therefore stabilised, autonomous elements (e.g. Hehl & Baker, 1989), or transposase genes or cDNAs fused to constitutive, inducible, or tissue-specific promoters (section 5.3). In this way, the level and timing of transposase production can be manipulated to best suit the parameters required for mutagenesis. Additional control can be afforded by engineering the non-autonomous element, as will be shown in the following sections.

5.3 Construct design for heterologous tagging

5.3.1 Transposase helpers

The most commonly used helper constructs have been fusions of the cauliflower mosaic virus (CaMV) 35S promoter to *Ac* transposase to drive transposition of *Ds* in *Arabidopsis* and tobacco (e.g. Swinburne et al., 1992; Fitzmaurice et al., 1992; Grevelding et al., 1992). Such experiments have shown that *Ds* excision is much more frequent when driven by a CaMV 35S-*Ac* helper than by wild-type or deleted *Ac*. Similarly, fusions of the *Spm* element *TnpA* and *TnpD* genes to CaMV 35S have been used to drive transposition of *dSpm/I* elements in transgenic tobacco and *Arabidopsis* (Cardon et al., 1993a,b). Following the observation that increased transposase expression does not necessarily lead to an increased excision frequency, however, recent work has demonstrated that there is a threshold level for *Ac* transposase expression, above which there is an inhibition of *Ds* excision (Scofield et al., 1993). Another potential problem is the finding that the high excision frequencies afforded by the strong CaMV 35S promoter are countered by an accompanying reduction in reinsertion frequency (Long et al., 1993a), which may be due to continued transposition and eventual loss of *Ds* elements.

A range of other promoters have been tested in transposase helper constructs, with varying degrees of success. These include chalcone synthase and chalcone isomerase, which are active during another development, in an unsuccessful attempt to increase germinal transposition (Rommens et al., 1992b), rbc-S (Honma et al., 1993), and a heat shock promoter which confers inducible transposition and is particularly effective during embryogenesis (Balcells et al., 1994). Germinal transposition might be more successfully increased using promoters active during early gametophyte development, rather than promoters active in maturing pollen like chalcone synthase and isomerase. To this end, our laboratory has isolated such a promoter (from the *Arabidopsis apg* gene), and experiments using *apg–Ac* transposase fusions demonstrate microspore-specific *Ds* transposition in tobacco (M. R. Roberts and J. Draper, unpublished).

A recent development which combines the benefits of controllable transposase helpers with the genetic simplicity of a one-element system is that of self-stabilising elements (Schmitz & Theres, 1994). Here, the promoter driving transposase expression resides outside a promoterless *Ac* derivative, such that read-through transcription can cause transposition, leaving the reinserted element in a stabilised condition separated from the promoter.

5.3.2 Excision markers

In addition to the ability to control transposase producton, methods for assaying excision have also been developed further since the original *nptII* excision marker of Baker et al. (1987). Most widely used has been a strepto-mycin-resistance (*spt*) gene, first employed by Jones et al. (1989) to assay *Ac* excision in tobacco. Streptomycin resistance is a cell-autonomous visual pheno-type (observed as greeen resistant cells in a background of bleached sensitive tissue) permitting detailed examinations of somatic excision in seedlings. This marker has proved to be a powerful tool in comparing the characteristic excision patterns driven by different promoter–transposase fusions (e.g. Sco-field et al., 1992, 1993). Since streptomycin resistance is cell-autonomous, germinally inherited events can be distinguished from somatic events by screening for fully green (non-variegated) seedlings. Other selectable markers such as hygromycin and basta (a herbicide) resistance (Haring et al., 1989; Cardon et al., 1993b) and screenable markers such as *gus* and luciferase (Haring et al., 1989; Cardon et al., 1993a; Charng & Pfitzner, 1994) have also proved successful in monitoring excision. Herbicide resistance is particularly useful as an excision marker, since herbicides can be applied to large popula-tions of plants in soil.

5.3.3 Markers for the selection of reinserted elements

Transposable elements containing marker genes to help enable selection and detection of reinsertion have also been used sucessfully, e.g. dihydrofolate reductase (Masterson et al., 1989), hygromycin resistance (Bancroft et al., 1992) and acetolactate synthase (Honma et al., 1993). As a further mechanism for control, screenable or negative selection marker genes have been included on transposase helper T-DNAs. For example, the transposase helper described by Bancroft et al. (1992) includes the *iaaH* gene, which converts externally applied auxin analogues to toxic derivatives. Segregation of the transposase helper and the transposable element can thus be followed or directly selected so that plants containing stabilised insertions can be obtained.

5.4 Gene tagging in transgenic plants

The availability of such a complex array of technologies should have provided enough information on the behaviour of *Ac/Ds* and *En/Spm* to assess the feasibility of full-scale tagging experiments. What, then, is the current status? A number of laboratories have produced collections of several hundreds of *Arabidopsis* families containing germinally transposed *Ds* or *dSpm/I* elements and screened these for visible mutants. Numerous mutations have been identified in this way, though, surprisingly, only a minority of these have been shown to be linked to transposed elements. This may be due to multiple transposition events where CaMV 35S transposase fusions are employed, with

some mutations being caused by transposon footprints (small sequence alterations resulting from imperfect transposon excision). The first genes to be
isolated using heterlogous tagging systems include genes involved in chloroplast development (Long et al., 1993b; Sundberg et al., 1994), pollen development (Aarts et al., 1993) and the regulation of flower pigmentation (Chuck et
al., 1993). Many more tagged mutants exist (e.g. Bancroft et al., 1993), and, at
the time of writing, several of the associated genes are in the process of being
cloned.

Most importantly, specific targeted tagging has also been achieved in transgenic plants. Transactivation of a linked *Ds* element was successfully used to
isolate the *Cladosporium fulvum* resistance gene, *Cf9*, from tomato (Jones et al.,
1994), whilst another notable success was the isolation of the tobacco *N* gene,
which affords resistance to tobacco mosaic virus, by tagging with *Ac* (Whitham
et al., 1994).

6 NOVEL APPLICATIONS FOR TRANSPOSABLE ELEMENTS

6.1 Gene identification

In addition to straightforward insertional mutagenesis, transposable elements
are being used in various other applications in molecular genetics. A range of
ideas based on the ability of transposons to carry sequences from one location
to another within the genome have recently been tested. For example, it
should be possible to create dominant mutations caused by ectopic gene
expression resulting from the introduction near open reading frames of constitutive promoters. Constructs consisting of the CaMV 35S promoter facing
out from the terminus of a *Ds* element are being employed for this purpose
(e.g. Grevelding et al., 1992). This strategy has already shown to be feasible in
a T-DNA tagging vector (Hayashi et al., 1992). Similarly, the promoter trapping
strategy discussed in Chapter 11 has also been adopted by workers using
transposons. The transposition of promoteless reporter genes such as *gus* to
new locations around the genome greatly reduces the number of independent
transformants required in comparison to the T-DNA-based experiments.

6.2 Identification of transposon insertions into known genes

One of the major problems in reverse genetics is that whilst genes active with
particular expression patterns can readily be isolated and sequenced, very
often no biological function can be attributed to these sequences. Strategies to
identify alleles of such genes which contain transposon insertions were originally devised for work on *Drosophila* (Ballinger & Benzer, 1989; Clark et al.,
1994), and are generally applicable to any organism containing cloned transposable elements. They depend on the power of the PCR to detect adjacent
sequences in large pools of DNA. Using PCR primers from the termini of a
transposon and primer(s) specific for the target, it is possible to detect and

identify plants in which the transposon has inserted into or very near the target gene. Such plants can then be analysed with respect to gene expression and phenotype in an attempt to link gene expression and function. This strategy, known as 'site-selected mutagenesis', has recently been used to isolate new mutant alleles of the maize *hcf106* locus (Das & Martienssen, 1995). This technique should also prove a useful alternative to homologous recombination to eliminate the expression of undesirable gene products.

6.3 Genome manipulation

A widely conceived application of linked transposition is the ability to create defined chromosome segments bordered by a transposed element and the T-DNA from which it excised, which can be manipulated either *in vivo* or *in vitro*. For example, the inclusion of target sequences for recombinase enzymes in the T-DNA and the transposon enables inversion or deletion of the intervening chromosomal DNA following the introduction of the recombinase enzyme into plants showing linked transposition events (e.g. van Haaren & Ow, 1993). Possibilities for isolating large sections of the genome exist in the case of linked transposition of elements carrying yeast artificial chromosome (YAC) or P1 phage sequences, such that the DNA intervening between a transposon and T-DNA might be cloned. These types of strategy may prove useful in fine mapping of chromosomes where different transposition events from the same T-DNA are analysed.

A novel use of the ability to cause excision of sequences from a particular location is in the elimination of genes from T-DNAs following transformation (Russell et al., 1992). For example, inclusion of the selectable marker gene within a transposable element allows relocation (or loss if reinsertion fails) and then segregation of the marker and T-DNA, a process which might be commercially valuable where the inclusion of antibiotic-resistance genes in transgenic crops is a public concern.

7 SOURCES OF MATERIAL

7.1 *Arabidopsis*

Transgenic *Arabidopsis* lines containing *Ac* helper T-DNAs and marked *Ds* T-DNAs (many of which are mapped) and a collection of transposed *Ds* elements are available from the Nottingham Arabidopsis Stock Centre in the UK. The transposed *Ds* collection consists of over 200 lines selected on the basis of germinal excision of *Ds* from a streptomycin-resistance gene, with hygromycin resistance being carried by *Ds*. Transposition was driven by a stabilised *Ac* element on a T-DNA marked with a *gus* gene. The parental lines containing these constructs, which were described by Bancroft et al. (1992), are also available from the stock centre. Lines containing a different but similar

Ds(Hyg):spt construct (Long et al., 1993a) and a CaMV 35S *Ac* transposase helper (Swinburne et al., 1992) which have been used successfully for tagging are also available. The stock centre runs a worldwide web (WWW) server permitting browsing of the seed catalogue and direct ordering of seed: the internet address is http://nasc.nott.ac.uk/ The centre can also be contacted by E-mail: the address is arabidopsis@nottingham.ac.uk

The endogenous *Arabidopsis* transposon *Tag1* is present in Landsberg *erecta*, but not in the Columbia or Wassilewskija ecotypes. There are three related *Tag1* elements in Landsberg *erecta*.

7.2 Maize

A comprehensive maize genome database administered in the USA is accessible through the internet: the address is gopher://teosinte.agron.missouri.edu/

Information on stocks and mutants can also be obtained from the Maize Genetics Cooperation: Maize Genetics Cooperation—Stock Center, S-123 Turner Hall, 1102 S. Goodwin Avenue, Urbana, IL 61801, USA.

7.3 Plasmids

A number of available plasmids containing *Ac* and *Ds* elements and selectable marker genes suitable for construction of new *Ac* helpers and *Ds* constructs along with binary transformation vectors are described by Jones et al. (1992).

8   CONCLUDING REMARKS

The study of mutable alleles in maize and *Antirrhinum* has led to the development of a valuable method of plant gene isolation: that of transposon tagging. Until recently, the scope of this technique was limited by the hosts available for transposon mutagenesis, but we have now seen the fruition of a concerted effort to engineer and transfer transposable elements to a range of species. Transposable elements have thus become firmly established as tools for gene isolation and manipulation, in both transposon tagging and the range of other applications being developed that utilise mobile DNA. With the increases in the numbers of hosts used and genes cloned by transposon tagging, mutant collections will grow and become better characterised. Consequently, it will become ever more straightforward to both clone important new genes and isolate valuable new alleles of known genes.

ACKNOWLEDGMENTS

I would like to acknowledge the support and tuition of John Draper and Rod Scott, in whose laboratory I first encountered plant transposons, and the insights freely proffered by George Coupland, Caroline Dean and Jonathan Jones.

ESSENTIAL READING

Aarts, M.G.M., Dirkse, W.G., Stiekema, W.J. and Pereira, A. (1993) Transposon tagging of a male-sterility gene in *Arabidopsis*. *Nature*, **363**, 715–717.

Fedoroff, N.V., Furtek, D.B. and Nelson, O.E. Jr (1984) Cloning the *bronze* locus in maize by a simple and generalizable procedure using the transposable controlling element *Activator* (*Ac*). *Proc. Natl. Acad. Sci. USA*, **81**, 3825–3829.

Jones, D.A., Thomas, C.M., Hammond-Cossack, K.E. et al. (1994) Isolation of the tomato *Cf-9* gene for resistance to *Cladosporium fulvum* by transposon tagging. *Science*, **266**, 789–793.

Scofield, S.R., Harrison, K., Nurrish, S.J. and Jones, J.D.G. (1992) Promoter fusions to the *Activator* transposase gene cause distinct patterns of *Dissociation* excision in tobacco cotyledons. *Plant Cell*, **4**, 573–582.

Walbot, V. (1992a) Strategies for mutagenesis and gene cloning using transposon tagging and T-DNA insertional mutagenesis. *Annu. Rev. Plant Physiol. Plant Mol. Biol.*, **43**, 49–82.

REFERENCES

Aitken, E.A.B., Callow, J.A. and Newbury, H.J. (1992) Mutagenesis of a race-specific rust resistance gene in *Antirrhinum majus* using a transposon-tagging protocol. *Plant J.*, **2**, 775–782.

Atkinson, P.W., Waren, W.D. and Obrochta, D.A. (1993) The *Hobo* transposable element of *Drosophila* can be cross-mobilized in houseflies and excises like the *Ac* element of maize. *Proc. Natl. Acad. Sci. USA*, **90**, 9693–9697.

Baker, B., Schell, J., Lorz, H. and Fedoroff, N. (1986) Transposition of the maize controlling element '*Activator*' in tobacco. *Proc. Natl. Acad. Sci. USA*, **83**, 4844–4848.

Baker, B., Coupland, G., Fedoroff, N. et al (1987) Phenotypic assay for excision of the maize controlling element *Ac* in tobacco. *EMBO J.*, **6**, 1547–1554.

Balcells, L., Sundberg, E. and Coupland, G. (1994) A heat-shock promoter fusion to the *Ac* transposase gene drives inducible transposition of a *Ds* element during *Arabidopsis* embryo development. *Plant J.*, **5**, 755–764.

Ballinger, D.G. and Benzer, S. (1989) Targeted gene mutations in *Drosophila*. *Proc. Natl. Acad. Sci. USA*, **86**, 9402–9406.

Bancroft, I., Bhatt, A.M., Sjodin, C. et al. (1992) Development of an efficient two-element transposon tagging system in *Arabidopsis thaliana*. *Mol. Gen. Genet.*, **233**, 449–461.

Bancroft, I., Jones, J.D.G. and Dean, C. (1993) Heterologous transposon tagging of the *DRL1* locus in Arabidopsis. *Plant Cell*, **5**, 631–638.

Bennetzen, J.L. (1984) Transposable element *Mu1* is found in multiple copies only in Robertson's *Mutator* maize lines. *J. Mol. Appl. Genet.*, **2**, 519–524.

Bhattacharyya, M.K., Smith, A.M., Ellis, T.H.N. et al. (1990) The wrinkled-seed character of pea described by Mendel is caused by a transposon-like insertion in a gene encoding starch-branching enzyme. *Cell*, **60**, 115–122.

Bonas, U., Sommer, H., Harrison, B.J. and Saedler, H. (1984) The transposable element *Tam1* of *Antirrhinum majus* is 17 kb long. *Mol. Gen. Genet.*, **194**, 138–143.

Brown, W.E., Robertson, D.S. and Bennetzen, J.L. (1989) Molecular analysis of multiple *Mutator*-derived alleles of the *Bronze* locus of maize. *Genetics*, **122**, 439–445.

Capel. J., Montero, L.M., Martinez-Zapater, J.M. and Salinas, J. (1993) Non-random distribution of transposable elements in the nuclear genome of plants. *Nucleic Acids Res.*, **21**, 2369–2373.

Cardon, G.H., Frey, M., Saedler, H. and Gierl, A. (1993a) Mobility of the maize transposable element *En/Spm* in *Arabidopsis thaliana*. *Plant J.*, **3**, 773–784.

Cardon, G.H., Frey, M., Saedler, H. and Gierl, A. (1993b) Definition and characterization of an artificial *En/Spm*-based transposon tagging system in transgenic tobacco. *Plant Mol. Biol.*, **23**, 157–178.

Carpenter, R. and Coen, E.S. (1990) Floral homeotic mutations produced by transposon mutagenesis. *Genes Dev.*, **4**, 1483–1493.

Carpenter, R., Martin, C. and Coen, E.S. (1987) Comparison of genetic behaviour of the transposable element *Tam3* at two unlinked pigment loci in *Antirrhinum majus*. *Mol. Gen. Genet.*, **207**, 82–89.

Charng, Y.C. and Pfitzner, A.J.P. (1994) The firefly luciferase gene as a reporter for *in-vivo* detection of *Ac* transposition in tomato plants. *Plant Sci.*, **98**, 175–183.

Chen, C.-H., Oishi, K.K., Kloeckener-Gruissem, B. and Freeling, M. (1987) Organ-specific expression of maize *Adh1* is altered after a *Mu* transposon insertion. *Genetics*, **116**, 469–477

Chuck, G., Robbins, T., Nijjar, C. et al. (1993) Tagging and cloning of a Petunia flower color gene with the maize transposable element *Activator*. *Plant Cell*, **5**, 371–378.

Clark, A.G., Silveria, S., Meyers, W. and Langley, C.H. (1994) Nature screen: an efficient method for screening natural populations of *Drosphila* for targeted *P*-element insertions. *Proc. Natl. Acad. Sci. USA*, **91**, 719–722.

Coen, E.S., Carpenter, R. and Martin, C. (1986) Transposable elements generate novel spatial patterns of gene expression in *Antirrhinum majus*. *Cell*, **47**, 285–296.

Coen, E.S., Romero, J.M., Doyle, S. et al. (1990) *Floricaula*—a homeotic gene required for flower development in *Antirrhinum majus*. *Cell*, **63**, 1311–1322.

Das, M. and Martienssen, R. (1995) Site-selected transposon mutagenesis at the *hcf106* locus in maize. *Plant Cell*, **7**, 287–294.

Dean, C., Sjodin, C., Page, T. et al. (1992) Behaviour of the maize transposable element *Ac* in *Arabidopsis thaliana*. *Plant J.*, **2**, 69–81.

Delgiudice, L., Manna, F., Massardo, D.R. et al. (1991) The maize transposable element *Mu1* does not transpose in tobacco. *Maydica*, **36**, 197–203.

Dellaporta, S.L., Greenblatt, I., Kermicle, J.L. et al. (1988) Molecular cloning of the maize *R-nj* allele by transposon tagging with *Ac*. In *Chromosome Structure and Function: Impact of New Concepts* (eds J.P. Gustafson and R. Appels), pp. 263–282. Plenum, New York.

Dennis, E.S., Sachs, M.M., Gerlach, W.L. et al. (1988) The *Ds1* transposable element acts as an intron in the mutant allele *Adh1-Fm335* and is spliced from the message. *Nucleic Acids Res.*, **16**, 3185–3828.

Dooner, H.K. and Belachew, A. (1989) Transposition pattern of the maize element *Ac* from the *bz-m2* (*Ac*) allele. *Genetics*, **122**, 447–457.

Earp, D.J., Lowe, B. and Baker, B. (1990) Amplification of genomic sequences flanking transposable elements in host and heterologous plants: a tool for transposon tagging and genome characterization. *Nucleic Acids Res.*, **18**, 3271–3279.

Fitzmaurice, W.P., Lehman, L.J., Nguyen, L.V. et al. (1992) Development and characterization of a generalized gene tagging system for higher plants using an engineered maize transposon *Ac*. *Plant Mol. Biol.*, **20**, 177–198.

Flavell, A.J., Smith, D.B. and Kumar, A. (1992) Extreme heterogeneity of *Ty1-copia* group retrotransposons in plants. *Mol. Gen. Genet.*, **231**, 233–242.

Frey, M., Tavantzis, S.M. and Saedler, H. (1989) The maize *En-1/Spm* element transposes in potato. *Mol. Gen. Genet.*, **217**, 172–177.

Frey, M., Reinecke, J., Grant, S. et al. (1990) Excision of the *En/Spm* transposable element of *Zea mays* requires two element-encoded proteins. *EMBO J.*, **9**, 4037–4044.

Gerats, A.G.M., Huits, H., Vrijlandt, E. et al. (1990) Molecular characterization of a nonautonomous transposable element (*dTph1*) of *Petunia*. *Plant Cell*, **2**, 1121–1128.

Grandbastien, M.-A., Spielmann, A. and Caboche, M. (1989) *Tnt1*, a mobile retroviral-like transposable element of tobacco isolated by plant cell genetics. *Nature*, **337**, 376–380.

Grappin, P. and Grandbastien, M.A. (1994) Isolation of a new mobile *Ac*-type transposable element in tobacco. In *Abstracts of the 4th International Congress on Plant Molecular Biology*, International Society for Plant Molecular Biology, Amsterdam.

Greenblatt, I.M. (1984) A chromosome replication pattern deduced from pericarp phenotypes resulting from movements of the transposable element, *modulator*, in maize. *Genetics*, **108**, 471–485.

Grevelding, C. Becker, D., Kunze, R. et al. (1992) High rates of *Ac/Ds* germinal transposition in *Arabidopsis* suitable for gene isolation by insertional mutagenesis. *Proc. Natl. Acad. Sci. USA*, **89**, 6085–6089.

Hardeman, K.J. and Chandler, V. (1989) Characterization of *bz1* mutants isolated from mutator stocks with high and low numbers of *Mu1* elements. *Dev. Genet.*, **10**, 460–472.

Haring, M.A., Gao, J., Volbeda, T. et al (1989) A comparative study of *Tam3* and *Ac* transposition in transgenic tobacco and petunia plants. *Plants Mol. Biol.*, **13**, 189–201.

Haring, M.A., Teeuwen-De Vroomen, M.J., Nijkamp, J.J. and Hille, J. (1991) Trans-activation of an artificial *dTam3* transposable element in transgenic tobacco plants. *Plant Mol. Biol.*, **16**, 39–47.

Hayashi, H., Czaja, I., Lubenow, H. et al. (1992) Activation of a plant gene by T-DNA tagging—auxin-independent growth *in-vitro*. *Science*, **258**, 1350–1353.

Hehl, R. and Baker, B. (1989) Induced transposition of *Ds* by stable *Ac* in crosses of transgenic tobacco plants. *Mol. Gen. Genet.*, **217**, 53–59.

Hirochika, H. (1993) Activation of tobacco retrotransposons during tissue culture. *EMBO J.*, **12**, 2521–2528.

Hirochika, H., Sugimoto, K., Kanda, M. and Otsuki, Y. (1994) Retrotransposons of rice activated by tissue culture. In *Abstracts of the 4th International Congress on Plant Molecular Biology*. International Society for Plant Molecular Biology, Amsterdam.

Honma, M.A., Baker, B.J. and Waddell, C.S. (1993) High-frequency germinal transposition of *Ds*^{ALS} in *Arabidopsis*. *Proc. Natl. Acad. Sci. USA*, **90**, 6242–6246.

Johns, M.A., Mottinger, J. and Freeling, M. (1985) A low copy number, *copia*-like transposon in maize. *EMBO J.*, **4**, 1093–1102.

Jones, J.D.G., Carland, F.M., Maliga, P. and Dooner, H.K (1989) Visual detection of transposition of the maize element *Activator* (*Ac*) in tobacco seedlings. *Science*, **244**, 204–207.

Jones, J.D.G., Carland, F., Lim, E. et al. (1990) Preferential transposition of the maize element *Activator* to linked chromosomal locations in tobacco. *Plant Cell*, **2**, 701–707.

Jones, J.D.G., Shlumukov, L., Carland, F. et al. (1992) Effective vectors for transformation, expression of heterologous genes, and assaying transposon excision in transgenic plants. *Transgenic Res.*, **1**, 285–297.

Knapp, S., Coupland, G., Uhrig, H. et al. (1988) Transposition of the maize transposable element *Ac* in *Solanum tuberosum*. *Mol. Gen. Genet.*, **213**, 285–290.

Knapp, S., Larondelle, Y., Roßberg, M. et al. (1994) Transgenic tomato lines containing *Ds* elements at defined genomic positions as tools for targeted transposon tagging. *Mol. Gen. Genet.*, **243**, 666–673.

Lassner, M.W., Palys, J.M. and Yoder, J.I. (1989) Genetic transactivation of *Dissociation* elements in transgenic tomato plants. *Mol. Gen. Genet.*, **218**, 25–32.

Li, H.M., Altschmied, L. and Chory, J. (1994) *Arabidopsis* mutants define downstream branches in the phototransduction pathway. *Genes Dev.*, **8**, 339–349.

Long, D., Swinburne, J., Martin, M. et al. (1993a) Analysis of the frequency of inheritance of transposed *Ds* elements in *Arabidopsis* after activation by a CaMV 35S promoter fusion to the *Ac* transposase. *Mol. Gen. Genet.*, **241**, 627–636.

Long, D., Martin, M., Sundberg, E. et al. (1993b) The maize transposable element system *Ac/Ds* as a mutagen in *Arabidopsis*—identification of an albino mutation induced by *Ds* insertion. *Proc. Natl. Acad. Sci. USA*, **90**, 10370–10374.

Marion-Poll, A., Marin, E., Bonnefoy, N. and Pautot, V. (1993) Transposition of the maize autonomous element *Activator* in transgenic *Nicotiana plumbaginifolia* plants. *Mol. Gen. Genet.*, **238**, 209–217.

Martin, C., Carpenter, R., Sommer, H. et al. (1985) Molecular analysis of instability in flower pigmentation of *Antirrhinum majus*, following isolation of the *pallida* locus by transposon tagging. *EMBO J.*, **4**, 1625–1630.

Martin, C.R., Prescott, A., Lister, C. and MacKay, S. (1989) Activity of the transposon *Tam3* in *Antirrhinum* and tobacco: possible role of DNA methylation. *EMBO J.*, **8**, 997–1004.

Masson, P. and Fedoroff, N.V. (1989) Mobility of the maize *Suppressor-mutator* element in transgenic tobacco cells. *Proc. Natl. Acad. Sci. USA*, **86**, 2219–2223.

Masterson, R.V., Furtek, D.B., Grevelding, C. and Schell, J. (1989) A maize *Ds* transposable element containing a dihydrofolate reductase gene transposes in *Nicotiana tabacum* and *Arabidopsis thaliana*. *Mol. Gen. Genet.*, **219**, 461–466.

McClintock, B. (1946) Maize genetics. *Carnegie Inst. Washington Year Book*, **45**, 176–186.

McClintock, B. (1948) Mutable loci in maize. *Carnegie Inst. Washington Year Book*, **47**, 155–169.

McLaughlin, M. and Walbot, V. (1987) Cloning of a mutable *bz2* allele of maize by transposon tagging and differential hybridization. *Genetics*, **117**, 771–776.

Meyer, C., Pouteau, S., Rouze, P. and Caboche, M. (1994) Isolation and molecular characterization of *dTnp1*, a mobile and defective transposable element of *Nicotiana plumbaginifolia*. *Mol. Gen. Genet.*, **242**, 194–200.

Motto, M., Marotta, R., Difonzo, N. et al. (1986) *Ds*-induced alleles at the *opaque-2* locus of maize. *Genetics*, **112**, 121–133.

Murai, N., Li, Z.J., Kawagoe, Y. and Hayashimoto, A. (1991) Transposition of the maize *Activator* element in transgenic rice plants. *Nucleic Acids Res.*, **19**, 617–622.

Nash, J., Luehrsen, K.R. and Walbot, V. (1990) *Bronze-2* gene of maize: reconstruction of a wild type allele and analysis of transcription and splicing. *Plant Cell*, **2**, 1039–1049.

O'Reilly, C., Shepherd, N.S., Pereira, A. et al. (1985) Molecular cloning of the *a1* locus of *Zea mays* using the transposable elements *En* and *Mu1*. *EMBO J.*, **4**, 877–882.

Osborne, B.I., Corr, C.A., Prince, J.P. et al. (1991) *Ac* transposition from a T-DNA can generate linked and unlinked clusters of insertions in the tomato genome. *Genetics*, **129**, 833–844.

Paz-Ares, J., Wienand, U., Peterson, P.A. and Saedler, H. (1986) Molecular cloning of the *c* locus of *Zea mays*: a locus regulating the anthocyanin pathway. *EMBO J.*, **5**, 829–833.

Peleman, J., Cottyn, B., van Camp, W. et al. (1991) Transient occurrence of extrachromosomal DNA of an *Arabidopsis thaliana* transposon-like element, *Tat1*. *Proc. Natl. Acad. Sci. USA*, **88**, 3618–3622.

Peschke, V.M., Phillips, R.L. and Gengenbach, B.G. (1987) Discovery of transposable element activity among progeny of tissue culture-derived maize plants. *Science*, **238**, 804–807.

Peterson, P.A. (1970) The *En* mutable system in maize. III. Transposition associated with mutational events. *Theor. Appl. Genet.*, **40**, 367–377.

Roberts, M.R., Kumar, A., Scott, R. and Draper, J. (1990) Excision of the maize transposable element *Ac* in flax callus. *Plant Cell Rep.*, **9**, 406–409.

Robertson, D.S. (1978) Characterization of a mutator system in maize. *Mutat. Res.*, **51**, 21–28.

Robertson, D.S. (1985) Differential activity of the maize mutator *Mu* at different loci and in different cell lineages. *Mol. Gen. Genet.*, **200**, 9–13.

Rommens, C.M.T., Rudenko, G.N., Dijkwel, P.P. et al. (1992a) Characterization of the *Ac/Ds* behaviour in transgenic tomato plants using plasmid rescue. *Plant Mol. Biol.*, **20**, 61–70.

Rommens, C.M.T., van Haaren, M.J.J., Buchel, A.S. et al. (1992b) Transactivation of *Ds* by *Ac*-transposase gene fusions in tobacco. *Mol. Gen. Genet.*, **231**, 433–441.

Russell, S.H., Hoopes, J.L. and Odell, J.T. (1992) Directed excision of a transgene from the plant genome. *Mol. Gen. Genet.*, **234**, 49–59.

Schmidt, R.J., Burr, F.A. and Burr, B. (1987) Transposon tagging and molecular analysis of the maize regulatory locus *opaque-2*. *Science*, **238**, 960–963.

Schmidtrogge, T., Weber, B., Borner, T. et al. (1994) Transposition and behaviour of the maize transposable element *Ac* in transgenic haploid *Datura innoxia* Mill. *Plant Sci.*, **99**, 63–74.

Schmitz, G. and Theres, K. (1994) A self-stabilizing *Ac* derivative and its potential for transposon tagging. *Plant J.*, **6**, 781–786.

Scofield, S.R., English, J.J. and Jones, J.D.G. (1993) High level expression of the *Activator* transposase gene inhibits the excision of *Dissociation* in tobacco cotyledons. *Cell*, **75**, 507–517.

Smyth, D.R. (1991) Dispersed repeats in plant genomes. *Chromosoma*, **100**, 355–359.

Sommer, H., Carpenter, R., Harrison, B.J. and Saedler, H. (1985) The transposable element *Tam3* of *Antirrhinum majus* generates a novel type of sequence alteration upon excision. *Mol. Gen. Genet.*, **199**, 225–231.

Sommer, H., Beltran, J.P., Huijser, P. et al. (1990) *Deficiens*, a homeotic gene involved in the control of flower morphogenesis in *Antirrhinum majus*: the protein shows homology to transcription factors. *EMBO J.*, **9**, 605–613.

Stucka, R., Schwarzlose, C., Lochmuller, H. et al. (1992) Molecular analysis of the yeast *Ty4* element—homology with *Ty1*, *copia*, and plant retrotransposons. *Gene*, **122**, 119–128.

Sundberg, E., Engstrom, P. and Coupland, G. (1994) Isolation of an *Arabidopsis thaliana* gene causing an albino phenotype when mutated by a transposon. In *Abstracts of the 4th International Congress on Plant Molecular Biology*. International Society for Plant Molecular Biology, Amsterdam.

Swinburne, J., Balcells, L., Scofield, S. et al. (1992) Elevated levels of *Activator* transposase mRNA are associated with high frequencies of *Dissociation* excision in Arabidopsis. *Plant Cell*, **4**, 583–595.

Taylor, L.P., Chandler, V. and Walbot, V. (1986) Insertion of 1.4 kb and 1.7 kb *Mu* elements into the *Bronze1* gene of *Zea mays*. *Maydica*, **31**, 31–45.

Thomas, C.M., Jones, D.A., English, J.J. et al. (1994) Analysis of the chromosomal distribution of transposon-carrying T-DNAs in tomato using the inverse polymerase chain reaction. *Mol. Gen. Genet.*, **242**, 573–585.

Tsay, Y.-F., Frank, M.J., Page, T. et al. (1993) Identification of a mobile endogenous transposon in *Arabidopsis thaliana*. *Science*, **260**, 342–344.

van Haaren, M.J.J. and Ow, D.W. (1993) Prospects of applying a combination of DNA transposition and site-specific recombination in plants—a strategy for gene identification and cloning. *Plant Mol. Biol.*, **23**, 525–533.

van Sluys, M.A., Tempe, J. and Fedoroff, N. (1987) Studies on the introduction and mobility of the maize *Activator* element in *Arabidopsis thaliana* and *Daucus carota*. *EMBO J.*, **6**, 3881–3889.

Walbot, V. (1988) Reactivation of the *Mutator* transposable element system following gamma irradiation of seed. *Mol. Gen. Genet.*, **212**, 259–264.

Walbot, V. (1992b) Reactivation of *Mutator* transposable elements of maize by ultraviolet light. *Mol. Gen. Genet.*, **234**, 353–360.

Weil, C.F. and Wessler, S.R. (1990) The effects of plant transposable element insertion on transcription initiation and RNA processing. *Annu. Rev. Plant Physiol. Plant Mol. Biol.*, **41**, 527–552.

Weil, C.F., Marillonnet, S., Burr, B. and Wessler, S.R. (1992) Changes in state of the *Wx-m5* allele of maize are due to intragenic transposition of the *Ds* element. *Genetics*, **130**, 175–185.

Whitham, S., Dinesh-Kumar, S.P., Choi, D. et al. (1994) The product of the tobacco mosaic virus resistance gene *N*: similarity to Toll and the interleukin-1 receptor. *Cell*, **78**, 1101–1115.

Yang, C.H., Ellis, J.G. and Michelmore, R.W. (1993a) Infrequent transposition of *Ac* in lettuce, *Lactuca sativa*. *Plant Mol. Biol.*, **22**, 793–805.

Yang, C.H., Carroll, B., Scofield, S. et al. (1993b) Transactivation of *Ds* elements in plants of lettuce (*Lactuca sativa*). *Mol. Gen. Genet.*, **241**, 389–398.

Yoder, J.I., Palys, J., Alpert, K. and Lassner, M. (1988) *Ac* transposition in transgenic tomato plants. *Mol. Gen. Genet.*, **213**, 291–296.

Zhang, H. and Sommerville, C.R. (1987) Transfer of the maize transposable element *Mu1* into *Arabidopsis thaliana*. *Plant Sci.*, **48**, 165–173.

Zhou, J.H. and Atherly, A.G. (1990) In situ detection of transposition of the maize controlling element (*Ac*) in transgenic soybean tissues. *Plant Cell Rep.*, **8**, 542–545.

V   PCR-based Cloning

13 PCR Techniques

CHRIS THOMAS

Advanced Technologies (Cambridge) Limited, Cambridge, UK

The polymerase chain reaction (PCR) has become an invaluable tool for molecular biologists. DNA sequences can be amplified in vitro from picogram quantities of DNA or even single copies of a gene within a matter of hours. The versatility of the process means that the researcher can use PCR to not only isolate DNA fragments but also modify them to aid cloning or to obtain mutants. PCR can be used to screen colonies for recombinants, for characterisation by sequencing and to screen for expression of endogenous or recombinant genes in plants. This chapter outlines the following:

(1) The principles of PCR and basic protocols.
(2) Preparation for PCR such as primer design and obtaining suitable templates.
(3) A range of PCR applications from cloning, screening and sequencing to quantitation.
(4) Some of the potential problems of the method and how to tackle them.

1 INTRODUCTION

The story of the invention of the polymerase chain reaction (PCR), a simple *in vitro* protocol for amplifying large amounts of DNA, is one of a dream coming true for a scientist. Kary B. Mullis tells of how he was driving through a star-spangled Friday night in April 1983, on the US 101, to a cabin in Mendocino County (Mullis, 1990). He was thinking about a proposed sequencing experiment involving the use of primers, DNA template, a DNA polymerase and deoxynucleotides when suddenly he realised that he had developed a method for doubling the amount of a specific DNA in his reaction. The process could be repeated, resulting in an exponential increase in the amount of product; a 1000-fold after 10 cycles, a 1 000 000-fold after 20. Measurable amounts of DNA could be generated from a single original copy of template, such as a gene in a single cell. The method was published (Saiki et al., 1985) and has since spawned a flood of applications and modifications expressed in an exponentially growing list of publications using PCR (14 000 in 1992–1993 alone). Mullis was subsequently awarded a Nobel prize for his invention.

The basic principle of PCR is straightforward (Fig. 1). The template DNA is denatured into single strands. Two oligonucleotide primer molecules are annealed to the templates, one primer to each strand. A DNA polymerase then

Plant Gene Isolation: Principles and Practice. Edited by G. D. Foster and D. Twell.
© 1996 John Wiley & Sons Ltd.

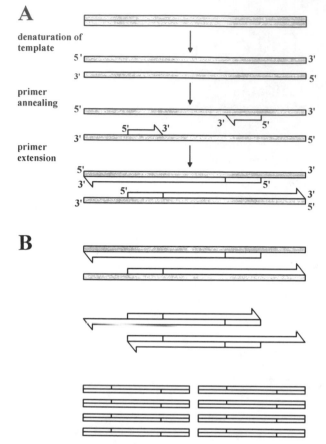

Figure 1. The principle of PCR. (A) A template such as double-stranded DNA is heat denatured to give single-stranded DNA. The sample is then cooled in the presence of primers specific to part of the DNA. Note that one primer binds to one DNA strand and the other primer binds to the complementary DNA strand. The binding sites of the primers are chosen such that they will prime DNA synthesis along the DNA strands towards each other. The reaction temperature is raised to allow DNA synthesis from the primers to occur using a thermostable DNA polymerase. Double-stranded DNA is produced in which the new strands re-create the priming region for the opposite primer. This effectively constitutes one PCR cycle. (B) After a set number of cycles, generally 25–40, the following reaction products have been generated: (1) the original template DNA strands with newly synthesised complements; (2) a small population of *linearly* amplified long complements to the original template DNA; (3) a large population of *logarithmically* amplified PCR fragments which contain the sequences to the two PCR primers at their ends. This latter population is generally amplified a 1 000 000-fold or more, resulting in microgram quantities of product from picogram quantities or less of the original target sequence

extends the primers by incorporating deoxyribonucleotides, and synthesises a complementary strand. The complementary strand must extend over the region primed by the other primer; the two priming sites can be separated by tens of bases to several thousand base pairs. These products are in turn denatured, and more primers are annealed and extended, thus doubling the amount of DNA bordered by the primers with each cycle until it becomes the dominant nucleic acid in the reaction tube. The use of thermostable DNA polymerases, initially *Taq* polymerase, has permitted the automation of the procedure (Saiki et al., 1988), as the DNA molecules can be heat denatured between each cycle without inactivating the polymerase. The repeated temperature cycling between denaturation, primer annealing and extension is provided by commercially available programmable heating blocks. Given the above requirements, even the smallest laboratory can isolate large quantities of a specific clonable product in a matter of hours.

Mullis was working for Cetus at the time of the invention of PCR, with the result that the process, with the use of *Taq* polymerase and cycling machines, was patented right from the start (US patents 4683195, 4683202, 4965188, 4889818, 5079352 and 5075216, and patents derived from these in other countries and the EU). The patents are currently held by Hoffmann La Roche, and it is worth checking that chosen suppliers for enzymes, kits and equipment relating to the PCR process do hold licences from Hoffmann La Roche to avoid any problems should work using PCR result in a commercially viable product, unintentionally or not.

The aim of this chapter is to cover the use of PCR for cloning plant sequences, highlighting any peculiarities specific to plants and demonstrating how techniques used on other organisms could be used in this field. Because of the vast literature on PCR, choice of references will be selective and sometimes arbitrary. There are many publications comprehensively reviewing or detailing the principles and applications of PCR in general (e.g. Erlich, 1989; Innis et al., 1990; McPherson et al., 1991a; Mullis et al., 1994; Newton & Graham, 1994; Sambrook et al., 1989; Thomas, 1993). These include detailed protocols for many of the procedures and are of great practical use. This chapter will give particular emphasis to the design of primers used for PCR, as experience has shown this to be a key factor in the success of any PCR experiment.

2   SETTING UP A PCR

2.1   Template

The template for a PCR is DNA. However, the purity and form of the template are flexible, ranging from caesium chloride purified genomic DNA to crude cell extracts, as long as the addition of the template does not inhibit the thermostable polymerase. A variety of methods for obtaining PCR templates will be described in section 3.1.

2.2   Polymerases

Any DNA polymerase capable of extending a DNA strand from a primer can be used for a PCR. Indeed, the first PCRs were conducted with Klenow (Saiki et

al., 1985; Scharf et al., 1986). Nowadays, however, thermostable DNA polymerases are preferred, as fresh enzyme does not need to be added after each
heat-denaturation step, as is the case with thermolabile enzymes (Saiki et al.,
1988).

The most frequently used and preferred enzyme is *Taq* polymerase (Saiki et
al., 1988). The enzyme was isolated from *Thermus aquaticus* and has been well
characterised (Gelfand, 1989). Cloned versions are now also available, e.g.
Amplitaq from Perkin Elmer. The optimum temperature range for the enzyme
is 70–80 °C, at which it has an extension rate of 35–100 nucleotides per second.

Taq polymerase does not have a 3'–5' exonuclease activity, which has two
consequences. First, in common with many polymerases, *Taq* polymerase
exhibits a non-template-specific addition of a deoxyribonucleotide at the 3' end
of double-stranded DNA molecules (Clark, 1988) which is not removed due to
the absence of the exonuclease activity. The added base tends to be adenine,
which must either be removed prior to blunt-end cloning by polishing the PCR
product with T4 DNA polymerase or can be exploited by cloning PCR products
directly into a TA vector (see section 4.8). Second, the lack of 3'–5' exonuclease
activity means that *Taq* polymerase does not correct for the incorporation of
mismatched bases and is therefore more likely to introduce single-base changes
(Tindall & Kunkel, 1988; Krawczac et al., 1989; Kwok et al., 1990; Eckert &
Kunkel, 1991). This is of particular concern where products are amplified from
just a few original template molecules, such as in a single cell. For most
purposes *Taq* polymerase is perfectly adequate.

Perkin Elmer also provide r*Tth* polymerase, isolated from *Thermus thermo-
philus*, which has reverse transcriptase activity, therefore enabling the synthesis
of cDNA from mRNA and subsequent PCR in the same tube (Myers &
Gelfand, 1991). The r*Tth* polymerase also lacks the 3'–5' exonuclease activity
and has the same limitations that this creates for *Taq* polymerase. Amersham
supply TET-z, which also has reverse transcriptase activity.

Other thermostable enzymes with proofreading ability and therefore a reduced error rate during replication have been isolated by various suppliers.
These include *Pfu* polymerase from Stratagene (Lundberg et al., 1991), Vent™
polymerase from New England Biolabs (Mattila et al., 1991) and UlTma™ DNA
polymerase.

Taq DNA polymerase can also be used in conjunction with *Pfu* DNA
polymerase at low concentrations to improve proofreading and removal
of non-specifically added nucleotides. The unexpected side-effect is that
molecules up to 10 kbp or more in length can be amplified (Barnes, 1994),
whereas the individual enzymes on their own cannot successfully amplify the
same large fragments. Stratagene is marketing its *Taq* PCR extender based on
this principle.

2.3 Reagents

The PCR reaction mix of 5–100 µl will contain the following:

- 1× polymerase buffer, provided as 10× stock by supplier

- 200 μM solution of each deoxynucleotide phosphate
- 1.2–2.5 mM magnesium chloride
- 0.1–1 μM primer 1
- 0.1–1 μM primer 2
- 25 units DNA polymerase per millilitre
- Template

A typical buffer for *Taq* polymerase is:

- 10 mM Tris-Cl, pH 8.3
- 50 mM potassium chloride
- 0.01% gelatin
- 0.01% nonidet P-40
- 0.01% Tween

Caution should be exercised when using *Taq* DNA polymerase and stock buffer from different sources, as the nonionic detergents may be supplied either with the enzyme or with the buffer. Other polymerases have their own buffer requirements. With *Taq* DNA polymerase, the potassium chloride concentration can be altered or reduced to facilitate PCR of long fragments (Ponce & Micol, 1992). Alternatively, a proprietary buffer system supplied, for example, with the *Taq* PCR Extender of Stratagene can be employed (Barnes, 1994).

The magnesium chloride concentration can be varied. The usual concentration is 1.5 mM but some products are only obtained at higher magnesium chloride concentrations. This is particularly the case for RAPDs, mentioned later in the text. The magnesium chloride is mopped up to some degree by the deoxyribonucleotides (dNTPs) in solution, so that if the dNTP concentration is altered from 200 μM, this will affect the available magnesium chloride.

The suggested concentration of 200 μM for each dNTP is a compromise between ensuring fidelity and adequate yields of PCR product. Higher concentrations can lead to more non-specific PCR products arising. It is important that the dNTPs are at equimolar concentrations, as imbalances also lead to increased misincorporation of bases.

The primers will be dealt with in more detail in section 3.2. In our hands, concentrations of approximately 5 μg of a 20-mer primer per millilitre give good results.

There is also a range of additives that have been reported to enhance specificity or yield of the inevitably more recalcitrant product. These include formamide (Sarkar et al., 1990), dimethylsulphoxide (Winship, 1989) and tetramethylammonium chloride (Hung et al., 1990). Newton & Graham (1994) report that such enhancers do not seem to be universally applicable and that their mode of action is not known. In our experience, primer design and optimisation of PCR conditions achieve a higher degree of success.

2.4 PCR machines

Strictly speaking, there is no real requirement for a PCR machine. Ample time and three water baths at the denaturation, annealing and extension temperatures will suffice. The repetitive nature of the task has thankfully been taken over by a variety of programmable heating blocks from a multitude of suppliers. However, many of these were developed before the PCR patents were issued and it is therefore worth enquiring whether the supplier of a chosen design has a licence for use of the machines with PCR.

Designs vary in the mode of heating and cooling the samples. The major points for the potential user are:

(1) The rate at which samples are heated and cooled. The current standard appears to be a rate of at least 1 °C per second. The length of time taken for a given PCR can be significantly longer if the samples do not reach the desired temperatures rapidly.
(2) The uniformity of temperature across a heating block (Hoelzel, 1990; Van Leuven, 1991). The simplest way to test for this is to run a machine full of identical reactions to reveal the reproducibility across the heating block.

Other factors to take into account are the number of samples that can be handled, whether a few for cloning or many for screening, ease of programming a PCR machine, serviceability, flexibility and the type of work to be done. A beneficial feature is the ability to actually measure the temperature changes in a special sample mimic tube in the block and to output the data to a data recorder. This will give a more reliable indication of the actual temperatures and durations of the various PCR steps during a PCR run and allow comparisons between different machines.

2.5   A basic PCR protocol

The example given is a protocol for screening 18 different cDNA preparations for the presence of a known sequence of 1 kbp length. The PCR is with *Taq* DNA polymerase and two 20-mer primers.

(1) Prepare 1.2 ml master mix containing the following:

   120 μl of 10× Taq buffer
   120 μl of 2 mM dNTP mixture
   6 μg of primer 1
   6 μg of primer 2
   942 μl of water

(2) Mix well, and then add 15 units of *Taq* DNA polymerase at 2.5 units per μl (6 μl) and mix gently.
(3) Dispense 20 × 50 μl aliquots into PCR vessels (either 0.5 ml thin-walled Eppendorf tubes or vessels recommended by the manufacturer of the PCR machine to be used).
(4) Overlay each aliquot with 50 μl of mineral oil or liquid paraffin.
(5) To 18 tubes add approximately 1 ng of cDNA in 1 μl.
(6) To tube 19 add 1 μl of a known positive control if available.
(7) To tube 20 add 1 μl of water (negative control).

(8)   Place tubes in PCR machine and run the following program:

94 °C   30 s
50 °C   1 min
72 °C   1 min
Cycles   30

Note that times given are for when the sample has reached the desired temperature. Time must be allowed for heating and cooling with some PCR machine programs.

(9)   At the end of the run, visualise samples by electrophoresis of a 5–10 μl aliquot on an agarose gel and staining with ethidium bromide (Sambrook et al., 1989). Fig. 2 shows a typical PCR gel.

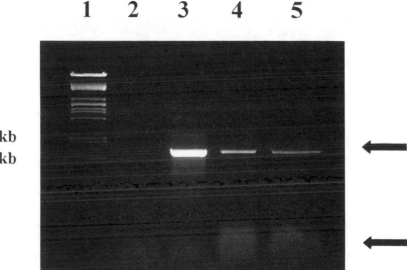

Figure 2. An example of PCR reactions visualised on an agarose gel with ethidium bromide. Lane 1 is a lambda *Pst*I digest used as a molecular weight marker. Lane 2 contains a negative PCR control where no template was included in an otherwise complete PCR. Lane 3 shows a positive PCR control where a PCR was spiked with 1 ng of plasmid known to contain the target sequence of 850-bp length (indicated by top arrow). Lane 4 shows a positive result from a test sample in which the product has been amplified specifically. Lane 5 shows a positive result from a test sample where additional, lower molecular weight products also appear on the gel. This could be due to the presence of truncated template or to PCR conditions that permit some less specific priming at other sequences. Nested or hot start PCR could be tried to resolve this question. The lower arrow indicates the presence of low molecular weight products in all PCR lanes. These occur in most PCRs to a greater or lesser degree and are referred to as 'primer dimers', although they do not only contain the products of self-priming events from the primers. 'Primer dimers' need to be removed from PCR reactions prior to cloning, by spun column chromatography, for example

2.6   Adapting the basic protocol

To prevent cross-contamination of PCRs, always add the template last. The one exception to this is where 'hot start' PCR (D'Aquila et al., 1991) is recommended.

In hot start PCR, the reaction mix with the template but minus the polymerase is heated to denaturing temperature and the polymerase is then added prior to annealing. This prevents the generation of non-specific PCR products from primers which anneal to the template and extend during the lower temperatures experienced by the reaction in the very first heating step (Fig. 2, lane 5).

The polymerase can be added manually, whilst the samples are held at a high temperature. Alternatively, the polymerase can be separated by a wax layer which melts above the annealing temperature, allowing enzyme and sample to mix (Bassam & Caetano-Anolles, 1993; Wainwright & Seifert, 1993). Perkin Elmer supplies 'Ampliwax PCR gems' for this purpose. A more sophisticated alternative is to embed the polymerase in trehalose inside a wax bead. When heated, the bead melts, releasing the polymerase (Kaijalainen et al., 1993).

The amount of template added depends upon its nature. For total cDNA preparations, 1 ng is ample. Picogram quantities of pure plasmids give strong products. With genomic DNA, up to 100 ng can be added to a reaction. Nested PCR with a second set of specific primers is generally recommended for genomic DNA (see section 4.1).

For plasmid and genomic DNA templates, the yield of product is improved by extending the denaturation time to 90 s for the first three cycles.

Whilst the annealing temperature for a given primer combination can be calculated (see section 3.2), it is worth trying a range of temperatures in optimisation experiments if enough template is available. Hamill et al. (1991) conducted a trial to compare theoretical and empirically determined annealing temperatures and found differences of varying degrees between the values, depending on the primers used. The annealing temperature should be as high as possible. If the annealing temperature is 68–72 °C, then the annealing and extension temperatures can be combined to reduce the overall run time.

A rough guide to the extension time is to allow 1 min for every kilobase of sequence. It is also worth increasing the extension time in later cycles as the yield of product increases and the amount of enzyme available becomes limiting.

3   PREPARATION FOR PCR

3.1   Generation of template

There are a variety of templates that can be used for PCR. This section will deal with genomic plant DNA, cDNA, mRNA, plasmid and phage libraries, colonies and templates immobilised to beads.

The best plant genomic DNA for PCR is the very pure material obtained by methods such as caesium chloride gradient purification exemplified in Chapter 3. The thermostable polymerases will tolerate some level of impurities, permitting the use of less labour-intensive protocols that are also more amenable to processing larger numbers of samples. Relatively pure DNA can be obtained by the cetyl triethylammonium bromide (CTAB)-based method of Dellaporta et al. (1983) and the method described by Gawel & Jarret (1991). Simpler methods use ethanol precipitation of DNA from crude extracts (Lassner et al., 1989; McGarvey & Kaper, 1991) or proteinase K digestion of ground plant material (Brunel, 1992; Guidet, 1994; Fangan et al., 1994). The tissue is ground to a powder either under liquid nitrogen in a pestle and mortar or (and this is for the adventurous) by vortexing with ballbearings (Colosi & Schaal, 1993), or ground straight into the extraction buffer.

Even cruder preparations can be made by grinding the plant tissue in 0.5 M sodium hydroxide solution, neutralising and diluting 100–1000-fold with Tris-HCl, pH 8. One microlitre of this preparation can then be used for PCR (Wang et al., 1993). Alternatively, a method has been published where a tissue squash on nylon membrane provides PCR-able template (Langridge et al., 1991). The simplest protocol of all is to utilise potassium ethyl xanthogenate to liberate DNA from leaves without the need for tissue homogenisation (Williams & Ronald, 1994).

Young or tissue-cultured leaf material provides an ideal source of genomic DNA for PCR, whilst tissues rich in oils or polyphenols can be less tractable. The method of Gawel & Jarret (1991) has proven successful with a wide variety of tree leaves, from oak to eucalyptus at ATC (M. Maunders, personal communication).

cDNA is a particularly useful PCR template for isolating expressed genes or parts of coding sequences without introns. Since only single-stranded template is required for priming by at least one oligonucleotide, the cDNA does not even have to be double-stranded. The cDNA can be synthesised by conventional methods from purified mRNA (Sambrook et al., 1989) and then used for PCR (Frohman et al., 1988; Sheardown, 1992). However, total RNA itself can be used for the initial cDNA synthesis, as the PCR amplifies any low levels of template generated (Carothers et al., 1989; Doherty, et al., 1989) and companies such as Pharmacia provide special kits for first-strand synthesis specifically for PCR (see also cDNA synthesis on magnetic beads below). Several methods have been published for performing cDNA synthesis and PCR in the same tube, simplifying the process (Goblet et al., 1989; Myers & Gelfand, 1991), although avian myeloblastosis virus (AMV) reverse transcriptase has been shown to inhibit *Taq* polymerase (Sellner et al., 1992). The problem can be circumvented by using r*Tth* polymerase, which has reverse transcriptase activity (Myers & Gelfand, 1991).

Companies such as Stratagene provide genomic and cDNA libraries in plasmids or phage for a variety of plant species, and aliquots of these can be used directly as templates for PCR to isolate a specific gene if some sequence information is available (Friedman et al., 1988). Alternatively, PCR is used as a screen of subsections of a library to facilitate isolation of a longer clone without going through successive lengthy hybridisation screens (Bloem & Yu, 1990; Kwiatkowski et al., 1990; Isola et al., 1991).

Bacterial colonies and phage plaques can be utilised directly in PCR reactions if very small amounts are added to the mixture so as not to poison the polymerase (Güssow & Clackson, 1989; Runnebaum et al., 1991). Genes can be identified with gene-specific primers or insert sizes determined by using vector-specific primers.

PCR can also be used to amplify DNA from bacteria immobilised onto a substrate such as nitrocellulose (Maskell et al., 1993). Maskell et al. (1993) needed to avoid the regrowth of immunoblotted bacterial colonies and yet detect the regulatory DNA sequences that they were investigating. The positive bacterial colonies were excised with a scalpel from the nitrocellulose filter and the DNA eluted by boiling in a small volume of water. The supernatant was then used to prime PCR reactions with primers flanking the regulatory sequences, and the resultant product was used for sequencing.

The introduction of nucleic acids immobilised to paramagnetic beads has opened another source of useful templates for PCR. Several companies (Dynal and Promega, for example) provide kits for purifying mRNAs directly from ground tissue using oligo(dT) bound to beads. The oligo(dT) can be used to prime the synthesis of cDNA on the beads. This single-stranded template is used to generate a complementary strand with suitable priming sites incorporated at the termini, by polyadenylation for example. The complementary strand is eluted and used for PCR (Lambert & Williamson, 1993; Raineri et al., 1991). Since the first strand is immobilised, it can be used repeatedly to generate fresh templates for PCR. Large pieces of DNA immobilised to filters (Lovett et al., 1991) or magnetic beads (Korn et al., 1992) can also be used to hybridise their respective cDNAs, which are eluted and amplified for cloning and analysis. It is possible to utilise immobilised template for solid-phase gene assembly; the small quantities of product are then amplified by PCR (Dombrowski & Wright, 1992). Chapter 14 deals with subtraction techniques in more detail.

3.2 Primer design

Oligonucleotide primers are on made on DNA synthesisers, which can be purchased from Applied Biosystems or Pharmacia, for example. A suitable alternative is to have oligonucleotide primers made to order by such companies as Promega or Pharmacia. Where possible, purified full-length oligonucleotides should be obtained for longer (50 bp plus) primers, as the proportion of full-length primers gradually decreases from 94% for a 20-mer to less than 50% for a 200-mer, even when the synthesis efficiencies are 99.7% per base.

The design of the best primers for the desired PCR reaction is perhaps the most crucial step for guaranteeing a successful outcome. Bad primers lead to inconclusive results, and too many or too few PCR products, and can waste valuable time. It is therefore worth making the effort to design suitable primers. This section will deal with fundamental features of primers, such as sequence specificity and stability.

The most fundamental feature of a PCR primer is its specificity, which is governed by the oligonucleotide sequence, primer length and internal stability. Most workers recommend a 50–60% G + C ratio, as well as the avoidance of GC-rich and repetitive sequences to reduce the risk of non-specific priming (Innis & Gelfand, 1990; Saiki, 1990). In most cases the primer sequence needs to be unique to the fragment to be amplified, e.g. to clone a specific gene. Primers that are deliberately less specific are used for methods such as differential display and RAPDS, described in section 4.3, where a range of products is required.

Sequence specificity increases with length. A 16-mer primer should theoretically have only one target copy in a genomic template of 10^9 bp (Sambrook et al., 1989). However, most specific primers start at about 18 bases and generally do not need to exceed 30 bases. Above 20 bases, primers will tolerate single-base mismatches. Binding of a primer to the template is influenced by an initial binding via a short run of bases within the oligonucleotide, with the rest of the oligonucleotide then snapping into place (Lathe, 1985). Primers can be designed so that annealing is promoted by stable sequences at the 5' end with more emphasis on specificity residing in the 3' end of the oligonucleotide (Rychlik, 1993).

Particular attention is given to the 3' terminal base, as studies on *Taq* and other polymerases have shown that extension can occur to varying degrees with primers that have a single base pair mismatch at the terminal base (Mendelman et al., 1989; Kwok et al., 1990). In fact, mismatch using dTTP at the 3' end results in PCR products almost as efficiently as if there were no mismatch. Mendelman et al. (1989) demonstrated that primers ending in either CC, AG or GA are 100 times less likely to result in extension at mismatched termini, and these should be used where possible.

Both the length and the sequence of a primer also affect another important feature, the annealing temperature (T_a) at which the primer binds optimally and specifically to the template. Ideally, the T_a should be as high as possible. This ensures fewer chances of random binding to other sequences and also reduces cycling times during PCR by minimising the cooling and heating times between temperatures. The annealing temperature is generally chosen to be 15 °C below the melting temperature (T_m) of the primer–template duplex (Sambrook et al., 1989). The simplest method for approximating the T_m of a primer is to ascribe a temperature of 2 °C for every A or T in the sequence and 4 °C for every G or C (Sambrook et al., 1989). A 20-mer with a 50% G + C content would have a T_m of 60 °C. It is then obvious that increasing primer length and/or having a higher G + C content increases the T_m. Certain sequences are more stable than others and this can also be taken into consideration (Rychlik, 1993). More sophisticated methods for determining the T_m and hence the T_a are available (Sambrook et al., 1989; Rychlik, 1993; Rychlik et al., 1990) and have been incorporated into computer programs to aid primer design (e.g. Oligo^{TM}, Amplify and Gene Jockey).

Primer sequences should be checked for possible self-complementarity, formation of hairpin loops and ability to hybridise preferentially with each other (Saiki, 1990) rather than the template. Such unwanted interactions occur to a greater or lesser degree in most PCRs and are generally referred to as primer-dimers (Fig. 2). Primers are short and abundant relative to the template, and any artefacts can rapidly take over a PCR and reduce the yield of the desired product (Mullis, 1991).

3.3 Degenerate primers

There are many instances where similar but not identical sequences need to be amplified; examples are the cloning of different members of gene families, cloning of identical genes from different species, and cloning of genes whose sequences are deduced with some ambiguity from amino acid sequence data. Chapter 5 discusses storage protein gene families, for example. The primers required for PCR of such products have to be designed to guarantee a high specificity for the desired product and yet allow for expected base differences.

The following considerations for designing degenerate primers are based on the method for designing degenerate oligonucleotides for hybridisation in Sambrook et al. (1989) and on primer design in McPherson et al. (1991b).

The number of variable bases required in a primer should be kept to a minimum. Where sequence information from a variety of sources is already available, this can be used to find conserved regions from which to design the primers. Teichmann et al. (1994), for example, designed primers to homologous 5' and 3' ends of tRNA^{ser} genes to isolate and clone eight different tRNA^{ser} genes from *Nicotiana rustica*. Tang et al. (1993) cloned and sequenced several cDNAs of the ACC oxidase gene family in *Petunia hybrida* using conserved primers.

Where such information is not available and the sequence of interest is part of a coding region, the amino acid sequence should be determined to aid primer design. Sequences rich in the amino acids serine, arginine or leucine are best avoided, because these are encoded by a wide range of different codons. The amino acid sequence is then back-translated using a codon table, and the variable bases highlighted. Codon usage does vary between different organisms and this can be exploited to further reduce redundancy. Murray et al. (1989) have published codon usage tables for monocots and dicots which, for example, demonstrate that AAT can be omitted as a codon for asparagine in monocots and that plants generally do not use the codon ATA for isoleucine. Alternatively, species-specific codon tables can be prepared from published sequences of other genes from the organism of interest.

Mixtures of primers can be synthesised encompassing all potential variants of the anticipated sequence, as only those that can bind specifically to the template will result in a visible PCR product. This approach has been applied successfully in plants with primers ranging from 2-fold to 1152-fold degeneracy (e.g. Bucciaglia & Smith, 1994; Lindstrom et al. 1993; Muench & Good, 1994;

Thangstad et al., 1993; Villand et al., 1993). However, the affinity of the primer to the template can be enhanced by using two further strategies. The first is based on the work of Uhlenbeck et al. (1971), who observed that uracil can base pair with both adenosine and guanosine in RNA. The principle is transferred to DNA. Any codon wobble which could be an A or G is encoded as G in the primer, and any wobble that could be C or T is coded as T. The second strategy is to use deoxyinosine at redundant bases, as deoxyinosine will pair with any base. Any sequence divergence in the 3' terminal four or five bases of the primer is retained to ensure maximum specificity (Sommer & Tautz, 1989).

3.4   Add-on features

The advantage of using PCR is not only the ease with which a sequence can be isolated relative to other methods but also that the termini of the PCR product can be modified to aid further cloning and detection (Fig. 3).

Most oligonucleotide primers are not phosphorylated at the 5' ends, in contrast to natural DNA fragments. Therefore, the simplest modification which aids blunt-end or TA cloning (section 4.8) is to synthesise primers with a 5' phosphate group. Adding the phosphate at the end of the oligonucleotide synthesis step is far more efficient than trying to phosphorylate primers or PCR products at a later stage.

Priming regions can be chosen to cover useful restriction sites or single-base changes can be incorporated to introduce a particular restriction site (section 4.5). Where no convenient sites are available, the sequence for a desired restriction site can be tagged onto the 5' end of the priming sequence in an oligonucleotide (Scharf et al., 1986; Jung et al., 1990), as shown in Fig. 3. Toguri et al. (1993), for example, included EcoRI sites in their primers to facilitate cloning of chalcone synthase from Petunia hybrida, whilst Firek et al. (1993) incorporated NotI, KpnI, HindIII, XbaI and BamHI sites into their primers to clone Nicotiana tabacum pathogenesis-related protein PR1 from genomic DNA. Many restriction enzymes do not cut sites that are at the 5' end of a PCR product but will cut when a few extra bases are added to extend the 5' end (Kaufman & Evans, 1990). Therefore, when including restriction sites to primers, it is advisable to add an additional four bases at the 5' end to ensure that restriction will occur. The New England Biolabs Catalogue for 1995 includes a table of restriction efficiencies for 35 restriction enzymes with various oligonucleotides.

The addition of extra sequences to the 5' end of primers can be taken even further to include promoters such as the T7 promoter with untranslated leader sequences (Browning, 1989; Kain et al., 1991). This allows in vitro transcription to be conducted directly from the PCR product. Alternatively, translational control sequences and restriction sites can be added so that a complete expression cassette is generated by PCR, ready for cloning into a bacterial overexpression vector (MacFerrin et al., 1990).

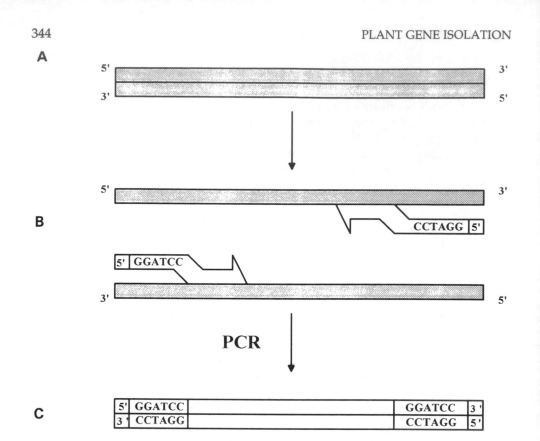

Figure 3. Incorporation of additional sequences into a PCR product. Additional sequences such as restriction sites or transcription initiation sites as well as phosphate groups or other ligands can be incorporated at the termini of sequences by PCR. The example in this figure shows the addition of *Bam*HI sites to facilitate cloning. (a) A DNA template where restriction sites are to be added at the termini of a desired sequence. (b) Primers designed to the termini of the desired sequence which are non-homologous to the template at their 5' ends due to the addition of a *Bam*HI restriction site and a few extra 5' bases. (c) The product of a PCR with the primers containing the desired sequence and *Bam*HI restriction sites at both ends

Primers can also be modified to aid their detection or to capture PCR products. Fluorescent groups are now routinely being added to primers to aid detection (Mayrand et al., 1992) and permit the detection of products from different PCRs in single reactions or gel lanes. They are also used extensively in large sequencing programmes (section 4.7). Biotin groups can also be used, both for detection using non-isotopic methods and for capture of specific PCR products or strands (see also Chapter 14). Primers can be made with non-nucleosidic reagents incorporated in the sequence. The polymerase is unable to extend the sequence at the point of inclusion, resulting in PCR products with single-stranded tails. These tails can be used for capture or detection purposes (Newton et al., 1993).

4 PCR APPLICATIONS

4.1 PCR of known sequences with two specific primers and reverse transcriptase PCR

There are many instances where it is desirable or just easier to use PCR to amplify a known coding sequence or promoter for characterisation, cloning or generating constructs. In such cases, the procedure is quite straightforward.

First, obtain the sequence on paper or as a data file. The sequence may originate from in-house sequencing work, from published papers or from a DNA databank such as EMBL or Genbank. The area to be amplified is located on the sequence. Primers are then designed to the ends of the chosen sequence, taking into account the considerations described in section 3.2, and any extra sequences, restriction sites or groups incorporated as required for downstream applications. Where possible, also design a second set of primers whose 3' termini extend beyond those of the first pair (see below). Calculate the annealing temperatures (T_a) of the sequences homologous to the template and the T_as of the full primers if different. Purchase or synthesise the primers.

Next, prepare the template for the PCR, which could be a bacterial clone or plasmid, a DNA library, plant genomic DNA or cDNA (section 3.1).

Set up the PCR as described in section 2.5 but with the following modifications to the PCR program. The first three cycles are conducted using an extended denaturation time of 1.5 min, and the T_a chosen is that for the sequence homologous to the template. This ensures that the initial template is denatured sufficiently and that the primers will bind. The extension time should be approximately 1 min per kilobase (times given refer to the time the sample actually spends at the desired temperature). The next 27 cycles are conducted with the denaturation time of 30 s, and the T_a calculated for the full-length primers if different to the previous value.

Upon completion of the PCR, electrophorese 10 μl of the samples on an agarose gel and visualise the bands using either ethidium bromide or fluorescence if suitable groups have been incorporated into the primers (Fig. 2). In most cases, with well-chosen primers and sequence, the product should be visible as a single band, with some 'primer-dimers' apparent at the electrophoresis front. 'Primer-dimers' refers to small non-specific PCR products which are seen with most reactions (Fig. 2). They can arise by interaction of the primers with themselves or one another and by fortuitous priming events on the template. Where too few or too many products are visible, the PCR should be repeated with different annealing temperatures until the optimum settings have been identified, or 'hot start' PCR should be tried (section 2.6).

With genomic DNA as a template in particular, the desired band may not be apparent after the first PCR, or may be one of several products even after optimisation. Nested PCR (Jackson et al., 1991) can be applied successfully in such cases (Fig. 4). A PCR is prepared using a second set of primers which code for sequence beyond the 3' ends of the first set. One microlitre of the primary PCR is used as a template for the secondary PCR and amplified for

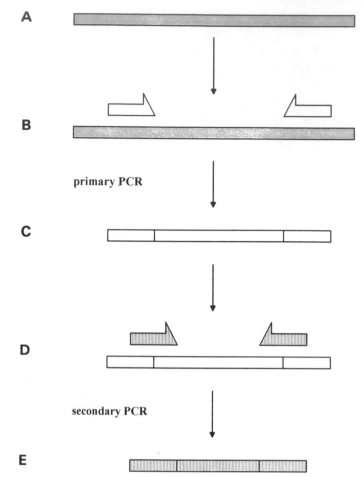

Figure 4. Nested PCR. Nested PCR is used where there is either no detectable product from an initial PCR or where several products have been generated when only one was desired. The products of a PCR are taken through a second PCR with a new set of primers, which provides additional specificity. (A) A template DNA containing a sequence of interest. (B) Primers are designed to the borders of the region of the DNA sequence desired and used in a PCR reaction. (C) The product of the PCR reaction terminated at either end by the primer sequences. (D) A PCR is performed on the products of the first reaction using a second set of primers, designed to prime internally to the first set. This is the nested PCR. (E) The PCR product of the second reaction

15–17 cycles. The use of a second set of primers, even if they extend as little as three bases beyond the first pair, introduces an additional level of sequence specificity. Ten microlitres of the secondary PCR are then visualised on an agarose gel.

Purify the PCR products away from the primers, primer-dimers and Taq buffer either by gel purification, spun column chromatography (Sambrook et al., 1989) or any other preferred method before using for downstream applications. Pirovano et al. (1994) provide a recent example for the isolation and analysis of an *Opaque-2*-related gene from sorghum. Using information from the sorghum genomic gene sequence that had been compared to the maize *Opaque-2* gene, primers were designed that would specifically amplify the sorghum gene from cDNA. The priming regions were made to different exon regions of the gene, thus allowing differentiation between true cDNA products and any possible products from genomic DNA contamination. Poly(A) RNA was isolated from endosperm tissue and 1 μg was reverse transcribed with Maloney murine leukaemia virus (M-MLV) reverse transcriptase in a total reaction volume of 20 μl. Five microlitres were used in 50-μl PCR reactions using *Taq* polymerase and amplified through 30 cycles of 90 °C for 30 s, 55 °C for 30 s, and 72 °C for 120 s. PCR products were visualised on an agarose gel, electro-eluted and subcloned into pGEM vectors for sequence analysis. Li et al. (1994) used PCR to isolate promoters and untranslated 3' regions of the small auxin up RNAs (SAURs) and the rubisco small-subunit gene of soybean. Fragments were fused to either *GUS* reporter gene or SAUR coding sequence in a variety of chimeric constructs designed to study their induction and stability in transgenic plants.

4.2 PCR with limited sequence information, inverse PCR, anchored PCR and RACE

In many cases, there is only limited sequence information available for a gene or DNA fragment to be isolated. Several innovative approaches have been devised to use PCR to amplify fragments containing uncharacterised DNA beyond a known sequence.

In inverse PCR (Ochman et al., 1988; Triglia et al., 1988), two primers are designed to the available known sequence which face *away* from each other (Fig. 5). The template is cut with a restriction enzyme that has a site close to one primer and another site at a distance from the other primer in the region to be characterised. The restriction fragments are circularised and ligated. The primers now face each other around the circular molecule, which is used as a template for PCR of the intervening unknown sequence with the two primers. The method has been used, for example, to obtain additional cDNA sequences (Huang et al., 1990), to construct hopping and linking libraries (Kandpal et al., 1990), and to isolate end probes from cosmids (Byth et al., 1994) and from T-DNA inserts (see Chapter 11 for protocol).

Anchored PCR covers a range of methods in which known sequence is added to one end of the template so that a universal primer can be used with a sequence-specific primer to amplify the intervening undefined DNA (Fig. 6). One such method is RACE (rapid isolation of cDNA ends), which is used where some central sequence information is available for a cDNA and the user wishes to obtain either the 3' or 5' terminal sequence (Belyavsky et al., 1989; Frohman et al., 1988; Struck & Collins, 1994). For 3' cDNA sequence, a specific primer is designed and used against a universal cDNA primer comprised of an oligo(dT) (17–30) sequence with additional 5' bases, generally restriction sites,

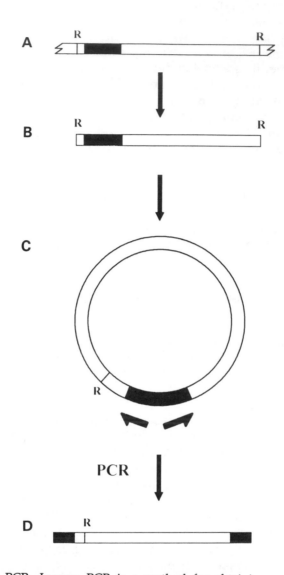

Figure 5. Inverse PCR. Inverse PCR is a method for obtaining amplified sequences outside a known DNA sequence. (A) A region of DNA containing known sequence is marked in black, and bordered by unknown sequence in white. 'R' marks restriction sites that have been identified from digests where one site is either within or at least close to the known sequence. (B) The DNA is digested with the appropriate restriction enzyme. (C) The fragments are self-ligated to form circular molecules. PCR is then performed using primers that face outwards from the known sequence into the un-characterised sequence. As the template molecules are circular, the region between the primers, consisting of uncharacterised sequence, can be amplified. (D) The product of the PCR showing the priming regions at the termini, the internal uncharacterised sequence and the restriction site

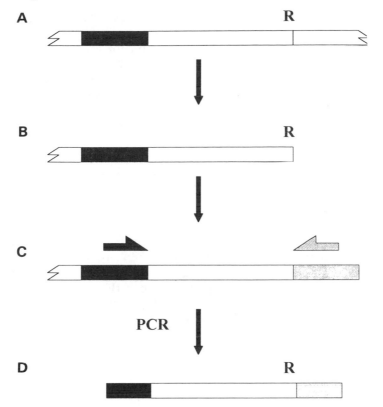

Figure 6. PCR with limited sequence information. Where inverse PCR is not possible or has been unsuccessful, uncharacterised sequence can be amplified from beyond a known sequence by adding a defined sequence, such as an oligonucleotide or vector, at a restriction site some distance from the known area. (A) A DNA template comprising a known sequence in black, uncharacterised sequence in white and a restriction site 'R'. (B) The template is digested with the appropriate enzyme. (C) A known oligonucleotide or vector sequence is ligated to the restriction fragment. PCR is performed using a primer specific to the known sequence and another primer to the added defined sequence. (D) The final PCR product with known sequences at the termini and the desired uncharacterised sequence internally

for increased stability. To obtain the 5' sequence of a cDNA, the product of first-strand cDNA synthesis is tailed with a string of dATPs or dGTPs using terminal transferase. A universal primer to this sequence is then used versus the specific cDNA primer in PCR. Companies such as Clonetech and Pharmacia provide kits specifically for RACE.

Sequences recognised by universal primers can also be generated by ligating linkers onto the ends of cDNAs or genomic DNA fragments (Loh et al., 1989; Riley et al., 1990). Alternatively, the DNA can be ligated into vectors so that a vector-specific primer close to the cloning site is used against the sequence-specific primer (Hovens & Wilks, 1989; Rasmussen et al., 1989; Rosenberg et al., 1991). This method is particularly useful in laboratories where libraries are being or have been generated, as the ligation mix prepared can be used as template.

Screaton et al. (1993) describe a rapid method for achieving short chromosomal walks in yeast artificial chromosomes (Chapter 10). One sequence-specific primer is used in a PCR under conditions that favour mispriming. Amongst the products generated, some will contain the true priming site at one end and a false priming site at the other. A second, specific primer downstream of the first is then used for sequencing into the region of unknown sequence.

All these methods are liable to generate additional PCR products to those desired, a problem that can be circumvented by nested PCR.

4.3   PCR of unknown sequences, RAPDS and differential display

The emphasis so far has been on maximising specificity to ensure amplification of one desired product from known sequence. In order to amplify unknown sequences, PCR conditions are actually altered to ensure that sequence specificity is reduced and multiple priming is promoted (Fig. 7). Typical differences from specific PCRs are the use of short primers, high magnesium chloride conditions and low annealing temperatures. The resultant products give multiple banding patterns on agarose or acrylamide gels. Several methods have been devised which employ what is in effect a random sampling of a large template to find polymorphisms between DNA from different sources. Such information can be used in genome mapping projects or to find differentially expressed genes from different tissues or stages of development.

The use of randomly amplified polymorphic DNA sequences (RAPDS) (Welsh & McClelland, 1990, 1991; Williams et al., 1990) has been a major success in the search for polymorphic markers for plant genome fingerprinting. RAPDS have been used for analyses from *Arabidopsis* (Reiter et al., 1992) to Red Pine (Mosseler et al., 1992). DNA is obtained from a number of different samples with one of the semipure template preparation methods outlined earlier and used in PCR with a single arbitrary primer, generally 10 bases long. PCR is conducted in high magnesium chloride buffer, generally 2.5 μM magnesium chloride, using a low annealing temperature of 35 °C, for 45 cycles or more. The products are visualised on agarose gels with either ethidium bromide or fluorescence technology. By the use of a range of different primers, a portfolio of up to 100 polymorphisms can be amassed in a relatively short time. Companies such as Genosys (Cambridge, UK) and Operon (California) provide kits of several thousand primers specifically for this purpose.

RAPDS has been adapted for use on cDNA from different tissues or stages of development, a method known as differential display (Liang & Pardee, 1992; Liang et al., 1993) (see also Chapter 14). The method is more technically demanding, as messenger RNA must be isolated and cDNA obtained to act as the template for the PCR. However, with suitable replication and controls, differential display promises to be a powerful technique in plant molecular biology. Bauer et al. (1993) describe modifications in procedures and mathe-

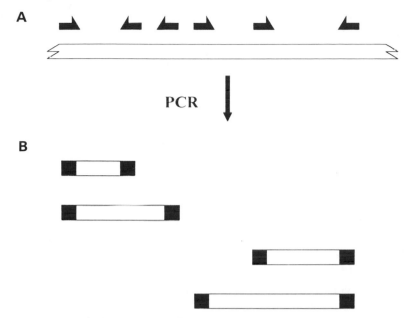

Figure 7. Amplification in the absence of specific information: RAPDS and differential display. (A) One or two primers are used in PCRs where non-specific priming is promoted, e.g. by reduced primer length, low annealing temperatures and high magnesium chloride concentrations. Some of the priming regions will be close enough and correctly oriented to permit amplification of the intervening DNA. In RAPDS, the template is genomic DNA. In differential display, the template is cDNA and one of the primers is anchored in the oligo(dT) sequence. (B) The products resulting from the PCR

matical considerations for choosing the number of primers for differential display to ensure that every messenger RNA in a population can be sampled. The PCR reactions are performed using an oligo(dT) primer versus a random primer. Radiolabel is included in the reaction and the products are visualised on sequencing gels. Polymorphic bands that are identified can be isolated by excision from the gel, elution and PCR, and the product used either for screening or for cloning.

4.4   cDNA libraries using PCR

The methods described for obtaining cDNA termini of specific cDNAs, that is RACE and ligation of universal primers to the cDNA, can also be used to amplify cDNA populations. cDNA is synthesised using an oligo(dT) primer with additional bases added at the 5' end. The universal priming region is generated at the 3' end of the synthesised cDNA and the population is amplified using the universal and oligo(dT) primer. This makes it possible to amplify sufficient cDNA for cloning from small amounts of tissue (Belyavsky et al., 1989; Dresselhaus et al., 1994; Gurr & McPherson, 1991). By judicious inclusion of restriction sites, the library can, for example, be directionally cloned into expression vectors. Dresselhaus et al. (1994) used this method to generate a cDNA library from the egg cells of maize. The major problems with this method are the preferential amplification of small fragments and the possibility that the resultant cDNA population is not representative. The former can be overcome by including size-fractionation steps at intervals in the PCR. The method can also be used to generate substantial amounts of cDNA, either for subtractive hybridisation or for screening (Herfort & Garber, 1991). Chapter 14 deals with making subtraction libraries.

4.5 PCR-based mutagenesis

The primers used in PCR can be exploited to create mutations. The simplest form concerns the introduction of single-base changes in the 5' end of one of the primers. The PCR product generated will contain the mutant sequence and can be cloned by any of the methods described below. More usually, single-base changes need to be introduced within a sequence rather than at the termini. Several solutions can be applied.

One method is to design two complementary primers with the required base change. Each primer is then used in PCR with its respective primer for either the 3' or 5' terminus of the sequence. The two separate PCR products, which have a common sequence in the region to be mutated, are purified and combined to act as a template for a second PCR with the 5' and 3' primers for the sequence required (Higuchi et al., 1988). The resultant full-length mutant molecule can then be cloned. A simplified version of the protocol with three primers has been developed (Landt et al., 1990; Barettino et al., 1993). The mutant primer is used with a terminal primer to amplify a mutant fragment. This fragment is purified and then used in a PCR as a long primer against the primer for the other terminus of the sequence. Again, the full-length product can then be cloned.

Frequently, the site to be mutated is in sequence that is already in a vector. In such instances, PCR-based unique site elimination (USE) can be applied (Ray & Nickoloff, 1992). This requires a primer with the required mutation and a second mutant primer which will remove a unique restriction site in the vector. The two primers are used for PCR on the vector with the target DNA, and the product is purified. The PCR product, with its two mutations, is annealed to denatured vector with the target sequence and the DNA is extended with a DNA polymerase. The result is a mixture of original rehybridised vector plus target DNA and hybrid vectors, where one strand contains the two mutations. Restriction with the unique restriction enzyme will linearise the original vector but not the hybrids. The digest is transformed into an E. coli *mutS* strain which is unable to repair mismatches in the hybrid vector. The plasmid DNA from the transformed population is isolated, restricted to completion with the targeted restriction site and used to transform a desired E. coli host. This step removes the non-mutated vector plus target which arises during hybrid plasmid replication in the first E. coli host. The resultant colonies should predominantly contain the mutated plasmid.

4.6 Deletions and insertions by PCR

Sequences can be deleted and inserted using chimeric primers which mark the boundaries of the mutation site. The simplest case is where the deletion is to be close (within 50 bp) to one end of the sequence. A mutant primer is designed which contains the terminal sequence in front of the desired deletion at its 5′ end and the sequence downstream of the deletion at its 3′ end. The primer is used for PCR on the template to be mutated with a suitable primer to the other end of the sequence. The PCR product contains the deletion and can be cloned as required.

Where the deletion is required internal to a linear sequence, the two parts of the DNA either side of the deletion are amplified. By ensuring that suitable restriction sites or regions of homology to each other are included at the 5′ ends of the primers either side of the deletion site, the two fragments can be combined by ligation or fusion PCR.

Deletion can be performed easily on circular sequences such as those in plasmids. Inverse PCR primers are designed at either side of the deletion target. The entire plasmid bar the deletion target is amplified and the linear product circularised to produce the deletion construct (Imai et al., 1991).

Insertional mutagenesis with a few bases can again be achieved by including the insertion in suitable primers. In order to insert larger fragments, the sequence to be inserted can be amplified with suitable primers and cloned by ligation if suitable restriction sites are available. Sticky feet PCR is an elegant method developed by Clackson & Winter (1989) which can be used to clone an insert into a circular single-stranded vector. The fragment to be inserted is amplified with phosphorylated primers that have homologous regions to the insertion site in a single-stranded DNA vector such as M13 at their 5′ ends. The fragments are used to prime second-strand synthesis on the single-stranded vector which has been grown in an *E. coli* strain (*dut ung*) which incorporates uracil instead of thymine. The second-strand circle is closed by ligation and the hybrid molecules are transformed into a *dut⁺ung⁺* host. The host degrades the uracil-containing template and only the desired mutant replicates.

4.7 Sequencing and characterisation of PCR products

The product of a PCR reaction is a population of molecules that may not all be identical even if only one band is visible on an agarose gel. One potential source of variation is the template, which may contain several different forms of the same sequence as, for instance, in a gene family. The other source of variation, particularly with *Taq* polymerase which lacks proofreading ability, is mutation occurring during the actual PCR (Tindall & Kunkel, 1988; Eckert & Kunkel, 1991). With *Taq* polymerases, an error rate of about 1 base in every 10 000, for 20 cycles of PCR, can lead to an accumulated error frequency of 1 in a 1000 bases, and for 50 cycles, to an error frequency of 1 in 400 (Eckert & Kunkel, 1991). It therefore makes sense to characterise the PCR product *prior* to cloning.

Digestion of a PCR product with a panel of restriction enzymes can reveal mixtures of closely related genes by the presence of additional restriction fragments to those expected either from the size of the product or previous sequence information. The restriction maps will also aid future cloning strategies.

Sequencing of a single PCR product has the advantage of providing a consensus sequence for the most abundant PCR fragments in the population. Any clone obtained later can then be compared to see if it reflects the expected sequence or is a minor variant. The main requirement for sequencing PCR products is that they are reasonably pure bands and that the primers have been removed, e.g. by spun column chromatography (Sambrook et al., 1989). The PCR products can be sequenced using standard double-stranded DNA-sequencing techniques; however, particularly with short products, rapid reannealing of the template can be a problem. There are a variety of methods for obtaining single-stranded template, from asymmetric PCR where one primer is present in limiting amounts (Gyllensten & Erlich, 1988), to using exonucleases to digest one strand of the PCR product preferentially (Higuchi & Ochman, 1989).

There are two methods to be particularly recommended for sequencing PCR products. The first uses magnetic beads (Green et al., 1990). The PCR product is generated, with one of the PCR primers being biotinylated at the 5' end. The resultant double-stranded DNA is then captured on streptavidin-coated magnetic beads (available from Promega or Dynal, for example). Denaturation of the bound DNA allows the elution of the non-biotinylated DNA strand. The two separated strands can then be sequenced independently using standard, commercially available sequencing protocols for single-stranded templates. Bhattacharya et al. (1993) used this method for sequencing amplified centrin cDNAs from green algae.

The second sequencing method utilises thermostable polymerase and a thermocycler (Connell et al., 1987; Gocayne et al., 1987). A single labelled sequencing primer is used on a small amount of the double-stranded DNA template and the sequencing reactions are repeated for approximately 20 cycles in a PCR machine with a thermostable polymerase. Sufficient single-stranded termination products are generated by this linear amplification for electrophoresis on a sequencing gel.

Both methods are particularly suited for large sequencing programs if used with fluorescent dye labels (Smith et al., 1986), as large numbers of samples can be sequenced and analysed on automated DNA sequencers. Chapter 15 deals with sequencing in more detail.

4.8   Cloning PCR products

PCR products are no more difficult to clone than other DNA fragments. There is the additional advantage that the PCR primers can be designed to accommodate an anticipated cloning strategy. Primers and unwanted primer-dimers should be removed prior to cloning by spun column chromatography (Sambrook et al., 1989). Where necessary, single PCR products can be separated out of a mixture of products by agarose gel electrophoresis or HPLC.

The simplest cloning method is blunt-end ligation into a suitable vector. Where PCR is conducted with *Taq* polymerase or other thermostable polymerases without 3'–5' exonuclease activity, the PCR products are not blunt-ended. This is due to an extendase activity of many enzymes which adds an additional 3' base to the termini of the DNA fragments (Clark, 1988). The ends of the DNA molecules can be polished with T4 or *Pfu* polymerase prior to cloning to dramatically improve the cloning efficiency (Costa & Weiner, 1994). PCR products made using enzymes with proofreading ability, such as *Pfu* polymerase, can be cloned directly by blunt-end ligation (Lohff & Cease, 1992). To improve the cloning efficiency further, ensure that the PCR primers are phosphorylated and clone into dephosphorylated vector.

Since *Taq* polymerase is still the most frequently used polymerase, the extendase activity described above can actually be exploited for cloning using TA vectors. *Taq* polymerase preferentially adds adenine to the 3' ends of PCR fragments and these can be ligated to vectors with complementary thymine overhangs (Kovalic et al., 1991). TA vectors are commercially available or can readily be constructed. At ATC we designed a linker adapted from Kovalic et al. (1991) to be cloned into an *Eco*RI/*Hind*III-digested vector. The following two phosphorylated complementary oligonucleotides were synthesised: 5'-AATTCCCATGGTTTTGTTGGCCATGTTATCCATGGA-3' and 5'-AGCTTCC-ATGGATAACATGGCCAACAAAACCATGGG-3'. The oligonucleotides were annealed and ligated into *Eco*RI- and *Hind*III-digested vector such that the vector still retained its blue–white selection. Linearised vector with single T overhangs is generated by digestion with *Xcm*I and is purified away from a small excised fragment of 15 bp by spun column chromatography. The vector is then ready for ligation to PCR products. Cloning of PCR fragments into TA vectors can be improved if the final incubation at 72 °C after 30 cycles of PCR is extended for 5 min or more to ensure that all fragments are adenylated at the 3' termini.

Dietmaier et al. (1993) describe a cloning strategy which similarly depends on generating compatible sticky termini in vector and PCR products but which can also be used with polymerases having proofreading ability and has the added advantage of permitting directional cloning. In their DISEC-TRISEC cloning method, the vector is digested with two restriction enzymes, e.g. *Eco*RI and *Hind*III, and partially filled in with Klenow in the presence of dATP. The vector then has a 2-bp and a 3-bp sticky end. The PCR primers are designed to contain two or four extra bases at their 5' ends. The PCR products are trimmed with the 3'–5' exonuclease activity of T4 DNA polymerase in the presence of dTTP to generate sticky termini complementary to the vector into which they are ligated. The method was used to clone cDNA fragments from *Chlamydomonas reinhardtii* and *Volvox carteri* (Dietmaier et al., 1993). The method is particularly attractive where suitable restriction sites are unavailable or cannot be used for cloning.

The more conventional method for generating sticky ends in PCR fragments to aid cloning is to include restriction sites at the 5′ ends of the primers (Scharf et al., 1986) with an additional four bases to ensure that digestion can occur (see Fig. 3 and section 3.4). The PCR fragment is digested with the appropriate restriction enzyme, the excised tails are removed by spun column chromato-graphy, and the fragment is ligated into suitably cut vector. As with blunt-end ligation, dephosphorylation of the vector or use of two non-complementary sites at either end of the fragment can enhance the cloning efficiency.

The above strategies should suffice for most cloning experiments; however, there are other cloning strategies, e.g. those that rely on vector homologies being introduced into the PCR primers. Sticky feet PCR (Clackson & Winter, 1989) has already been mentioned in section 4.6). PCR products can even be cloned in vivo in *E. coli* using homologous recombination to combine vector and PCR product (Oliner et al., 1993).

4.9   Screening libraries, colonies and plaques

PCR is an invaluable technique in the isolation and identification of positive clones from libraries or transformation experiments in bacteria. DNA purifica-tion is not required, as the PCR buffers and enzymes will withstand the addition of small quantities of impurities such as a minute aliquot of bacteria or phage. The DNA is released from the added organisms during the denatura-tion step of the PCR cycle, which should be extended to last about 90 s for the first three cycles. The released DNA then acts as the template for the PCR.

Screening of libraries has been extensively described (Bloem & Yu, 1990; Kwiatkowski et al., 1990; Isola et al., 1991; Israel, 1993). The library is sub-divided into a convenient number of subsections, which are amplified separ-ately. A 0.5–1-µl aliquot of each subsection is then used as the template for PCR with primers specific to the desired fragment. Ideally the fragment size should not exceed 500 bp, to minimise amplification of artefacts created by false priming. Subsections giving a positive signal can then be diluted further, subdivided, amplified and screened by PCR until a preparation sufficiently enriched in the desired clone can be screened by hybridisation or even colony PCR. The procedure can be performed in microtitre plates. Israel (1993) divided a murine genomic library in lambda into 64 wells in a grid of eight rows and eight columns, each well containing 1000 clones. Aliquots of the rows and of the columns were pooled separately and used for PCR, so that 16 PCRs could identify the 1 in 64 positives in the grid. The positive PCR product was verified by hybridisation to a third oligonucleotide probe. The procedure was repeated on dilutions and aliquots of the positive well and performed a third time on the positives from that screen. Individual plaques from a plating of the final positive well were screened by colony PCR and the majority contained the desired insert. Garbarino & Belknap (1994) used a PCR library screen to isolate a potato genomic ubiquitin-ribosomal protein gene.

Güssow & Clackson (1989) demonstrated that it was possible to screen individual bacterial colonies or plaques quite simply by PCR. The colony or plaque is diluted in a small aliquot of water and 1 µl of the suspension is used to prime the PCR. With bacterial colonies, an alternative method used at ATC is to touch the edge of a colony with a toothpick, make a short streak on a gridded selective agar plate and then tap the toothpick several times into a PCR reaction mix. Any positives identified by PCR can then be easily located on the gridded streaks. Colonies can be screened specifically with primers to the desired clone or with primers for the vector sequence either side of the cloning site (Güssow & Clackson, 1989; Runnebaum et al., 1991) to characterise insert size. Hundreds of colonies can be screened in this manner. With the inclusion of 12% sucrose and 1% tartrazine in the PCR mix prior to performing the amplification, the PCR reactions can be loaded directly onto a gel when amplification is completed (Hoppe et al., 1992). The method was developed with *E. coli* but we have had equal success with plant transformation vectors in *Agrobacterium tumefaciens* and *A. rhizogenes*.

4.10   Screening transgenic plants

Putative transgenic plants can be screened at a very early stage using PCR, with the added benefit of screening larger numbers of plants before they begin to occupy too much growth room or greenhouse space (Lassner et al., 1989; Hamill et al., 1990, 1991; McGarvey & Kaper, 1991). Several DNA extraction methods for plants have been described in section 3.1; however, the method of Edwards et al. (1991) has been successfully used on *Brassica napus* by Turgut et al. (1994), and on several thousand potato transgenics at ATC. The method is rapid and avoids the use of hazardous organic solvents.

A small leaf or leaf disc from in vitro plant material is macerated in an Eppendorf tube with a disposable grinder for 15 s at room temperature. Four hundred microlitres of extraction buffer (200 mM Tris–HCl, pH 7.5, 250 mM NaCl, 25 mM EDTA, 0.5% sodium dodecyl sulphate (SDS)) is added and the sample is vortexed. The samples are microfuged for 1 min and the DNA is precipitated from the supernatant with an equal volume of isopropanol. The DNA pellet is dried and dissolved in 50 μl of water, and 2.5 μl are used for PCR.

When generating transgenic plants, the sequences incorporated can be entire coding sequences with novel promoters, several kilobase-pairs in length and often with homologous genes naturally present in the plant. Therefore, rather than attempting to amplify the entire cloned sequence, one should target both junctions of the cloned fragment with either the transferred T-DNA or with any chimeric sequence, such as a 35S or tissue-specific promoter. The risk of non-specific or spurious amplification is reduced by keeping the size of the diagnostic fragments to about 500 bp. Screening can be further simplified by amplifying the two border fragments simultaneously in the same PCR (Multiplex PCR) (Lassner et al., 1989; Chamberlain et al., 1990), though primers have to be designed to avoid cross-hybridisation (section 3.2).

The transgenic plants can also be screened for transcription of the novel sequences by performing reverse transcriptase (RT) PCR on their RNA. Brown et al. (1993) describe how to detect antisense transcripts in transgenic plants by

RT PCR. Frequently, one also wishes to determine the magnitude of transcription, and this is dealt with in the next section on quantitative PCR.

4.11 Quantitative PCR

This section will deal with PCR quantitation of mRNA transcripts, though all considerations after the generation of cDNA are equally valid for quantitating DNA. The method of Shirzadegan et al. (1991) provides good total RNA preparations; alternatively, commercially available mRNA direct purification kits, e.g. from Dynal or Pharmacia, can be used. Chapter 10 deals with RNA purification in more detail.

Quantitative PCR is an indirect assay, as PCR is an amplification process. Quantitation relies on relating the yield of amplified product to the amount of starting template. Initially, amplification occurs linearly during PCR until the reaction components such as primers, dNTPs and enzyme become limiting and a final plateau concentration of product is reached. If all PCR reactions were identical, then the concentration of product after a set number of cycles during the linear amplification range could be related directly to template concentration when compared to a standard curve. However, in reality there will be differences in PCR ability between template preparations due to minor impurities and synthesis efficiencies. Methods for quantitative PCR therefore aim to take account of these differences or to include internal controls. Two different procedures are currently in use, with the second predominating in the methods literature.

The first method, described by Dallman & Porter (1991) and Golde et al. (1990), involves taking test samples of total or mRNA, synthesising cDNA which is then used for PCR. Aliquots or replicates are removed at five-cycle intervals during PCR, and the amount of product is measured to find the linear amplification range of the PCR. Quantitation is achieved by comparison with an external standard curve obtained using known concentrations of the target sequence in identical PCRs. The PCR products are assayed by incorporating either radiolabel or fluorescent bases into the primers.

The second method uses an internal RNA standard that is added to the samples prior to cDNA synthesis, thus providing a control for both cDNA synthesis and PCR efficiencies. Chang et al. (1993) demonstrated the applicability of the method to plant tissues. Several replicates are set up for each sample. Each replicate is spiked with an increasing series of mRNA standard concentrations before being utilised for cDNA synthesis and PCR for 20–40 cycles. The resulting products are run on an agarose gel and visualised by either ethidium bromide, fluorography or radiolabelling. Where the yield of products from the test template and the internal standard is identical, the amount of initial test template equals the amount of standard added. In order to function properly, the internal standard should be as similar to, and be primed by, the same primers as the test sample. The standard must differ slightly from the test RNA in size or by the presence of an additional restriction site in order to distinguish between the two products. Celi et al. (1993) describe a simple method for creating deletion mutants for use as internal standards. The standards can then be cloned into a vector such as pBluescript, which permits the generation of run-off RNA transcripts (Chang et al., 1993).

Where large numbers of samples need to be assayed regularly, consideration ought to be given to detection methods where the PCR products are bound to a solid support such as a membrane or microtitre plate rather than electrophoresed. The immobilised samples are then screened by hybridisation with test or standard specific probes (Kohsaka et al., 1993; Khudyakov et al., 1994; Rhoer-Moja et al., 1994).

5 TROUBLESHOOTING

PCR is a deceptively simple technique that has its own problems unique to the principle of amplification. The following points may be of use to the user.

5.1 Controls

For every PCR, *always* include a negative control, that is, a PCR reaction without template. If available, also include a positive control.

These two controls alone will demonstrate whether the PCR conditions are correct and whether contamination of any of the reagents has occurred. Particularly when dealing with plant genomic DNA, where the large genome size increases the chance of priming at several sites, it is also worth including control PCRs with template and each primer individually. This ensures that the products observed are really the result of two specific primers.

5.2 Too many or too few bands

Where too many or no bands are observed, the PCRs should be repeated at various annealing temperatures from 35 °C upwards at 5 °C intervals. With negative results, spiking the test sample with a known amplifiable template should also be tried to check whether the reaction has been poisoned by reagents carried over from DNA or RNA purification. PCR specificity is also affected by the magnesium chloride concentration in the reaction buffer, so this may be varied if necessary. Nested PCR can also result in improved results by adding an additional specificity and amplification step. Hot start PCR (D'Aquila et al., 1991) should be tried where too many bands are present. Synthesising a new set of primers with a shift of a few bases can also be tried if the above methods do not work.

Manufacturers of PCR machines are now producing machines to a high specification. However, if consistently variable results are obtained, it is also worth checking for PCR reproducibility across the heating block to see if certain positions give poor results (Hoelzel, 1990; Van Leuven, 1991). There may also be differences in results between PCRs conducted on machines of different origin or batch number.

5.3 Contamination

PCR amplifies product from very small quantities of template and is very susceptible to false results from minute traces of contaminants. Contamination is therefore almost inevitable where PCR is conducted regularly with the same template.

In most cases, the problem is better solved by starting with fresh template and reagents on a cleaned work area, rather than wasting valuable time trying to identify the source of contamination. The risk of contamination can be minimised by purchasing the basic reagents such as PCR buffers and dNTPs and by having the primers synthesised at a separate location by people not directly involved in the PCR work. Store DNA templates and PCR reagents at separate locations and always add the template last to a PCR mix. Use disposable plasticware where possible and clean reusable materials with a 1-h soak in 0.2 M sodium hydroxide followed by 0.2 M HCl, and then rinse with sterile water.

PCR can also be conducted with dUTP. To prevent carry-over of contaminant PCR products, future PCRs are treated with uracil-N-glycosylase, which degrades the dUTP-modified DNA (Longo et al., 1990). The method has been adapted for RT PCR by Udaykumar et al. (1993). Perkin-Elmer provides a PCR carry-over prevention kit. Alternatively, the PCR primers can be designed to contain a 3' terminal ribose residue which permits degradation by ribonuclease A or sodium hydroxide (Walder et al., 1993). Other methods, such as the use of UV irradiation (Sarkar & Sommer, 1990), have also been used.

6 CONCLUDING REMARKS

PCR is a valuable tool for the plant molecular biologist. This is evidenced by the increasing number of papers which routinely use PCR-based methodologies to isolate and clone genes, to create chimeric constructs, to sequence them and to screen transformants. This chapter aims not only to show some of the versatility of PCR in plants but also to emphasise the importance of experimental design and the inclusion of suitable controls to guarantee reproducible, problem-free results.

New applications are continually arising and PCR is tending towards faster throughput and greater automation. PCR-based methods and protocols are being published regularly, for example in *Plant Molecular Biology Reporter*, the methods section of *Nucleic Acids Research* and *Biotechniques*. The journal *PCR Methods and Applications* is devoted entirely to the subject.

Finally, if you are considering purchasing a PCR machine for the first time, buy two, because you will probably end up needing them! PCR is compulsive.

ACKNOWLEDGMENTS

I would like to thank my friends and colleagues at ATC, who have wholeheartedly adopted PCR to great effect and made the writing of this chapter possible. Jane Dawson gave invaluable assistance in the typing of the references. My thanks go to Keith Blundy, who allowed me to write half the chapter at work, and to my wife Jane and the children, who let me complete the other half in their prime time.

ESSENTIAL READING

Hoelzel, R. (1990) The trouble with 'PCR' machines. *Trend. Genet.*, **6**, 237.

Mullis, K.B. (1990) The unusual origin of the polymerase chain reaction. *Sci. Am.*, April, 36–43.

Newton, C.R. and Graham, A. (1994) *PCR*. BIOS, Oxford.

Rychlik, W. (1993) Selection of primers for polymerase chain reaction. In *Methods in Molecular Biology*, Vol. 15: *PCR Protocols: Current Methods and Applications* (Ed. B.A. White), pp. 31–40. Humana Press, Totowa.

Saiki, R.K., Scharf, S., Faloona, F. et al. (1985) Enzymatic amplification of beta globin genomic sequences and restriction site analysis for diagnosis of sickle cell anemia. *Science*, **230**, 1350–1354.

REFERENCES

Barettino, D., Feigenbutz, M., Valcarcel, R. and Stunnenberg, H.G. (1993) Improved method for PCR-mediated site-directed mutagenesis. *Nucleic Acids Res.*, **22**, 541–542.

Barnes, W.B. (1994) PCR amplification of up to 35-kb DNA with high fidelity and high yield from λ bacteriophage templates. *Proc. Natl. Acad. Sci. USA*, **91**, 2216–2220.

Bassam, B.J. and Caetano-Anolles, G. (1993) Automated 'Hot Start' PCR using mineral oil and paraffin wax. *Biotechniques*, **14**, 31–33.

Bauer, D., Müller, H., Reich, J. et al. (1993) Identification of differentially expressed mRNA species by an improved display technique (DDRT-PCR). *Nucleic Acids Res.*, **21**, 4272–4280.

Belyavsky, A., Vinogradov, T. and Rajewsky, K. (1989) PCR-based cDNA library construction: general cDNA libraries at the level of a few cells. *Nucleic Acids Res.*, **17**, 2919–2932.

Bhattacharya, D., Steinkötter, J. and Melkonian, M. (1993) Molecular cloning and evolutionary analysis of the calcium-modulated contractile protein, centrin, in green algae and land plants. *Plant Mol. Biol.*, **23**, 1243–1254.

Bloem, L.J. and Yu, L. (1990) A time-saving method for screening cDNA or genomic libraries. *Nucleic Acids Res.*, **18**, 2830.

Brown, J.W.S., Simpson, G.G., Clark, G. et al. (1993) Detection of antisense transcripts in transgenic plants by RT-PCR. *Plant J.*, **4**, 883–885.

Browning, K.S. (1989) Transcription and translation of mRNA from polymerase chain reaction-generated DNA. *Amplifications*, **3**, 14–15.

Brunel, D. (1992) An alternative, rapid method of plant DNA extraction for PCR analyses. *Nucleic Acids Res.*, **20**, 4676.

Bucciaglia, P.A. and Smith, A.G. (1994) Cloning and characterization of *TAG 1*, a tobacco anther β-1,3-glucanase expressed during tetrad dissolution. *Plant Mol. Biol.*, **24**, 903–914.

Byth, B.C., Thomas, G.R., Hofland, N. and Cox, D.W. (1994) Application of inverse PCR to isolation of end probes from cosmids. *Nucleic Acids Res.*, **22**, 1766–1767.

Carothers, A.M., Urlaub, G., Mucha, J. et al. (1989) Point mutation analysis in a mammalian gene: rapid preparation of total RNA, PCR amplification of cDNA and Taq sequencing by a novel method. *Biotechniques*, **7**, 494–499.

Celi, F.S., Zenilman, M.E. and Shuldiner, A.R. (1993) A rapid and versatile method to synthesize internal standards for competitive PCR. *Nucleic Acids Res.*, **21**, 1047.

Chamberlain, J.S., Gibbs, R.A., Ranier, J.E. and Laskey, C.T. (1990) Multiplex PCR for the diagnosis of duchenne muscular dystrophy. In *PCR Protocols: A Guide to Methods and Applications* (eds M.A. Innis, D.H. Gelfand, J.J. Sninsky and T.J. White), pp. 272–281. Academic Press, San Diego.

Chang, P.-F., Narasimhan, M.L., Hasegawa, P.M. and Bressan, R.A. (1993) Quantitative mRNA-PCR for expression analysis of low-abundance transcripts. *Plant Mol. Biol. Rep.*, **11**, 237–248.

Clackson, T. and Winter, G. (1989) Sticky feet-directed mutagenesis and its application to swapping antibody domains. *Nucleic Acids Res.*, **17**, 10163–10170.

Clark, J.M. (1988) Novel non-templated nucleotide additions catalyzed by procaryotic and eucaryotic DNA polymerases. *Nucleic Acids Res.*, **16**, 9677–9686.

Colosi, J.C. and Schaal, B.A. (1993) Tissue grinding with ball bearings and vortex mixer for DNA extraction. *Nucleic Acids Res.*, **21**, 1051–1052.

Connell, C., Fung, S., Heiner, C. et al. (1987) Automated DNA sequence analysis. *Biotechniques*, **5**, 342–348.

Costa, G.L. and Weiner, M.P. (1994) Polishing with T4 or P*fu* polymerase increases the efficiency of cloning of PCR fragments. *Nucleic Acids Res.*, **22**, 2423.

Dallman, M.J. and Porter, A.C.G. (1991) Semi-quantitative PCR for the analysis of gene expression. In *PCR: A Practical Approach* (eds M.J. McPherson, P. Quirke and G.R. Taylor) pp. 215–224. Oxford University Press, Oxford.

D'Aquila, R.T., Bechtel, L.J., Videler, J.A. et al. (1991) Maximizing sensitivity and specificity of PCR by pre-amplification heating. *Nucleic Acids Res.*, **19**, 3749.

Dellaporta, S.L., Wood, J. and Hicks, J.B. (1983) A plant DNA minipreparation: version-II. *Plant Mol. Biol. Rep.*, **1**, 19–21.

Dietmaier, W., Fabry, S. and Schmitt, R. (1993) DISEC-TRISEC: di- and trinucleotide-sticky-end cloning of PCR-amplified DNA. *Nucleic Acids Res.*, **21**, 3603–3604.

Doherty, P.J., Huesla-Contreras, M., Dosch, H.M. and Pan, S. (1989) Rapid amplification of complementary DNA from small amounts of unfractioned RNA. *Anal. Biochem.*, **177**, 7–10.

Dombrowski, K.E. and Wright, S.E. (1992) Construction of a multiple mucin tandem with a mutation in the tumor-specific epitope by a solid-phase gene assembly protocol. *Nucleic Acids Res.*, **20**, 6743–6744.

Dresselhaus, T., Lörz, H. and Kranz, E. (1994) Representative cDNA libraries from few plant cells. *Plant J.*, **5**, 605–610.

Eckert, K.A. and Kunkel, T.A. (1991) The fidelity of DNA polymerases used in the polymerase chain reactions. In *PCR: A Practical Approach* (eds M.J. McPherson, P. Quirke and G.R. Taylor, pp. 225–244. Oxford University Press, Oxford.

Edwards, K., Johnstone, C. and Thompson, C. (1991) A simple and rapid method for the preparation of plant genomic DNA for PCR analysis. *Nucleic Acids Res.*, **19**, 1349.

Erlich, H.A. (ed.) (1989) *PCR Technology. Principles and Application for DNA Amplification.* Stockton Press, New York.

Fangan, B.M., Stedje, B., Stabbetorp, O.E. et al. (1994) A general approach for PCR-amplification and sequencing of chloroplast DNA from crude vascular plant and algal tissue. *Biotechniques*, **16**, 484–494.

Firek, S., Draper, J., Owen, M.R.L. et al. (1993) Secretion of a functional single-chain Fv protein in transgenic tobacco. *Plant Mol. Biol.*, **23**, 861–870.

Friedman, K.D., Rosen, N.L., Newman, P.J. and Montgomery, R.R. (1988) Enzymatic amplification of specific complementary DNA inserts from lambda gt-11 libraries. *Nucleic Acids Res.*, **16**, 8718.

Frohman, M.A., Dush, M.K. and Martin, G.R. (1988) Rapid production of full-length cDNAs from rare transcripts. Amplification using a single gene-specific oligonucleotide primer. *Proc. Natl. Acad. Sci. USA*, **85**, 8998–9002.

Garbarino, J.E. and Belknap, W.R. (1994) Isolation of a ubiquitin-ribosomal protein gene (*ubi3*) from potato and expression of its promoter in transgenic plants. *Plant Mol. Biol.*, **24**, 119–127.

Gawel, N.J. and Jarret, R.L. (1991) A modified CTAB DNA extraction procedure for *Musa* and *Ipomoea*. *Plant Mol. Biol. Rep.*, **9**, 262–266.

Gelfand, D. (1989) *Taq* DNA polymerase. In *PCR Technology: Principles and Applications for DNA Amplification* (ed. H.A. Erlich), pp. 17–22. Stockton Press, New York.

Goblet, C., Prost, E. and Whalen, R.G. (1989) One step simplification of transcripts in total RNA using the polymerase chain reaction. *Nucleic Acids Res.*, **17**, 2144.

Gocayne, J., Robinson, D.A., FitzGerald, M.G. et al. (1987) Primary structure of rat cardiac beta-adrenergic and muscarinic cholinergic receptors obtained by automated DNA sequence analysis; further evidence for a multigene family. *Proc. Natl. Acad. Sci. USA*, **84**, 8296–8300.

Golde, T.E., Estus, S., Usiak, M. et al. (1990) Expression of beta amyloid protein precursor mRNAs: recognition of a novel alternatively spliced form and quantitation in Alzheimer's disease using PCR. *Neuron*, **4**, 253–267.

Green, A., Roopra, A. and Vaudin, M. (1990) Direct single stranded sequencing from agarose of polymerase chain reaction products. *Nucleic Acids Res.*, **18**, 6163–6164.

Guidet, F. (1994) A powerful new technique to quickly prepare hundreds of plant extracts for PCR and RAPD analyses. *Nucleic Acids Res.*, **22**, 1772–1773.

Gurr, S.J. and McPherson, M.J. (1991) PCR-directed cDNA libraries. In *PCR: A Practical Approach* (eds M.J. McPherson, P. Quirke and G.R. Taylor), pp. 147–170. Oxford University Press, Oxford.

Güssow, D., and Clackson, T. (1989) Direct clone characterisation from plaques and colonies by the polymerase chain reaction. *Nucleic Acids Res.*, **17**, 4000.

Gyllensten, U.B. and Erlich, H.A. (1988) Generation of single stranded DNA by the polymerase chain reaction and its application to the direct sequencing of the HLA-DQA locus. *Proc. Natl. Acad. Sci. USA*, **85**, 7652–7656.

Hamill, J.D., Rounsley, S., Spencer, A. et al. (1990) The use of the polymerase chain reaction to detect specific sequences in transformed plant tissues. In *Progress in Plant Cellular and Molecular Biology* (eds H.J.J. Nijkamp, L.H.W. Van der Plas and J. Van Aartrijk), pp. 183–188. Kluwer Academic Publishers, Dordrecht.

Hamill, J.D., Rounsley, S., Spencer, A. et al. (1991) The use of the polymerase chain reaction in plant transformation studies. *Plant Cell Rep.*, **10**, 221–224.

Herfort, M.R. and Garber, A.T. (1991) Simple and efficient subtractive hybridization screening. *Biotechniques*, **11**, 598–604.

Higuchi, R. (1989) Using PCR to engineer DNA. In *PCR Technology: Principles and Applications for DNA Amplification* (ed. H.A. Erlich), pp. 61–70. Stockton Press, New York.

Higuchi, R.G. and Ochman, H. (1989) Production of single stranded DNA templates by exonuclease digestion following the polymerase chain reaction. *Nucleic Acids Res.*, **17**, 5865.

Higuchi, R., Krummel, B. and Saiki, R.K. (1988) A general method of *in vitro* preparation and specific mutagenesis of DNA fragments: studies of protein and DNA interactions. *Nucleic Acids Res.*, **16**, 7351–7367.

Hoppe, B.L., Conti-Tronconi, B.M. and Horton, R.M. (1992) Gel-loading dyes compatible with PCR. *Biotechniques*, **12**, 679–680.

Hovens, C.M. and Wilks, A.F. (1989) Rapid screening of highly complex cDNA libraries using the polymerase chain reaction. *Nucleic Acids Res.*, **17**, 4415–4416.

Huang, S., Yiyuan, H., Wu, C. and Holcenberg, J. (1990) A simple method for direct cloning cDNA sequence that flanks a region of known sequence from total RNA by applying the inverse polymerase chain reaction. *Nucleic Acids Res.*, **18**, 1922.

Hung, T., Mak, K. and Fong, K. (1990) A specific enhancer for polymerase chain reaction. *Nucleic Acids Res.*, **18**, 4953.

Imai, Y., Matsushima, Y., Sugimara, T. and Terada, M. (1991) A simple and rapid method for generating a deletion by PCR. *Nucleic Acids Res.*, **17**, 723.

Innis, M.A. and Gelfand, D.H. (1990) Optimization of PCRs. In *PCR Protocols: A Guide to Methods and Applications* (eds M.A. Innis, D.H. Gelfand, J.J. Sninsky and T.J. White), pp. 3–12. Academic Press, San Diego.

Innis, M.A., Gelfand, D.H., Sninsky, J.J. and White, T.J. (eds) (1990) *PCR Protocols: A Guide to Methods and Applications*. Academic Press, San Diego.

Isola, N.R., Harn, H.J. and Cooper, D.L. (1991) Screening recombinant DNA libraries: a rapid and efficient method for isolating cDNA clones utilizing the PCR. *Biotechniques*, 11, 580–582.

Israel, D.I. (1993) A PCR-based method for high stringency screening of DNA libraries. *Nucleic Acids Res.*, 21, 2627–2631.

Jackson, D.P., Hayden, J.D. and Quirke, P. (1991) Extraction of nucleic acid from fresh and archival material. In *PCR: A Practical Approach* (eds M.J. McPherson, P. Quirke and G.R. Taylor), pp. 29–50. Oxford University Press, Oxford.

Jung, V., Pestka, S.B. and Pestka, S. (1990) Efficient cloning of PCR generated DNA containing terminal restriction endonuclease recognition sites. *Nucleic Acids Res.*, 18, 6156.

Kaijalainen, S., Karhunen, P.J., Lalu, K. and Lindström, K. (1993) An alternative hot start technique for PCR in small volumes using beads of wax-embedded reaction components dried in trehalose. *Nucleic Acids Res.*, 21, 2959–2960.

Kain, K.C., Orlandi, P.A. and Lanar, D.E. (1991) Universal promoter for gene expression without cloning: Expression-PCR. *Biotechniques*, 10, 366–374.

Kandpal, R.P., Shukla, H., Ward, D.C. and Weissman, S.M. (1990) A polymerase chain reaction approach for constructing jumping and linking libraries. *Nucleic Acids Res.*, 18, 3081.

Kaufman, D.L. and Evans, G.A. (1990) Restriction endonuclease cleavage at the termini of PCR products. *Biotechniques*, 9, 304–306.

Khudyakov, Y.E., Gaur, L., Singh, J. et al. (1994) Primer specific solid-phase detection of PCR products. *Nucleic Acids Res.*, 22, 1320–1321.

Kohsaka, H., Taniguchi, A., Richman, D.D. and Carson, D.A. (1993) Microtiter format gene quantification by covalent capture of competitive PCR products: application to HIV-1 detection. *Nucleic Acids Res.*, 21, 3469–3472.

Korn, B., Sedlacek, Z., Manca, A. et al. (1992) A strategy for the selection of transcribed sequences in the Xq28 region. *Human Mol. Genet.*, 1, 235–242.

Kovalic, D., Kwak, J.-H. and Weisblum, B. (1991) General method for direct cloning of DNA fragments generated by the polymerase chain reaction. *Nucleic Acids Res.*, 19, 4560.

Krawczac, M., Reiss, J., Schmidke, J. and Rosler, U. (1989) Polymerase chain reaction: replication errors and reliability of gene diagnosis. *Nucleic Acids Res.*, 17, 2197–2201.

Kwiatkowski, T.J., Zoghbi, H.Y., Ledbetter, S.A. et al. (1990) Rapid identification of yeast artificial chromosome clones by matrix pooling and crude lysate PCR. *Nucleic Acids Res.*, 18, 7191.

Kwok, S., Kellog, D.E., Spasic, D. et al. (1990) Effects of primer–template mismatches on the polymerase chain reaction: human immunodeficiency virus type 1 model studies. *Nucleic Acids Res.*, 18, 999–1005.

Lambert, K.N. and Williamson, V.M. (1993) cDNA library construction from small amounts of RNA using paramagnetic beads and PCR. *Nucleic Acids Res.*, 21, 775–776.

Landt, O., Grunert, H.-P. and Hahn, U. (1990) A general method for rapid site-directed mutagenesis using the polymerase chain reaction. *Gene*, 96, 125–128.

Langridge, U., Schwall, M. and Langridge, P. (1991) Squashes of plant tissue as substrate for PCR. *Nucleic Acids Res.*, 19, 4954.

Lassner, M.W., Peterson, P. and Yoder, J.I. (1989) Simultaneous amplification of multiple DNA fragments by polymerase chain reaction in the analysis of transgenic plants and their progeny. *Plant Mol. Biol. Rep.*, 7, 116–128.

Lathe, R. (1985) Synthetic oligonucleotide probes deduced from amino acid sequence data. *J. Mol. Biol.*, 183, 1–12.

Li, Y., Strabala, T.J., Hagen, G. and Guilfoyle, T.J. (1994) The soybean SAUR open reading frame contains a *cis* element responsible for cycloheximide-induced mRNA accumulation. *Plant Mol. Biol.*, **24**, 715–723.

Liang, P. and Pardee, A.B. (1992) Differential display of eukaryotic messenger RNA by means of the polymerase chain reaction. *Science*, **259**, 967–971.

Liang, P., Averboukh, L. and Pardee, A.B. (1993) Distribution and cloning of eukaryotic mRNAs by means of differential display: refinements and optimization. *Nucleic Acids Res.*, **21**, 3269–3275.

Lindstrom, J.T., Chu, B. and Belanger, F.C. (1993) Isolation and characterization of an *Arabidopsis thaliana* gene for the 54 kDa subunit of the signal recognition particle. *Plant Mol. Biol.*, **23**, 1265–1272.

Loh, E., Elliot, J.F., Cwirla, S. et al. (1989) Polymerase chain reaction with single-sided specificity: analysis of 7. cell receptor and chain. *Science*, **243**, 217–220.

Lohff, C.J. and Cease, K.B. (1992) PCR using a thermostable polymerase with 3' to 5' exonuclease activity generates blunt products suitable for direct cloning. *Nucleic Acids Res.*, **20**, 144.

Longo, M.C., Berninger, M.S. and Hartley, J.L. (1990) Use of uracil DNA glycosylase to control carryover contamination in polymerase chain reactions. *Gene*, **93**, 125–128.

Lovett, M., Kere, J. and Hinton, L.M. (1991) Direct selection: a method for the isolation of cDNAs encoded by large genomic regions. *Proc. Natl. Acad. Sci. USA*, **88**, 9628–9632.

Lundberg, K.S., Schoemaker, D.D., Adams, M.W. et al. (1991) High fidelity amplification using a thermostable DNA polymerase isolated from *Pyrococcus furiosus*. *Gene*, **108**, 1.

MacFerrin, K.D., Terranova, M.P., Schreiber, S.L. and Verdine, G.L. (1990) Overproduction and dissection of proteins by the expression cassette polymerase chain reaction. *Proc. Natl. Acad. Sci. USA*, **87**, 1937–1941.

Maskell, D., Szabo, M. and High, N. (1993) PCR amplification of DNA sequences from nitrocellulose-bound, immunostained bacterial colonies. *Nucleic Acids Res.*, **21**, 171–172.

Mattila, P., Korpela, J., Tenkanen, T. and Pitkänen (1991) Fidelity of DNA synthesis by the *Thermococcus litoralis* DNA polymerase—an extremely heat stable enzyme with proofreading activity. *Nucleic Acids Res.*, **19**, 4967–4973.

Mayrand, P.E., Corcoran, K.P., Ziegle, J.S. et al. (1992) The use of fluorescence detection and internal lane standards to size PCR products automatically. *Appl. Theor. Electrophoresis*, **3**, 1–11.

McGarvey, P. and Kaper, J.M. (1991) A simple and rapid method for screening transgenic plants using the PCR. *Biotechniques*, **11**, 428–432.

McPherson, M.J., Quirke, P. and Taylor, G.R. (eds) (1991a) *PCR: A Practical Approach.* Oxford University Press, Oxford.

McPherson, M.J., Jones, K.M. and Gurr, S.J. (1991b) PCR with highly degenerate primers. In *PCR: A Practical Approach* (eds M.J. McPherson, P. Quirke and G.R. Taylor), pp. 171–186. Oxford University Press, Oxford.

Mendelman, L.V., Petruska, J. and Goodman, M.F. (1989) Base mispair extension kinetics. Comparison of DNA polymerase a and reverse transcriptase. *J. Biol. Chem.*, **265**, 2338–2346.

Mosseler, A., Egger, K.N. and Hughes, G.A. (1992) Low levels of genetic diversity in red pine confirmed by random amplified polymorphic DNA markers. *Can. J. Forest. Res.*, **22**, 1332–1337.

Muench, D.G. and Good, A.G. (1994) Hypoxically inducible barley alanine aminotransferase; cDNA cloning and expression analysis. *Plant Mol. Biol.*, **24**, 417–427.

Mullis, K.B. (1991) The polymerase chain reaction in an anemic mode: how to avoid cold oligodeoxyribonuclear fusion. *PCR Methods Applic.*, **1**, 1–4.

Mullis, K.B., Ferré, F. and Gibbs, R.A. (eds) (1994) *The Polymerase Chain Reaction.* Birkhaüser, Boston.

Murray, E.E., Lotzer, J. and Eberle, M. (1989) Codon usage in plant genes. *Nucleic Acids Res.*, **17**, 477–498.

Myers, T.W. and Gelfand, D.H. (1991) Use of rTth DNA polymerase for reverse transcription and PCR. *Biochemistry*, **30**, 7661–7666.

Newton, C.R. and Graham, A. (1994) *PCR*. BIOS, Oxford.

Newton, C.R., Holland, D., Heptinstall, L.E. et al. (1993) The production of PCR products with 5′ single-stranded tails using primers that incorporate novel phosphoramidite intermediates. *Nucleic Acids Res.*, **21**, 1155–1162.

Ochman, H., Gerber, A.S. and Hartl, D.L. (1988) Genetic applications of an inverse polymerase chain reaction. *Genetics*, **120**, 621–623.

Oliner, J.D., Kinzler, K.W. and Vogelstein, B. (1993) *In vivo* cloning of PCR in *E. coli*. *Nucleic Acids Res.*, **21**, 5192–5197.

Pirovano, L., Lanzini, S., Hartings, H. et al. (1994) Structural and functional analysis of an *Opaque*-2-related gene from sorghum. *Plant Mol. Biol.*, **24**, 515–523.

Ponce, M.R. and Micol, J.L. (1992) PCR amplification of long DNA fragments. *Nucleic Acids Res.*, **20**, 623.

Raineri, I., Moroni, C. and Senn, H.P. (1991) Improved efficiency for single-sided PCR by creating a reusable pool of first-strand cDNA coupled to a solid phase. *Nucleic Acids Res.*, **19**, 4010.

Rasmussen, U.B., Basset, P. and Daniel, J.-Y. (1989) Direct amplification of cDNA inserts from λ libraries using the cloning-adapter as primer for PCR. *Nucleic Acids Res.*, **17**, 3308.

Ray, F.A. and Nickoloff, J.A. (1992) Site-specific mutagenesis of almost any plasmid using a PCR-based version of unique site elimination. *Biotechniques*, **13**, 342–348.

Reiter, R.S., Williams, J.G.K., Feldman, K.A. et al. (1992) Global and local genome mapping in *Arabidopsis thaliana* by using recombinant inbred lines and random amplified polymorphic DNAs. *Proc. Natl. Acad. Sci. USA*, **89**, 1477–1481.

Rhoer-Moja, S., Bazin, H., Sauvaigo, S. et al. (1994) Solid support quantitation of c-myc PCR products using a cleavable reporter. *Nucleic Acids Res.*, **22**, 547–548.

Riley, J., Butler, R., Finniear, R. et al. (1990) A novel, rapid method for the isolation of terminal sequences from yeast artificial chromosome (YAC) clones. *Nucleic Acids Res.*, **18**, 2887.

Rosenberg, H.F., Corrètte, S.E., Tenen, D.G. and Ackerman, S.J. (1991) Rapid cDNA library screening using the polymerase chain reaction. *Biotechniques*, **10**, 53–54.

Runnebaum, I.R., Syka, P. and Sukumar, S. (1991) Vector PCR. *Biotechniques*, **11**, 446–447.

Rychlik, W. (1993) Selection of primers for polymerase chain reaction. In *Methods in Molecular Biology*, **15**, *PCR Protocols: Current Methods and Applications* (ed. B.A. White), pp. 31–40. Humana Press, Ottowa.

Rychlik, W., Spencer, W.J. and Rhoads, R.E. (1990) Optimization of the annealing temperature for DNA amplification *in vitro*. *Nucleic Acids Res.*, **18**, 6409–6412.

Saiki, R.K. (1990) Amplification of genomic DNA. In *PCR Protocols: A Guide to Methods and Applications* (eds M.A. Innis, D.H. Gelfand, J.J. Sninsky and T.J. White), pp. 13–20. Academic Press, San Diego.

Saiki, R.K., Gelfand, D.H., Stoffel, S. et al. (1988) Primer-directed enzymatic amplification of DNA with a thermostable DNA polymerase. *Science*, **239**, 487–491.

Saiki, R.K., Scharf, S., Faloona, F. et al. (1985) Enzymatic amplification of beta globin genomic sequences and restriction site analysis for diagnosis of sickle cell anemia. *Science*, **230**, 1350–1354.

Sambrook, J., Fritsch, E.F. and Maniatis, T. (eds) (1989) *Molecular Cloning: A Laboratory Manual*, 2nd edn. Cold Spring Harbor Laboratory Press, New York.

Sarkar, G. and Sommer, S.S. (1990) Shedding light on PCR contamination. *Nature*, **343**, 27.

Sarkar, G., Kapelner, S. and Sommer, S.S. (1990) Formamide can dramatically improve the specificity of PCR. *Nucleic Acids Res.*, **18**, 7465.

Scharf, S.J., Horn, G.T. and Erlich, H.A. (1986) Direct cloning and sequence analysis of enzymatically amplified genomic sequences. *Science*, **233**, 1076–1078.

Screaton, G.R., Bangham, C.R.M. and Bell, J.I. (1993) Direct sequencing of single primer PCR products: a rapid method to achieve short chromosomal walks. *Nucleic Acids Res.*, **21**, 2263–2264.

Sellner, L.N., Coelen, R.J. and Mackenzie, J.S. (1992) Reverse transcriptase inhibits Taq polymerase activity. *Nucleic Acids Res.*, **20**, 1487–1490.

Sheardown, S.A. (1992) A simple method for affinity purification and PCR amplification of poly(A)$^+$ mRNA. *Trends Genet.*, **8**, 121.

Shirzadegan, M., Christie, P. and Seemann, J.R. (1991) An efficient method for isolation of RNA from tissue cultured plant cells. *Nucleic Acids Res.*, **19**, 6055.

Smith, L., Sanders, J.Z., Kaiser, R.J. et al. (1986) Fluorescence detection in automated DNA sequence analysis. *Nature*, **321**, 674–679.

Sommer, R. and Tautz, D. (1989) Minimal homology requirements for PCR primers. *Nucleic Acids Res.*, **17**, 6749.

Struck, F. and Collins, J. (1994) Simple and rapid 5' and 3' extension techniques in RT-PCR. *Nucleic Acids Res.*, **22**, 1923–1924.

Tang, X., Wang, H., Brandt, A.S. and Woodson, W.R. (1993) Organisation and structure of the 1-aminocyclopropane-1-carboxylate oxidase gene family from *Petunia hybrida*. *Plant Mol. Biol.*, **23**, 1151–1164.

Teichmann, T., Urban, C. and Beier, H. (1994) The tRNA^{ser}-isoacceptors and their genes in *Nicotiana rustica*: genome organization, expression *in vitro* and sequence analyses. *Plant Mol. Biol.*, **24**, 889–901.

Thangstad, O.P., Winge, P., Husebye, H. and Bones, A. (1993) The myrosinase (thioglucoside glucohydrolase) gene family in Brassicaceae. *Plant Mol. Biol.*, **23**, 511–524.

Thomas, C.J.R. (1993) The polymerase chain reaction. *Methods Plant Biochem.*, **10**, 117–140.

Tindall, K.R. and Kunkel, T.A. (1988) Fidelity of DNA synthesis by the *Thermus aquaticus* DNA polymerase. *Biochemistry*, **27**, 6008–6013.

Toguri, T., Umemoto, N., Kobayashi, O. and Ohtani, T. (1993) Activation of anthocyanin synthesis genes by white light in eggplant hypocotyl tissues, and identification of an inducible P-450 cDNA. *Plant Mol. Biol.*, **23**, 933–946.

Triglia, T., Peterson, M.G. and Kemp, D.J. (1988) A procedure for *in vitro* amplification of DNA segments that lie outside the boundaries of known sequences. *Nucleic Acids Res.*, **16**, 8186.

Turgut, K., Barsby, T., Craze, M. et al. (1994) The highly expressed tapetum-specific A9 gene is not required for male fertility in *Brassica napus*. *Plant Mol. Biol.*, **24**, 97–104.

Udaykumar, Epstein, J.S. and Hewlett, I.K. (1993) A novel method employing UNG to avoid carry-over contamination in RNA-PCR. *Nucleic Acids Res.*, **21**, 3917–3918.

Uhlenbeck, O.C., Martin, F.H. and Doty, P. (1971) Self-complementary oligo-ribonucleotides: effects of helix defects and guanylic acid–cytidylic acid base pairs. *J. Mol. Biol.*, **57**, 217.

Van Leuven, F. (1991) The trouble with PCR machines: fill up the empty spaces! *Trends Genet.*, **7**, 142.

Villand, P., Olsen, O.-A. and Klecxkowski, L.A. (1993) Molecular characterization of multiple cDNA clones for ADP-glucose pyrophosphorylase from *Arabidopsis thaliana*. *Plant Mol. Biol.*, **23**, 1279–1284.

Wainwright, L.A. and Seifert, H.S. (1993) Use of oil overlays in 'oil-free' PCR technology. *Biotechniques*, **14**, 33–36.

Walder, R.Y., Hayes, J.R. and Walder, J.A. (1993) Use of PCR primers containing a 3'-terminal ribose residue to prevent cross-contamination of amplified sequences. *Nucleic Acids Res.*, **21**, 4339–4343.

Wang, H., Qi, M. and Cutler, A.J. (1993) A simple method of preparing plant samples for PCR. *Nucleic Acids Res.*, **21**, 4153–4154.

Welsh, J. and McClelland, M. (1990) Fingerprinting genomes using PCR with arbitrary primers. *Nucleic Acids Res.*, **18**, 7213–7218.

Welsh, J. and McClelland, M. (1991) Genomic fingerprinting using arbitrarily primed PCR and a matrix of pairwise combinations of primers. *Nucleic Acids Res.*, **19**, 5275–5279.

Williams, C.E. and Ronald, P.C. (1994) PCR template-DNA isolated quickly from monocot and dicot leaves without tissue homogenization. *Nucleic Acids Res.*, **22**, 1917–1918.

Williams, J.G.K., Kubelick, A.R., Livak, K.J. et al. (1990) DNA polymorphisms amplified by arbitrary primers are useful as genetic markers. *Nucleic Acids Res.*, **18**, 6531–6535.

Winship, P.R. (1989) An improved method for directly sequencing PCR amplified material using dimethyl sulphoxide. *Nucleic Acids Res.*, **17**, 1266.

14 cDNA and Genomic Subtraction

M. P. BULMAN
University of Exeter, Exeter, UK

S. J. NEILL
University of the West of England, Bristol, UK

This chapter will deal with the technique of subtractive hybridisation, which has been used successfully to isolate genomic and cDNA clones by the selective enrichment of genomic or cDNA that contains a particular population or sequence of cDNA. The chapter will highlight some of the successes to date, as well as providing typical experimental protocols. Other areas which are covered include:

(1) Outline of the basis of cDNA and genomic subtraction; situations where appropriate; overview of techniques.
(2) cDNA subtraction in detail. Preparation of driver and target nucleic acids; mRNA–cDNA and cDNA–cDNA hybridisations; use of polymerase chain reaction (PCR); hybridisation conditions; separation: hydroxylapatite chromatography, biotinylation/phenol/chloroform extraction; generation of subtracted libraries and hybridisation probes.
(3) cDNA subtraction protocol: adapter–primer ligation; PCR amplification; biotinylation; hybridisation; separation; generation of subtracted probe; controls.
(4) Use of subtractive hybridisation to isolate cDNA clones representing differentially expressed genes.
(5) Genomic subtraction; development; PCR-based approaches.

1 INTRODUCTION

A major goal of plant molecular biology is the isolation of genes and identification of the regulatory DNA sequences and proteins associated with them. Consequently, the objective of many cloning experiments is to isolate cDNA clones encoding mRNAs unique to specific tissues or whose abundance is increased in response to particular treatments. Furthermore, it is essential to isolate and analyse genes of developmental and regulatory significance, whose mRNAs may be relatively rare within the cell, in addition to those encoding more abundant mRNAs. However, despite considerable advances in the techniques available to the molecular biologist for the identification of clones in cDNA libraries (Chapters 5, 6 and 7), differential screening is unlikely to identify clones representing mRNAs of low abundance, i.e. those present at

Plant Gene Isolation: Principles and Practice. Edited by G. D. Foster and D. Twell.
© 1996 John Wiley & Sons Ltd.

lower than 0.05–0.01% in the cDNA library. In addition, with techniques such as chromosome walking (Chapter 10), and T-DNA and transposon tagging (Chapters 11 and 12), isolation of a gene when only the phenotype is known is a difficult task. An alternative approach is the technique of subtractive hybridisation, which has been used successfully to isolate genomic and cDNA clones by the selective enrichment of genomic or cDNA that contains a particular population or sequence of DNA.

In this technique, two populations of nucleic acid, of which one contains the unique or enriched sequence(s), are hybridised in such a way that hybrids are formed between sequences common to both populations. These hybrid sequences can then be removed in a number of ways to leave behind a population of nucleic acid enriched in the target sequence(s). The subtracted DNA ultimately obtained can be radiolabelled for use as a hybridisation probe or used to construct a subtractive library. Subtractive hybridisation is an old approach, first used to purify T4 phage mRNA (Bautz & Reilly, 1966), that has been advanced by a number of developments, including incorporation of the polymerase chain reaction (PCR) to facilitate the generation of large amounts of DNA for hybridisations and/or recovery of the very small amounts of target DNA remaining after hybridisations.

In this chapter we review the various subtractive cloning approaches that have been used to isolate specific genes from both plant and non-plant sources and highlight some of the successes to date. A number of reports have described the use of subtractive hybridisation, and it is now possible to purchase commercial cDNA subtraction kits. However, it should be stressed that subtractive hybridisation is technically demanding and prone to the generation of artefacts, particularly when PCR is involved. In addition, this technique may in fact be of little use for isolating clones whose mRNA abundance is only moderately increased in target tissues (Li et al., 1994). Undoubtedly, a number of unsuccessful attempts at subtractive cDNA and genomic cloning have not appeared in the literature! Nonetheless, we also describe some experimental protocols, based on our own work and that of others, to carry out subtractive hybridisations and recover subtractively enriched DNA for subsequent probing or library construction.

2   SUBTRACTIVE APPROACHES

Generalised schemes for cDNA and genomic subtractions are depicted in Fig. 1. In both cases, the driver and target nucleic acids (NAs) must be prepared and modified if necessary for subsequent manipulation. Such modifications include biotinylation for removal after hybridisation and adapter ligation for amplification by PCR. Driver and target NAs are then denatured where appropriate and hybridised. Following hybridisation, the duplexes formed, representing sequences common to both driver and target, are removed, and the subtracted DNA used in a number of ways: it may be amplified, re-subtracted, labelled for use as a probe or used to generate a subtracted library.

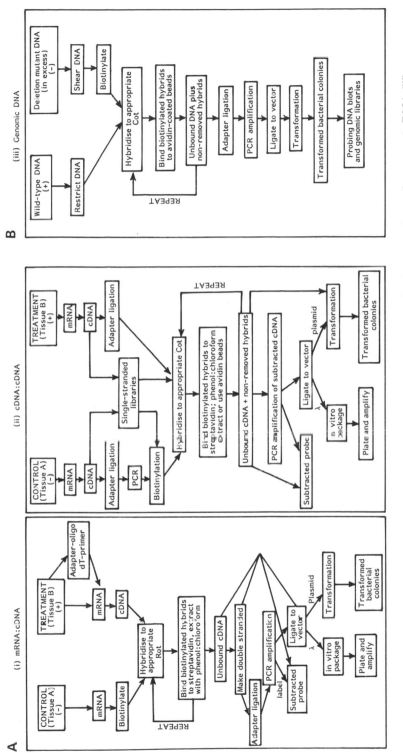

Figure 1. (A) Subtractive hybridisation schemes for cDNA (i, ii). (B) Subtractive hybridisation scheme for genomic DNA (iii)

2.1   cDNA subtraction

cDNA subtraction relies on the principle that only certain subpopulations of mRNA species are present, or present at increased concentrations, in specific tissue types or in tissues subjected to certain treatments, e.g. hormonal or environmental, but that certain subpopulations (i.e. those encoding 'house-keeping genes') are present in all cells.

2.2   Genomic subtraction

Genomic subtraction is dependent on the absence of a gene in a particular DNA population, e.g. genomic DNA prepared from deletion mutants in which a short section of DNA is missing, this being the only difference between DNA isolated from wild type and mutant.

2.3   Preparation of target and driver nucleic acids

In either genomic or cDNA subtractions, two populations of NA, the *driver* NA, deficient or depleted in the target sequence(s), and the *target* NA, enriched for the target sequences, are hybridised such that the concentration of the driver greatly exceeds that of the target. The rate of annealing or reanneal-ing is determined by the concentration of the driver NA which drives the reaction to completion. Driver NA is therefore either mRNA or cDNA prepared from tissue not expressing the gene of interest or genomic DNA isolated from the deletion mutant, and is ofter referred to as the minus sample. Target NA is either cDNA from tissues containing the mRNA species of interest or wild-type genomic DNA containing the gene which is missing from the deletion mutant, and is often referred to as the plus sample (Fig. 1).

2.4   Hybridisation

Subtractive hybridisations have been carried out using mRNA and single-stranded cDNA (Fig. 1A), between single-stranded cDNA and single-stranded cDNA, between double-stranded cDNA and single-stranded cDNA, between double-stranded cDNA and double-stranded cDNA (Fig. 1A) and between (double-stranded) genomic DNA (Fig. 1B). When the target NA is single-stranded, then after hybridisation those sequences unique to or enriched in the target will remain single-stranded; such single-stranded sequences can be separated from the hybrids. When double-stranded DNA populations are denatured and then hybridised, sequences present in equal abundance in both populations will cross-hybridise more rapidly than sequences less abundant in the driver can anneal to their complementary sequences. As the concentration of driver DNA is in excess over that of target DNA, sequences derived from the target population but common to both are more likely to hybridise to com-plementary sequences derived from the driver population. Driver DNA (and mRNA) can be 'tagged' with biotin, such that, after hybridisation, any tagged sequences can be separated from those that are not biotin-labelled.

Hybridisation between single-stranded molecules, e.g. mRNA/cDNA or single-stranded cDNA/single-stranded cDNA, is essentially a pseudo-first-order reaction (Sargent, 1987; Okamura & Goldberg, 1989), described by equation 1:

$$C/C_0 = \exp(-kR_0 t) \qquad (1)$$

where C is the concentration of single-stranded cDNA sequences at time t, C_0 is the initial concentration of cDNA, R_0 is the initial concentration of RNA (in moles of nucleotide per litre), t is the time (in seconds) and k is the pseudo-first-order rate constant. It has been calculated that hybridisation to a $R_0 t$ of approximately 1000 is required to achieve 99.9% completion, such that any single-stranded cDNA remaining behind is probably not homologous to driver mRNA (Sargent, 1987).

The reassociation of single-stranded DNA molecules to form double-stranded duplexes starting with double-stranded molecules is a second-order reaction described by equation 2:

$$C/C_0 = 1/(1 + kC_0 t) \qquad (2)$$

where C is the concentration of single-stranded DNA sequences at time t, C_0 is the initial concentration of single-stranded DNA (in moles of nucleotide per litre), t is the time (in seconds) and k is the second-order rate constant (Okamura & Goldberg, 1989). Detailed discussions of the kinetics of NA hybridisation can be found in Van Ness & Hahn (1982), Sargent (1987) and Okamura & Goldberg (1989), and references therein. Furthermore, Milner et al. (1995) have carried out a theoretical analysis of (double-stranded) DNA DNA subtractive hybridisation and formulated a model to predict the enrichment of target sequences during hybridisation of genomic or double-stranded cDNA.

The degree of subtraction is dependent on $R_0 t$ or $C_0 t$, i.e. the concentration of driver NA and the time of hybridisation, and the molar excess of driver over target NA. The requirement for a large excess of driver over target has led to various refinements of the hybridisation technique, in which the driver mRNA is first converted to cDNA, followed by cloning and/or amplification by PCR to produce a large and renewable source of driver cDNA.

2.5 Separation

After hybridisation of driver and target NAs, it is clearly essential to separate non-hybridised cDNA/DNA from hybridised duplexes in order to allow subsequent manipulation of the subtracted DNA, i.e. for rehybridisation, probing or library construction. In some protocols, the common and unique NAs are not physically separated but only the subtracted DNA is recovered. This has been achieved by specific cloning of the subtracted cDNA utilising clonable termini present only on plasmids containing target cDNA (Gulick & Dvorak, 1990; Aguan et al., 1991). Alternatively, crosslinking of RNA–cDNA duplexes has been used, such that only non-hybridised cDNA is available for subsequent labelling (Hampson et al., 1993). However, in most cases, the subtracted target single-stranded or single-stranded/double-stranded DNA is physically separated from the double-stranded hybrids, by hydroxylapatite chromatography or biotin labelling.

3   cDNA SUBTRACTION

3.1   Preparation of driver and target NAs

The use of RNA as the driver NA (Fig. 1A, i) has the advantage that hybridisation is between single-stranded RNA and cDNA. Thus, only inter-pool hybridisation will occur, as opposed to the intra-pool hybridisation that will also occur when populations of double-stranded DNA are melted and subsequently allowed to hybridise (Fig. 1A, ii). RNA can also be labelled with biotin for the subsequent removal of hybrids (Sive & St John, 1988). A number of groups have successfully used RNA as the driver NA in subtractive hybridisation reactions and identified several upregulated clones (Table 1). However, RNA–cDNA hybridisation has the disadvantage that large quantities of rRNA-free mRNA are required and the mRNA may be thermally degraded during long hybridisations. To circumvent this problem, some recent reports have described the use of oligo(dT)$_{30}$–latex and paramagnetic bead technology: mRNA or cDNA is coupled to the solid support and can then be hybridised repeatedly with cDNA or mRNA (Hara et al., 1991; Lopez-Fernandez & del Mazo, 1993; Sharma et al., 1993). Alternatively, in vitro transcription has been used to increase the availability of RNA for hybridisations. Directional cDNA libraries were first prepared from control tissues in vectors containing T3 or T7 promoters. Large amounts of RNA were then synthesised via T3 or T7 RNA polymerase (Swaroop et al., 1991; Gruber et al., 1993; Li et al., 1994).

Instead of using mRNA as the driver, several groups have instead used cDNA (Fig. 1A, ii). One approach has been to use single-stranded phagemids as both the target and driver. Duguid et al. (1988) developed a subtractive cloning procedure based on the hybridisation of single-stranded cDNA libraries in a vector containing the M13 phage origin of replication. Single-stranded cDNA libraries were produced from control and scrapie-infected hamster brain tissue. The single-stranded DNA from either could then be biotinylated and used as the driver to subtract common sequences from the other single-stranded DNA. A similar method has been described by Rubenstein et al. (1990), who improved the efficiency of subtractive hybridisation by using single-stranded phagemids with directional inserts. Herfort & Garber (1991) also used directional single-stranded cDNA libraries as the driver and target. However, inserts were first released from the target library prior to hybridisation with an excess of the biotinylated driver library. Directional single-stranded cDNA libraries were used as the target by Swaroop et al. (1991) and Li et al. (1994) but the driver NA was biotinylated RNA generated by in vitro transcription of the control library as described above. Recently, Wilson et al.

Table 1. Identification of cDNA clones via subtractive hybridisation

cDNA clones and expression	Species	Subtractive hybridisation method used	Potential function of gene products	Reference
Several, e.g.: E4, E8, E17. Ethylene inducible in fruit	*Lycopersicon esculentum* (tomato)	RNA–cDNA/HAP. Subtracted library screened with subtracted and self-subtracted control probes	Proteinase inhibitor; unknown	Lincoln et al. (1987)
cDef 1	*Antirrhinum majus* (snap-dragon)	RNA–cDNA/HAP. Subtracted flower-specific library screened with subtracted wild-type and mutant probes	Transcription factor	Sommer et al. (1990)
Several, e.g.: *ESI2, ESI18, ESI28, ESI32.* Salt-stress-induced in roots	*Lophopyrum elongatum* (wheatgrass)	cDNA library–cDNA library/FPERT/selective cloning of enriched fragments, differential screening	Unknown	Gulick & Dvorak (1990)
Several, e.g. *lip 5, lip 9.* Low-tempera-ture-induced	*Oryza sativa* (rice)	cDNA–cDNA. Selective cloning of adapter-ligated target cDNA	Unknown	Aguan et al. (1991)
Several. Cytokinin-inducible	*Physcomitrella patens* (a moss)	PCR amplification of adapter-ligated cDNA. Biotinylation, cDNA–cDNA hybridisa-tion, streptavidin treat-ment. Subtracted cDNA used as hybridisation probe	Unknown	Wang et al. (1992)
Imt1. Induced by osmotic stress	*Mesembryanthemum crystal-linum* (common ice plant)	RNA–cDNA/HAP. Subtracted library screened by differential hybridisation	Myoinositol-O-methyl-transferase	Vernon & Bohnert (1992)

continued overleaf

Table 1. (*cont.*)

cDNA clones and expression	Species	Subtractive hybridisation method used	Potential function of gene products	Reference
NGSUB7, NGSUB8. Expression increased during callus formation	*Nicotiana glauca*	PCR amplification of adapter-ligated cDNA. Biotinylation, cDNA–cDNA hybridisation, streptavidin treatment, construction of subtracted library	Unknown; DNA-binding protein	Cecchini et al. (1993)
GmN 36, GmN 70, GmN 93, GmN 315. Nodule-specific or nodule-enhanced expression	*Glycine max* (soybean)	RNA–cDNA/HAP. Subtracted probe used to screen a nodule cDNA library	Unknown; metal-binding protein; assimilate transport	Kouchi & Hata (1993)
DR 57. Induced by infection by *Pseudomonas syringae* pathover tomato	*L. esculentum* (tomato)	RNA–cDNA/HAP. Subtracted probe used to screen a cDNA library prepared from infected leaves	Leucine aminopeptidase	Pautot et al. (1993)
Several, e.g. 2A, 12A, 18A, 102B, 108A. Expression induced or repressed by silicon treatment	*Cylindrotheca fusiformis* (a marine diatom)	cDNA library–cDNA library/FPERT. Selective cloning of enriched fragments, differential screening	Various, e.g. tubulin, ADP/ATP translocator, adenosylhomocysteinase	Hildebrand et al. (1993)
Several. Induced by infection of host roots by *Pythium dimorphum*	*Picea abies* (Norway spruce)	cDNA–RNA hybridisation on magnetic beads PCR amplification of subtracted cDNA. Differential screening of subtracted library	Unknown	Sharma et al. (1993)

pAt 1892, pAt 1894. Gibberellin-induced	Arabidopsis thaliana (thale-cress)	RNA–cDNA. cDNA synthesised from mRNA using adapter–oligo(dT) primers for subsequent PCR. Biotinylation, streptavidin treatment, PCR amplification of subtracted cDNA. Subtracted library screened by differential hybridisation	γ-TIP (tonoplast intrinsic protein); unknown	Phillips & Huttly (1994)
Many, e.g. DB103, DB117, DB226, DB265. Giant cells in root induced by infection by the nematode Meloidogyne incognita	L. esculentum	PCR amplification of adapter-ligated cDNA. ss cDNA–library–cDNA hybridisation, biotinylation, streptavidin treatment, construction of subtracted library	E2 enzyme (protein ubiquitination), RNA polymerase II, proton-ATPase, antisense-Tnt1-94 transposon	Wilson et al. (1994) Bird & Wilson (1994)
Several, wilt-induced	A. thaliana	PCR amplification of adapter-ligated cDNA. cDNA–cDNA–hybridisation, biotinylation, avidin bead treatment. Subtracted probe used to screen cDNA library prepared from wilted leaves	Unknown; monooxygenase	Bulman (1994)

HAP, hydroxylapatite; FPERT, formamide-phenol emulsion reassociation technique.

(1994) have similarly utilised a single-stranded phagemid cDNA library for subtractive hybridisation; in this case, the target library was hybridised with an excess of PCR-generated biotinylated double-stranded cDNA as driver.

Another approach has been to hybridise double-stranded libraries. Gulick & Dvorak (1990) constructed plasmid libraries from the roots of both salt-stressed (target) and control (driver) wheatgrass (*Lophopyrum elongatum*). Sonicated plasmid DNA from the driver library was then hybridised in excess with inserts from the target library. After hybridisation (in a formamide–phenol emulsion), only sequences unique to or highly enriched in the target library were expected to reassociate with themselves and reconstitute their cloning ends (the hybridisation mix was cloned into the *Xba*I and *Sst*I-digested pJET2 vector). Hildebrand et al. (1993) used the same technique to clone silicon-responsive cDNA clones from the marine diatom *Cylindrotheca fusiformis*. Aguan et al. (1991) hybridised control (driver) cDNA (2 µg) with adapter-ligated target cDNA (0.1 µg). The driver cDNA was *Alu*I-digested prior to hybridisation to increase greatly the molar ratio, and the target cDNA ligated to an adapter, such that it possessed a 5′ TT overhang. The hybridisation mix was then cloned into a pUC18 vector treated so as to possess a 5′ AA overhang. Finally, Travis & Sutcliffe (1988) used cDNA–cDNA hybridisation to generate a subtracted probe. A plasmid cDNA library was first made from monkey brain cerebellum. Inserts from this library were then hybridised in excess with cortex single-stranded cDNA to generate the probe.

The development of PCR technology has obviated the need to produce cDNA libraries in order to meet the requirement for large amounts of driver cDNA (Fig. 1A, ii). Further, incorporation of PCR into subtraction protocols has also facilitated the generation of workable quantities of subtracted cDNA from the nanogram amounts remaining after subtraction and can even be used to incorporate biotin into the driver cDNA. In recent years, PCR has been used to generate cDNA libraries from minute quantities of RNA isolated from very small amounts of tissue or even single cells. A PCR amplification sequence is attached to the cDNA either via inclusion of the sequence in an adapter–oligo(dT) primer used for reverse transcription (Belyavsky et al., 1989; Domec et al., 1990; Gurr & McPherson, 1991) or via ligation of specific double-stranded adapters to the cDNA, i.e. ligation-mediated PCR (Jepson et al., 1991; Neill et al., 1993; Wilson et al., 1994). The resulting cDNA can then be amplified using a primer complementary to one strand of the adapter. Both approaches have been used to generate sufficient cDNA following subtraction. Timblin et al. (1990) used the latter approach to amplify cDNA remaining after exhaustive subtraction. A synthetic oligonucleotide containing a number of restriction sites was ligated to the cDNA (first made double-stranded). This was then amplified by PCR using a single primer complementary to one strand of the adapter to generate a renewable source of cDNA for probing or library construction. Sharma et al. (1993) and Hara et al. (1991) have also used PCR of adapter-lig-ated cDNA to recover sufficient DNA for cloning. Phillips & Huttly (1994) prepared a subtracted library from gibberellin (GA)-treated *Arabidopsis* using the adapter–oligo(dT) method. Following cDNA synthesis, G-tailing (to pro-vide a primer site for second-strand formation) and subtractive hybridisation, they converted the cDNA to a double-stranded form using an oligo(dC) adapter–primer, and amplified the cDNA using a primer complementary to a sequence in both adapter–primers. The adapters contained an EcoRI site for subsequent cloning.

PCR has also been used to great advantage to generate large amounts of driver cDNA for use in hybridisation reactions, so that C_0t and driver/target ratios can be maximised. Lebeau et al. (1991) used PCR to amplify the insert sequences from a cDNA library prepared from control tissue. They used PCR primers corresponding to the T7 and SP6 promoters of the λGem4 vector and included biotin-16-dUTP in the PCR reaction mix, so that the resulting amplified cDNA was also biotinylated. This driver was then used to subtract single-stranded cDNA from target tissue to generate a subtracted probe.

Duguid & Dinauer (1990) developed a system in which double-stranded cDNA could be amplified by PCR to generate cDNA libraries in vitro. These libraries could then be subtracted and the subtracted cDNA reamplified using PCR such that the subtraction could be repeated *ad infinitum*. In their method, cDNA was prepared from both target and control tissues and ligated to a PCR adapter containing an EcoRI restriction site. The adapter was blunt-ended at one end but had a four-nucleotide overhang at the other, such that it would ligate to both ends of the cDNA via its blunt end and the same adapter-complementary primer could be used to amplify the cDNA sequences internal to the two adapters. The control cDNA was then amplified and EcoRI-restricted (to reduce the chances of later amplification of non-subtracted cDNA) before being biotinylated and used to subtract the target cDNA. After being subtracted and separated, the subtracted target cDNA was amplified by PCR and re-subtracted. This sequential subtraction and amplification was repeated several times before the subtracted target cDNA was EcoRI-restricted and cloned into a plasmid vector. To increase further the chances of detecting induced clones, the control library was subtracted against itself to generate a subtracted library enriched for low-abundance sequences. The rate of annealing depends on, amongst other factors, sequence abundance. Thus, low-abundance sequences would anneal more slowly and therefore the representation of such sequences in the non-biotinylated pool would increase. Consequently, their representation in a PCR-generated library should be increased relative to that in a non-subtracted one. The same in vitro library amplification and subtraction has been used by Ace et al. (1994), except that in their case driver cDNA was digested with AluI and Rst prior to hybridisation. Wang et al. (1992) have also outlined a similar strategy. However, driver and target cDNAs were ligated to different adapters, and different primers were used to amplify target or driver cDNA, thereby reducing the chance of amplifying any driver cDNA that may have sneaked through the subtraction. Cecchini et al. (1993) have recently described an improved method based on that of Duguid & Dinauer (1990). In this protocol, cDNA was first synthesised in such a way as to generate asymmetric ends, which could be amplified using two specific primers complementary to each end. Target cDNA was subtracted six times, with the subtracted cDNA being reamplified after every second hybridisation cycle. Finally, in a combination of methods, Wilson et al. (1994) have constructed a subtractive cDNA library from tomato roots enriched in nematode-induced sequences. Excess biotinylated driver cDNA was prepared by PCR using a protocol similar to that of Duguid & Dinauer (1990) and used to subtract a M13 single-stranded cDNA library prepared from target tissue. However, this cDNA library had itself been generated by PCR of adapter-ligated cDNA.

3.2   Hybridisation

Whatever means have been employed to generate target and driver NAs, the two populations must then be hybridised in such a way as to maximise annealing of complementary sequences and therefore the degree of subtraction. Thus it is desirable to hybridise to as high a R_0t or C_0t value as possible and to maximise the molar excess of driver over target. R_0t can be increased by reducing the hybridisation volume, and hybridisation accelerated by increasing the temperature and salt concentration (Van Ness & Hahn, 1982; Sargent, 1987). Mixing of the reaction also increases the extent of hybridisation (Van Ness & Hahn, 1982). Increasing the hybridisation time increases the apparent R_0t but can lead to thermal degradation of RNA, and, to a lesser extent, DNA, such that the true R_0t value is actually decreased (Travis & Sutcliffe, 1988). The figures reported in the literature for R_0t (in $mol\,l^{-1}\,s^{-1}$) vary considerably, from ~ 1000 (Phillips & Huttly, 1994) to 3000–4000 (Lincoln et al., 1987, Timblin et al., 1990, Pautot et al., 1993), whereas those for C_0t range from 500 (Wilson et al., 1994) to 1000 (Duguid et al., 1988; Rubenstein et al., 1990). Often the values are not reported, but clearly the important point is that the concentration of driver and the excess over target be as high as possible. Ratios of driver to target are usually in the range 10–100, and the molar ratio can be increased by reducing the size of the driver cDNA, either by shearing (e.g. Gulick & Dvorak, 1990) or by treatment with a restriction enzyme such as AluI (Aguan et al., 1991; Ace et al., 1994). Usually, reactions are carried out in 5–50 μl volumes at approximately 65 °C (after first boiling if necessary to denature double-stranded molecules), in 0.5–1.0 M NaCl for 20–50 h. Hybridisation mixtures are often buffered with 50 mM HEPES (4-(2-hydroxyethyl)piperazine-1-ethanesulphonic acid)-KOH or phosphate buffer at pH values between 6.8 and 8.0 (usually pH 7.5 or 7.6), although some workers have used EPPS (N-(2-hydroxyethyl)piperazine-N'-3-propanesulphonic acid), for which the pK_a varies less with temperature (Straus & Ausubel, 1990). Sodium dodecyl sulphate (SDS) and Na_2EDTA are usually included in the buffer, at 0.1–0.2% (w/v) and 0.5–2 mM respectively. Some workers have also included formamide in the reaction buffer so that the hybridisation temperature can be reduced, thereby reducing the chances of thermal degradation of the hybridising NAs during the long reaction times (Aguan et al., 1991; Herfort & Garber, 1991; Gruber et al., 1993). In addition, PERT (phenol emulsion reassociation technique) and variations thereof have also been used (Travis & Sutcliffe, 1988; Gulick & Dvorak, 1990; Hildebrand et al., 1993); in this method, hybridisation is carried out in a phenol–water emulsion, in which the rate of DNA reassociation is increased many-fold (Kohne et al., 1977). PERT has the advantage that in addition to the very great increases in the rate of hybridisation, the reduced temperatures and shorter incubation times lessen the chances of thermal degradation.

3.3 Separation

The first subtraction protocols with RNA–cDNA hybridisation used hydroxylapatite chromatography to separate double-stranded hybrids from non-reacted single-stranded molecules (e.g. Sargent, 1987). Hydroxylapatite is a form of insoluble calcium phosphate that binds NAs. Hybridisation mixtures are applied to hydroxylapatite columns at an elevated temperature (usually 60 °C) in low-phosphate buffer (e.g. 10–30 mM sodium phosphate). At this concentration of phosphate, both double-stranded and single-stranded NAs bind to the column. Single-stranded molecules can then be eluted with 0.12–0.14 M phosphate; double-stranded molecules are retained on the column but can be eluted with 0.4–0.5 M phosphate (Britten et al., 1974). This method has been incorporated into several successful subtractive cloning schemes (e.g. Lincoln et al. 1987; Sommer et al., 1990; Vernon & Bohnert, 1992; Kouchi & Hata, 1993; Pautot et al., 1993) but does have a number of disadvantages (Britten et al., 1974) and has largely been superseded in the literature by the use of biotin labelling and removal of labelled complexes using avidin or streptavidin. Sive & St John (1988) were the first to describe this technique. Driver RNA was first photobiotinylated. Following hybridisation of the biotinylated RNA and cDNA, streptavidin was added to the reaction mixture. Streptavidin has a high affinity for biotin, and the streptavidin–biotin–RNA–cDNA complexes could then be removed from the hybridisation mixture by precipitation and extraction with phenol/chloroform. Biotin labelling and removal with streptavidin are efficient and easy to perform and have been used a number of times with both RNA and cDNA as the driver (e.g. Phillips & Huttly, 1994; Rubenstein et al., 1990; Swaroop et al., 1991; Wang et al., 1992; Ace et al., 1994). DNA can be labelled with biotin either via photobiotinylation (perhaps 1 in 200 residues labelled (Sun et al., 1992)) or via PCR in the presence of biotin–dUTP (Lebeau et al., 1991; Wilson et al., 1994; Bulman, 1994), which should increase the degree of labelling and therefore maximise the efficiency with which biotin-labelled hybrids are subsequently removed. Biotin-labelled complexes can also be removed from reaction mixes using avidin–resin (Duguid et al., 1988) or with streptavidin- or avidin-coated beads (Wilson et al., 1994; Bulman, 1994).

A number of workers have completely avoided the separation of hybridised and non-hybridised cDNA. Gulick and Dvorak (1990) and Aguan et al. (1991) devised strategies that should have permitted the subsequent cloning of only subtracted (non-hybridised) sequences. Recently a technique has been described in which mRNA–cDNA hybrids are irreversibly crosslinked, such that only non-hybridised single-stranded cDNA can be subsequently radiolabelled for use as a probe (Hampson et al., 1993). However, it is not known how much non-subtracted DNA may contaminate the subtracted cDNA in these situations.

3.4 Uses of the subtracted cDNA

Subtracted cDNA remaining after hybridisation should be enriched for those cDNAs representing mRNAs whose abundance is increased in target tissue. Thus, the cDNA can be used either as a hybridisation probe to screen conventional cDNA libraries, or to construct a subtracted cDNA library, or a combination of the two (Fig. 1A). Screening with a subtracted probe has the advantage that the conventional library should contain full-length sequences and can be used for other purposes. The library can be differentially screened using the subtracted probe, and potentially positive plaques rescreened and their inducibility checked by northern analysis or RNA dot-blotting (e.g. Travis & Sutcliffe, 1988; Wang et al., 1992; Kouchi & Hata, 1993; Ace et al., 1994). The use of subtracted libraries, generated either pre- or post-hybridisation, has the advantage that they are very much smaller than conventional libraries (e.g. Cecchini et al., 1993; Phillips & Huttly, 1994), so that the chances of detecting induced/specific clones should be increased considerably. The chances can be increased even further by screening subtracted libraries by differential hybridisation (e.g. Sharma et al., 1993) or even with a subtracted probe (e.g. Lincoln et al., 1987; Sommer et al., 1990). Potential disadvantages of subtracted libraries include difficulties in cloning and the generation of artefacts, particularly if PCR is involved (Bulman, 1994), and the substantially reduced length of the clones, due to thermal degradation during hybridisation and preferential amplification of short fragments during PCR. However, the latter problem can be circumvented by size fractionation prior to cloning (Ace et al., 1994; Bulman, 1994) or by rescreening conventional libraries with subtracted clones (Phillips & Huttly, 1994).

4 cDNA SUBTRACTION PROTOCOL

Which one to use? There are clearly a large number of subtraction protocols available in the literature, developed for various reasons outlined in the text, and the choice of which to use is a difficult one. The main factors to be considered include the ease with which the procedure can be performed, the species and tissue to be used and the availability of NA for hybridisations; successful cloning requires large amounts of driver NA so that R_0t/C_0t values and driver/target ratios can be maximised. An ideal system would maximise the number of subtracted clones identified whilst minimising the number of technically demanding steps involved, but it is not always easy to discern from the literature just how successful a subtractive cloning procedure has been. However, a number of groups have attempted to determine the success of their subtractive hybridisations, either by following the removal of cDNA during hybridisation or, more directly, by estimating the representation within the cDNA of specific clones. Estimates of subtractive enrichment vary from 4–16-fold (Gulick & Dvorak, 1990) to 100–150-fold (Phillips & Huttly, 1994), to 200–300-fold (Hampson et al., 1993; Li et al., 1994) and even up to ~ 5000-fold (Rubenstein et al., 1990; Wilson et al., 1994). Generally, > 90% removal of common sequences is achieved during hybridisation (based on loss of target cDNA). Li et al. (1994) reported > 99% removal of actin sequences from their subtracted cDNA, and Cecchini et al. (1993) could not detect any actin hybridisation to their subtracted cDNA. Similarly, Ace et al. (1994) found that the representation of housekeeping genes fell dramatically in their subtracted cDNA library, whereas that of an inducible mRNA remained constant. In most cases the judgment must be empirical: a cDNA library is of little use unless the desired clones can be identified. Conversely, the subtractive strategy can be deemed successful if the desired clones are identified, even though they may well have been isolated as efficiently by conventional means.

We chose a PCR-based approach, in which cDNA is amplified in vitro to generate large amounts of cDNA that can be subsequently hybridised and separated. Such PCR-generated cDNA can be manipulated and stored conveniently, biotin-labelled by PCR and used both to construct libraries and to generate renewable probes. This approach has been used successfully to generate subtracted cDNA highly enriched for specific sequences (Duguid & Dinauer, 1990; Cecchini et al., 1993; Ace et al., 1994; Wilson et al., 1994). We have used it to isolate cDNAs whose expression is induced by wilting and abscisic acid (ABA) in *Arabidopsis*, as described in sections 4.2–4.9.

4.1 cDNA preparation

As with all cloning strategies, good quality mRNA, free from rRNA, is desirable. We isolate RNA using guanidine hydrochloride (Logemann et al., 1987) and purify mRNA using two rounds of oligo(dT)–cellulose treatment or with oligo(dT)-linked beads (Dynal or Promega). Double-stranded, blunt-ended cDNA is prepared using commercially available kits (e.g. Amersham, Stratagene). It is essential at this stage that the cDNA is blunt-ended, so that the next step of adapter ligation is efficient. Target cDNA was synthesised from mRNA isolated from 4-h-wilted leaves and driver cDNA from mRNA isolated from turgid leaves.

4.2 Adapter–primer ligation

We have used several different adapter–primers that all have certain features in common: they are blunt-ended at one end but possess an overhang at the other, so that they will ligate to both ends of the cDNA via their blunt ends; they also have an inbuilt restriction site to facilitate subsequent cloning. We synthesised a number of oligonucleotides based on sequence provided by M. Leech (IPSR, Norwich) (Wang et al., 1992) and also used a commercially available PCR adapter, the Uni-Amp adapter from Clontech. The homemade adapter we used was prepared from 25-mer and 13-mer oligonucleotides, termed M3 and M4 respectively. The non-kinased oligonucleotides were annealed by boiling together at equimolar concentrations (10^{-4} M) followed by slow cooling to room temperature. The annealed adapter–primer was aliquotted and stored at $-20\,°C$. This adapter–primer had a 12-nucleotide 5′

overhang and an *EcoRI* restriction site in the primer region. The 13-mer M4 oligonucleotide was subsequently used as the PCR primer. The Uni-Amp adapter is a 33-mer oligonucleotide with a 3' overhang, that has been phosphorylated, and, in our case, contained an *EcoRI* restriction site within the adapter region. PCR amplification uses the 25-mer Uni-Amp primer. Formation of adapter-dimers during ligation is eliminated when non-phosphorylated adapters are used; in this case the PCR conditions must be such that the unligated strand of the adapter falls off and is filled in before the actual PCR amplification cycle. However, we have not encountered any problems with the Uni-Amp adapter. We have used the M3/M4 adapter for driver cDNA and the Uni-Amp adapter for the target, to avoid any subsequent amplification of driver cDNA that might remain after subtraction. However, the use of different adapters and primers has the disadvantage that the PCR process itself may introduce artefactual differences into the two cDNA populations. An alternative strategy, therefore, is to use the same adapter for both driver and target, but to ensure that the driver cDNA is restriction-enzyme-treated to remove the PCR amplification sequences.

Determine the amounts of adapters and cDNA to be used by UV spectrometry and by an ethidium bromide plate assay. To carry out ligation reactions on a (relatively) large scale (for example, in our case, using the M3/M4 adapters), use the following reaction conditions: 1.5 μl 10× standard ligase buffer (containing ATP; supplied), cDNA (~ 300 ng), M3/M4 adapter–primer (500 pmol), T4 DNA ligase (3 U, Pharmacia), final volume 15 μl. For small-scale reactions (e.g. using the Uni-Amp adapter) use the following conditions: 1 μl 10× ligase buffer, cDNA (~ 25 ng), 1 μl Uni-Amp adapter (as supplied), T4 DNA ligase (2 U), final volume 10 μl. Incubate reactions at 16 °C overnight. Remove excess adapters by passage down a Spin-400 column (Clontech). This step has the added advantage that the cDNA is size-fractionated, with only molecules > 400 bp being retained.

4.3   PCR amplification

Different PCR conditions will be required depending on which adapter has been ligated to the cDNA. For M3/M4-adapted driver cDNA, the reaction mix contains 5 μl 10× PCR buffer, 8 μl dNTP mix (1.25 mM, final concentration 0.2 mM), adapted cDNA (1–10 ng), M4 primer (500 pmol), *Taq* polymerase (2.5 U, SuperTaq, HT Biotechnology, UK), final volume 50 μl. Cap the mixture with mineral oil (Sigma) and subject to the following PCR programme: 72 °C for 3 min, followed by 30 cycles of 94 °C for 1 min, 55 °C for 1 min, 72 °C for 10 min, with a final extension at 72 °C for 20 min. For Uni-Amp-adapted target cDNA, the reaction mix contains 5 μl 10× PCR buffer, 8 μl dNTP mix (1.25 mM, final concentration 0.2 mM), adapted cDNA (~ 1 ng), 1 μl Uni-Amp primer (as supplied), *Taq* polymerase (1.5 U), final volume 50 μl. Cap with mineral oil and subject to the following thermal cycle: 94 °C for 1 min, 60 °C for 1 min, and 72 °C for 2 min, for 30 cycles, followed by 72 °C for 10 min. Typically, several (2–10) micrograms should be obtained. PCR cDNA generated in this way is shown in Fig. 2A.

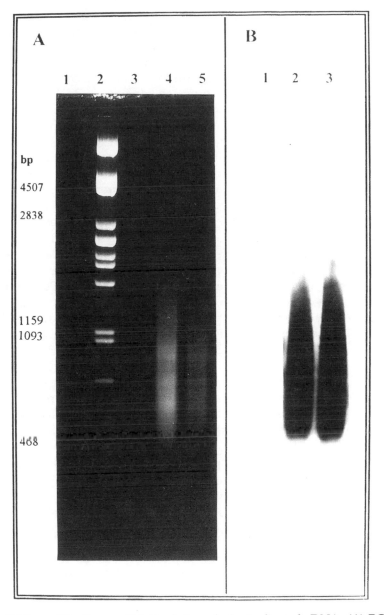

Figure 2. PCR amplification and biotinylation of adapter-ligated cDNA. (A) PCR amplification products–DNA detection: lane 2, MW markers; lane 1, control (no primers); lane
3, control (non-adapted cDNA); lane 4, M3/M4-adapted cDNA; lane 5, Uni-Amp-
adapted cDNA. (B) PCR amplification products—biotin detection: lane 1, non-biotinylated cDNA; lane 2, photobiotinylated; lane 3, PCR-biotin-labelled

4.4 Biotinylation of cDNA

We have used two methods of incorporating biotin into driver cDNA—photobiotinylation and PCR. We use photoprobe-biotin from Vector Laboratories,

although photobiotin is available from other sources, such as Sigma. Photobio-tinylation is relatively straightforward and should be performed as per the manufacturer's instructions. Mix 10 μg of cDNA (in water) with an equal volume of photoprobe-biotin stock (a 2-ml microcentrifuge tube with open lid is convenient) and irradiate in an ice bath for 10 min, 15 cm below a sunlamp. We used a 350-W mercury vapour/tungsten filament lamp from Vector Labs. The optimal wavelengths for photobiotinylation are between 350 and 370 nm, although the lamp has a wider range. Low UV wavelengths should not be used, as this may result in nicking of the DNA. After irradiation, dilute the solution, extract exhaustively with butanol to remove the photobiotin and ethanol precipitate the biotinylated cDNA. To incorporate biotin directly into the driver cDNA, it is possible to use PCR in the presence of biotin-21-dUTP. The reaction mix contains 5 μl 10× PCR buffer, adapted cDNA (1–10 ng), 1 μl dNTPs minus dTTP (10 mM, final concentration 0.2 mM), 1 μl dTTP (7.5 mM, final concentration 0.15 mM), 5 μl biotin-21-dUTP (Clontech, final concentra-tion 0.05 mM), primer (complementary to one strand of the adapter; 500 pmol), *Taq* polymerase (2.5 U, SuperTaq), final volume 50 μl. Use the same PCR conditions as above and remove excess primers on a Spin-400 column (Clon-tech). The success of biotinylation is shown in Fig. 2B. A DNA gel similar to that in Fig. 2A was Southern-blotted and the blot developed with the 'Southern Light' chemiluminescence biotin–avidin assay system (Tropix).

4.5 Hybridisation

We have only carried out hybridisations using aqueous buffers, but the use of PERT seems likely to increase greatly the efficiency of hybridisation (e.g. Travis & Sutcliffe, 1988). Co-precipitate target cDNA (250 ng) and biotinylated driver cDNA (20 μg) and redissolve in 4 μl of 1.25× hybridisation buffer. Add 1 μl of 5 M NaCl to give a final buffer composition of 10 mM EPPS, pH 8.0, 1 M NaCl, 1 mM Na_2EDTA, 0.1% (w/v) SDS. Cap the hybridisation mix with mineral oil, heat to 100 °C for 3 min and incubate at 65 °C for 48 h.

4.6 Separation of biotinylated and non-biotinylated hybrids

We have used two methods, both successfully, to remove biotin-labelled hybrids from reaction mixes. The first involves binding to streptavidin followed by precipitation and extraction with phenol/chloroform (Sive & St John, 1988). To do this make up the hybridisation mix to 50 μl with water and add 20 μg of streptavidin (Gibco BRL, prepared in 0.01 M phosphate buffer, pH 7.2, 0.15 M NaCl, 0.05% (w/v) sodium azide). After incubation at room temperature for 10 min, extract the mixture with an equal volume of phenol/chloroform (1 : 1) and remove the aqueous phase, avoiding the white precipitate at the interface. Back-extract the organic phase with 50 μl of water and repeat the streptavidin and phenol/chloroform steps. Combine the aqueous phases and ethanol preci-pitate the cDNA (with more driver cDNA if the subtraction is to be repeated). The second method utilises avidin-coated beads (Straus & Ausubel, 1990). We have tried streptavidin-coated paramagnetic beads (Dynal) but could not get any appreciable binding of ^{32}P-labelled hybrids. However, Wilson et al. (1994) have used streptavidin-coated paramagnetic beads from Promega with some success. The avidin beads must be washed in EE buffer (10 mM EPPS, pH 8.0, 1 mM Na$_2$EDTA) before use. Pellet 2 ml of 5% avidin beads (Baxter Health Corporation, USA) in a microcentrifuge, resuspend in 2 ml of EE and repeat the centrifugation and washing steps. Resuspend the beads in 2 ml of EEN (EE plus 0.5 M NaCl). After subtractive hybridisation, make up the reaction mix to 100 μl with EEN and add 100 μl of washed beads. After 1 h of incubation at room temperature, collect the beads by centrifugation through a microcentrifuge cup (Millipore, UK). Wash the reaction tube with 50 μl of EEN and repeat the bead treatment. Ethanol precipitate the residual target cDNA (with excess driver cDNA if the subtraction is to be repeated).

It is advisable to determine the efficiency with which biotin-labelled molecules can be removed. We tested our system with biotinylated restricted lambda DNA (by UV) and by using random-primed ^{32}P-labelled biotinylated plasmids and cDNA. In the latter case, it is essential that unincorporated nucleotides are removed (e.g. by passage down a Nick column (Pharmacia)) before use. Typically, we find that both streptavidin and avidin bead treatments are very efficient, with > 90% of biotinylated molecules being removed.

4.7 Library construction

The subtraction cycles can be repeated several times, although too many may be detrimental, as the cDNA will suffer from thermal degradation. Moreover, there may be sequences that do not subtract efficiently (e.g. Van Brunt, 1992). If the target cDNA has been adapter-ligated, it can be amplified after each subtraction if desired. The target cDNA can then be restricted and cloned into a suitable vector, such as lambdaZAPII, which has a number of useful characteristics. We carried out four subtraction cycles and amplified the residual target cDNA (ligated to the Uni-Amp adapter). However, we have not been successful at cloning this DNA. Both driver and target PCR cDNAs were hybridised with rbcS and A1494 probes (see section 4.9) prior to subtraction and gave strong signals. After subtraction, the amplified target cDNA no longer hybridised to these probes, or to ^{32}P-labelled first-strand cDNA. Clearly, the target cDNA had been contaminated with extraneous DNA and so was not cloned. In our second attempt, we recovered PCR-amplified target cDNA that still hybridised to Arabidopsis cDNA probes. This time, however, we could not clone the DNA, possibly because it would not restrict prior to ligation to the lambdaZAP vector. Consequently, we instead generated a subtracted cDNA probe.

4.8 Screening of conventional library with subtracted probe

Radiolabelled first-strand cDNA was synthesised from 2 μg of mRNA prepared from target tissue (in this case Arabidopsis leaves wilted for 4 h). This cDNA was hybridised as before with biotinylated driver cDNA to a C_0t value of

~ 1100 (assuming an average cDNA length of 900 bp). Approximately 90% of target cDNA was removed by hybridisation and avidin bead treatment (as determined by Cerenkov counting) and the subtraction was repeated, resulting in a subtractive enrichment of ~ 100-fold. At this point, the remaining cDNA can be labelled to high specific activity by random priming (e.g. with the Amersham Megaprime kit). We removed a portion of the target cDNA prior to subtractive hybridisation and screened plaque lifts of a cDNA library (prepared from 4-h-wilted *Arabidopsis* leaves using a Stratagene lambdaUniZAP kit) using equivalent amounts of radioactive non-subtracted and subtracted probes. The corresponding autoradiograph is shown in Fig. 3. Clearly, subtraction resulted in a dramatic reduction in the number of cDNAs which hybridised to plaques in the library.

Plaques which did hybridise to the subtracted probe were picked off into 250 μl of SM buffer (Sambrook et al., 1989) and their inducibility confirmed by PCR amplification and RNA dot-blot analysis (Mutchler et al., 1992). Twenty-five-microlitre aliquots were freeze-thawed and amplified by PCR using T3 and T7 primers corresponding to lambdaZAP sequences outside the cloning region (see Chapter 13). Inserts were recovered from the gel, ^{32}P-labelled and then used to probe RNA and mRNA dot-blots of RNA prepared from turgid and wilted leaves. Typical results, showing one clone that was induced by wilting and one that was not, are shown in Fig. 3. To confirm the inducibility of clones identified in this way and to determine transcript size, northern analyses were carried out (Fig. 3). In this particular case, the subtracted probe has clearly identified a clone encoding a mRNA whose transcript levels increase dramatically following wilting and low-temperature treatment.

4.9 Controls

Any protocol involving PCR amplification is prone to the generation of arte-facts, and it is wise therefore to include a number of control steps to ensure that such artefacts can be spotted early on. Furthermore, wherever possible, the success of the subtractive hybridisation should be assessed by the use of clones representing housekeeping and inducible genes. The PCR process may itself introduce bias into the cDNA populations, as some sequences, e.g. relatively short ones, may be amplified more efficiently than others. Amplifica-tion of contaminating DNA can only be avoided by scrupulous cleanliness and care when carrying out all steps. The lack of amplified contaminants can be checked by setting up control PCR reactions (see Fig. 2), e.g. minus primers, minus cDNA and minus adapted cDNA but plus non-adapted cDNA (to discount non-specific amplification). Additionally, the PCR-generated cDNA can be probed with known sequences from the species being investigated. For example, we probed PCR cDNA with the ATS17 probe, representing *Arabidop-sis rbcS* mRNA (Krebbers et al. 1988); a band of the expected size was present on Southern blots (Bulman, 1994). We have also used this probe as a negative control for the subtractive hybridisations. The amount of *rbcS* mRNA declines dramatically following wilting of *Arabidopsis* (Williams et al., 1994a). Thus the

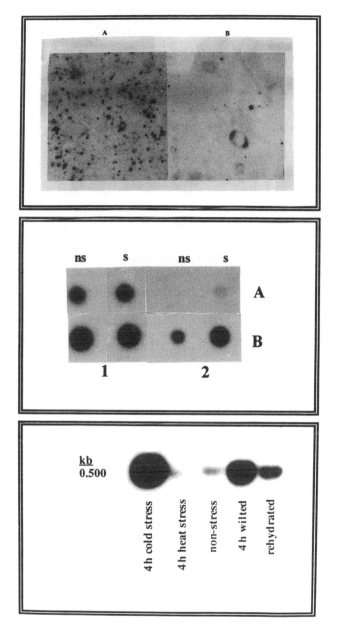

Figure 3. Identification of a wilting-stress-inducible cDNA using a subtracted probe. Upper: screening of a cDNA library with equivalent amounts of radioactivity of (A) cDNA probe; (B) subtracted cDNA probe. Middle: diagnostic RNA dot-blot with (1) non-inducible cDNA clone; (2) stress-inducible cDNA clone. s, stressed; ns, non-stressed; A, 5 μg RNA; B, 250 ng mRNA. Lower: northern analysis of mRNA isolated from *Arabidopsis* shoots exposed to various treatments and probed with wilt-induced cDNA clone identified by subtractive hybridisation

representation of *rbcS* cDNA in the subtracted cDNA should be far less than in the non-subtracted cDNA. Conversely, as a positive control, we have used A1494, a cDNA whose expression is strongly upregulated following wilting (Williams et al., 1994b).

5   IDENTIFICATION OF cDNAs USING SUBTRACTIVE HYBRIDISATION

The cDNAs isolated using subtractive hybridisation are shown in Table 1. Clearly, there have been some considerable successes in identifying inducible and tissue-specific clones. In many cases, the functions of the proteins encoded by the cDNAs still remain unknown. However, in at least two cases, the clones so isolated were found to encode proteins of great developmental importance. The first of these was cDef1, encoding a transcription factor containing a new DNA-binding motif (subsequently termed the MADS box) (Sommer et al., 1990). The other was *Imt1*, a salt-stress-inducible gene encoding myoinositol-*O*-methyltransferase, an enzyme likely to be required for osmotic adaptation (Vernon & Bohnert, 1992). A number of other groups have also adopted the subtractive approach to search for environmentally induced clones (Table 1). Bird and his colleagues (Wilson et al. 1994; Bird & Wilson, 1994) have isolated more than 200 clones that appear to represent mRNAs upregulated in response to nematode infection. It is likely that sequence and expression analyses of these clones will illuminate those cellular processes initiated during the formation of giant cells. Similarly, cDNA clones have been isolated from host tissues following invasion by pathogenic or symbiotic micro-organisms (Table 1); again, analysis of these clones will be informative with regard to the processes of plant–microbe interaction. Finally, several cDNA clones have been isolated representing mRNAs whose abundance is increased following treatment with hormones such as ethylene, cytokinins, gibberellins and ABA.

6   GENOMIC SUBTRACTION

In contrast to cDNA subtraction, genomic subtraction (Fig. 1B) has rarely been used to isolate specific sequences, and in fact only one plant gene has so far been cloned in this way (Sun et al., 1992). However, the method developed in Ausubel's laboratory (Straus & Ausubel, 1990) may well find application for plants with relatively small genomes, such as *A. thaliana*, provided that suitable deletion mutants can be generated. Unlike cDNA hybridisation, where the cDNA populations are of limited complexity, genomic subtraction involves the hybridisation of whole genomes, and therein lies the problem. The more complex the DNA population, the lower the concentration of particular DNA sequences and thus the slower the rate of reannealing. Single rounds of subtraction therefore result in only a relatively low degree of enrichment. On the other hand, methods described by Ausubel and others employ multiple rounds of subtraction coupled to PCR to amplify the subtracted DNA and can effect much greater levels of enrichment.

In essence, genomic subtraction involves the hybridisation of two genomic DNA populations, of which one is deficient in a sequence or sequences for various reasons (e.g. deletions, different strains). Lamar & Palmer (1984) were the first to utilise genomic subtraction, to clone sequences from the Y chromosome of mice. Target DNA was prepared by cleavage with Sau3A and hybridised with an excess of randomly sheared driver DNA. Double-stranded DNA was then cloned into the BamHI site of a cloning vector, assuming that only target DNA that had reannealed to itself would have generated clonable ends. Similar methods have been used to clone DNA fragments which become deleted in certain human conditions (Kunkel et al., 1985; Nussbaum et al., 1987). Although this technique has been successful, the yield is poor, as it is limited by the rate of annealing of dilute target DNA, and the degree of enrichment cannot exceed the mass ratio of driver to target DNA. Thus the 100-fold or so enrichments achieved are of limited use in cloning fragments corresponding to small deletions (Chasan, 1992).

Wieland et al. (1990) and Straus & Ausubel (1990) both described genomic subtraction protocols which employed multiple rounds of subtraction and recovery of target DNA by PCR. Wieland et al. (1990) used their protocol in a model situation, in which lambda DNA had been added to human DNA to act as the target sequence. This DNA was Sau3A digested, size-fractionated and biotinylated before being ligated to an oligonucleotide adapter–primer. The target DNA was then subtracted with sheared driver DNA (human DNA minus lambda DNA) and the non-annealed single-stranded sequences recovered with hydroxylapatite. After several rounds of subtraction and recovery, single-stranded target DNA was recovered by avidin/biotin chromatography, rendered double stranded and amplified by PCR (with one strand of the adapter as primer), and finally cloned. Plasmid colonies were pooled and analysed for the presence of lambda DNA: enrichments of 300–700-fold were achieved.

The protocol developed by Straus and Ausubel is similar but more efficient and straightforward. In their method, wild-type yeast DNA (target) was Sau3A digested and hybridised with an excess of randomly sheared photobiotinylated DNA from a deletion mutant (driver). Following denaturation and hybridisation, hybrids formed between DNA from the target and corresponding biotinylated driver sequences were subtracted with avidin beads. After several rounds of subtraction the residual target DNA was ligated to an oligonucleotide adapter and PCR amplified using an adapter-complementary primer. The resulting DNA was radiolabelled and used to probe a yeast genomic library. Straus and Ausubel isolated a 5-kb fragment corresponding to the deletion and estimated that they had achieved a 1000-fold enrichment. This technique has recently been used by Darrasse et al. (1994) to isolate DNA probes specific for the bacterium Erwinia carotovora subsp. atroseptica. DNA from this potato pathogen was subtracted with DNA from subsp. carotovora CH26, a weaker pathogen, with the aim of identifying DNA related to pathogenicity and for use as a diagnostic probe. The technique has also been used to isolate mating-type genes from the rice-blast fungus *Magnaporthe grisea* (Kang et al., 1994).

Sun et al. (1992) have applied genomic subtraction to clone the *GA1* locus in *Arabidopsis*. Mutations at this locus result in gibberellin-deficient dwarf plants, due to a lack of *ent*-kaurene synthesis. The same protocol was used as that described by Straus & Ausubel (1990): *Arabidopsis* wild-type DNA served as target, and DNA isolated from the presumptive deletion mutant *ga1-3* as driver. Sun et al. also performed a reconstruction experiment, in which adenovirus (Ad-2) DNA was added as target DNA to *Arabidopsis* wild-type DNA. The results of this reconstruction are shown in Fig. 4. Each round of subtraction resulted in a 10-fold enrichment. Sun et al. then did the *Arabidopsis* *ga1-3*/wild-type subtraction, carrying out five rounds of subtraction prior to adapter ligation and PCR amplification. Unlike the situation with yeast, however, the target DNA was not sufficiently enriched to screen an *Arabidopsis* genomic library. Instead, six cloned PCR fragments were ^{32}P-labelled and used to probe DNA gel blots of wild-type and *ga1* mutants (Fig. 4). One of these

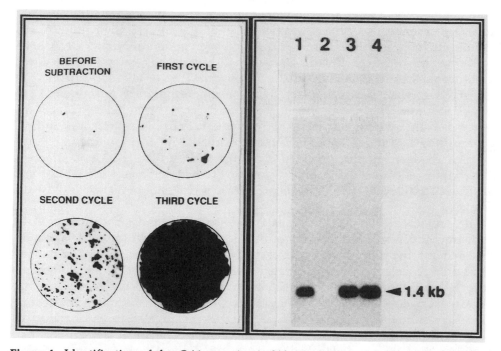

Figure 4. Identification of the *GA1* gene in *Arabidopsis*. Left: enrichment of target gene (Ad-2) during three subtraction cycles in a genomic subtraction reconstruction experiment. PCR-amplified subtracted DNA was cloned and the resulting colonies probed with Ad-2. Right: detection of deletion in *GA1* gene in *Arabidopsis* genomic DNA. The blot contains 1 µg of *Hind*III-digested DNA from *A. thaliana* Landsberg *erecta* (1), *ga1-3* (2), *ga1-4* (3) and *ga1-2* (4). The blot was probed with a 250-bp *Sau*3A fragment from pGA1-1, cloned via subtractive hybridisation of wild-type and *ga1-3* DNA. Reproduced by permission of the American Society of Plant Physiologists from Sun et al. (1992)

pGA1-1, detected a 1.4-kb fragment present in wild-type, *gal-2* and *gal-4* DNA, but missing from *gal-3* DNA. This clone was then used to isolate genomic clones and show that the *gal-3* mutation was due to a 5-kb deletion, corresponding to 0.005% of the genome. In this case, genomic subtraction had again resulted in an enrichment of target sequence of ~ 1000-fold. Unfortunately, this is unlikely to be sufficient to identify deletions in plants with more complex genomes than *Arabidopsis* (Chasan, 1992), even if well-defined mutants were available. Improvements in genomic subtraction may be achieved by increasing the rate of hybridisation of target and driver DNA, e.g. by the use of PERT, and by improving the recovery of target DNA following subtraction. The Straus and Ausubel method requires adapter ligation to target DNA *after* subtraction, thus relying on the self-annealing of dilute target sequences. If the PCR adapter were ligated to the target DNA prior to subtraction, then PCR would generate double-stranded DNA for amplification. In fact, Bjourson et al. (1992) have described a combined subtraction and PCR amplification scheme (to generate strain-specific *Rhizobium* sequences) that does just that. In their protocol, both driver and target DNAs were restricted and adapter-ligated prior to subtraction, thus providing a means to both recover subtracted target and regenerate further driver. The driver DNA was amplified using dUTP instead of dTTP and then photobiotinylated. After hybridisation, common hybrids were removed by streptavidin treatment. Any residual streptavidin–biotin–DNA complexes were removed on a NENSORB cartridge and by treatment with uracil-DNA glycosylase, which destroys any traces of dUTP-containing driver DNA. Finally, residual target DNA was PCR amplified using a different primer to that required for driver DNA. Two rounds of subtraction–amplification were sufficient to generate a highly specific probe; this method is clearly applicable to both cDNA and genomic cloning of plant genes.

7   CONCLUDING REMARKS

There is no doubt that there have been some remarkable successes achieved by subtraction cloning, such as isolation of the *Arabidopsis GA1* locus by Sun et al. (1992) and generation of the nematode-induced subtracted cDNA library by Bird and his colleagues (Wilson et al., 1994). Many cDNA clones have been isolated which encode mRNAs whose expression is upregulated in specific tissues or in response to various hormonal or environmental treatments. However, in many cases a key task remains the identification of genes of regulatory significance, i.e. genes whose cellular function may be to initiate and co-ordinate the cascades of gene expression associated with cellular differentiation and adaptation. The products of such genes may include transcription factors, enzymes and regulators of hormone metabolism, receptor proteins and other proteins involved in cell signalling, and it may be that their mRNAs are of low abundance within the cell. Thus the advantages of subtractive technology offer an attractive means of fishing out such genes. On the other hand, only one plant gene has so far been isolated using genomic subtraction. The utility of this technique is likely to be limited by genomic complexity, although the incorporation of PERT may permit its use for species with greater genome complexity than *Arabidopsis*. However, a key requirement is the existence of well-characterised deletion mutants. If such mutants can be generated for *Arabidopsis*, then the use of genomic subtraction may well be extended. A further development likely to make an impact in plant molecular genetics is the technique of restriction fragment length polymorphism (RFLP) subtraction recently developed by Straus and his colleagues (Rosenberg et al., 1994).

ACKNOWLEDGMENTS

Funding for some of the experimental work described in this chapter was provided by the SERC and Nuffield Foundation, and we thank Ms J. Williams, UWE, Bristol for advice and technical assistance. We are grateful to Dr E. Krebbers, Plant Genetic Systems, Gent, Belgium for the pATS17 clone and to Dr N.P. Minton, CAMR, Porton Down, UK for oligonucleotide synthesis. We are also grateful to a number of people for information, advice, pre-prints and unpublished personal communications. In particular, we thank Dr A. Phillips (LARS, Bristol, UK), Drs D. Baulcombe and M. Leech (IPSR, Norwich, UK), Dr J.J. Milner (Department of Botany, University of Glasgow, Scotland, UK), Dr A. Lonneborg (Norwegian Forest Research Institute, As, Norway) and Dr D. Bird (Department of Nematology, University of California, Riverside, California, USA).

ESSENTIAL READING

Ace, C.I., Balsamo, M., Le, L.T. and Okulicz, W.C. (1994) Isolation of progesterone-dependent complementary deoxyribonucleic acid fragments from rhesus monkey endometrium by sequential subtractive hybridisation and polymerase chain reaction amplification. *Endocrinology*, **134**, 1305–1309.
Duguid, J.R. and Dinauer, M.C. (1990) Library subtraction of *in vitro* cDNA libraries to identify differentially expressed genes in scrapie infection. *Nucleic Acids Res.*, **18**, 2789–2792.
Straus, D. and Ausubel, F.M. (1990) Genomic subtraction for cloning DNA corresponding to deletion mutations. *Proc. Natl. Acad. Sci. USA*, **87**, 1889–1893.
Sun, T.p., Goodman, H.M. and Ausubel, F.M. (1992) Cloning the *Arabidopsis GA1* locus by genomic subtraction. *Plant Cell*, **4**, 119–128.
Wilson, M.A., Bird, D.McK. and van der Knapp, E. (1994) A comprehensive cDNA cloning approach to identify nematode-induced transcripts in tomato. *Phytopathology*, **83**, 299–303.

REFERENCES

Aguan, K., Sugawara, K., Suzuki, N. and Kusano, T. (1991) Isolation of genes for low-temperature-induced proteins in rice by a simple subtractive method. *Plant Cell Physiol.*, **32**, 1285–1289.
Bautz, E.K.F. and Reilly, E. (1966) Gene-specific messenger RNA: isolation by the deletion method. *Science*, **151**, 328–330.

Belyavsky, A., Vinogradova, T. and Rajewsky, K. (1989) PCR-based cDNA library construction: general cDNA libraries at the level of a few cells. *Nucleic Acids Res.*, **17**, 2919–2932.

Bird, D. McK. and Wilson, M.A. (1994) DNA sequence and expression analysis of root-knot nematode-elicited giant cell transcripts. *Mol. Plant-Mic. Interact.*, **7**, 419–424.

Bjourson, A.J., Stone, C.E. and Cooper, J.E. (1992) Combined subtraction hybridisation and polymerase chain reaction amplification procedure for isolation of strain-specific *Rhizobium* DNA sequences. *Appl. Environ. Microbiol.*, **58**, 2296–2301.

Britten, R.J., Graham, D.E. and Neufeld, B.R. (1974) Analysis of repetitive DNA sequences by reassociation. *Methods Enzymol.*, **29**, 363–418.

Bulman, M.P. (1994) The development of subtractive cDNA technology in the identification of wilt- and ABA-inducible genes in *Arabidopsis thaliana*. PhD thesis, University of the West of England, Bristol.

Cecchini, E., Dominy, P.J., Geri, C. et al. (1993) Identification of genes up-regulated in dedifferentiating *Nicotiana glauca* pith tissue, using an improved method for constructing a subtractive cDNA library. *Nucleic Acids Res.*, **21**, 5742–5747.

Chasan, R. (1992) When more is less: using deletions to clone genes. *Plant Cell*, **4**, 111–112.

Darrasse, A., Kotoujansky, A. and Bertheau, Y. (1994) Isolation by genomic subtraction of DNA probes specific for *Erwinia carotovora* subsp. *atroseptica*. *Appl. Environ. Microbiol.*, **60**, 298–306.

Domec, C., Garbay, B., Fournier, M. and Bonnet, J. (1990) cDNA library construction from small amounts of unfractionated RNA: association of cDNA synthesis with polymerase chain reaction. *Anal. Biochem.*, **188**, 422–426.

Duguid, J.R., Rohwer, R.G. and Seed, B. (1988) Isolation of cDNAs of scrapie-modulated RNAs by subtractive hybridisation of a cDNA library. *Proc. Natl. Acad. Sci. USA*, **85**, 5738–5742.

Gruber, C.E., Li, W.B., Lin, J.-J. and D'Alessio, J.M. (1993) Subtractive cDNA hybridization using the multifunctional plasmid vector pSPORT 2. *Focus*, **15**, 59–65.

Gulick, P.J. and Dvorak, J. (1990) Selective enrichment of cDNAs from salt-stress-induced genes in the wheatgrass, *Lophopyrum elongatum*, by the formamide-phenol emulsion reassociation technique. *Gene*, **95**, 173–177.

Gurr, S.J. and McPherson, M.J. (1991) PCR-directed cDNA libraries. In *PCR, A Practical Approach* (eds M.J. McPherson, R. Quirke and G.R. Taylor), pp. 147–170. IRL Press, Oxford.

Hampson, I.N., Pope, L., Cowling, G.J. and Dexter, T.M. (1993) Chemical cross-linking subtraction (CCLS): a one-tube protocol for producing subtractive hybridization probes. *Focus*, **15**, 50–51.

Hara, E., Kato, T., Nakada, S. et al. (1991) Subtractive cDNA cloning using oligo(dT)$_{30}$-latex and PCR: isolation of cDNA clones specific to undifferentiated embryonal carcinoma cells. *Nucleic Acids Res.*, **19**, 7097–7104.

Herfort, M.R. and Garber, A.T. (1991) Simple and efficient subtractive hybridisation screening. *Biotechniques*, **11**, 598–604.

Hildebrand, M., Higgins, D.R., Busser, K. and Volcani, B.E. (1993) Silicon-responsive cDNA clones isolated from the marine diatom *Cylindrotheca fusiformis*. *Gene*, **132**, 213–218.

Jepson, I., Bray, J., Jenkins, G. et al. (1991) A rapid procedure for the construction of PCR cDNA libraries from small amounts of plant tissue. *Plant Mol. Biol. Rep.*, **9**, 131–138.

Kang, S., Chumley, F.G. and Valent, B. (1994) Isolation of the mating-type genes from the phytopathogenic fungus *Magnaporthe grisea* using genomic subtraction. *Genetics*, **138**, 289–296.

Kohne, D.E., Levison, S.A. and Byers, M.J. (1977) Room temperature method for increasing the rate of DNA reassociation by many thousandfold: the phenol emulsion reassociation technique. *Biochemistry*, **16**, 5329–5341.

Kouchi, H. and Hata, S. (1993) Isolation and characterisation of novel nodulin cDNAs representing genes expressed at early stages of soybean nodule development. *Mol. Gen. Genet.*, **238**, 106–119.

Krebbers, E., Seurinck, J., Herdies, L. et al. (1988) Four genes in two diverged families encode the ribulose-1,5-bisphosphate carboxylase small subunit polypeptides of *Arabidopsis thaliana*. *Plant Mol. Biol.*, **11**, 745–759.

Kunkel, L.M., Monaco, A.P., Middlesworth, W. et al. (1985) Specific cloning of DNA fragments absent from the DNA of a male patient with an X-chromosome deletion. *Proc. Natl. Acad. Sci. USA*, **82**, 4778–4782.

Lamar, E.E. and Palmer, E. (1984) Y-encoded, species-specific DNA in mice: evidence that the Y chromosome exists in two polymorphic forms in inbred strains. *Cell*, **37**, 171–177.

Lebeau, M.-C., Alvarez-Bolado, G., Wahli, W. and Catsicas, S. (1991) PCR driven DNA–DNA competitive hybridisation: a new method for sensitive differential cloning. *Nucleic Acids Res.*, **19**, 17.

Li, W.-B., Gruber, C.E., Lin, J.-J. et al. (1994) The isolation of differentially expressed genes in fibroblast growth factor stimulated BC$_3$H1 cells by subtractive hybridisation. *BioTechniques*, **16**, 722–729.

Lincoln, J.E., Cordes, S., Read, E. and Fischer, R.L. (1987) Regulation of gene expression by ethylene during *Lycopersicon esculentum* (tomato) fruit development. *Proc. Natl. Acad. Sci. USA*, **84**, 2793–2797.

Logemann, J., Schell, J. and Willmitzer, L. (1987) Improved method for the isolation of RNA from plant tissues. *Anal. Biochem.*, **163**, 16–20.

Lopez-Fernandex, L.A. and Mazo, J. del (1993) Construction of subtractive cDNA libraries from limited amounts of mRNA and multiple rounds of subtraction. *BioTechniques*, **15**, 654–659.

Milner, J.J., Cecchini, E. and Dominy, P.J. (1995) A kinetic model for subtractive hybridisation. *Nucleic Acids Res.*, **23**, 176–187.

Mutchler, K.J., Klemish, S.W. and Russo, A.F. (1992) A rapid PCR protocol for identification of differentially expressed genes from a cDNA library. *PCR Methods Applic.*, **1**, 195–198.

Neill, S.J., Hey, S.J. and Barratt, D.H.P. (1993) Analysis of guard cell gene expression in *Pisum sativum*. *J. Exp. Bot.*, **44S**, 3.

Nussbaum, R.L., Lesko, J.G., Lewis, R.A. et al. (1987) Isolation of anonymous DNA sequences from within a submicroscopic X chromosome deletion in a patient with choroideremia, deafness, and mental retardation. *Proc. Natl. Acad. Sci. USA*, **84**, 6521–6525.

Okamura, J.K. and Goldberg, R.B. (1989). Regulation of gene expression: general principles. In *The Biochemistry of Plants* Vol. 15 (ed. A. Marcus), pp. 1–76. Academic Press, NY.

Pautot, V., Holzer, F.M., Reisch, B. and Walling, L.L. (1993) Leucine aminopeptidase: an inducible component of the defense response in *Lycopersicon esculentum* (tomato). *Proc. Natl. Acad. Sci. USA*, **90**, 9906–9910.

Phillips, A.L. and Huttly, A.K. (1994) Cloning of two gibberellin-regulated cDNAs from *Arabidopsis thaliana* by subtractive hybridisation: expression of the tonoplast water channel, γ-TIP, is increased by GA$_3$. *Plant Mol. Biol.*, **24**, 603–615.

Rosenberg, M., Przybylska, M. and Straus, D. (1994) RFLP subtraction: a method for making libraries of polymorphic markers. *Proc. Natl. Acad. Sci. USA*, **91**, 1706–1710.

Rubenstein, J.L.R., Brice, A.E., Ciaranello, R.D. et al. (1990) Subtractive hybridisation system using single-stranded phagemids with directional inserts. *Nucleic Acids Res.*, **18**, 4833–4842.

Sambrook, J., Fritsch, E.F. and Maniatis, T. (1989) *Molecular Cloning, A Laboratory Manual*, 2nd edn. Cold Spring Harbor, NY.

Sargent, T.D. (1987) Isolation of differentially expressed genes. *Methods Enzymol.*, **152**, 423–432.

Sharma, P., Lonneborg, A. and Stougaard, P. (1993) PCR-based construction of subtractive cDNA library using magnetic beads. *BioTechniques*, **15**, 610–611.

Sive, H.L. and St John, T. (1988) A simple subtractive hybridization technique employing photoactivatable biotin and phenol extraction. *Nucleic Acids Res.*, **16**, 10937.

Sommer, H., Beltran, J.-P., Huijser, P. et al. (1990) *Deficiens*, a homeotic gene involved in the control of flower morphogenesis in *Antirrhinum majus*: the protein shows homology to transcription factors. *EMBO J.*, **9**, 605–613.

Swaroop, A., Xu, J., Agarwal, N. and Weissman, S.M. (1991) A simple and efficient cDNA library subtraction procedure: isolation of human retina-specific cDNA clones. *Nucleic Acids Res.*, **19**, 1954.

Timblin, C., Battey, J. and Kuehl, W.M. (1990) Application for PCR technology to subtractive cDNA cloning: identification of genes expressed specifically in murine plasmacytoma cells. *Nucleic Acids Res.*, **18**, 1587–1593.

Travis, G.M. and Sutcliffe, J.G. (1988) Phenol emulsion-enhanced DNA-driven subtracted cDNA cloning: isolation of low abundance monkey cortex-specific mRNAs. *Proc. Natl. Acad. Sci. USA*, **85**, 1696–1700.

Van Brunt, J. (1992) Fishing for genes. *Bio/Technology*, **10**, 852–853.

Van Ness, J. and Hahn, W.E. (1982) Physical parameters affecting the rate and completion of RNA-driven hybridisation of DNA: new measurements relevant to quantitation based on kinetics. *Nucleic Acids Res.*, **10**, 8061–8077.

Vernon, D.M. and Bohnert, H.J. (1992) A novel methyl transferase induced by osmotic stress in the facultative halophyte *Mesembryanthemum crystallinum*. *EMBO J.*, **11**, 2077–2085.

Wang, T.L., Leech, M.J., Martin, C.R. et al. (1992) Molecular genetic approaches to studying cytokinin action in moss. In *Physiology and Biochemistry of Cytokinins in Plants* (eds M. Kaminek, D.W.S. Mok and E. Zazimalova), pp. 149–155. SPB Academic Publishing bv, The Hague.

Wieland, I., Bolger, G., Asouline, G. and Wigler, M. (1990) A method for difference cloning: gene amplification following subtractive hybridisation. *Proc. Natl. Acad. Sci. USA*, **87**, 2720–2724.

Williams, J., Bulman, M.P. and Neill, S.J. (1994a) Wilt-induced ABA biosynthesis, gene expression and down-regulation of rbcS mRNA levels in *Arabidopsis thaliana*. *Physiol. Plant.*, **91**, 177–182.

Williams, J., Bulman, M., Huttly, A. et al. (1994b) Characterisation of a cDNA clone from *Arabidopsis thaliana* encoding a potential thiol protease whose expression is induced independently by wilting and abscisic acid. *Plant Mol. Biol.*, **25**, 259–270.

VI   Sequencing Projects

15 EST and Genomic Sequencing Projects

RICHARD COOKE
Université de Perpignan, Perpignan, France

RÉGIS MACHE
Laboratoire de Biologie Moleculaire vegetale, Grenoble, France

HERMAN HÖFTE
INRA, Versailles, France

(1) Genomic and cDNA sequencing is under way for several organisms, particularly humans, *Caenorhabditis elegans*, *Arabidopsis thaliana* and rice.

(2) More than 12 000 partial single-run sequences (expressed sequence tags: ESTs) of cDNAs for rice and *Arabidopsis* are available in the EST databank, dbEST. Sequence from the 5' extremity allows identification of putative protein products, while 3' sequences can provide gene-specific probes. Information on putative protein products is available either in dbEST or directly in EMBL, PIR or GenBank files.

(3) Tags to more than 1000 *Arabidopsis* genes coding for known proteins have been identified. More than half of these proteins had not yet been identified in *Arabidopsis*. Two-thirds of the ESTs show no similarity to known sequences and are probably plant or species specific.

(4) ESTs can be used directly or through identification of primers for PCR amplification to isolate the corresponding genes in other species. Comparison of rice and *Arabidopsis* EST sequences will be particularly useful in this respect. Information on protein function can be obtained by the use of sense or antisense copies of translated sequences to turn off gene expression and by correlation of mutant loci with the position of ESTs on the genetic map.

(5) Information on ESTs is available using a variety of network tools. In addition, specific information on *Arabidopsis* sequences, clones, mutant lines, genetic and physical maps and much more is directly accessible.

(6) Genomic sequencing will rapidly provide information on gene structure and chromosome organisation. Coupled with sequencing of cognate cDNAs, it will allow rapid and precise identification of structural and functional motifs on the *Arabidopsis* genome.

1 INTRODUCTION

The development in recent years of automatic high-throughput sequencing methods has made it possible to determine the sequence of complete genomes.

Plant Gene Isolation: Principles and Practice. Edited by G. D. Foster and D. Twell.
© 1996 John Wiley & Sons Ltd.

For the plant community, such an invaluable source of biological information is being uncovered as the sequencing of the *Arabidopsis thaliana* genome proceeds. While awaiting the completion of the genomic sequence, a much more rapid and cost-effective way to identify new genes has been developed in a number of laboratories with the systematic sequencing of anonymous cDNAs. Single-run sequences of 250–400 bases (expressed sequence tags; EST) are determined on randomly picked cDNA clones. This provides sufficient information to unambiguously identify the corresponding gene, and comparisons with sequence databases frequently allow the assignation of potential functions to the corresponding gene products. Large collections of ESTs for several plant species are accumulating in public databases, providing a wealth of information for the plant biologist. In this chapter we will review the ESTs and genomic sequencing projects for plants. In addition, we will provide information on how to access the sequences and how they can be used in your research projects.

2   cDNA SEQUENCING PROJECTS

The accumulation of data from sequencing projects has been extremely rapid and has led to the creation at the National Centre for Biotechnology Information at the National Institutes of Health of a new database, dbEST (Boguski et al., 1993), into which all expressed sequence tags are incorporated, either by direct submission or by transfer from EMBL or GenBank. As for these other databases, it is updated regularly. Fig. 1 shows the accumulation of ESTs for three plant species compared with those from the human sequencing projects since mid-1993. The latest version of dbEST (16 August 1994) contains more than 50 000 tags from 23 different organisms. However, the vast majority of these tags are derived from only four organisms: *Homo sapiens*, *Arabidopsis thaliana*, *Caenorhabditis elegans* and *Oryza sativa*. Reviews have been published for humans (Adams et al., 1991, 1992, 1993), mouse (Höög, 1991), *C. elegans* (Waterston et al., 1992), maize (Keith et al., 1993), rice (Uchimiya et al., 1993; Umeda et al., 1994) and *Arabidopsis* (Höfte et al., 1993; Newman et al., 1994).

2.1   Systematic sequencing in plants

Table 1 shows the libraries and number of sequences for plant species in dbEST. While they have only been underway for three years, the rice and *Arabidopsis* projects have already contributed more than 12 000 sequences to the international databases. In addition, some 10 000 rice ESTs are still unpublished (T. Sasaki, personal communication) (Table 1). These projects are of particular interest because very few data on coding sequences from plants were available when sequencing began. In mid-1992 there were less than 500 protein-coding sequences in the EMBL database for the whole plant kingdom, whereas the number of genes expressed during the lifetime of a plant is estimated to be between 16 000 and 33 000 (Gibson & Somerville, 1993). In addition, many of

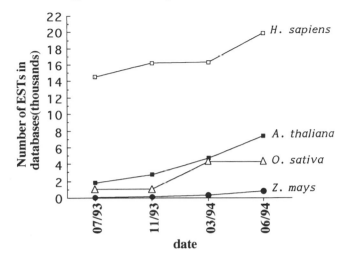

Figure 1. Comparison of the accumulation from mid-1993 to mid-1994 of ESTs in the dbEST database for *Homo sapiens* and the three plants for which significant sequencing data are available

the sequences available were those of abundantly expressed genes in multi-gene families. It was thus inevitable that systematic cDNA sequencing would lead to the identification of a considerable number of genes and that the benefits of such an approach would extend well beyond the direct exploitation within the framework of a genome sequencing project. The particular interest of the rice and *Arabidopsis* projects is that these two organisms represent the division of the plant kingdom into monocotyledonous and dicotyledonous species, and results obtained for these two organisms are likely to be directly applicable to many other species. We will discuss the results from these two projects, with particular emphasis on those obtained for *Arabidopsis*.

2.2   Plant cDNA libraries

Rice was chosen as a model species because many morphological and isozyme markers have already been mapped to their respective chromosomes (Kinoshita, 1990) and restriction fragment length polymorphism (RFLP) maps are available (McCouch et al., 1988; Saito et al., 1991). In addition, it has the smallest genome among the cereals (6×10^8 bp) and its agronomical importance makes it an obvious candidate for detailed study. The first cDNA libraries sequenced for this organism were prepared from cells cultured in different concentrations of sucrose (Uchimiya et al., 1993). More recently, libraries have been prepared from cells cultured under stress (saline or nitrogen starvation conditions), with the aim of studying stress-related gene expression and signal transduction (Umeda et al., 1994). A large amount of unpublished data has also been obtained from roots and shoots of seedlings and from leaves (T. Sasaki,

Table 1. Number of EST sequences from plants

Organism	Organ of library	Number of sequenced clones	
Rice	Callus with 2, 4-D	2655	
	Callus with BA	960	
	Callus with NAA	553	
	Callus with heat shock	626	
	Young root	2232	
	Shoot	4069	
	Etiolated shoot	3401	
	Adult leaf	499	
	Developing seed	488	
	Total	15483	
Maize	Etiolated seedlings	166	
	Membrane-free polysomes (endosperm)	426	
	Leaf	189	
	Total	781	
Arabidopsis	Mixed library	4163	
	Cell suspension culture[a]	535	(649)
	Xanthomonas-treated cell suspension[a]	99	(102)
	Green siliques[a]	379	(441)
	Flowering tips[a]	31	(53)
	Green shoots[a]	254	(421)
	Flower buds[a]	477	(543)
	Whole etiolated seedlings[a]	903	(1547)
	Dry seeds[a]	211	(414)
	Sliced leaves[a]	267	(290)
	Total	3156	(4460)
Tobacco	Three-day-old cells	13	
Rapeseed	Leaf	39	

[a]Non-redundant sequences submitted to dbEST through EMBL. Figures in brackets show the total number of sequences for these libraries.
2,4-D, 2,4-dichlorophenoxyacetic acid; BA, 6-benzylaminopurine; NAA, 1-naphthalene acetic acid.

personal communication) (Table 1). These libraries are oriented, and sequencing has been from the 5′ extremity to facilitate the identification of the putative protein products.

Arabidopsis thaliana was chosen several years ago as a model species for sequencing projects for, among other reasons, its small genome, short generation time and ease of transformation, and the fact that physical mapping was already under way (Redei & Koncz, 1992). An integrated genetic map of visual and RFLP markers has recently been published (Hauge et al., 1993), but very few genes had been identified and analysed when systematic sequencing began. Sequencing of *Arabidopsis* cDNAs in the USA is being carried out on an oriented, mixed library from the 5′ extremity. The library was prepared using equal amounts of poly(A)⁺ RNA from four tissue types: seven day germinated etiolated seedlings, tissue-culture-grown roots, rosettes of plants of all ages grown either in continuous light or with a 16-h photoperiod with harvesting in the dark, and, finally, aerial tissues of plants grown under the same two sets of conditions (Newman et al., 1994). The French *Arabidopsis* sequencing group, representing nine different laboratories, sequences cDNAs from separate libraries, almost all of which are oriented, prepared from different organs or material grown under different culture conditions (Höfte et al., 1993; Cooke et al., 1996). Among others, cDNA libraries constructed from developing or mature siliques, suspension-cultured cells, flower buds, leaf strips cultured in the presence of 2,4-D to induce cell division and etiolated seedlings are being used (Table 1). The aim is to obtain information on the mRNA population in the individual libraries by random sequencing, allowing an overall view of the genes transcribed in the different organs and conditions. For example, sequencing of the dry seed library has shown the presence of mRNAs for storage and late embryogenesis abundant proteins previously undetected in *Arabidopsis*, the relative frequency of these transcripts suggesting that the abundancy of the corresponding proteins varies between different plant species. Sequencing in maize is also carried out on libraries corresponding to particular tissues (Table 1). An analysis of 130 clones from endosperm polysomes has been published (Keith et al., 1993).

2.3 Sequencing strategies

Single-run sequences of 250–400 bp are determined in most cases using an automated sequencer. Almost all of the libraries are oriented, and the sequencing is mostly carried out from the 5′ extremity. Currently, in the framework of the European *Arabidopsis* sequencing project, the 5′ extremity is sequenced first and all non-redundant clones are subsequently sequenced from the 3′ end. The 3′ end not only provides gene-specific probes for closely related members of gene families, but also provides an invariant tag (the libraries were primed from the poly(A) tail) for cDNAs derived from the same mRNA but which are truncated at different positions at the 5′ end. Non-overlapping sequences from the 5′ ends of such clones may otherwise be identified as corresponding to different mRNAs. Finally, in a number of cases, sequences from both extremities have allowed the identification of chimeric clones containing fragments derived from two or, very rarely, more mRNAs. As many ESTs are only sequenced from the 5′ extremity, the existence of this type of artefact should be borne in mind if the clone is to be used for more detailed study.

2.4 Quality of EST data

2.4.1 *Accuracy*

Since most EST sequences submitted to dbEST are obtained by a single sequencing run on an automatic sequencer, it is important to have an idea of the accuracy of the sequences obtained. Comparison of our raw sequences from the ABI 373A sequencer (after removal of vector and linker sequences) between redundant clones or with sequences which are already known in *Arabidopsis* show that the accuracy is usually greater than 95% over more than 300 bases. The majority of uncertainties correspond to unattributed bases or spurious bases inserted towards the end of the exploitable data. Visual examination of the graphs can increase the accuracy of the data to up to 98% with no further treatment. Even without manual correction the accuracy is highly acceptable for the purposes of homology searches, for both nucleotide and amino acid alignments. The accuracy of the sequences is especially important if they are to be of use in defining PCR primers for mapping purposes and the identification of homologous sequences in other organisms.

2.4.2   Vector contamination

Vector contamination in ESTs is a serious problem in sequencing projects, and it is necessary to be vigilant in analysing the sequences in order to eliminate even short portions of vector or cloning linkers. However, even before EST projects started, the databanks already contained many sequences with considerable portions of vector sequences. An alignment of the commonly used vector pBluescript with either Genbank or dbEST reveals the extent of contamination: more than 400 EST sequences show significant similarity to the cloning vector (our own unpublished data). These two considerations of accuracy and contamination are of great importance for the correct use of databanks.

2.4.3   Redundancy

While the knowledge of the redundancy of a specific tag in a given cDNA library gives an indication of the level of expression of the corresponding gene, redundancy in the public EST databases (more than one EST corresponding to identical mRNAs) is becoming an increasingly serious problem as random sequencing continues. The actual percentage of redundancy is very difficult to estimate, since two ESTs are very rarely identical, even if they are derived from identical mRNAs, as they almost always vary in length and contain sequencing errors. For example, only 121 out of 51 595 (0.23%) sequences in the EST section of GenBank are truly redundant (Cooke, unpublished data). However, a search for sequences encoding one particular plant protein, the ribulose bisphosphate carboxylase small subunit, in dbEST reveals 63 tags in the *Arabidopsis* mixed library alone, while there are at least 140 tags to ubiquitin-coding sequences in the database. Only in the early phases of the French project were attempts made to estimate the redundancy by looking for overlaps between ESTs using an arbitrary cut-off score of 90% nucleotide identity in a stretch of 50 nucleotides. This led to an estimated redundancy of 31% for the first 1152 ESTs, which is probably close to the true figure. The presence of tens and sometimes hundreds of copies of the same sequence in databases can interfere with the detection of significant alignments with a low level of similarity (e.g. homologous sequences in other organisms), since usually only a limited number of significant alignments are provided in a search output.

As a result of this redundancy, the some 12 000 expressed sequence tags to plant genes in dbEST seem to represent only a fraction of the estimated 16 000–33 000 genes present in a plant genome. To get an idea of what percentage of all genes already has been tagged, one can search for 'ancient conserved regions' (ACRs) (Claverie, 1993). ACRs are protein segments whose sequence is highly conserved between phyla, and include structural proteins such as actin and tubulin, ribosomal proteins, kinases and many enzymes involved in the principal metabolic pathways. Green et al. (1993) concluded from their work tht currently known proteins already include representatives of most of the ACRs, and that moderately expressed genes are more likely to contain ACRs than genes expressed at a low level. In 1993, they had identified 509 different ACRs. Preliminary analysis of 2361 sequences in the French project using tFASTA (Pearson & Lipman, 1988) showed that only 104 of the 509 ACRs were represented (our unpublished observations). This suggests that the vast majority of genes, even those which are moderately expressed, has yet to be identified.

2.5 Data analysis

Similarity searches between ESTs and the international databanks may provide clues on the potential function of the corresponding gene. The methods of analysis used and the citation of the similarities observed differ widely between different projects. These differences probably account for the great variation observed in the percentage of ESTs for which a database match was found. *Arabidopsis* ESTs from the American project are submitted directly to dbEST. Automatic alignment of all sequences in dbEST against nucleotide and protein sequence databases is carried out regularly, and the best alignments are incorporated into the sequence files. While no similarities are cited in the GenBank files, a list of the results of the dbEST BLAST (Altschul et al., 1990) searches is regularly posted to the *Arabidopsis* electronic newsgroup (see section 5.1 for information on joining). Since the annotations with the results of the database comparisons are done automatically, the similarities given in these lists should be treated with caution but nonetheless constitute a valuable source of information.

For the French collaborative project, sequence similarities were detected by BLASTX alignment (Altschul et al., 1990) of amino acid translations of the six possible reading frames against the NCBI non-redundant protein databank, which contains sequences from SWISS-PROT, PIR and translations of coding sequences cited in GenBank. These alignments are carried out either locally or using the network NCBI BLAST server. As a general rule, only alignments with a similarity score greater than 80 are further considered, and results are inspected to eliminate high-scoring but insignificant alignments, such as those with proteins containing repetitive motifs (e.g. proline- or glycine-rich stretches). The filter option in BLAST, which replaces low-complexity sequences by Xs, usually allows the distinction to be made between these alignments and cDNAs corresponding to bona fide proteins. The BLAST algorithm is particularly useful in this type of project, as it does not allow gaps in alignments, and interruptions in these can thus often be traced to errors (insertions or deletions) in the sequence which subsequently can be corrected. This preliminary analysis is often followed by a more sensitive alignment of the correct amino acid translation using FASTA (Pearson & Lipman, 1988). This detailed analysis allows citation in the files submitted to the EMBL data library not only of the similar sequences detected in the databanks but also of the limits of the open coding region. The result of this is that the translations of the coding regions of the cDNAs appear in the GenPept and PIR databanks.

The rice ESTs were first analysed by alignment with GenBank using the program FASTA (Pearson & Lipman, 1988). Only sequences showing similarity scores greater than 160 were subsequently analysed for amino acid homology (Uchimiya et al., 1993; Umeda et al., 1994). This method has a serious drawback in that similarity searches between protein sequences are much more sensitive than between nucleic acid sequences. If we consider that many proteins which are known to be homologous have less than 30% identity between species, and taking into account the degeneracy of the genetic code, this can lead to very low levels of similarity at the nucleotide level. This probably explains why only 8% (Uchimiya et al., 1993) and 15% (Umeda et al., 1994) of the rice ESTs were cited as showing similarity to known sequences. The number of tags to genes coding for known proteins is probably greater than these figures suggest. Amino acid sequence comparisons with 130 maize ESTs uncovered significant similarities to known genes in 20% of the cases (130).

3   GENE IDENTIFICATION

3.1   ESTs similar to known genes

In a first global analysis of 1152 *Arabidopsis* cDNAs, Höfte et al. (1993) found that 375 (32%) were similar to known proteins and presented a list of these similarities. Information for more recently sequenced tags is directly available in the EMBL sequence files. Of these 375 sequences, 35% were identical to known *Arabidopsis* genes. Among the remaining 65%, 33% were similar to genes from other plants and, in most cases, can be considered to represent the homologous gene in *Arabidopsis*. The majority of genes of this class represent highly expressed genes which were either found in several libraries or frequently in individual libraries corresponding to particular culture conditions or a specific tissue (rubisco small subunit, seed storage proteins). However, 7% of the clones sequenced correspond to new members of previously characterised gene families which had not been identified by classical methods, and 25% to genes which had never been sequenced in plants. Examples of the former are ESTs identifying new members of a family of plant K^+ channel proteins or a family of NO_3^- transporters, a fifth member of the *Arabidopsis* 2S seed storage protein (napin) gene family which had not been described in this or any other organism, a meri-5 homologue and three new oleosins which correspond to those described in other crucifers. In fact, one of the surprising findings is that many proteins are encoded by multigene families which may often contain more members than their animal counterparts. For instance, five thioredoxin genes were identified in *Arabidopsis* compared with three in yeast and only one in humans (Y. Meyer, personal communication).

In fact, as shown in Table 2, the global figures for similarities to known proteins given above mask quite large differences between the individual libraries, although the overall percentage showing significant similarity is around 30% for all the libraries studied. The figures in the table differ slightly from those given above, as we have only considered the non-redundant sequences in this case. Thus, among the suspension-cultured cell ESTs, 55% of the similarities detected are to proteins previously unidentified in plants, including structural proteins, enzymes involved in intermediary metabolism and a considerable number of ribosomal proteins. On the other hand, more than half of the similarities found for the flower bud library are with proteins already identified in other plants, and the percentage is also high for the developing silique and etiolated seedling libraries. These similarities are to a wide variety of proteins: heat shock proteins, histones, photosystem subunits, proteins involved in the translational apparatus and many enzymes. Thus, these libraries are likely to form an excellent source of ESTs for workers wishing to isolate *Arabidopsis* genes homologous to a particular plant protein. Finally, ESTs representing new members of gene families are particularly abundant in the etiolated seedling and developing silique libraries.

Table 2. Sequence similarities between amino acid translations of *Arabidopsis* ESTs and known protein sequences

cDNA library	Number of ESTs	Identical to *Arabidopsis* sequences	Similar to *Arabidopsis* sequences	Similar to other plant sequences	Similar to non-plant sequences	Total
Flower buds	144	10 (23%)	1 (2%)	23 (53%)	9 (21%)	43
Developing siliques	226	15 (20%)	9 (12%)	32 (42%)	20 (26%)	76
Suspension culture	321	15 (19%)	5 (6%)	15 (19%)	43 (55%)	78
Cultured leaf strips	53	4 (29%)	0 (0%)	9 (64%)	1 (7%)	14
Etiolated seedlings	144	8 (20%)	6 (15%)	16 (40%)	10 (25%)	40
Total	888	52 (21%)	21 (8%)	95 (38%)	83 (33%)	251

The figures are derived from the 903 non-redundant ESTs discussed in Höfte et al (1993). Similarities were detected by BLAST (Altshcul et al., 1990) or FASTA (Pearson & Lipman, 1988) alignments as described in the text. For each library, the number of similarities in each category is given as a percentage of the total number of ESTs showing significant similarity to known proteins. (Figures for the cultivated leaf strip library should be considered with caution, as the number of sequences considered is low.)

In addition to the large number of *Arabidopsis* ESTs to genes coding for metabolic enzymes and proteins involved in translation, significant similarities are detected to a certain number of 'exotic' sequences, as in other systematic sequencing projects. These may simply be proteins which have been given names which correspond to the organism in which they were first detected, such as the abnormal wing disc protein of *Drosophila*, which is in fact a nucleoside diphosphate kinase (Kimura et al., 1990). However, others are more intriguing, like the hemoglobinase or urease homologues and the human laminin homologue, which suggests the presence in plants of this class of cytoskeletal elements. It should be emphasised that many of these genes for which no known plant homologue was known may have been difficult to identify by other methods.

In contrast to the results obtained for the *Arabidopsis* sequences, the vast majority of the similarities detected between rice ESTs and known genes correspond to previously characterised plant proteins (Uchimiya et al., 1993; Umeda et al., 1994), which is more probably due to the fact that these authors searched the databases with nucleotide sequences than to a difference between the mRNA populations in the two organisms. This is corroborated by the fact that 20% of maize ESTs were found to show significant similarity to known genes and, more particularly, 8 of the 18 similarities cited (44%) are to non-plant genes (Keith et al., 1993). Taking this into account, these figures are comparable with those observed in *C. elegans*, where 31% of the sequences show similarity to known genes (Waterston et al., 1992). Higher percentages were observed for human brain cDNAs (52%) (Adams et al., 1991) and mouse ESTs (44%) (Höög, 1991), but this probably reflects the much greater number of sequences already available for animals before systematic sequencing began.

3.2   Identification of new genes

Only about one-third of *Arabidopsis* and 20% of maize ESTs encode known proteins, and the cited percentage for rice ESTs is even lower (Uchimiya et al., 1993; Umeda et al., 1994). Thus, more than two-thirds of ESTs sequenced correspond to genes coding for proteins which have no known counterpart in other organisms or whose sequence has diverged to the extent that the homology is no longer detectable. This is in fact the corollary of the definition of the ACRs (Green et al., 1993). The fact that new sequences showing no similarity to known proteins are unlikely to contain ACRs implies that little or no sequence similarity should be detected between these new sequences and those already in the databanks. The unidentified ESTs from systematic sequencing projects therefore represent a colection of sequences which are at least specific to plants and, in a number of cases, maybe even only to closely related species. As is the case for the ESTs corresponding to known proteins, some of the ESTs without database matches are also highly represented in the cDNA libraries (Höfte et al., 1993), indicating the existence of abundantly expressed genes or gene families which have not yet been identified in plants. These unidentified proteins fall into two classes: those which were only found in one library and therefore probably represent proteins which are only present in the particular tissue or culture conditions used, and those which are found in several libraries and probably correspond to proteins which play a more general role in cell function and which may be specific to plants.

3.3 Abundantly expressed genes

Because sequencing has so far been carried out with little or no elimination of redundancies, it has been possible to establish a catalogue of abundantly expressed genes in these organisms and, in those projects in which individual libraries have been sequenced, to obtain an overview of the mRNA population at a given stage or under particular conditions. The degree of redundancy varies considerably between the different cDNA libraries, as has been shown for *Arabidopsis*, in which internal redundancy is only 9% in suspension-cultured cells, whereas it is greater than 30% in more specialised tissues such as developing siliques (Höfte et al., 1993). In order to identify genes which are expressed at a lower level it will soon be necessary to eliminate redundancies from the latter libraries. Several strategies are being adopted in order to achieve this goal: hybridisation with probes corresponding to the most abundant gene classes, with a probe prepared from a pool of all clones which have already been sequenced or simply by hybridising the library against itself. An alternative, and eventually a more efficient, approach would be the use of a normalised library in which all cDNA clones should be represented at a similar abundance (Sankhavaram et al., 1991).

4   HOW CAN ESTs BE USED?

4.1   Isolation of genes from other plant species corresponding to ESTs

For closely related organisms, it should be possible in many cases to use the ESTs as classical hybridisation probes to detect homologous genes. This is the case for rice and maize sequences, where similarity between ESTs is very high (T. Sasaki, personal communication) and should also be the case between *Arabidopsis* and related crucifers. For other plants, the information obtained by comparing EST sequences between rice and *Arabidopsis* will permit the identification of highly conserved nucleotide sequences which can subsequently be used to define unique or degenerate primers to isolate homologous sequences by PCR. An example of this possibility is given in Fig. 2, which shows the results of a FASTA (Pearson & Lipman, 1988) alignment of an *Arabidopsis* EST against dbEST. The two sequences are tags to one of the glutamine synthetase genes, and it can clearly be seen that, whereas the non-coding regions show only very limited similarity, there are blocks of highly conserved nucleotide sequence within the coding region against which specific PCR primers can be made. As the two species concerned here are representatives of monocots and dicots, it is most probable that these sequences are at least equally well conserved in other plant species.

```
D15766|RIC1212A Rice cDNA, partial sequence  610   547   635
75.6% identity in 279 nt overlap


ATTS1966  CAT-CTTCTTTCTCTGTCGGAACAAAAATGACGTCTCCTCTCTCAGATCTCCTAAAGCGG
          : : :::: : ::          :      ::: :  :: ::::: : :::::: : :: :
D15766    CTTGCTTCCTCCTCCTCATCGTCCGCCATG---GCTTCTCTCACCGATCTCGTCAACCTC


ATTS1966  GATCTATCAGACACCA---AGAAAATCATCGCTGAATACATATGGATCGGGGGCTCTGGA
          : :: :: :::::::   :::: X::::::: :: ::::::::::::::: :: :::::
D15766    AACCTCTCCGACACCACGGAGAAGATCATCGCCGAGTACATATGGATCGGTGGATCTGGC


ATTS1966  ATGGATATTAGAAGCAAAGCCAGGACATCACCAGGCCCAGTAAGTAATCCAACAGAGCTT
          ::::::: : :: ::::: :: :::::       : ::::: :: : : :::: :   ::::
D15766    ATGGATCTCAGGAGCAAGGCTAGGACTCTCTCCGGCCCTGTGACTGATCCCAGCAAGCTG


ATTS1966  CCAAAATGGAACTACGATGGCTCTAGCACCGCACAAGCTGCCGGAGATGATAGTGAAGTC
          :: ::  :::::::::::::::::::: :::::::  :: ::  ::::.:: :: :::.: :::
D15766    CCCAAGTGGAACTACGATGGCTCCAGCACCGCCCAGGCCCCCGGNGAGGACAGTNAGGTC


ATTS1966  ATTCTTTATCCACAGGCAATATTTACGGACCCATTCAGGAAGGGGAACAACATTCTGGTG
          :: :: :: :::::::: :: :: : :::::::::::::::::: :::::::: :: ::
D15766    ATCCTGTACCCACAGGCTATCTTCAAGGACCCATTCAGGAAGGGAAACAACATCCTTGTC


ATTS1966  ATGTGTGATGCTTACAGCCCG
          ::::X :::
D15766    ATGTGCGATTG
```

Figure 2. Use of *Arabidopsis* and rice ESTs to define highly conserved sequences which can be used to define PCR primers for cloning of homologous sequences. The figure shows the result of a FASTA alignment of an *Arabidopsis* EST to a glutamine synthetase gene against the dbEST database. The ATG start codon is underlined and highly conserved regions are boxed

4.2 Correlation of map positions of ESTs with the position of mutant loci

With the availability of partial sequences for the majority of all genes in the foreseeable future, efficient methods for analysis of the in vivo function of genes need to be developed. In the absence of an efficient system for gene replacement by homologous recombination in plants, strategies in which the endogenous gene product is turned off through ectopic expression of antisense (Pnueli et al., 1994) or sense (cosuppression) (Brusslan et al., 1993) copies are most commonly used. Alternatively, information on the potential function of cloned genes is provided by the correlation of their location on the genetic map with the position of mutant loci. For *Arabidopsis*, the number of genetic loci marked by mutation (already more than 800) is rapidly growing (Dennis et al., 1993). The mapping of ESTs can be done by RFLP analysis using a collection of recombinant inbred lines (Lister & Dean, 1993). A more rapid and precise procedure is becoming available with the construction of a physical map of ordered overlapping YAC clones. The map position of cDNAs can then be determined through hybridisation to filters of overlapping YACs or through PCR on pools of YAC clones using primers based on the partial sequences. In a collaborative effort, four laboratories (Ecker, Goodman, Dean and Bouchez/ Caboche) are using a new library (the CEPH-INRA-CNRS (CIC) library) with large inserts (average size 420 kb, ±4 genome equivalents) and with a low percentage of chimerism to line up contigs of overlapping YAC clones with the genetic map using mapped DNA fragments as anchored probes. In France, a project funded by the GREG (Groupement de Recherche pour l'Etude des Génomes) has begun to maps ESTs using a PCR strategy. B. Lemieux (York University, Ontario) has also started a similar project in collaboration with D. Lashkari and R. Davis (Stanford, California). The latter group plans to map 100–200 ESTs per month. Also in rice a considerable effort is being put into mapping ESTs: 1626 DNA markers have been mapped by RFLP, among which 1085 are ESTs (T. Sasaki, personal communication; Antonio et al., 1993).

5 EST DATABASES AND STOCK CENTRES

5.1 Network search tools

As partial sequences for plant genes accumulate at a tremendous pace, the need increases for efficient procedures for retrieving information from databases. Most of these are available on the Internet, and various servers and tools exist which allow researchers to retrieve information. In the past years new transfer protocols such as WAIS (Wide Area Information Server) and Gopher (Parker, 1993) have become available, to retrieve information in a flexible way through keyword searches. A major improvement was the development of the World Wide Web (WWW) initiative, which is a global information retrieval system linking up worldwide networks using hypertext and multimedia. A number of programs exist for all widely distributed computer operating systems that allow the user to wander through the WWW. The most powerful browser is MOSAIC from the National Center for Supercomputing Applications (NCSA). It can be obtained by anonymous FTP from 'ftp.ncsa.uiuc.edu'. Another browser that does not require access to the network in a graphical mode is LYNX, obtainable from ftp2.cc.ukans.edu.

Comprehensive guidelines for consulting sequence databases are described by Newman et al. (1994). Instructions can also be obtained using the NCBI World Wide Web (WWW) server at http://www.ncbi.nlm.nih.gov. You can access information in this server with the MOSAIC software by clicking on the blue words, which are hypertext links. Subscriptions to the *Arabidopsis* electronic newsgroup can be requested by contacting one of the following two addresses, depending upon one's location: biosci@daresbury.ac.uk for Europe, Africa and Central Asia, or biosci@net.bio.net for the Americas and the Pacific Rim.

5.2   Searching for ESTs using keywords

The simplest way to investigate whether any ESTs homologous to your favourite gene have been identified is to search dbEST on the NCBI WWW server with a text string using the MOSAIC software. To identify only *Arabidopsis* sequences, use the word 'cress', which is specific to *Arabidopsis* files. The word 'Arabidopsis' is also found in other files containing ESTs similar to *Arabidopsis* sequences. For instance, look for *Arabidopsis* peroxidases using the search string 'cress and peroxidase'. Alternatively, you can retrieve individual records with keywords using the RETRIEVE e-mail server. For more information, send the following e-mail message to retrieve@ncbi.nlm.nih.gov:
DATALIB DBEST
HELP

5.3   Searching for ESTs matching a DNA or a protein sequence

ESTs matching a DNA or a protein sequence can be found using the BLAST e-mail server at blast@ncbi.nlm.nih.gov. A BLAST manual can be obtained also by putting the word HELP in the e-mail message.

For the vast majority of ESTs, no translated protein sequences are stored in the databases. As nucleotide sequences are very rarely well conserved, even between relatively closely related organisms, similarity searches at the nucleotide level of the EST database will only provide useful information for a small number of sequences. For instance, among 29 ESTs from rice and *Arabidopsis* for which significant amino acid similarity to the same protein has been detected, similarity at the nucleotide level is only seen for 18 (62%). It is thus often necessary to adopt another approach in order to determine whether an EST corresponding to a known protein or peptide has already been sequenced.

tBLASTN or tFASTA programs of the BLAST (Altschul et al., 1990) and FASTA (Pearson & Lipman, 1988) families, respectively, allow one to look for ESTs corresponding to known protein sequences. These options align an amino acid sequence against a nucleotide sequence database dynamically translated in the six possible reading frames and, as such, provide an extremely sensitive means of detection of sequences coding for a given protein. For example, a tobacco thioredoxin amino acid sequence (Marty & Meyer, 1991) was used to screen the *Arabidopsis* EST database using tFASTA, allowing the identification of five sequences corresponding to thioredoxin mRNAs. Complete sequencing of these five clones showed that they correspond to five different thioredoxin genes, some of which have very divergent amino acid sequences (Y. Meyer, personal communication). A second approach using the same search tools is to look for diagnostic 'signatures' which are specific to proteins executing similar functions. It is often difficult to detect short, local homologies of this kind in proteins whose sequence is otherwise very different by alignment of a complete protein sequence with databanks. However, by using either only the signature for a given protein or a complete sequence in which the divergent residues have been 'masked' (replaced by Xs), it is possible to limit alignment only to the conserved regions. For example, ESTs corresponding to protein kinases can be detected using a masked protein kinase sequence, in which only the kinase-specific motif remained unmasked.

5.4 Obtaining sequenced cDNA clones

Most cDNA clones corresponding to ESTs in the databases are available from the Arabidopsis Biological Resource Center (ABRC) at Ohio State University (1735 Neil Avenue, Columbus OH 43210 USA). DNA stocks can be ordered by fax (614-292-0603), by e-mail (dna@genesys.cps.msu.edu) or on-line through AIMS (Arabidopsis Information Management database: for instructions send an e-mail message to inquire-aims@genesys.cps.msu.edu). For e-mail orders, type STOCK ORDER in the subject line.

5.5 Obtaining mapping information

Mapping information on ESTs can be obtained in several ways (Dennis et al., 1993):

(1) From AAtDB: ('An *Arabidopsis thaliana* Data Base'). The software and database files are available via anonymous FTP from genome-ftp.stanford.edu in the AAtDB directory.
(2) Via the internet: from the dbSTS database of sequence-tagged sites database on the NCBI WWW server (http://www.ncbi.nlm.nih.gov), which contains 315 entries for rice sequences, or from the AIMS.

6 SEQUENCING GENOMIC DNA

While EST sequencing, in combination with genetic and physical mapping, has already provided important spinoffs to the scientific community, in the long run it is desirable to have access to the complete genetic information of a model plant. At the moment there is one programme (ESSA) (European Scientists Sequencing Arabidopsis) for the large-scale sequencing of the *Arabidopsis* genome, involving a network of European laboratories, co-ordinated by Dr Michael Bevan (Norwich, UK), and supported by the EU. For the first three years (1993–1996), the aim is to sequence 2 Mb of chromosome IV centred on the *FCA* and *AP2* genes. In addition, DNA fragments of 75 kb will be sequenced around genes of interest, giving information scattered throughout the genome and representing a further 525 kb. In the future, this programme should be extended in order to fully sequence chromosomes IV and V.

6.1 Why sequence the genome?

In addition to the sequence and physical position of genes, genome sequencing will provide valuable information on many other aspects of chromosome organisation, including sequences regulating transcription and splicing, repetitive sequences, telomeric sequences and structural elements of chromosomes (sequences involved in replication, recombination and association with the cytoskeleton and membranes) as well as the distribution of mobile elements. It will also allow the identification of new genes which, due to their low expression level or their expression during a restricted phase of development, are not represented in the cDNA libraries used for EST sequencing.

6.2   Genomic sequencing strategies

Due to the limited resources, sequencing strategies and technologies are mostly borrowed from other genome projects (human, *C. elegans* and yeast). Starting from the partial physical map of overlapping YAC clones (see above), ordered cosmid contigs are constructed, from which sequencing templates are prepared. As an alternative, interest exists for constructing libraries in bacteriophage P1, which accepts fragments in the 100-kb size range. The advantage of this vector is the ease with which libraries can be manipulated and insert DNA recovered. For sequencing, most laboratories use a 'shotgun' approach to generate random libraries of overlapping fragments. Sequences are determined using the Sanger method and automated fluorescent DNA sequencers. Once 85–90% of the final sequence is obtained, a directed sequencing method using specifically designed primers allows the remaining gaps to be closed.

6.3   Genome analysis

As the systematic sequencing of the *Arabidopsis* genome has only just begun, only very limited information is available. At the end of the first year of the ESSA project, 82 kbp of sequence had been established. Sequencing of regions around known genes has shown that gene density in these regions can be as high as four or five genes within 20 kbp (our unpublished results). One of the priorities in this preliminary stage has been to accumulate information on cDNA sequences corresponding to the new genes which are identified. The algorithms which are currently used for analysis of human genomic sequences, such as GRAIL (Shah et al., 1994), while usually providing approximate locations for introns, are very inaccurate in predicting intron–exon boundaries in plants and often do not detect small introns. For instance, a region of the *Arabidopsis thaliana* chromosome III containing the *GapA* gene was analysed using two GRAIL programs (Fig. 3). The GRAIL 1a program recognises *GapA* exon 4 quite well (although with no accuracy at its 3' end) but recognises only part of exon 5 and none of exons 1 to 3. The GRAIL 2 program (Xu et al., 1994) is more efficient, as it recognises exons 3, 4 and 5, although with a number of uncertainties at the extremities. Comparison between genomic sequences and their cognate cDNAs allows the precise determination of the position of intron–exon boundaries, which will permit the adaptation to *Arabidopsis* of existing algorithms for automatic analysis and should eventually lead to the establishment of rules for the automatic assignation of intron positions in genomic sequences.

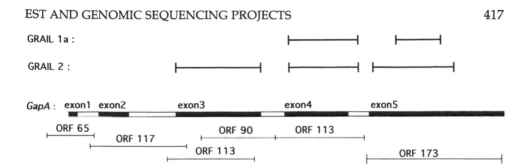

Figure 3. Exon prediction by GRAIL 1a and GRAIL 2 in the *Arabidopsis thaliana* DNA region containing the *GapA* gene encoding the A subunit of the glyceraldehyde-3-phosphate dehydrogenase. The *GapA* gene is located on chromosome III. Black bars represent exons whose positions have been determined by sequencing the corresponding cDNA (Shih et al., 1991). White bars represent introns. ORFs (minimum of 50 codons) are indicated below the *GapA* gene. Exon prediction using the GRAIL 1a and the GRAIL 2 algorithms are indicated above the *GapA* gene

7 CONCLUDING REMARKS

In the last two years, our knowledge of plant genes has increased dramatically. At the current rate of progress it is probable that tags to the majority of *Arabidopsis* and rice genes will be available in 2–3 years, while genomic sequencing in the framework of ESSA should provide 2.25 Mb within the next two years. It is clear that systematic sequencing of plant DNA will rapidly provide a wealth of information of great biological, technological and most probably agronomical interest.

ACKNOWLEDGMENTS

We thank G. Clabault (Grenoble) and Dr S. Klosterman (Martinsried Institute for Protein Sequence, Martinsried, Germany) for their help in the computer analysis of the *GapA* gene and T. Sasaki and B. Lemieux for communicating unpublished data. The French *Arabidopsis* cDNA sequencing project is supported by the CNRS (GDR-ACS 1003), the European project by BIOTECH and the French Groupement de recherche et d'etudes des genomes (GREG).

ESSENTIAL READING

Adams, M.D., Kelley, J.M., Gocayne, J.D. et al. (1991) Complementary DNA sequencing: expressed sequence tags and human genome project. *Science*, **252**, 1651–1656.

Höfte, H., Desprez, T., Amselem, J. et al. (1993) An inventory of 1152 expressed sequence tags obtained by partial sequencing of cDNA from *Arabidopsis thaliana*. *Plant J.*, **4**, 1051–1061.

Newman, T., De Bruijn, F., Green, P. et al. (1994) Genes galore: a summary of the methods for accessing the results of large scale partial sequencing of anonymous *Arabidopsis* cDNA clones. *Plant Physiol.*, **106**, 1241–1255.

Uchimiya, H., Kido, S., Shimazaki, T. et al. (1993) Random sequencing of cDNA libraries reveals a variety of expressed genes in cultured cells of rice (*Oryza sativa* L.). *Plant J.*, **2**, 1005–1009.

Cooke, R., Raynal, M., Laudie, M., et al. (1996) Further progress towards a catalogue of all *Arabidopsis* genes; analysis of a set of 5000 non-redundant ESTs. *The Plant J.*, **9**, 101–124.

REFERENCES

Adams, M.D., Dubnick, M., Kerlavage, A.R. et al. (1992) Sequence identification of 2,375 human brain genes. *Nature*, **355**, 632–634.

Adams, M.D., Kerlavage, A.R., Fields, C. and Venter, J.C. (1993) 3,400 new expressed gene sequence tags identify diversity of transcripts in human brain. *Nature Genet.*, **4**, 256–267.

Altschul, S.F., Gish, W., Mileer, W. et al. (1990) Basic local alignment search tool. *J. Mol. Biol.*, **215**, 403–410.

Antonio, B.A., Yamamoto, K., Harushima, Y. et al. (1993) More than one thousand markers mapped by RFLP. *Rice Genome*, **2**, 3–8.

Boguski, M.S., Lowe, T.M.J. and Tolstoshev, S.H. (1993) dbEST—database for 'expressed sequence tags'. *Nature Genet.*, **4**, 332–333.

Brusslan, J.A., Karlin-Neumann, G.A., Huang, L. and Tobin, E.M. (1993) An Arabidopsis mutant with a reduced level of cab140 RNA is a result of cosuppression. *Plant Cell*, **5**, 667–677.

Claverie, J.-M. (1993) Database of ancient sequences. *Nature*, **364**, 19–20.

Dennis, L., Dean, C., Flavell, R. et al. (1993) The multinational coordinated *Arabidopsis thaliana* genome research project progress report: year three. Publication NSF 93-173. US National Science Foundation, Washington DC.

Gibson, S. and Somerville, C. (1993) Isolating plant genes. *Trends Biotechnol.*, **11**, 306–313.

Green, P., Lipman, D., Hillier, L. et al. (1993) Ancient conserved regions in new gene sequences and the protein databases. *Science*, **259**, 1711–1716.

Hauge, B.M., Hanley, S.M., Cartinhour, S. et al. (1993) An integrated genetic/RFLP map of the *Arabidopsis thaliana* genome. *Plant J.*, **3**, 745–754.

Höög, C. (1991) Isolation of a large number of novel mammalian genes by a differential cDNA screening strategy. *Nucleic Acids Res.*, **19**, 6123–6127.

Keith, C.S., Hoang, D.O., Barrett, B.M. et al. (1993) Partial sequence analysis of 130 randomly selected maize cDNA clones. *Plant Physiol.*, **101**, 329–332.

Kimura, N., Shimada, N., Nomura, K. and Watanabe, K. (1990) Isolation and characterisation of a cDNA clone encoding rat nucleoside diphosphate kinase. *J. Biol. Chem.*, **265**, 15744–15749.

Kinoshita, T. (1990) Report of the committee on gene symbolisation, nomenclature and linkage groups. *Rice Genetics Newslett.*, **7**, 22–23.

Lister, C. and Dean, C. (1993) Recombinant inbred lines for mapping RFLP and phenotypic markers in *Arabidopsis thaliana*. *Plant J.*, **4**, 745–750.

Marty, I. and Meyer, Y. (1991) Nucleotide sequence of a cDNA encoding a tobacco thioredoxin. *Plant Mol. Biol.*, **17**, 143–147.

McCouch, S.R., Kochert, G., Yu, Z.H. et al. (1988) Molecular mapping of rice chromosomes. *Theor. Appl. Genet.*, **76**, 815–829.

Parker, M. (1993) Biological data access through gopher. *TIBS*, **18**, 485–486.

Pearson, W.R. and Lipman, D.J. (1988) Improved tools for biological sequence comparison. *Proc. Natl. Acad. Sci. USA*, **85**, 2444–2448.

Pnueli, L., Hareven, D., Rounsley, S.D. et al. (1994) Isolation of the tomato AGAMOUS gene TAG1 and analysis of its homeotic role in transgenic plants. *Plant Cell*, **6**, 163–173.

Redei, G.P. and Koncz, C. (1992) Classical mutagenesis. In *Methods in Arabidopsis Research* (eds C. Koncz, N.-H. Chua and J. Schell), pp. 16–82. World Scientific, London.

Richards, E.J., Chao, S., Vongs, A. and Yang, J. (1992) Characterization of *Arabidopsis thaliana* telomeres isolated in yeast. *Nucleic Acids Res.*, **20**, 4039–4046.

Saito, A., Yano, M., Kishimoto, N. et al. (1991) Linkage map of restriction fragment length polymorphism loci in rice. *Japan. J. Breed.*, **41**, 665–670.

Sankhavaram, R.P., Parimoo, S. and Weissman, S.M. (1991) Construction of a uniform abundance (normalized) cDNA library. *Proc. Natl. Acad. Sci. USA*, **88**, 1943–1947.

Shah, M.B., Guan, X., Einstein, J.R. et al. (1994) User's guide to GRAIL and GEN-QUEST (Sequence analysis, gene assembly and sequence comparison systems) E-mail servers and XGRAIL(Version 1.2) and XGENQUEST (version 1.1) client-server systems. Available by anonymous FTP to arthur.epm.ornl.gov

Shih, M.C., Heinrich, P. and Goodman, H.M. (1991) Cloning and chromosomal mapping of nuclear genes encoding chloroplast and cytosolic glyceraldehyde-3-phosphate dehydrogenase from *Arabidopsis thaliana*. *Gene*, **104**, 133–138.

Umeda, M., Hara, C., Matsubayashi, Y. et al. (1994) Expressed sequence tags from cultured cells of rice (Oryza sativa L.) under stressed conditions: analysis of transcripts of genes engaged in ATP-generating pathways. *Plant Mol. Biol.*, **25**, 469–478.

Waterston, R., Martin, C., Craxton, M. et al. (1992) A survey of expressed genes in *Caenorhabditis elegans*. *Nature Genet.*, **1**, 114–123.

Xu, Y., Mural, R.J., Shah, M.B. and Uberbacher, E.C. (1994) Recognizing exons in genomic sequence using GRAIL II. In *Genetic Engineering: Principles and Methods* (ed. J. Setlow), Vol. 15, pp. 241–253. Plenum Press.

Index